Turquoise in Mexico and North America

Science, Conservation, Culture and Collections

Turquoise in Mexico and North America

Science, Conservation, Culture and Collections

Edited by J.C.H. King, Max Carocci, Caroline Cartwright,
Colin McEwan and Rebecca Stacey

Archetype
Publications

in association with

The British
Museum

First published 2012 by Archetype Publications Ltd. in association with the British Museum

Archetype Publications Ltd.
6 Fitzroy Square
London W1T 5HJ
www.archetype.co.uk

ISBN: 978-1-904982-79-1

Cover illustration: Aztec/Mixtec wooden mask from Mexico covered in turquoise mosaic with scattered turquoise cabochons, mother-of-pearl eyes and conch shell teeth. Am,St.400.
© Trustees of the British Museum.

Designed by Marcus Nichols at PDQ Digital Media Solutions Ltd.
Typeset by PDQ Digital Media Solutions Ltd., Bungay
Printed and bound in Great Britain by Latimer Trend & Co. Ltd., Plymouth

Contents

Introduction and acknowledgements vii
J.C.H. King, Max Carocci, Caroline Cartwright, Colin McEwan and Rebecca Stacey

Prologue 1
J. C.H. King

Science and Conservation

Mastering materials: comparative properties of raw materials on turquoise mosaics and their significance for mosaic technology 7
Caroline Cartwright, Rebecca Stacey and Colin McEwan

Mineralogy and manufacturing technique in a group of archaeological greenstone mosaics from Classic period Mesoamerican sites 15
Laura Filloy Nadal

Cracking the code of pre-Columbian turquoise trade networks and procurement strategies 29
Sharon Hull and Mostafa Fayek

Conservation of a turquoise mosaic disk from Tula, Mexico 41
Patricia Meehan and Valerie Magar

An integrated approach to understanding the selection and fate of turquoise on Mexican mosaics in the British Museum 55
Rebecca Stacey, Caroline Cartwright, Giovanni Verri and Colin McEwan

An alternative approach to the prehispanic turquoise trade 65
Alyson M. Thibodeau, John T. Chesley, Joaquin Ruiz, David J. Killick and Arthur Vokes

The turquoise disk from Offering 99 at the Templo Mayor in Tenochtitlan 75
Adrián Velázquez Castro, María Eugenia Marín Benito, Emiliano Ricardo Melgar Tísoc, Reyna Beatriz Solís Ciriaco and José Luis Ruvalcaba Sil

Culture and Collections

Turquoise in the Aztec imperial world 91
Frances Berdan

Mexican turquoise mosaics in Vienna 103
Christian Feest

The symbolism of turquoise in ancient Mesoamerica 117
Karl A. Taube

Teoxihuitl: turquoise in Aztec thought and poetry 135
Patrick Johansson K.

Turquoise and squash blossom: a Pueblo dialogue of the long run 145
Peter M. Whiteley

The exotica of the Settala Museum and other northern Italian collections 155
Antonio Aimi

A new glance at Bologna's sixteenth- and seventeenth-century museums and their Mexican items 165
Laura Laurencich-Minelli

Progress and evolution and changes to the practice of collecting in the mid-nineteenth century 173
Michael Thompson

In pursuit of the unity of the human race: Henry Christy and Mexico 177
Jill Cook

Henry Christy, A.W. Franks and the British Museum's turquoise mosaics 183
Marjorie Caygill

On the eternal and the elusive: an appreciation of the work of Johnson and Bird 197
Henrietta Lidchi

Shared images: continuity and innovation in turquoise jewellery 203
Gail Bird and Yazzie Johnson

We wear our wealth 213
Shane Hendren

Eighteenth-century economy, twentieth-century merchandising: the market for turquoise in the American Southwest, 1900–1940 215
Cheri Falkenstien-Doyle

The turquoise mines of the Cerrillo Hills Mining District, gemstone materials, and their usage: ancient and modern 221
Douglas Magnus

Index 225

Introduction and acknowledgements

J.C.H. King, Max Carocci, Caroline Cartwright, Colin McEwan and Rebecca Stacey

The papers in this volume are the outcome of the conference 'Turquoise, Henry Christy and European Collections' organised by the Department of Africa, Oceania and the Americas and held at the British Museum in December 2009. The idea of the conference was inspired in the first place by a widening and accelerating interest in the nine sixteenth-century Mexica mosaics in the British Museum. Earlier publications (Read 1895; Carmichael 1970) had already drawn attention to the historical, cultural and artistic interest of these exceptional objects. The results of the initial scientific studies of the mosaics (from A. Middleton, C.R. Cartwright and R. Stacey) subsequently appeared in *Turquoise Mosaics from Mexico* (McEwan et al. 2006). The conference took place during the exhibition *Moctezuma: Aztec Ruler* at the British Museum as a scholarly adjunct to the project celebrating the 200th anniversary of Mexican Independence in 2010 (McEwan and López Luján 2009). Some of the iconic mosaic objects featured in the exhibition including those brought together by the renowned nineteenth-century collector Henry Christy (1810–1865). One outstanding mask was used as the lead image for the exhibition poster and also graces the cover of the present volume.

The purpose of the conference was to enable a lively interdisciplinary exchange and dialogue to take place and was thematically structured to focus on reporting further scientific analyses, conservation and interpretation, geology and provenance studies, collections histories and not least Native American turquoise use and commentary. These themes inform the organisation of this book and the contributors embrace a wide variety of academic and non-academic backgrounds, all drawing on their respective expertise to create a new knowledge and deeper understanding of the subject. This is preceded by a discursive introduction that aims to offer useful background where this is not adequately covered in the individual contributions.

We extend appreciative thanks to our colleagues who attended the conference, the session chair-people, all those who have contributed to the volume as well as the anonymous reviewers. In particular, we would like to thank Antony Simpson, Catherine Higgitt, Henrietta Lidchi, Norman Hammond, Warwick Bray, Elizabeth Baquedano, Joan Caballero and Fiona Grisdale.

The conference and the publication have been generously supported by the Eugene V. and Clare E. Thaw Charitable Trust, Santa Fe, for which the British Museum is most grateful; we extend our special thanks to Eugene V. Thaw and Sherry Thompson.

References

Carmichael, E. (1970) *Turquoise Mosaics from Mexico*. London: British Museum Press.

McEwan, C. and López Luján, L. (eds) (2009) *Moctezuma: Aztec Ruler*. London: The British Museum Press.

McEwan, C., Middleton, A., Cartwright, C. R. and Stacey, R. (2006) *Turquoise Mosaics from Mexico*. London: The British Museum Press.

Read, C.H. (1895) 'On an ancient Mexican headpiece coated with mosaic', *Archaeologia* **14**: 383–398.

Prologue

J.C.H. King

Turquoise, $Cu(Al,Fe^{3+})_6(PO_4)_4(OH)_8{\cdot}4H_2O)$, is an extraordinary mineral, of exceptional colours, varieties and forms. It is a secondary mineral occurring in the potassic alteration zone of hydrothermal porphyry copper deposits or formed by the action of meteoric waters, usually in arid regions, on aluminous igneous or sedimentary rocks (as vein filling in volcanic rocks and phosphatic sediments). Of compelling beauty, part of the attractiveness of turquoise comes from its opaque colour and glassy sheen, qualities that contrast markedly with the transparent and refractive qualities of most gemstones, although its diaphaneity (transparency) ranges from opaque to translucent or even transparent. Its lustre, that is surface finish, ranges from the vitreous to waxy and to dull. In terms of solidity, it ranges from moderately soft to moderately hard (Mohs scale 5–6), exhibiting conchoidal fracture. The colour is also very varied and includes bright blue, sky blue, pale green, blue-green, turquoise blue, apple green and green-grey (Kolitsch and Giester 2000).

Turquoise thus comes in many forms and demonstrates a variety of properties and is demanding of those who work on its scientific and cultural analysis, because of its peculiarities and mutability. This was recognised in the first paper in English published in the eighteenth century by Cromwell Mortimer (ca.1692–1752; Mortimer 1746–1747). Physician and antiquary, he was a friend and colleague of Sir Hans Sloane (1660–1753), founder of the British Museum. In his paper, presented at the Royal Society, Mortimer examined the continental literature, looked for classical antecedents, described his experiments, and showed his audience a large specimen of turquoise. He also mentions the examples in Sloane's collection – from China and Spain. These would have entered the founding collection of the British Museum, although some were removed in the 1880s to the new sister museum in South Kensington, the Natural History Museum. Mortimer describes the hybrid nature of turquoise, the practice of faking it by colouring ivory, and the propensity of the mineral to change colour. His chemical experiments proved that it was indeed a mineral containing copper. Intriguingly he says 'there are two Sorts of [turquoise], the oriental, from the *East Indies* and *Persia*; and the occidental, from *Spain, Germany, Bohemia, Silesia* & c.' (Mortimer 1746–1747). While Mortimer does not distinguish satisfactorily between eastern and western turquoise, we can follow this categorisation and speculate that perhaps some of the 'occidental' turquoise he saw in Sloane's collection may have come from Mexico and North America.

Turquoise is a complex material (see Hull and Fayek, in this vol., pp. 29–40), and the word 'turquoise' also has contradictory overlapping meanings and usage that are perhaps best explored through etymology. As Mortimer put it 'This Stone has received its Name of Turchesia, and Turquoise, from its being most commonly brought from Turky' (Mortimer 1746–47: 429). The definitions of 'turquoise', even when used as a generic term, are as problematic as the material itself, and consequently provide multiple avenues for varied interpretations. These explanations act as a counterpart to the physical mutability of the substance. The origin of the English word turquoise lies in the medieval French word meaning 'turkish'. Larousse defines the stone as: 'ancien français *turcois*, turc, parce que cette pierre, provenant de Perse, transitait par la Turquie', which translates as 'the stone, though from Persia, was first traded through Turkey to Europe' (Larousse 2010).

For Europeans, particularly between the conquest of Byzantium/Constantinople in 1453, through the Battle of Lepanto in 1571 to the second Siege of Vienna of 1683, Turkey was the place and country which represented the exotic. The Ottoman Empire was the troubling, threatening 'other' that also defined the different and the unknown, and it was this vision of Turkey that would have prevailed in the late fifteenth and early sixteenth century at the time of the European arrival in the Americas and the invasion of Mexico. It was in this period that the mineral turquoise starts to be described in the English language as a powerful stone called a 'turkey' and using other words that would later evolve to become 'turquoise'. The *Oxford English Dictionary* includes descriptions, dating from the fifteenth century onwards, of rings set with diamonds and other stones including 'turkeys' and the modern turquoise: 'a ryng gold with a torcas' (1478), or 'A ryng of gold, sett wt. a turques, a dymamaunt...' (1513). By the middle of the eighteenth century the use of 'turkey' had been entirely replaced by the word 'turquoise' (*OED* 2011).

An important early description of turquoise as a jewel is found in Shakespeare's *Merchant of Venice*, in which Shylock laments the loss of his ring set with a turquoise: 'it was my Turkies, I had it of Leah when I was a Batcheler' (Act III.i.126). This dialogue indicates that he was given the ring by his wife, perhaps as a betrothal gift, suggesting it was understood as a symbol of constancy. From the late sixteenth and early seventeenth century the meaning of turquoise changed and developed. It came to be believed that the stone brightened and faded as the health of the wearer improved or declined. The presence of turquoise was also said to remove enmity,

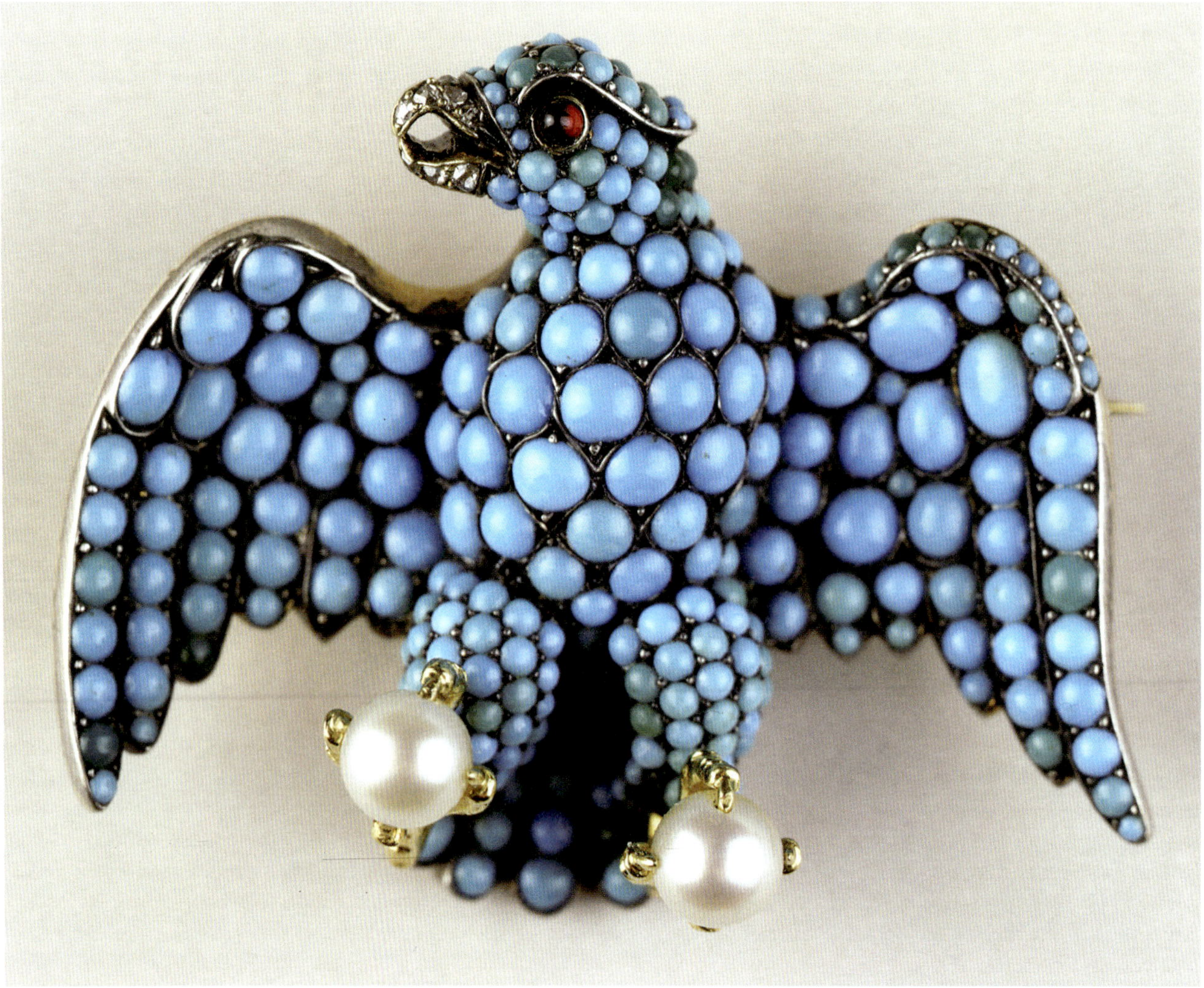

Figure 1. Eagle brooch with pavé-set turquoises in silver with ruby eyes, diamond beak and a pearl in each claw, backed in gold. One of the series given to train bearers by Prince Albert after his marriage to Queen Victoria in 1840. Made by Charles Du Vé, for R.& S. Garrard. Presented by Mrs Mary Anna Marten to mark the directorship of Dr Robert Anderson, 1992–2002 (Department of Prehistory and Europe, British Museum. 2002,03001.1, © Trustees of the British Museum).

and in medieval Europe was associated with horses and with travelling. It was particularly fashionable for men, and in one seventeenth-century association turquoise was linked, when worn by women, with sterility (Boswell 1963).

By the early nineteenth century the meaning of turquoise had changed again, and the stone was used by both genders. It remained commonly associated with the east, with Persia, and also with Russia and with Russian jewellery. However the use of turquoise occurs across the world and is of great antiquity, for example with material mined in the Sinai peninsula being used in ancient Egypt. Despite this, it was not until the nineteenth century that the west became aware of the use of turquoise in China and Tibet (Pogue 1915). Recent accounts revisit both the Old and New World production and use of the stone (Vigil 2005; Lowry and Lowry 2010). In the early nineteenth century Russian use and exploitation of turquoise from Persia was dominant, but European fashion for turquoise was more widespread. In 1840 the young Queen Victoria presented her bridesmaids with turquoise brooches, one of which is in the British Museum (Gere and Rudoe 2010) (Fig 1.) Later in the 1840s Albert acquired turquoises from Russia for Victoria of a paler colour and lesser quality that were set in a suite by Garrard, the then crown jewellers (Gere and Rudoe 2010: 158–159). Within a few years they had, however, turned green; the problem was ascribed by Victoria not to any supposed inconstancy of Albert's but to the use of pomade. In the 1830s and 1840s – when the nineteenth-century manufacturer, philanthropist and collector Henry Christy (1810–1865) was in his 20s and 30s and beginning to move in society – turquoise became a highly fashionable jewel (Gere and Rudoe 2010). What is less clear is how North American turquoise entered the consciousness of Henry Christy and when turquoise then came to be associated with the Americas, the mines of New Mexico, and the mosaics of the early sixteenth-century Mexica. It was Christy who initiated the gathering together of the British Museum's nine sixteenth-century Mexica mosaics (Caygill, in this vol., pp. 183–196).

In Europe, in the middle of the nineteenth century, turquoise jewels were also worn by commoners, in relatively prosaic formats such as forget-me-knot themed jewels, or exotic archaeological jewellery. One display of turquoise at the Great Exhibition in 1851 may in fact have been seen by Christy. A large Russian floral brooch, set with diamonds supported by turquoise, was exhibited by the St Petersburg court jewellers Kämmerer & Zeftigen (Gere and Rudoe 2010: 254–255). Priced at £750, this may give some rough indication of the economics in Western Europe of using Mexica mosaics as a source of raw materials. In the 1860s, during

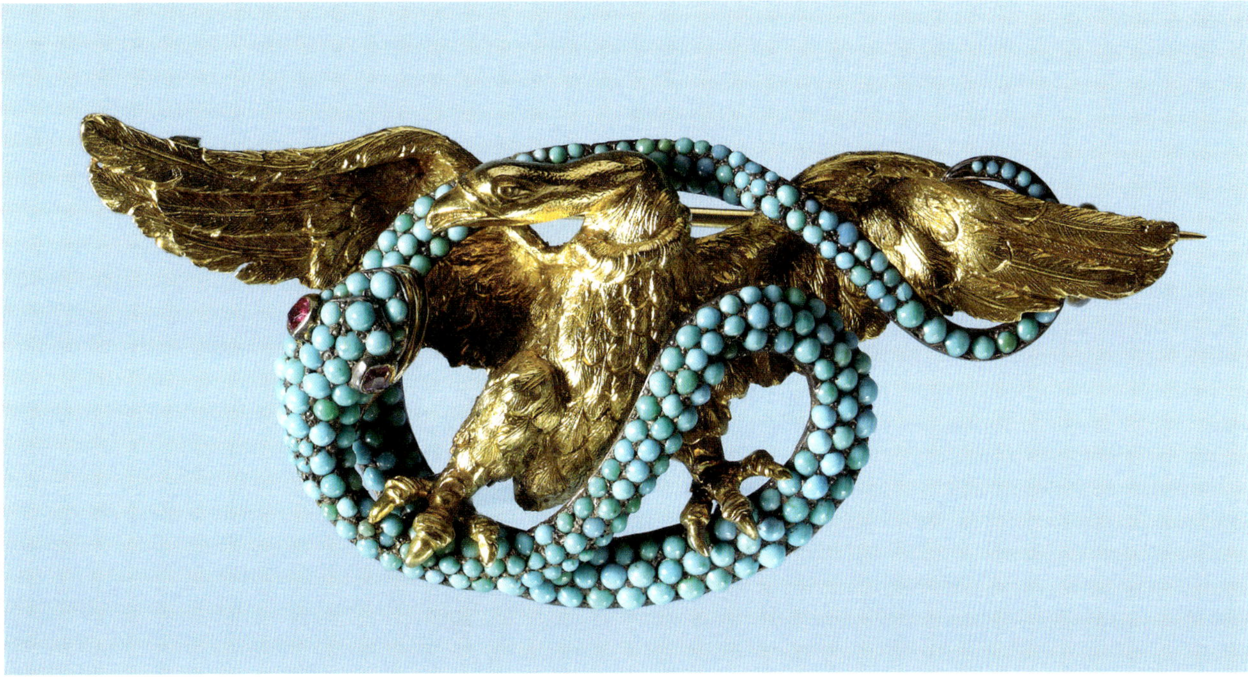

Figure 2. French cast and chased gold brooch in the form of an eagle fighting a snake. The snake is pavé-set with turquoises and has cabochon ruby eyes. It was probably made in Paris during the French intervention in Mexico of the 1860s to represent the mythological founding of Tenochtitlan in 1325, at the spot where an eagle was seen eating a serpent. Hull Grundy Gift (Department of Prehistory and Europe, British Museum, 1978,1002.189, © Trustees of the British Museum).

the French period in Mexico, Paris jewellers created brooches using turquoise representing the founding of Tenochtitlan and Mexico (Fig. 2). None of the Mexica turquoise mosaics bought by Christy and A.W. Franks (1826–1897), one of the Museum's curators, in the nineteenth century cost more than £100. A large Mexica mosaic, such as the British Museum's double-headed serpent with 15,000 tesserae, in the wrong hands could have provided raw materials for numerous European jewels. It is extraordinary that any of the Mexica mosaics survived at all, but one should also wonder how many other Mexica mosaics were cannibalised and the turquoise reused in European jewellery. From 1860 onwards books about the symbolism of gems and jewels confirmed the relationship of constancy of colour in turquoise to constancy in love (Gere and Rudoe 2010). Another factor that may have made second-hand tesserae from Mexica and other items attractive to nineteenth-century jewellers is that they would have been certain of the quality and permanence of colour. By the late nineteenth century, interest in Native North America turquoise jewellery was also developing in the United States. Design of Native pieces changed in response to this interest, with the reduction in the size of jewellery, and the creation of lighter more wearable forms. New mines for the production of turquoise were also opened up at this period, especially in Nevada and Colorado (Vigil 2005: 19–20).

Concerning the instability and inconstancy of turquoise two further points may be made. The first is that it is frequently copied and indeed faked, and the copies and reproductions overlap, as emphasised nearly a century ago (Pogue 1915). The attempt to reproduce turquoise in the Neolithic in Syria, seven or eight thousand years ago, may have given rise to 'The First Fake?' turquoise beads (Taniguchi et al. 2002), long before Mortimer reported on the dyeing of ivory to mimic turquoise (Mortimer 1746–1747). This suggests that even in antiquity the mineral was regarded and understood in a highly sophisticated manner. The study of prehistoric aesthetics, including colour, is a rapidly expanding field. If we reframe this in the historic and modern context as a question of taste, we can see the way in which in Iran, for instance, purely non-variegated blue turquoise is preferred, whereas in North America matrix turquoise with contrasting black or brown inclusions is probably the most sought after (Lowry and Lowry 2010: 13). Finally it should be noted that the tertiary etymologies of 'turkey' provide an intriguing commentary on this stone – which is even now quite difficult to characterise. The acclamatory positive use of 'turkey', which 500 years ago was an expression of the great, the exotic and wonderful, is best demonstrated by the misnaming of the great North American bird the turkey, as such, on the presumption that it too came from Asia Minor; and in the misnaming of maize as Turkish wheat or Turkish (eNotes 2010). Further, by the mid-twentieth century, 'turkey' became common American slang for a dud, ironic given that today, in describing American jewellery, so many turquoises do indeed turn out to be duds – that is fakes, or of low quality and chemically enhanced.

Turquoise, Henry Christy and European collections

The conference, 'Turquoise, Henry Christy and European Collections', allowed a group of experts from a wide variety of academic and non-academic backgrounds to come together in order to open up new insights in the understanding of turquoise. The research presented was based both on material in museum collections and the study of more recent finds, whether from archaeological contexts or the art market. It also explored the challenges of the scientific analysis, conservation, presentation and interpretation of such objects. Many of the approaches and issues discussed are also applicable to the study of collections of Native American jewellery – where of course changing taste and the availability of Asian and other turquoise means an extraordinary diversity in the use of original, adapted and adopted materials.

The conference also explored the early history of collecting of Mexica mosaics in Europe and in Britain, where Christy is a central figure. The association with European collections has a long history, with some mosaics having been in these collections since the sixteenth century or tentatively identified as part of the group of 16 shields given to Cortés in 1519 (Feest, in this vol., pp. 103–116). A central theme of the conference was the British Museum's nineteenth-century collection of Mexica mosaics and the role of Henry Christy in assembling this collection and understanding the broader intellectual environment in which he was operating, and networks of scholarly relationships in which he was involved.

As discussed above etymology is key in exploring the idea of turquoise, and the Mexica and Maya mosaics, with their Nahuatl and other Middle and North American meanings. Of particular importance is the way in which very distinct Europeans terms such as, in English, 'blue' and 'green' are grouped together as single words. Etymology was therefore widely discussed at the conference, including the links between definitions and the symbolism of the materials being described, and the underlying Mexica belief systems.

The complex themes and literature explored for Middle American mosaics are very different, but of a similar complexity, in Native North America. The conference examined these themes and the degree of shared symbolism and belief systems between the Mexica and Native North America and the history of turquoise use in the southwest United States for Native North American and non-Native North American jewellers up to the present day.

References

Boswell, J.C. (1963) 'Shylock's turquoise ring', *Shakespeare Quarterly* **14**(4) Autumn: 481–483.

eNotes, (2010) http://www.enotes.com/food-encyclopedia/natural-history-maize (accessed 10 December 2010) (Christian Feest, pers. comm., 3 December 2010).

Gere, C. and Rudoe, J. (2010). *Jewellery in the Age of Queen Victoria: A Mirror to the World.* London: British Museum.

Kolitsch, U. and Giester, G. (2000) 'The crystal structure of faustite and its copper analogue turquoise locality: Iron Monarch iron ore deposit, South Australia', *Mineralogical Magazine* **64**: 905–913.

Larousse (2010) http://www.larousse.com/en/dictionaries/french/turquoise (accessed 10 December 2010).

Lowry, J, D. and J. P. Lowry (2010) *Turquoise: The World Story of a Fascinating Gemstone.* Layton, UT: Gibbs Smith.

Mortimer, C. (1746–1747) 'Some remarks on the precious stone called the turquoise', *Philosophical Transactions of the Royal Society* **44**: 429–432.

Pogue, J.E. (1915) 'The turquois: a study of its history, mineralogy, geology, ethnology, archaeology, mythology, folklore, and technology', in *Memoirs of the National Academy of Sciences,* Vol. XII, Part II. Washington, DC: National Academy of Sciences.

Taniguchi, Y.H., Shimadzu, Y. and Tsuneki, A. (2002) 'The first fake? Imitation turquoise beads recovered from a Syrian Neolithic site, Tell el- Kerkh', *Studies in Conservation* **47**(3): 175–183.

Vigil, A. (ed.) (2005) *The Allure of Turquoise,* 2nd edn. Sante Fe: New Mexico Magazine.

Science and Conservation

Mastering materials: comparative properties of raw materials on turquoise mosaics and their significance for mosaic technology

Caroline Cartwright, Rebecca Stacey and Colin McEwan

Introduction

Since 1992, the nine Mexican turquoise mosaics in the British Museum have undergone intensive scientific investigation to characterise different types of natural raw materials selected for their manufacture (McEwan et al. 2006; Stacey et al. 2006; Cartwright and Meeks 2007; Cartwright 2009; Stacey and Cartwright 2009; McEwan et al. forthcoming). Although the overwhelming impression is of turquoise, at least 18 other material categories are present, some of which consist of more than one type. These include wood, shell, resins, lignite, malachite, pyrite, human skull, flint, plant fibres, gold, pigments and deerskin, and some later additions such as glass, beeswax, brass wire, gemstones (garnet, beryl, emerald, spinel, zircon), shark teeth and pearls. Understanding why so many varieties of material other than turquoise have been used is key to our continued scientific investigation. Leaving aside later additions and modifications (see Stacey et al., in this vol., pp. 55–64), the original selection of raw material in order to achieve a diversity of colours and textures on these mosaics appears to have been paramount. This paper describes selected specific objectives within the most recent detailed scientific examination of many of the materials of the mosaics using optical and scanning electron microscopy. These results expand and enhance previously published references about the extent to which both the microstructural and decorative properties of different raw materials were understood by Mixtec specialist craftsmen and which factors governed their selection. The term 'turquoise mosaics' is used purely for convenience and because it is in common usage; it is very clear that many other materials besides turquoise were selected for crafting these objects.

Methods and objectives

Standard techniques of optical microscopy were used recently to re-examine most of the British Museum mosaics. The objectives were threefold:

1. to obtain a suite of digital images (to augment and enhance the 2006 slides series) that record the working techniques chosen to reflect the specific material properties exhibited in the different types of tesserae;
2. to identify materials that were not previously covered by Cartwright in McEwan et al. (2006), such as some shell tesserae on the animal head Am,St.400.a;
3. to further evaluate what were the principal factors governing the choice of raw materials for tesserae, such as colour, symbolism, longevity, artistic impact, economic constraints, and personal preferences on the part of the craftsmen (see also Stacey et al., in this vol., pp. 55–64).

Enhanced techniques of scanning electron microscopy (SEM) have been utilised to extend previous work on and experimentation with the various raw materials present on the turquoise mosaics. Subsequent to the publication of McEwan et al. (2006), the British Museum purchased a large chamber Hitachi S-3700N variable pressure scanning electron microscope (VP-SEM) and a high resolution Hitachi S4800 field emission scanning electron microscope (FE-SEM), both with energy-dispersive X-ray (EDX) analysis capabilities. These SEMs have enabled an even wider range of materials to be investigated through the application of different techniques that were not possible with previous SEM instrumentation. The principal objectives were twofold:

1. to obtain a series of SEM micrographs (images) that display key properties of different types of materials chosen for the tesserae, many of which were identified in the earlier studies (listed above in the Introduction) but were not characterised in the SEM;
2. to provide a confirmed identification of the wood chosen for the animal head Am,St.400.a, which was previously characterised only with a suggested taxon.

Results and discussion

Wood

Following the protocol laid down by the International Association of Wood Anatomists (IAWA), wood identification was carried out by examining three different sections – transverse, radial longitudinal and tangential longitudinal (frequently abbreviated to TS, RLS and TLS).

VP-SEM micrographs of *Cedrela odorata* (Meliaceae family) wood, selected for six of the nine Mexican turquoise

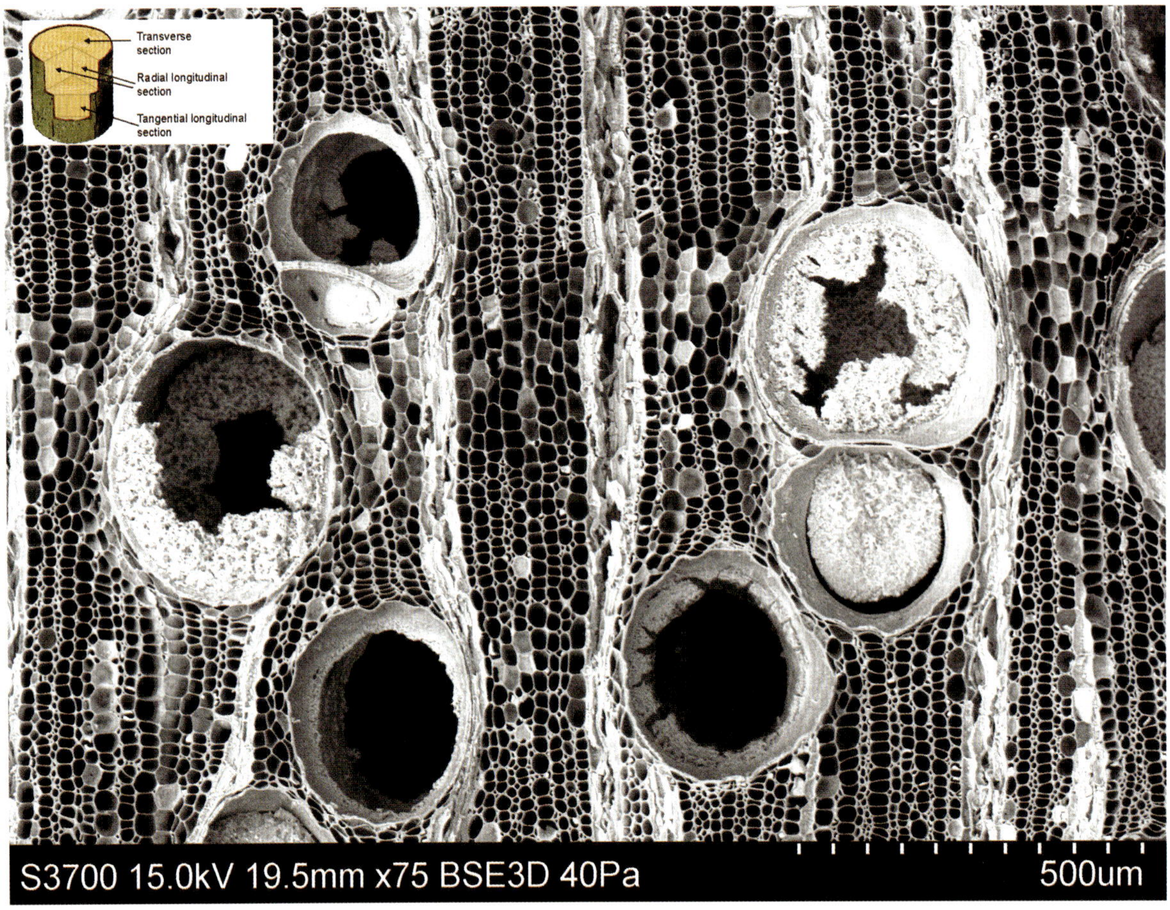

Figure 1. VP-SEM micrograph of a transverse section of *Cedrela odorata* wood showing resin in many of the vessels (© Trustees of the British Museum).

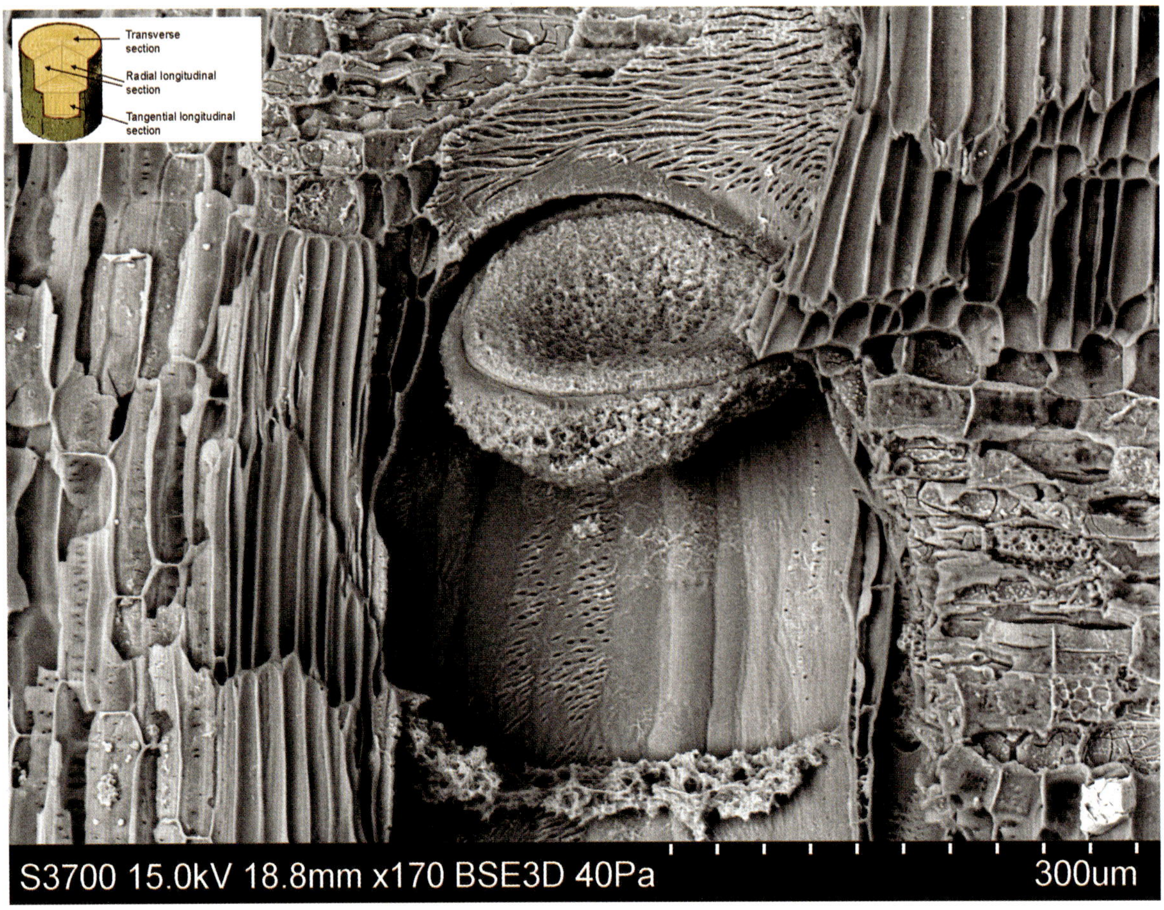

Figure 2. VP-SEM micrograph of a radial longitudinal section of *Cedrela odorata* wood showing a distinct resin plug in one of the vessels (© Trustees of the British Museum).

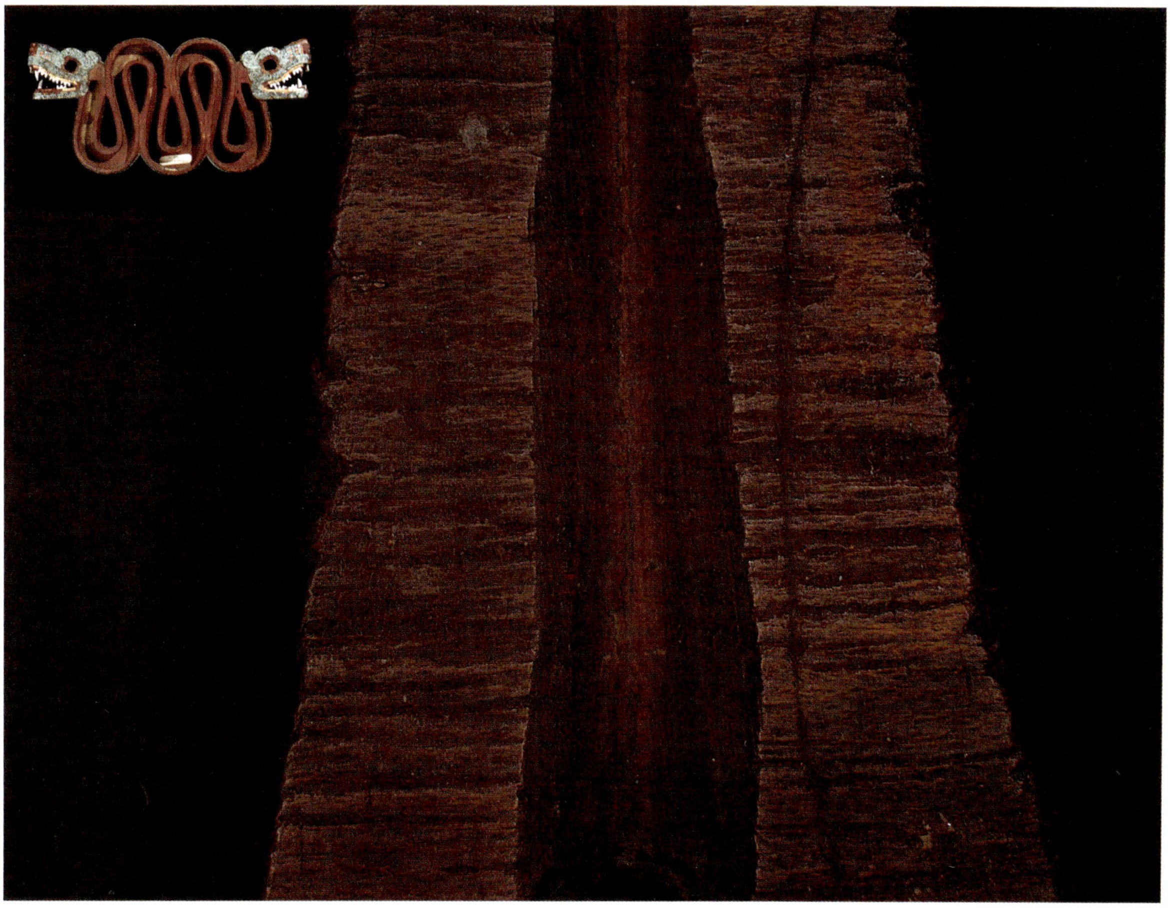

Figure 3. Optical microscope image of the reverse of the double-headed serpent Am1894.634 showing how the *Cedrela odorata* wood has been shaped and planed along the tangential grain (© Trustees of the British Museum).

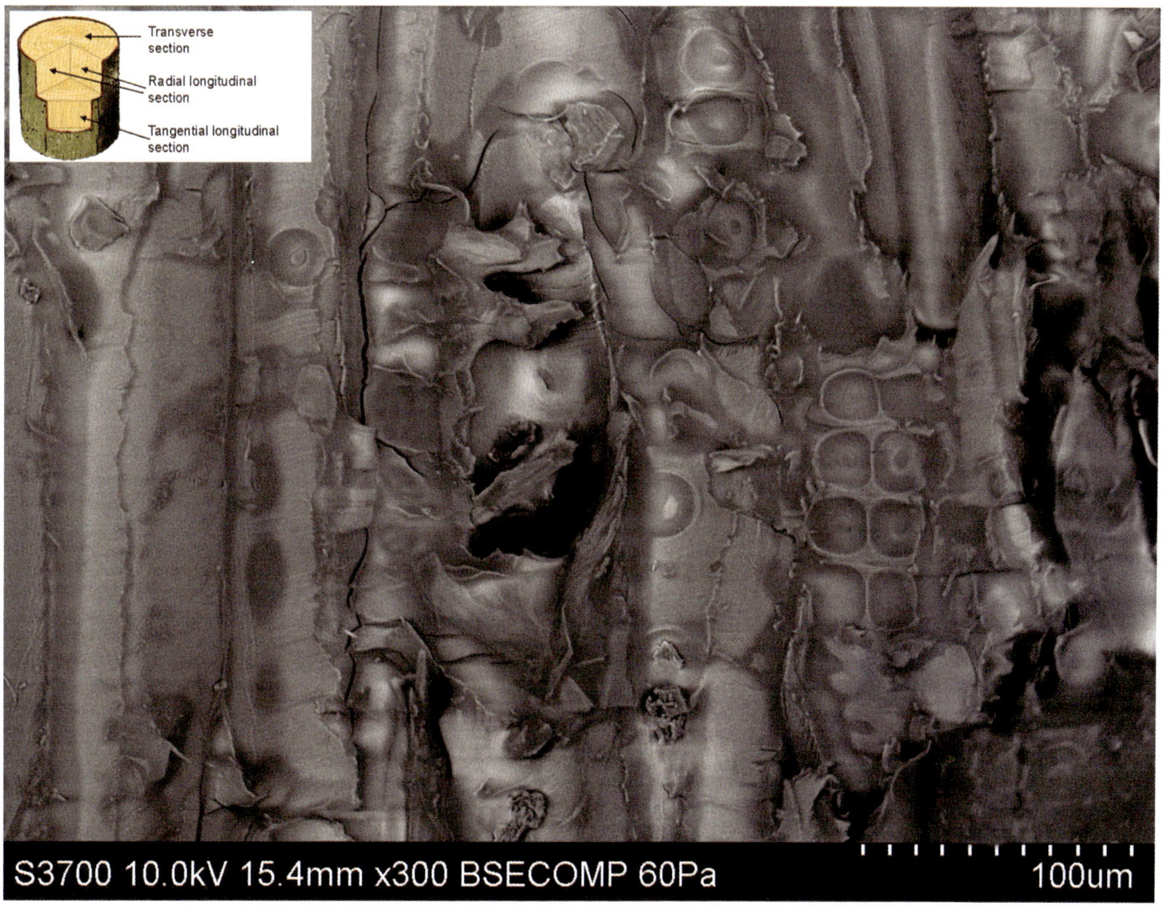

Figure 4. VP-SEM micrograph of the radial longitudinal section of *Pinus* sp., pine wood used to make the shield, Am,St.397.a (© Trustees of the British Museum).

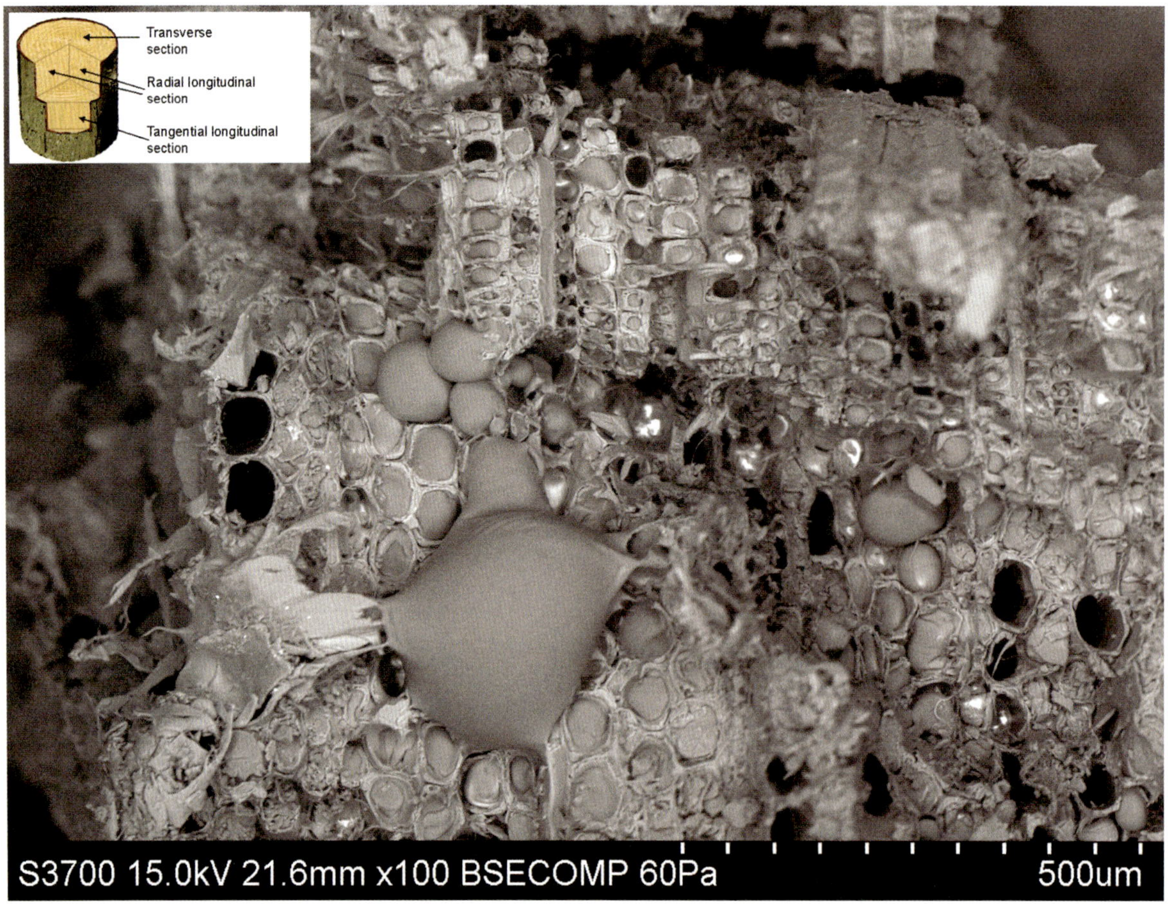

Figure 5. VP-SEM micrograph of the transverse section of *Pinus* sp., pine wood used to make the shield, Am,St.397.a. Resin is clearly present in and exuding from many of the tracheids (© Trustees of the British Museum).

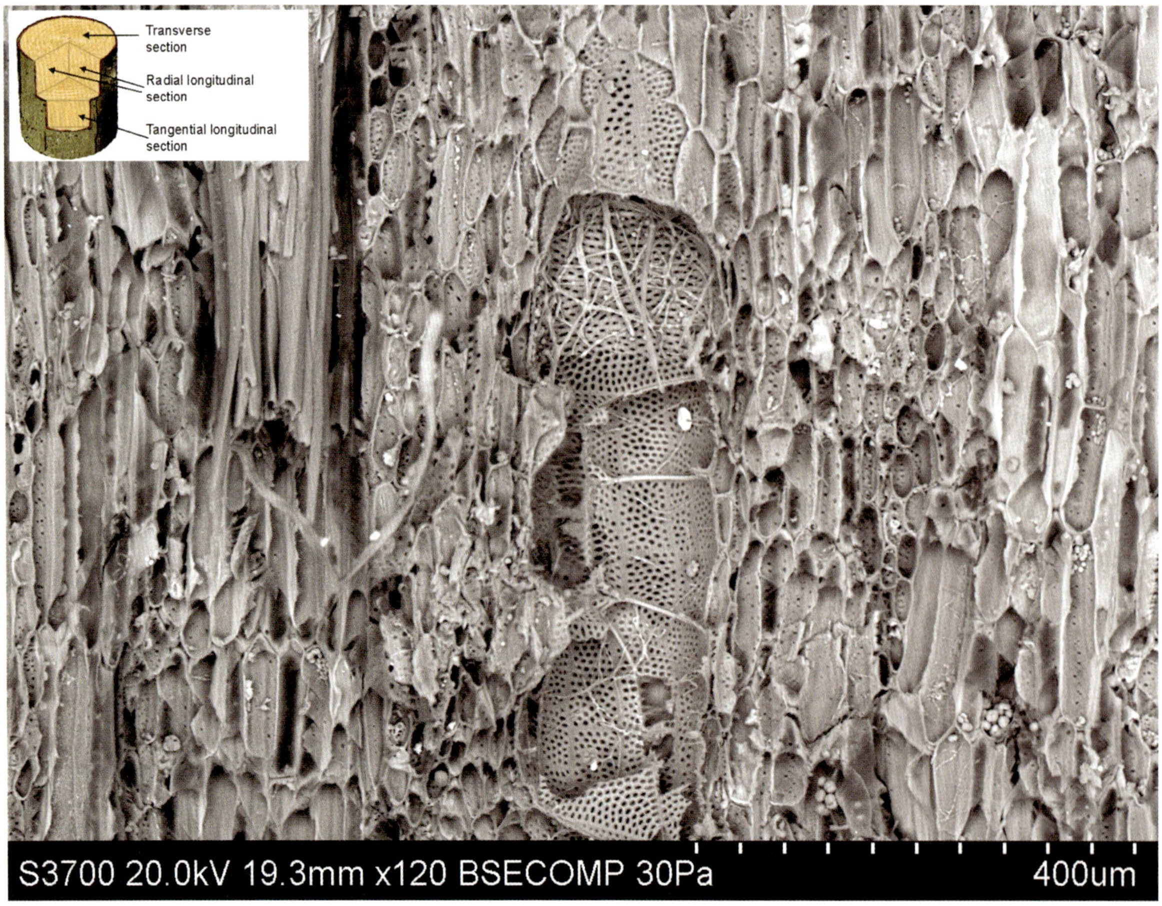

Figure 6. VP-SEM micrograph of a tangential longitudinal section of a tiny sample of wood from the animal head Am,St.400.a, the anatomical details of which confirm the *Erythrina coralloides* identification (© Trustees of the British Museum).

mosaics in the British Museum, reveal that in transverse section, the vessels are mostly full of resin (Fig. 1). The radial longitudinal section, Figure 2, shows a distinct resin plug in one of the vessels. The presence of resin in *C. odorata* makes the heartwood aromatic, rot-resistant and insect-repelling. It may also create an additional ability to bond well with the applied resin used to affix the tesserae (Stacey and Cartwright 2009). With a specific gravity of 0.4, this is a moderately lightweight wood that is easy to carve. Figure 3, an image of the reverse of the double-headed serpent, Am1894.634, shows how the wood has been shaped and planed along the tangential grain. The characteristic anatomical features of *C. odorata* wood include: simple perforation plates, polygonal, small, alternate intervessel pits, vessel-ray pits with distinct borders, fibres with simple to minutely bordered pits, septate and non-septate fibres, thin- to thick-walled fibres, axial parenchyma which can be scanty paratracheal, vasicentric, diffuse and in marginal bands, rays 1–3 cells wide, larger rays commonly 4–10 seriate, heterocellular rays, prismatic crystals present in upright or square ray cells and also in chambered and non-chambered axial parenchyma cells (Miller and Détienne 2001).

Cedrela odorata is of commercial importance and is a highly valued forest species because of its quality wood. Consequently, it has been overexploited in natural rainforest and is considered endangered in Mesoamerica (Navarro et al. 2004). Although replanting is taking place (e.g. Ramos and del Amo 1992), its distribution continues to be reduced without sustainable regeneration. It is problematic, therefore, to map its ecological and geographical distribution in Mexico at the time of the creation of the turquoise mosaics (particularly in the light of their unknown provenances), and equally difficult to determine whether or not *C. odorata* forests were then systematically managed on a regional basis for their wood resources (Xolocotzi 1998).

For the shield, Am,St.397.a, *Pinus* sp., pine wood was selected. This identification is confirmed by the VP-SEM micrograph (Fig. 4) of the cross-field pitting in the radial longitudinal section, which yields the most diagnostic anatomical information for coniferous woods. The shield was shaped and planed along the radial grain. Although pine wood is only considered to be of medium quality, the resins present in the timber (Fig. 5) might have helped protect against insect attack. Around 47 species of pines are distributed throughout Mexico (Farjon et al. 1997) but with most abundance in the central region. Their ability to grow successfully in a variety of conditions makes them valuable and commercially useful. Anatomically, pines species are virtually indistinguishable, hence the identification of the wood of Am,St.397.a being made to genus level only.

For the McEwan et al. (2006) publication, the available techniques for examination of the wood used for animal head Am,St.400.a could only result in a suggestion that the soft, rather pale yellow to white wood appeared similar to the coral tree *Erythrina americana* or *E. coralloides* (known locally as colorín). Optical microscopical examination revealed a spongy cellular structure but did not yield any diagnostic features to verify the identification as colorín wood. Recent re-examination of a tiny sample of the wood from Am,St.400.a in the VP-SEM confirmed the identification as *E. coralloides*, coral-tree (Fabaceae family), which is indigenous to the rocky cliffs and canyons of Mexico. Figure 6 shows a VP-SEM micrograph of a tangential longitudinal section of this sample that supports the *E. coralloides* identification.

Key features include: simple perforation plates; alternate intervessel pits; vestured pits; vessel-ray pits with distinct borders, similar to intervessel pits in size and shape throughout the ray cell; fibres with simple to minutely bordered pits; non-septate fibres; confluent/ banded axial parenchyma; fusiform parenchyma cells; larger rays commonly 4–10 seriate or more; storied parenchyma, vessel elements and axial parenchyma.

As the wood is so soft, with a low specific gravity of less than 0.4, one may speculate as to why it was chosen (apart from extreme ease of carving). Figure 7a and b show the pale trunk wood of *E. coralloides* and its scarlet flowers. Could

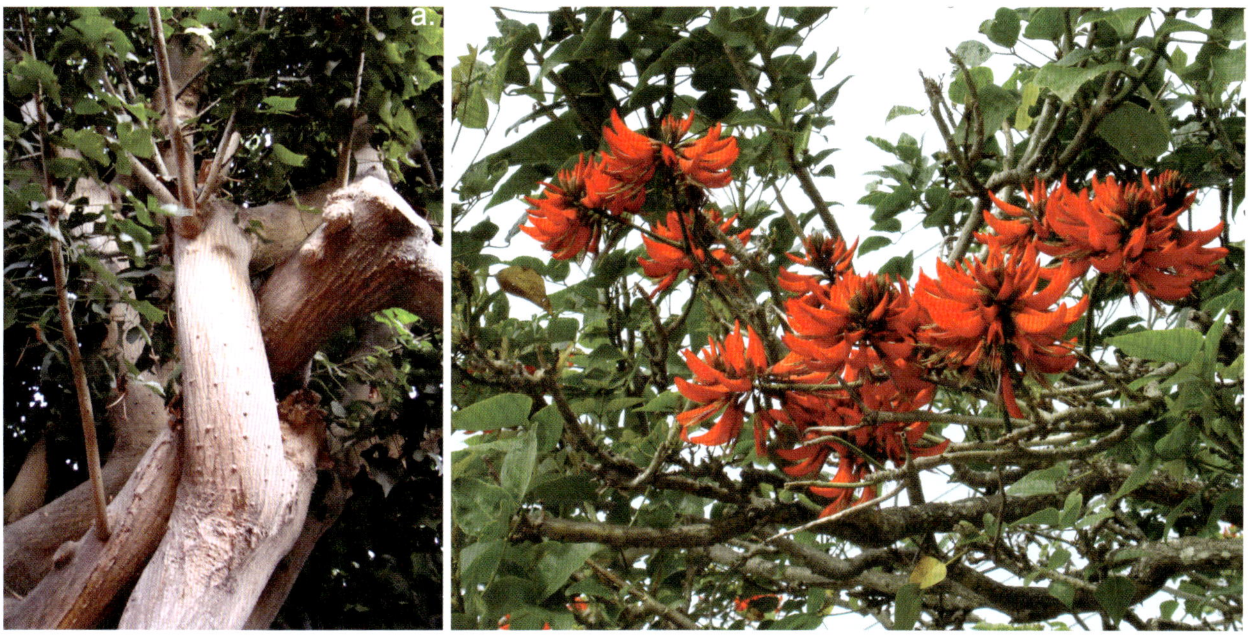

Figure 7. (a) The pale trunk wood of *Erythrina coralloides* and (b) its scarlet flowers (© Caroline R. Cartwright).

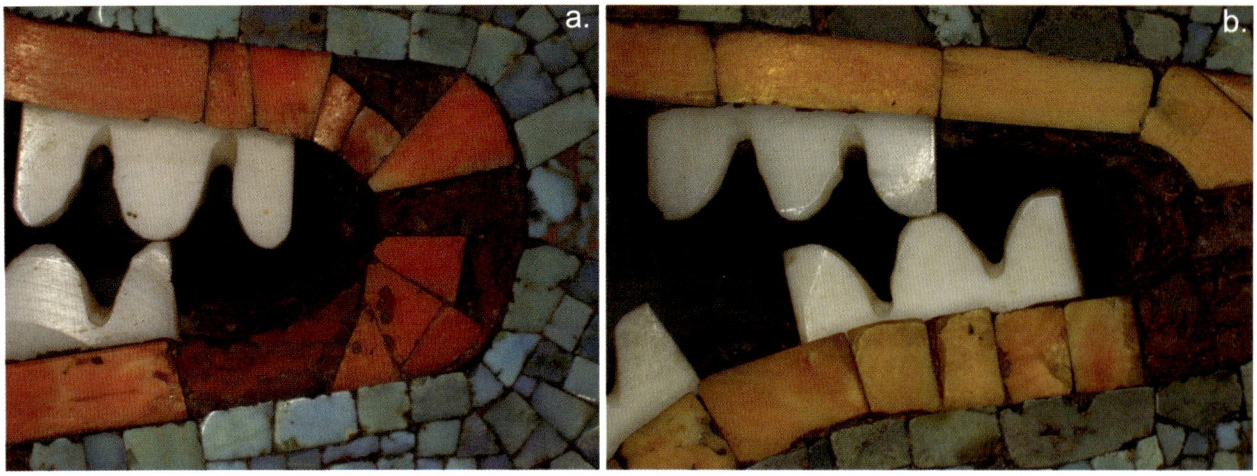

Figure 8. Optical microscope images: (a) red *Spondylus princeps* shell tesserae used to depict 'gums' in the mouths on the front of the heads of the double-headed serpent and (b) yellow *Spondylus princeps* shell tesserae used to depict 'gums' in the mouths on the rear of the heads of the same object (© Trustees of the British Museum).

the choice be related to these, possibly having a symbolic parallel in some way with the scarlet of *Spondylus princeps* (thorny oyster) shells and the significance of blood in Aztec/Mixtec rituals?

Shell

The contrasting material properties of the different types of shell for tesserae on the Mixtec turquoise mosaics such as conch, oyster and mother-of-pearl have already been published in some detail (Cartwright and Meeks 2007; Cartwright, 2009) and will not be repeated here. However, three updates to these publications are presented:

1. the selection for particular purposes or artistic effects of yellow varieties of *Spondylus*, rather than red;
2. recent taxonomic revisions of conch shell (and the importance of being aware of new taxonomic revisions);
3. new identifications of a shell taxon present on the animal head Am,St.400.a that has not yet been identified from any of the other nine turquoise mosaics in the British Museum collections.

Many of the turquoise mosaics from Mexico show the selection and use of gradations of scarlet, blood-red and orange coloured segments of *S. princeps*, possibly for symbolic and artistic effect. Red and orange-red *S. princeps* shell tesserae were used to great effect to depict 'gums' in the mouths on the front of the heads of the double-headed serpent (Fig. 8a). The deployment of yellow *Spondylus* tesserae for the 'gums' on the rear of the heads of the double-headed serpent (Fig. 8b) has long been the subject of speculation as to whether the yellow colour indicated weathering or fading through exposure to excessive sunlight at certain points in the object's history. Recent optical microscopical re-examination confirms, however, that yellow was specifically selected, not only on the double-headed serpent but also on other mosaics including the shield Am,St.397.a. Although trade networks could potentially have offered several species of *Spondylus* that display a preponderance of yellow on the shells, such as *S. calcifer*, *S. ictericus* and *S. americanus*, it is important to note (and see Pillsbury 1996) that *S. princeps* itself can occur in a variety of colours but mainly in red, orange, and purple; sometimes yellow and white. Such colour variations may be depth-related, for example between 18 and 27 metres of water white and yellow *S. princeps* specimens may be most common, whereas deeper than 27 metres, reds and oranges predominate.

Other shell taxa, gastropods as well as bivalves, could potentially have yielded raw material for yellow tesserae via long-distance trade. Examples, either striped yellow and white (e.g. conch) or with areas of yellow colouration, include: *Laevicardium elatum*, *Tegula regina*, *Thais speciosa*, *Cymatium parthenopeum keenae*, *Semele tortuosa*, *Cittarium pica* and *Fasciolaria* species. From the Sea of Cortès, Mexico to Ecuador, the orange prince cone shell, *Conus (Ductoconus) princeps apogrammatus*, is often found in vivid yellow (McEwan et al. forthcoming). However, none of these taxa has been used as a source for the yellow tesserae in Figure 8b on the rear of the double-headed serpent; these are all *S. princeps* colour variants. This does seem to reinforce the culturally significant and symbolic importance of the Aztec/Mixtec use of the thorny oyster, in both its red and yellow colourations.

For the second objective, recent taxonomic revisions of conch shell, this is an opportunity to record the current confusion and uncertainty over the Linnean binomial classifications for conch shells. *Strombus gigas* had been named by Linnaeus in 1758. There are many local and common names for this conch including: queen conch, pink conch, caracol reina, caracol rosa, caracol rosado, caracol de pala, cobo, botuto, guarura, and lambí. It is a species of very large edible marine gastropod molluscs in the family Strombidae, the true conches. It is indigenous to the tropical Caribbean faunal zone of the Western Atlantic Ocean, found at depths from 0.3 to 18 metres. In 2007, Petuch and Roberts reclassified *Strombus gigas* (queen conch) as *Eustrombus gigas*, only to be supplanted in 2008 by Landau et al. who reclassified this taxon as *Lobatus gigas*. As yet, there seems to be no completely unified application in the literature of this taxonomic reclassification and it is worth bearing in mind that all three terms may still be in current use for some time, i.e. *S. gigas*, *E. gigas* and *L. gigas*.

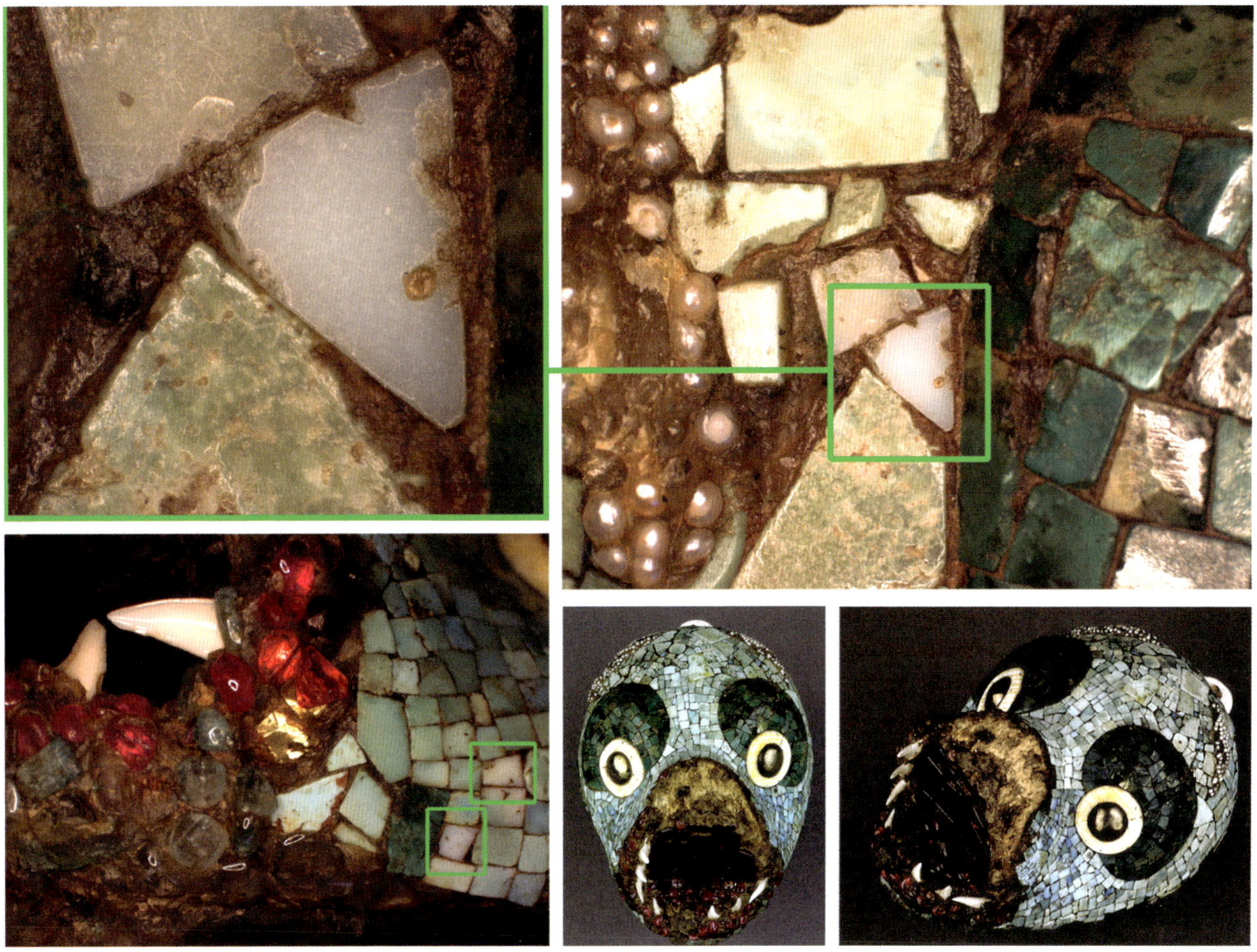

Figure 9. Optical microscope images of one of the *Cypraecassis* sp., helmet/bonnet shell white tesserae on the animal head Am,St.400.a (© Trustees of the British Museum).

Finally, it is important to record a new shell taxon present on the animal head Am,St.400.a; one that has not yet been identified from any of the other nine turquoise mosaics in the British Museum. Given the complex history of this particular object and the likelihood of extensive reworking and additions once in Europe (see McEwan et al. 2006), the recent recognition of small white shell tesserae other than from conch shell stimulated much interest. Had these shell tesserae been sourced to Mediterranean species, for example, there would have been a specifically European category of later material to be added to those of the glass, gemstones, brass wire, possibly the pearls and shark's teeth. However, the taxon identified as the source of these partially translucent white tesserae (Fig. 9) was *Cypraecassis* sp., helmet/bonnet shell (family Cassidae), common in the Sea of Cortés, Mexico.

Conclusions

The interpretation of any artefact, archaeological or historical, is usually firmly embedded in the contextual information regarding its location and will be made in conjunction with associated finds, environmental evidence and a chronological and cultural framework. As the nine turquoise mosaics from Mexico have arrived in the British Museum collections through different routes and lack such a framework, it is necessary to base the scientific research on their component materials and, most importantly, material properties. Codices do offer some avenues for comparison; however it is not clear, for example, how literally depictions of turquoise tesserae, conch or *Spondylus* shells in the codices can serve as direct models for material selection or precise cultural significance in terms of their usage on the mosaics. It seems there is no consensus on whether such mosaics were, in fact, for active ritual or ceremonial use, or whether they were specifically created as showy objects to appease or impress the Spanish incomers. What is clear, however, is that long-distance trading of different qualities and types of symbolically significant materials was integral to the cultural systems in operation and that specialist traders as well as specialist craftsmen were necessary for the creation of these objects, each unique in its own way. The mastery of materials is self-evident; understanding disparate material properties and transforming them into a coherent, artistic and culturally reverberant whole can rarely have been equalled.

References

Cartwright, C.R. (2009) 'Scanning electron microscopy at the British Museum and the Aztec turquoise mosaics from Mexico', *Hitachi Electron Microscopy News* **3**: 38–42.

Cartwright, C.R. and Meeks, N.D. (2007) 'Aztec conch shell working: high-tech design', *British Museum Technical Research Bulletin* **1**: 35–42.

Farjon, A., Perez de la Rosa, J.A. and Styles, B.T. (1997) *A Field Guide to the Pines of Mexico and Central America.* Kew: Kew Publishing.

Landau, B.M., Kronenberg, G.C. and Herbert, G.S. (2008) 'A large species of *Lobatus* (Gastropoda: Strombidae) from the Neogene of the Dominican Republic, with notes on the genus', *The Veliger* **50**: 31–38.

McEwan, C., Middleton, A., Cartwright, C.R. and Stacey, R. (2006) *Turquoise Mosaics from Mexico.* London: British Museum Press.

McEwan, C., Stacey, R.J. and Cartwright, C.R. (forthcoming) 'The "Tezcatlipoca" skull mosaic in the British Museum collections: new insights and questions of identity', in E. Baquedano (ed.), *Tezcatlipoca: Trickster & Supreme Aztec Deity.* Niwot, CO.: University of Colorado Press, 119–127.

Miller, R.B. and Détienne, P. (2001) *Major Timber Trees of Guyana: Wood Anatomy*. Wageningen: Tropenbos International, 68–69.

Navarro, C., Montagnini, F. and Hernández, G. (2004) 'Genetic variability of *Cedrela odorata* Linnaeus: results of early performance of provenances and families grown in association with coffee', *Forest Ecology and Management* **192**: 217–227.

Petuch, E.J. and Roberts, C.E. (2007) *The Geology of the Everglades and Adjacent Areas*. Boca Raton, FL: CRC Press.

Pillsbury, J. (1996) 'The thorny oyster and the origins of empire; implications of recently uncovered *Spondylus* imagery from Chan Chan, Peru', *Latin American Antiquity* **7**: 313–340.

Ramos, J. and del Amo, S. (1992) 'Enrichment planting in a tropical secondary forest in Veracruz, Mexico', *Forest Ecology and Management* **52**: 289–304.

Stacey, R.J. and Cartwright, C.R. (2009) 'Construction, modification and repair of Mexican mosaics: evidence from the adhesives', in J. Ambers, D. Saunders and L. Harrison (eds), *Holding it all Together: Ancient and Modern Approaches to Joining, Repair and Consolidation.* London: Archetype Publications, 61– 67.

Stacey, R.J., Cartwright, C.R. and McEwan, C. (2006) 'Chemical characterisation of ancient Mesoamerican "copal" resins: preliminary results', *Archaeometry* **48**: 323–340.

Xolocotzi, E.H. (1998) 'Aspectos de la domesticación de plantas en México: una apreciación personal', in T.P. Rammoorthy, R. Bye, A. Lot and J. Fa (eds), *Diversidad biológica de México: orígenes y distribución.* Mexico City: Instituto be Biología, Universidad Nacional Autónoma de México (UNAM), 715–735.

Authors

- Caroline Cartwright, Department of Conservation and Scientific Research, The British Museum, London, UK
- Rebecca Stacey, Department of Conservation and Scientific Research, The British Museum, London, UK
- Colin McEwan, Department of Africa, Oceania and the Americas, The British Museum, London, UK

Mineralogy and manufacturing technique in a group of archaeological greenstone mosaics from Classic period Mesoamerican sites

Laura Filloy Nadal

Introduction

In Mesoamerica, there is abundant evidence to show the ways in which turquoise, together with obsidian, jade and seashells, was fashioned into exquisite mosaics and sacred objects destined for use by rulers, dignitaries and gods. Mosaics were complex pieces to execute and required specialised teamwork, time, trade networks and resources. The selection of colours, materials and iconography evident in mosaics was closely tied to the use they were to be given. In Mesoamerica, the art of mosaic work was used to create objects that served to mark the social status of high-ranking individuals. To create them, artists employed materials whose use was restricted and controlled by the ruling class, such as feathers, jade, hematite, *Spondylus* shell and turquoise. Mesoamerican mosaics were laid onto a remarkable range of 'backing' materials involved including shell, fine wood, metals, polished stone and even human bone. Turquoise and jade tesserae were applied to skulls, diadems, earflares, nose plugs, lip plugs, pectorals, masks, sceptres, knife handles, disks and plaques that propitiated the presence of the gods and promoted cosmic equilibrium.

Although during the Formative (or Preclassic) and Classic periods, stone ornaments in various hues of green predominated, the Toltecs and their successors added multiple shades of blue to their palette of stone from sources located far beyond the frontiers of Mesoamerica and thousands of kilometres away from the Mexica capital. These exotic materials were obtained primarily through tribute, exchange and long-distance trade. Contact and relations between the inhabitants of the American Southwest and their Mesoamerican counterparts during the Postclassic period are well documented in the archaeological record as many scholars have discussed (González and Olmedo 1986; Harbottle and Weigand 1992; Izeki 2008; Weigand 1968; Weigand and Weigand 2001). We know that the influences between both cultural 'super-areas' have always been reciprocal. However, even after decades of study and debate, there is still no scientific consensus to specifically determine the geological sources for the tesserae used to create Mesoamerican mosaics.

A similar phenomenon prevails in the study of intense green stones, so highly prized by Mesoamerican cultures of all eras. The debate continues over the location of mineral sources from which jade was obtained. Furthermore, this precious mineral was also traded over long distances to reach all corners of Mesoamerica. Detailed analysis conducted by several research institutions has opened a door into the private world of prehispanic artisans and begun to unravel the secrets of their trade. Specialised lapidary stone craftsmen worked directly in the court under the scrutiny of the ruling class, and no doubt, technical expertise was passed down from generation to generation. Clearly, turquoise mosaic flourished in Mesoamerica among the Toltecs, Mixtecs and Mexicas, but the lapidary tradition of mosaic work had its roots much earlier in time in the Classic period when it was used with great artistry by the people of Teotihuacan and by the Mayas.

In this paper, I will discuss a group of mosaics that was created during the Classic period. This corpus was studied and restored in recent years in the Conservation Laboratory at the Museo Nacional de Antropología (MNA) in Mexico City. Discussion will also cover a piece from the third century AD in the Central Highlands and a corpus of mosaics made in the Maya area during the seventh century AD. I would also like to devote a few lines to a greenstone, Teotihuacan-style mask, covered with turquoise and shell mosaic, because the mystery surrounding the time and place of its creation has been a matter of debate. For each case study, discussion will proceed from the archaeological context in which the objects were found, the raw materials employed, a formal description of the objects, and the techniques utilised by Mesoamerican lapidary artists to produce them.

The mask from Malinaltepec, Guerrero

This unusual Teotihuacan-style mask covered with splendid mosaic work was brought to the MNA in Mexico City in 1921 (Castillo Ledón 1922: 35) (Fig. 1). However, it did not end up at the Museum by chance. In the early years of Mexico's post-revolutionary government, there was a growing interest in acquiring archaeological objects that glorified the country's prehispanic past. It is no surprise then that the commission that was organised for the Celebration of the First Centennial of Mexican Independence gave funds to the Museum to undertake a study trip to Oaxaca and Guerrero aimed at collecting archaeological pieces to enrich the National Museum's collection (Archivo Histórico del Museo Nacional de Antropología, vol. 33, exp. 1480, f. 1–14: 10; vol. 39, exp. 1518, 9 f. 92–99: 93).

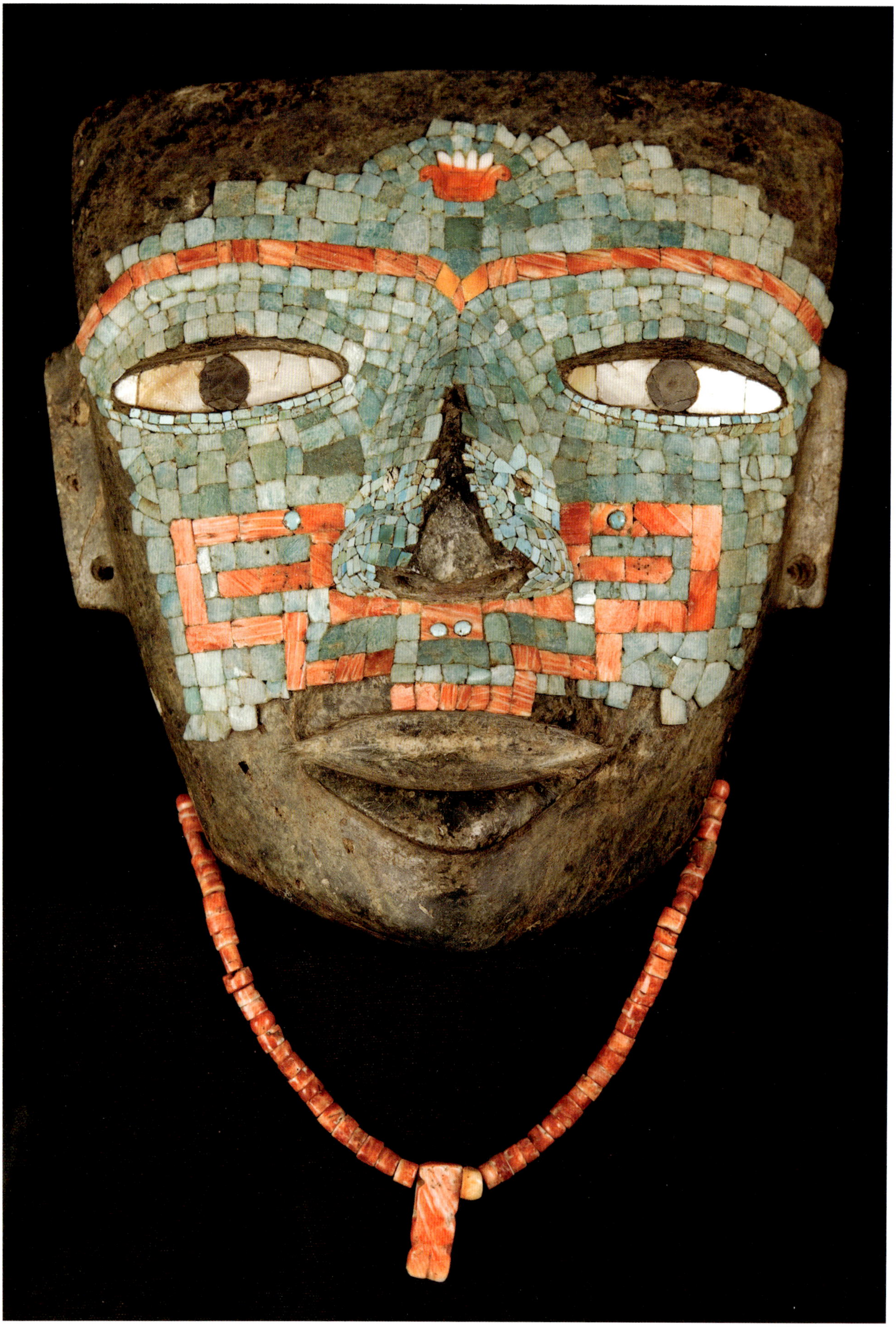

Figure 1. Teotihuacan-style mask from Malinaltepec, Oaxaca (MNA 10-9639). National Museum of Anthropology, courtesy of CONACULTA-INAH. (Photo: Martirene Alcántara.)

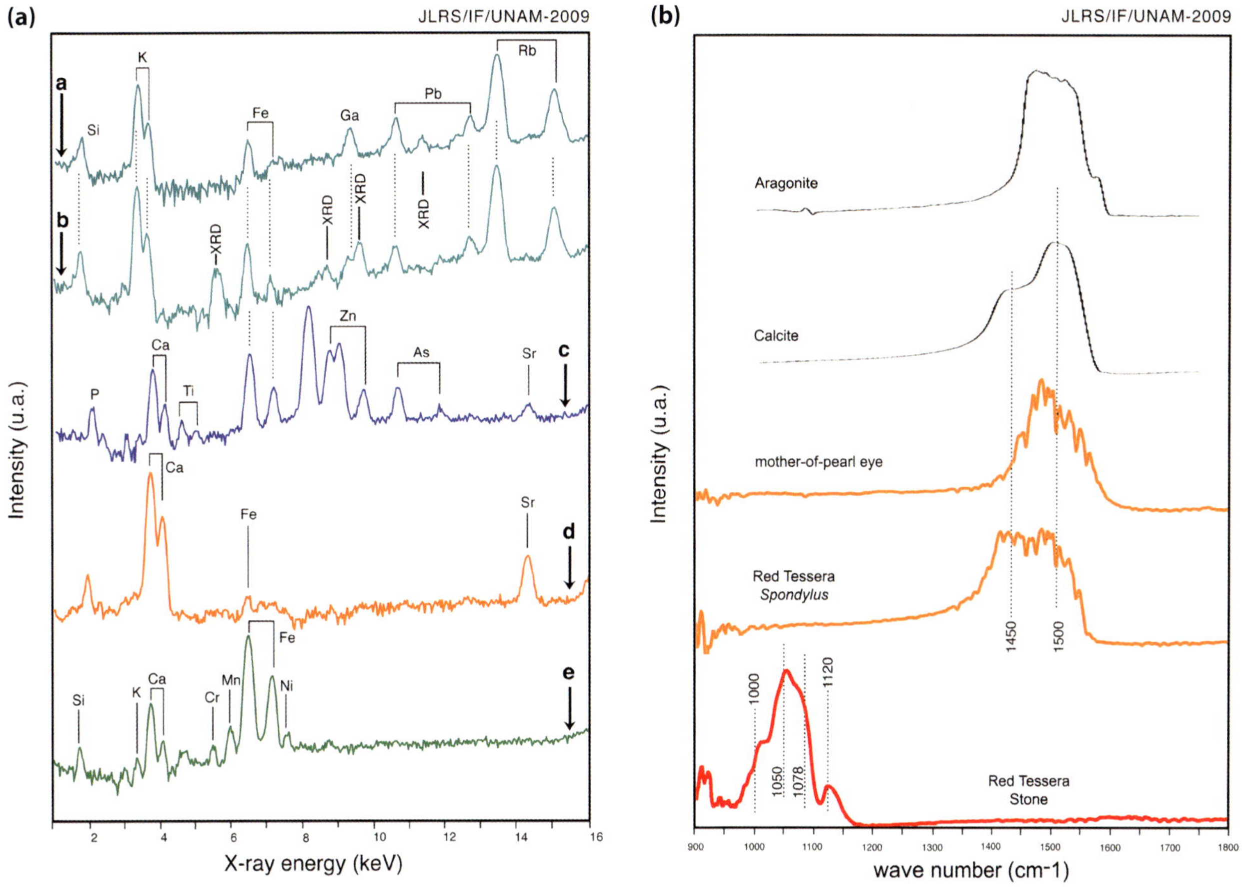

Figure 2. (a) XRD spectra of (**a–b**) amazonite tessera; (**c**) turquoise tessera; (**d**) applied marine shell and (**e**) the mask support. (b) FTIR spectra of the shell components of the mask (*Spondylus* tessera and the mother-of-pearl eye) and from other red tessera.

Little data is available on circumstances surrounding the mask's discovery. We know it was recovered from a funerary deposit found in the vicinity of the town of Malinaltepec in the state of Guerrero (Aguirre 1922). Porfirio Aguirre excavated it together with a series of fired clay fragments that were part of a large pot and some human bones that fell apart when exposed to the air (Aguirre 1922: 34). Further down, he found two stone masks, one of them covered with blue, white and red mosaic tesserae. On the same spot he also found a series of red shell beads from a necklace (Aguirre 1922: 35). The pieces were part of a funerary deposit in which the individual was placed in a ceramic pot and dressed in rich trappings created from materials not locally available in the mountainous region of Guerrero.

Once the excavations were concluded, the archaeologist returned to Mexico City. A couple of months later, when the mask was cleaned, it was put on display in the MNA. Almost immediately this delicate object provoked controversy about its authenticity (Castillo Ledón 1922). The Museum's director assembled a workgroup composed of diverse experts, from Mexico and abroad, to analyse the mask (Archivo Histórico del MNA, vol. 39, exp. 1518, 7f. 53–56). After two months of work it was concluded that the piece was a magnificent specimen from the prehispanic era crafted from polished greenstone, probably serpentine or serpentinite, and adorned with tesserae made of *Spondylus princeps*, turquoise, amazonite and specular hematite (Archivo Histórico del MNA, vol. 39, exp. 1518, 9f. 92–99: 95; Cummings 1922; Níven 1922; Estrada 1922; Waitz 1922). In addition, the adhesive used was identified as a resin, probably copal (Noriega 1922).

In 2009 we had the opportunity to study the Malinaltepec mask: a carved greenstone covered with 650 blue and 75 orange tesserae. The generous and invaluable collaboration of Dr José Luis Ruvalcaba Sil (Instituto de Física/Universidad Nacional Autónoma de México (IF/UNAM)), made it possible to conduct a series of non-invasive analyses on the piece in order to determine exactly what raw materials were used to make it. To study the object, the IF/UNAM team conducted X-ray fluorescence (XRF) and Fourier transform infrared spectroscopy (FTIR) analyses (Ruvalcaba Sil et al. 2010). Of the 152 blue tesserae analysed with XRF, 80 percent turned out to be amazonite and only 20 percent were made of mineral turquoise, i.e. turquoise that can be determined chemically (Ruvalcaba Sil et al. 2010: 162). It is also interesting to note that the tesserae made of mineral turquoise are located in the nose area, in the cabochons on the cheeks and upper lip of the mask, as well as on the contour of the eyes. This material was probably chosen for carving these delicate tesserae because it is easier to work than amazonite and could also be cut with greater precision to make the mosaic more homogeneous.

As for the white pieces to make the sclera of the eyes, they registered spectra that correspond to the composition of seashells (Fig. 2a); Velázquez Castro et al. (2010a: 171) identified the use of mother-of-pearl (*Pinctada mazatlanica*) to produce the sclera of the mask's eyes. In contrast, the appliqués for the iris of both eyes turned out to be rich in iron and are made of specular hematite, identified by XRF (Ruvalcaba Sil et al. 2010: 162) (Fig. 2b).

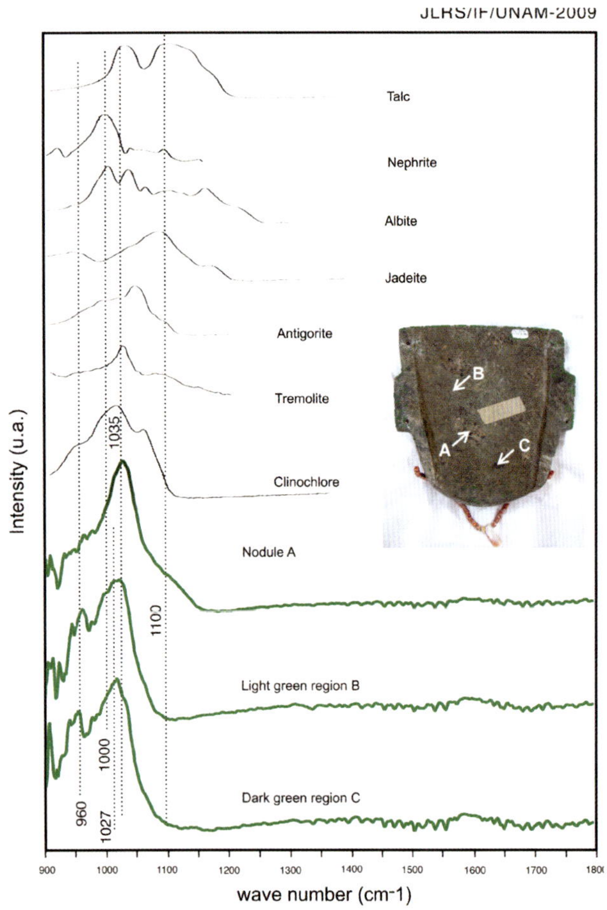

Figure 3. FTIR spectra of the support of the Malinaltepec mask.

The orange shell of the stepped fret and the necklace beads were made of *S. princeps* (Velazquez Castro et al. 2010a: 171–172), a species endemic to the Panamic province that spans the Sea of Cortés to northern Peru (Keen 1971). An exceptional case was a small orange tessera, located in the lower right corner of the stepped nose plug, whose XRF spectrum revealed it is a mineral rich in Si, Al and Fe with lesser amounts of K, Ca, Ti, Mn, Cr, Ni and Ga; however, we were unable to determine the rock type with precision.

For the greenstone support, mid-FTIR spectra were obtained that were similar to those of clinochlore (Ruvalcaba Sil et al. 2010: 166): a chloritite, a dark-coloured, metamorphic stone with cream, light brown or black veins and spots (Fig. 3). Although the trace elements that we obtained in our analysis were insufficient to securely identify the material used to make the mask, above all because a combination of elements was present, it was possible to reject the suggestion of the use of antigorite and serpentinite, which had been proposed earlier (Waitz 1922; Níven 1922; Sánchez and Robles 2002; Sánchez Hernández et al. 2002). In a recent study based on X-ray diffraction (XRD), Sánchez and Robles (2010: 129) were able to confirm the identification of the stone used in the support as, indeed, clinochlore.

Proposing the place of origin of the raw materials that form part of archaeological objects is no easy task. To date, relatively few sources with clear evidence of prehispanic exploitation have been studied. Nevertheless, we can offer some suggestions. The stone used for the support, the clinochlore, is a mineral commonly associated with serpentinite. The nearest sources are found on the slopes of the Sierra Madre del Sur in the modern-day states of Oaxaca and Guerrero (Sánchez and Robles 2005).

At present, only one source of amazonite has been detected in Mexico located in the south-southeastern part of the state of Chihuahua near Peñoles (Sánchez Hernández et al. 2002). However, amazonite mines have also been located in the American Southwest (USA Geological Survey 2002) and researchers from UNAM have referred to the presence of amazonite sources in the state of Oaxaca (Schmitter-Villada and del Campo 1980). To date, no mineral turquoise source has been located in Mesoamerican territory, but rather confined to American Southwest (for a detailed discussion of this topic, see the papers by Hull and Fayek, and Thibodeau et al., in this vol., pp. 29–40 and 65–74 respectively; in addition to González and Olmedo 1986; Harbottle and Weigand 1992; Izeki 2008; Weigand 1968; Weigand and Weigand 2001; Weigand et al. 1977).

The Malinaltepec mask forms part of a corpus of Teotihuacan-style masks in the MNA collections. Contrary to what might be believed, the corpus of Teotihuacan-style masks in the Museum is not very large. In only a few cases is there any information on the provenance of the pieces because almost all of them came to the Museum from private collections assembled in the first half of the twentieth century (INAH 2009); moreover, specific analyses have not yet been conducted to identify the stones used in their manufacture. Stylistically, the Teotihuacan-style masks adhere to a geometric pattern: the overall contour of the face is U-shaped; the forehead is flat and forms a narrow, horizontal band; the eyebrows are slightly marked by a curved bridge; the eyes are elliptical; the ears rectangular tabs; the nose broad; and the lips fleshy, clearly delineated and slightly parted (de la Fuente 2003a: 27; López Luján 1989: 19).

One might add that Teotihuacan-style masks are purely generic representations that do not represent specific individuals. They conform to a repetitive stylistic pattern, portraying a preconceived notion of representing the face through codified conventions, a kind of archetype of the community (de la Fuente 2003b: 43; de la Fuente 2003c: 224–225; Filloy Nadal 2011a: 164–165). The Malinaltepec mask fits in perfectly with these characteristics, a fact that suggests it was carved at a place under Teotihuacan influence. Recent technological studies of the Malinaltepec mask reinforce this idea, because the tools used to produce the mask are similar to those detected for other objects excavated at Teotihuacan and in Guerrero from the Classic period (Velázquez Castro et al. 2010a). We can therefore propose an approximate date of manufacture, but not its provenance.

As of now, the Malinaltepec mask is the only one in the MNA's collection that has a stone base covered with turquoise mosaic. In fact, the mask specimens with tesserae in blue tones –made of mineral turquoise as well as 'turquoise substitute' (so-called 'cultural turquoise') – in the Museum are much later in date. These are pieces from Oaxaca and central Mexico made during the Postclassic period and the support backing is carved in wood. This fact, obviously, was one of the factors that led to the controversy of 1921. Furthermore, no turquoise fragments have been found at Teotihuacan to date, whether in controlled excavations at different compounds or as part of surface collections (Turner 1992: 89). This suggests that turquoise was not used in Teotihuacan workshops.

It is worth considering the possibility that the mask might be a reused piece. In other words, the greenstone support might have been fashioned during the Classic period, which would explain it perfectly matching Teotihuacan stylistic canons, but the mosaic might have been added later. If this is the case, this magnificent specimen adorned with turquoise mosaic could be an example of the reuse of an object from the past by inhabitants of the prehispanic community near the modern-day town of Malinaltepec, Guerrero.

Although there is still not enough evidence to be able to speak of a turquoise mosaic tradition from the Classic period, the lapidary technique of assembling small tesserae was well known among Mesoamerican cultures during this period. Turning now to some examples of greenstone mosaic from the city of Teotihuacan and the Maya area, we shall see that greenstone mosaic work is a technology that dates back to the third century AD and its mastery requires a high degree of specialisation.

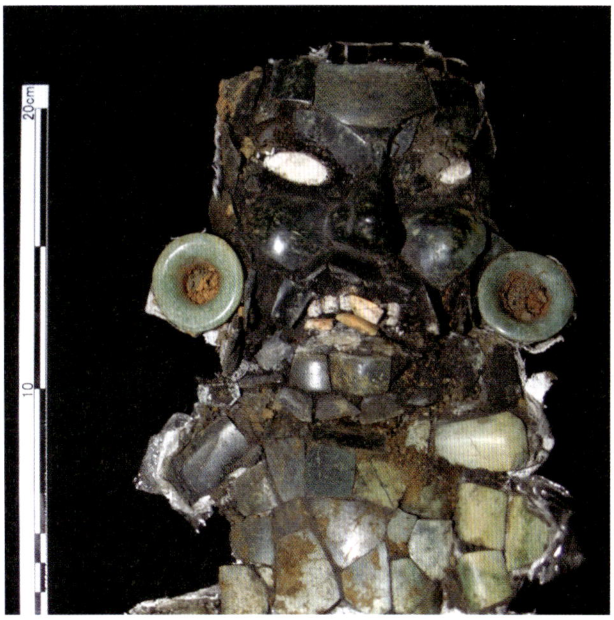

Figure 4. Teotihuacan mosaic before conservation. Courtesy of CONACULTA-INAH. (Photo: Saburo Sugiyama.)

Burial 6 from the Moon Pyramid

Since 1998, the Moon Pyramid Project, coordinated by Saburo Sugiyama and Rubén Cabrera, has systematically excavated the interior of this enormous structure (Sugiyama and Cabrera 2007). The complex of offerings known as Burial 6 was discovered on the north–south axis of Building 5. Based on recent research, this ritual deposit dates to roughly AD 300, when the fourth stage of the Moon Pyramid was built (Sugiyama and López Luján 2007: 138–141).

Burial 6 is a complex deposit containing human skeletons, bone remains of canines, felines and birds of prey, nine pairs of large obsidian eccentrics, a slate and pyrite mirror and two obsidian human figures. It also included an unusual concentration of small pieces of greenstone covered by loose sediment (Sugiyama and López Luján 2007: 138–141) that turned out to be a complete human figure composed of hundreds of tesserae (Fig. 4). Never before in the history of excavations at Teotihuacan had a similar piece been found. In 2005 we undertook the restoration of the mosaic (Filloy Nadal et al. 2006).

Although we can never fully know the production sequence and techniques used by the ancient people of Teotihuacan to produce this human figure in mosaic, the restoration of the sculpture enabled us to identify some of the materials used during the Classic period and to hypothesise the technical steps necessary to create the piece. Tracing the steps of a sculptor, one would suppose the first undertaking was to carve the support. For this the artist had to select a piece of wood and smooth it down bit by bit to produce the human form. Biologist Alejandra Quintanar Isaías (2005) was able to identify a local conifer of the Pinaceae family, probably of the *Pinus* genus, as the wood used to carve the support.

During the Classic period, to make the tesserae, the first step was to select a greenstone core or fragment of the right tonality. The Teotihuacan artist employed different types of stone when creating the mosaic. Results indicate that all the minerals sampled came from regions distant from the Teotihuacan Valley. The stones used by the people of Teotihuacan to produce the tesserae display a wide range of colour from greenish white to very dark green. Thanks to the use of XRD in the laboratories of the Instituto Nacional de Antropología e Historia (INAH), the following minerals were identified in the tesserae: antigorite, chrysotile and clinoclore (the same mineral used for the Malinaltepec mask, as discussed earlier), and components of serpentinite, a metamorphic rock characterised by exquisite green tonalities (Sánchez and Robles 2005). This information helps us to resolve two questions, one associated with the possible source or zone of stone extraction, and the other linked to the technique of manufacture.

The serpentine sources closest to the Teotihuacan Valley used for obtaining the raw material are located in the Tehuitzingo-Tecomatlán zone in the state of Puebla and on the spurs of the southern Sierra Madre (Sánchez and Robles 2005), all of which were evidently within the Teotihuacan sphere of influence. All the stones chosen by the Teotihuacanos are minerals of relatively low hardness, which facilitates the processes of cutting and smoothing, as well as producing the final lustre. Once the hue of the rock was selected, they proceeded to make the tesserae; through percussion and smoothing of the stone, the overall shapes were obtained. To identify the type of tools and the technical sequence employed to make the different pieces of the mosaic, we examined the pieces with an optical microscope and a scanning electron microscope (SEM).

For this purpose, first we studied the pieces with a ×20 hand lens to select the tesserae with the most visible traces. Then, a stereoscopic microscope coupled with a camera was used to examine the materials at magnifications of ×10 and ×30 and to take digital photos of the traces observed. This made it possible to compare the archaeological traces in greater detail with the modern traces of workmanship produced in the laboratory.

The third step consisted of making impressions of the traces left by different tools on the surface of the greenstone pieces. For this, we used a replication polymer tape, which was softened in acetone and pressed against the work surface to be analysed. Once dried, the strip was removed and it

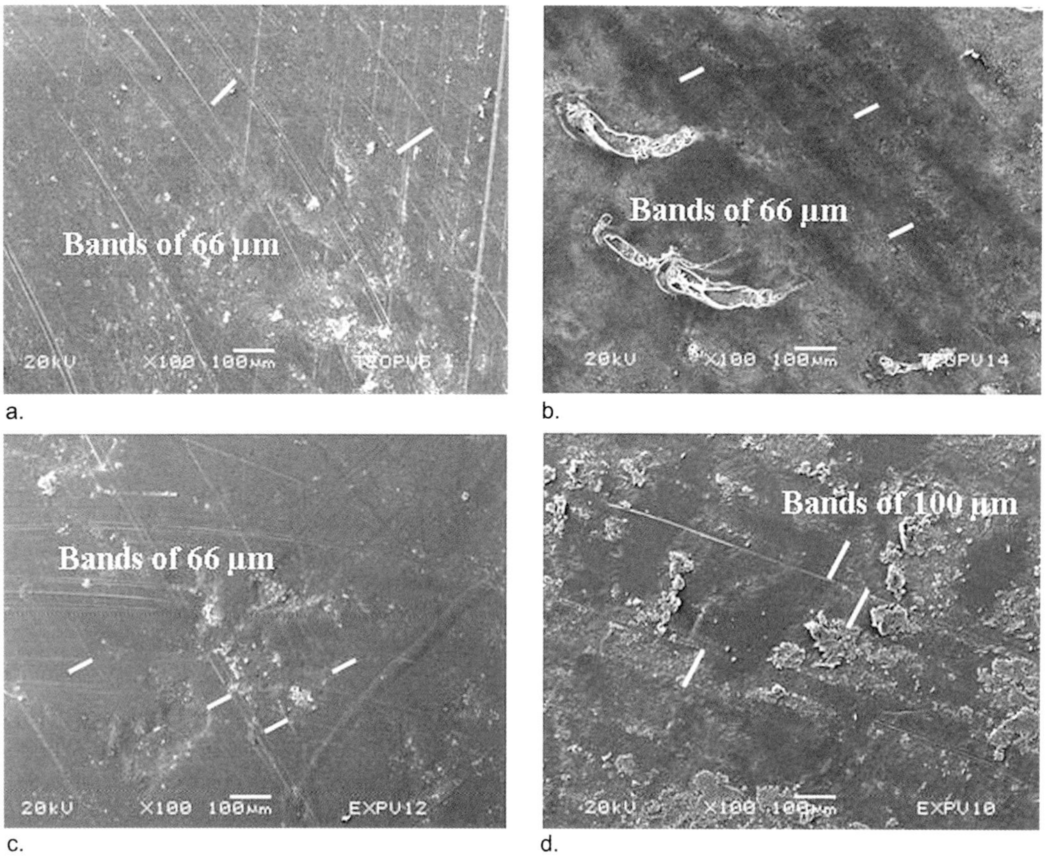

Figure 5. Surface of the tesserae in SEM. Shaping the tessera: (a) surface of the archaeological mosaic (front); (b) unpolished surface of the mosaic (back); (c) experimental abrasion made on andesite slab; (d) experimental abrasion made on basaltic slab. Bands on the mosaic are similar to those obtained when shaping a serpentinite rock with an andesite slab.

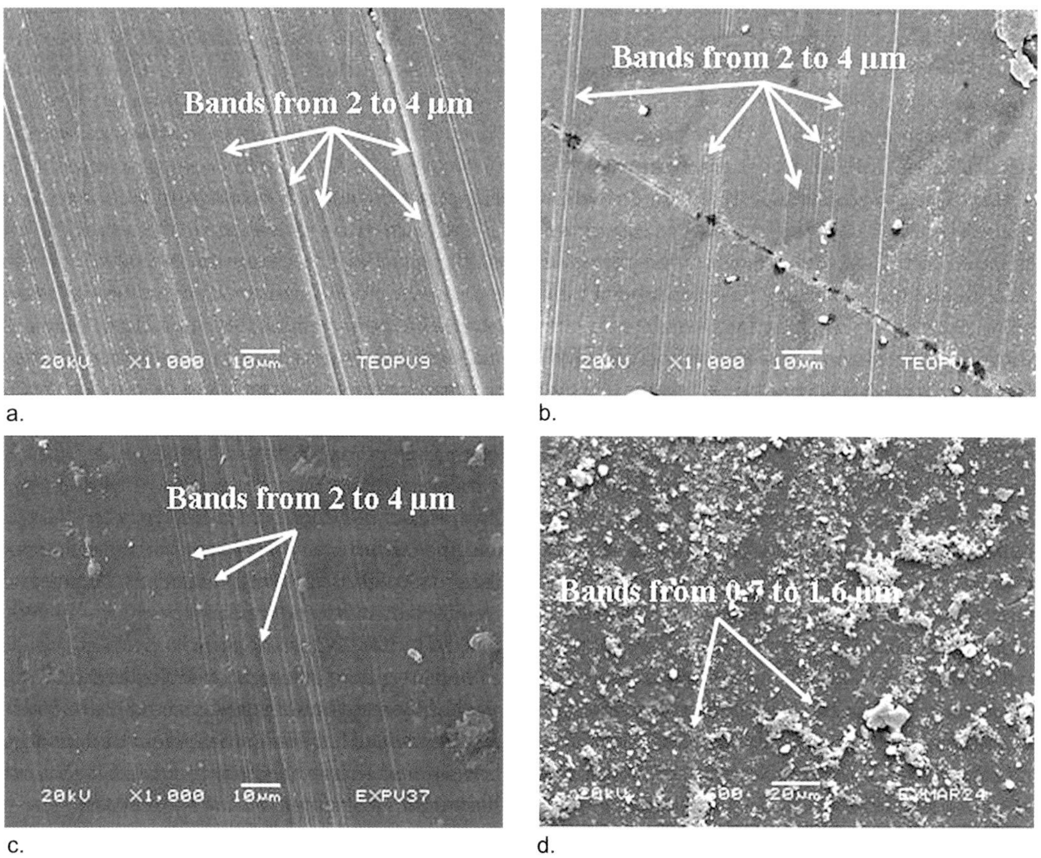

Figure 6. Surface of the tesserae in SEM. Polishing the tessera surface: (a and b) polished surface of the archaeological mosaic; (c) experimental abrasion made with flint and andesite slab; (d) experimental polishing made on basaltic slab and sand. Bands on the archaeological mosaic are similar to those obtained when polishing a serpentinite rock with flint and andesite slab.

produced a mould of the surface marked with manufacturing traces. The moulds were then coated in a gold ion bath to be examined using the high vacuum mode of the SEM. This technique facilitated analysis and made it possible to avoid moving the archaeological pieces to the laboratory. For the final analysis, we used one of INAH's SEMs. The samples were examined at magnifications of ×100, ×300, ×600 and ×1000, taking micrographs as digital images.

At the same time, in the lapidary workshop in the Museo Templo Mayor, the team led by archaeologist Emiliano Melgar replicated in the laboratory the modifications that can be observed on diverse archaeological objects, such as cuts, percussion, wearing, perforations, incising and surface finishes. To this end they used tools and processes that: (a) are mentioned in sixteenth-century historical sources (Sahagún 2000), (b) have been observed in the archaeological record (Domínguez Carrasco and Folan 1999: 643; Kovacevich 2007: 74–86; Rochette 2009: 209–216), (c) have been proposed by various researchers (e.g. Digby 1972: 16 , 20; Kidder 1947: 122–123; Mirambell 1968; Smith and Kidder 1951: 33–36), and that could have been used in prehispanic Mexico. At the end of the experimental phase, replication moulds were produced and the traces of manufacturing present on the pieces produced experimentally in the laboratory were obtained and analysed in the SEM. Finally, they were compared with those detected on the archaeological pieces. After examining the traces of the tesserae from the Teotihuacan mosaic in the SEM, we can propose that the initial shaping of the tesserae was probably carried out on andesite slabs, because they differ from the traces left by working on basalt. The surface was polished carefully with a flint nodule until a smooth finish was produced and the surface was burnished with leather to achieve a glossy appearance; in this case at higher magnification (×600 and ×1000), flat zones can be seen crossed by uninterrupted series of extremely fine, well-defined straight-parallel bands 2–4 μm in width; these are similar to those obtained by polishing with flint nodules and burnishing with leather produced in the laboratory, and they differ from other abrasives such as sand (Filloy Nadal and Melgar Tisoc 2009). To form the bevelled edges, the stone was cut with a flint tool and shaped on andesite slabs to produce slightly rough surfaces that could better adhere to adjacent tesserae (Filloy Nadal and Melgar Tisoc 2009) (Figs 5 and 6).

Sánchez and Robles (2005) also identified the rock employed in making the lips and iris of the eyes. The tesserae for the sculpture's lips were carved in small sections from a carbonate rock known as dolomite (polarised optical microscopy, microchemical tests and XRD were used for the identification). The pupil of the right eye was carved of a microcrystalline bituminous limestone. It is interesting to note that neither of these limestone varieties is from the Teotihuacan Valley, where minerals of volcanic origin predominate. The nearest outcrops of carbonate rocks are found to the northwest of the zone of Tula, Hidalgo, and to the southwest in the states of Morelos and Puebla (Sánchez and Robles 2005). Miniscule plaques of obsidian from nearby mines (Sierra de las Navajas) were used for the delicate diadem worn by the figure.

Putting this puzzle together would not have been easy for the Teotihuacan artist. The placement of the tesserae on the wood must have been extremely complex because among the 300 pieces, no two are alike. To resolve this problem, the artist employed an extremely sticky grey paste that served on the one hand as an adhesive and on the other as a material to compensate for differences in thickness of the tesserae. Once the 300 pieces were in place, the final operation must have been to fill in the tiny spaces that remained between the tesserae, using the same paste (Fig. 7). It is important to mention that the greenstone mosaic only covered the front part of the sculpture and that the back part was made solely of wood.

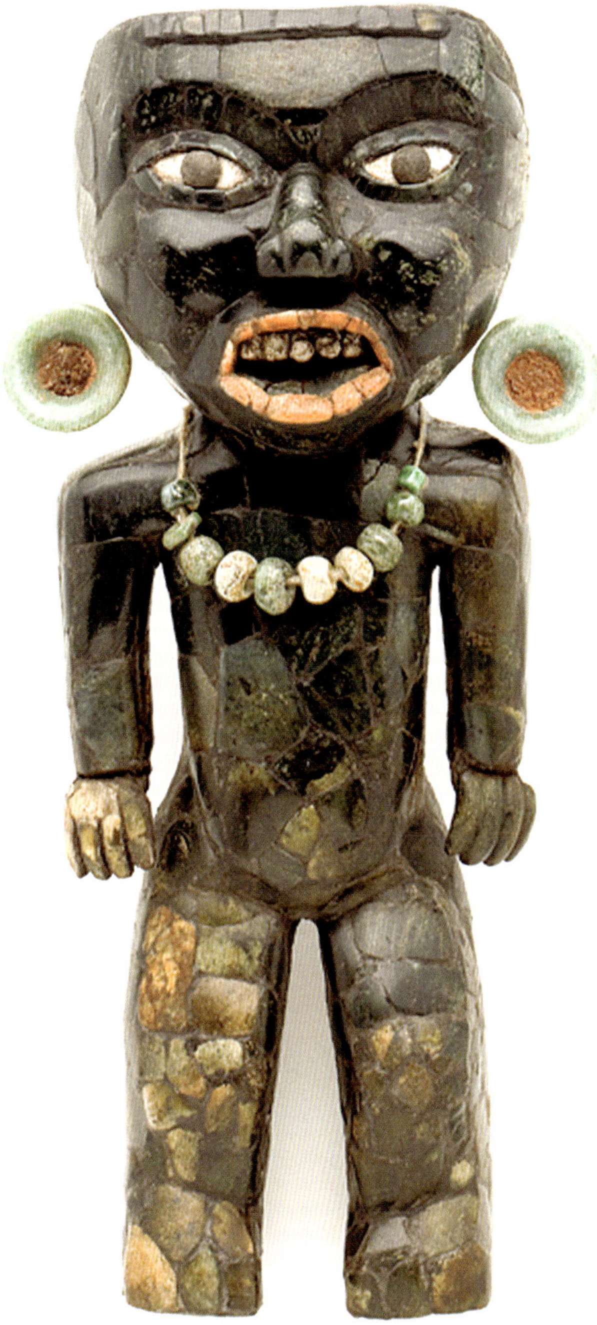

Figure 7. Serpentinite anthropomorphic mosaic after conservation. Courtesy of Estudio Zabé. (Photo: Michel Zabé.)

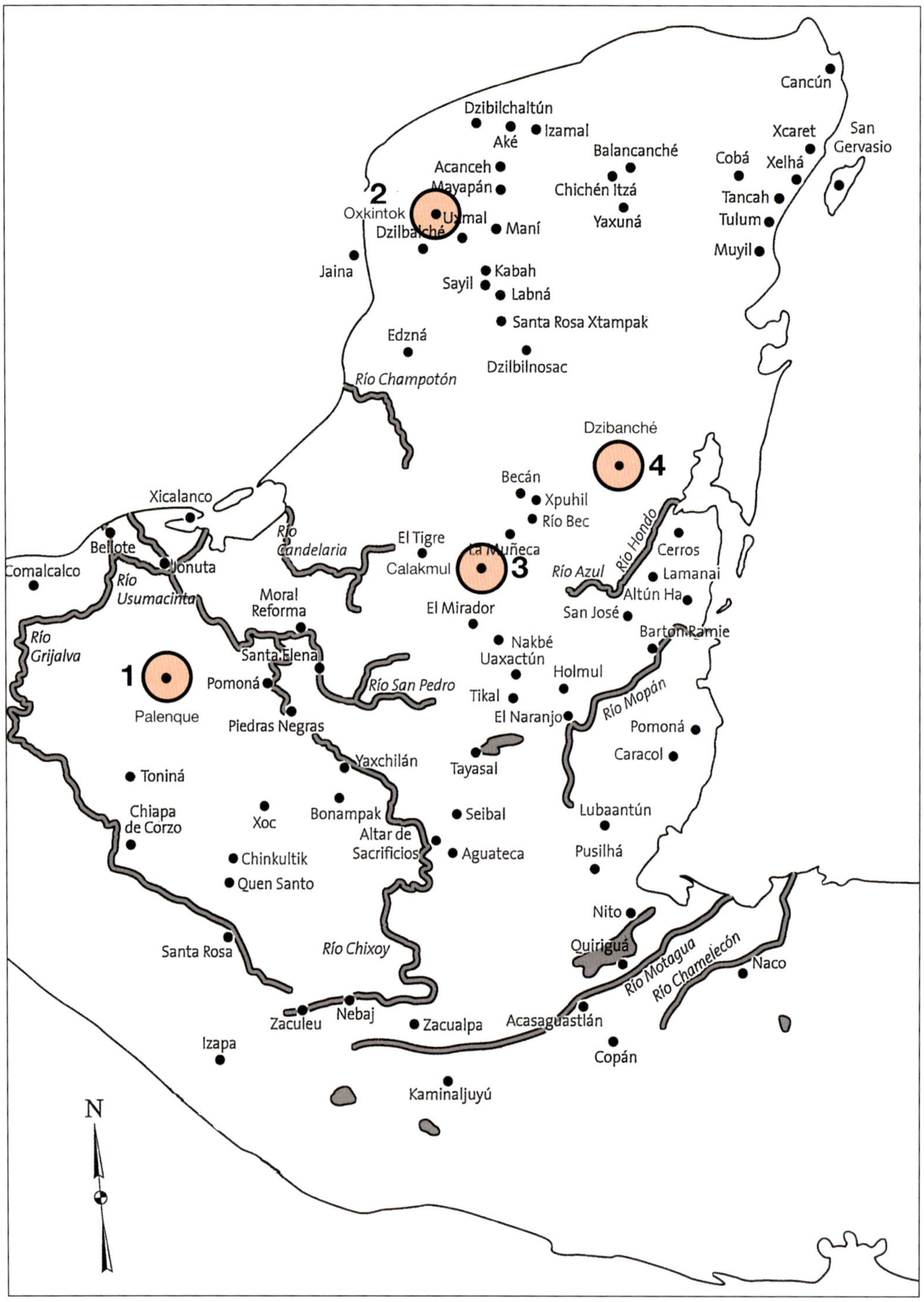

Figure 8. Map: archaeological sites with greenstone mosaic, Mayan area. Drawing by Ivette García Arreola, National Museum of Anthropology, CONACULTA-INAH.

Corpus of funerary masks from the Maya area

Until the mid-1980s, only two jade masks were known from the Maya area. The first was discovered in Burial 160 at Tikal (published by Harrison 2000). The second was the funerary mask of K'inich Janaab' Pakal of Palenque recovered in 1952 in the Temple of the Inscriptions (Ruz Lhuillier 1953, 1973). However, explorations undertaken by INAH at other sites in the Maya zone have led to the discovery of 12 other greenstone mosaic masks. The following discussion offers a brief overview of the different finds made at Oxkintok, Calakmul, Dzibanché and Palenque (Fig. 8).

Oxkintok, Yucatán

Two spectacular archaeological mosaics were found in Oxkintok, Yucatán, one made of shell and the other of greenstone. The first piece, a shell ornament in the shape of a bird's head, was discovered in a tomb from the Ah Canul group in a Late Classic context dated from AD 600 to 700 (Filloy Nadal 2004; Schmidt 2004: 83; Velázquez Valadez n.d.). The 130 shell pieces that formed the ornament were created from four different shell taxa: *S. princeps*, *Strombus gigas*, *P. mazatlanica* and one of the *Unio* genus (Velázquez Castro et al. 2004) (Fig. 9).

The jade mask from Oxkintok was recovered in Tomb 5 of Building CA-3, in the same Ah Canul group, but it comes from an earlier context dated from AD 550 to 630. This tomb contained an adult male together with a ceramic bowl, a jade and shell bead necklace, a stingray spine and a jade mask (Palomero Sánchez 1989: 105–107; Schmidt 2004: 82). The fact that the mask was restored at the MNA gave us an opportunity to analyse it carefully (Filloy Nadal 2004). To shape the face, Maya artists used 340 thin plaques of intense green jade: omphacite-cosmochlor pyroxene, identified by XRD (Sánchez and Robles 2003). The teeth and sclera of the eyes were made with pieces of *P. mazatlanica* shell (Velázquez Castro et al. 2003) and with two small grey obsidian disks. The lips were made with stucco painted a pinkish colour and the edge of the mask was of the same material, although dark red in colour.

This mask is entirely unique in the Maya zone because the mosaic was mounted on a support made of flint chips, ceramic fragments and stucco. Consequently, the facial features are extremely geometric and the surfaces tend to be flat. The little relief that exists was produced through the unevenness of the forehead, the eyelids, the nose and the cheekbones. However, the whole is harmonious and the countenance displays exaggerated cranial deformation and wears a headdress with a knot and an engraved, spiral-shaped, shell element, surrounded by semicircular pieces covered with jade. These elements denote the high status of the figure represented.

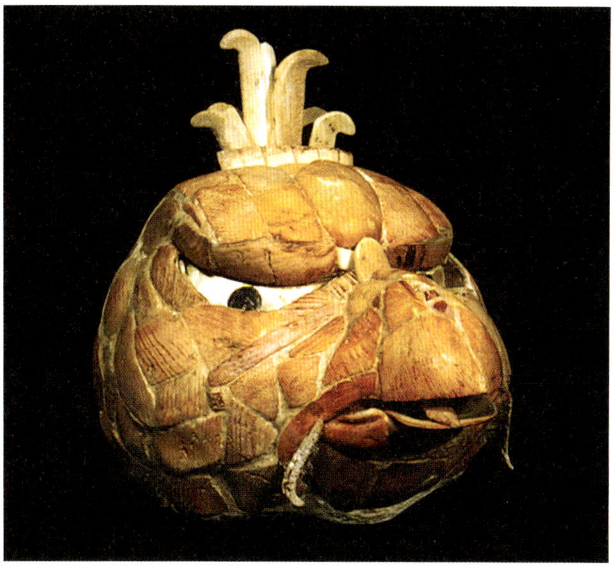

Figure 9. National Museum shell mosaic from Oxkintok, Yucatan. Courtesy of the Proyecto Fotográfico de la Colección del Museo Museo Nacional de Antropología, CONACULTA-INAH. (Photo: Luis Martín Martínez.)

Masks from the kingdom of *K'aan* (Calakmul and Dzibanché)

In the impressive city of Calakmul in Campeche, a total of five human masks made of greenstone has been found (to date). These examples were originally mounted on a tropical wood support. A magnificent face that formed part of a jade belt, a symbol of royal power, was also recovered. Three of the face masks were found in Early Classic royal tombs and the other two in Late Classic tombs (Carrasco 2004; Devendahl 2008).

In all cases, including the example from the belt, the shell taxon for the sclera of the eyes was *P. mazatlanica* or a species of the *Unio* genus (García Vierna 2007; Martínez del Campo Lanz 2004, 2007a). All the masks displayed grey obsidian appliqués to form the iris. In contrast, the materials used in the decorations are less homogeneous and they were made either with *S. princeps*, *S. calcifer* or *S. gigas* shell (Martínez del Campo Lanz 2004, 2007a). In one case, delicate pieces made of specular hematite were also used. It is interesting to note that the restorers managed to identify the wood used as the support in two of the masks (specimens from Tomb 4, Building II and Tomb 1, Building II-D of Calakmul); in both cases mahogany was used (García Vierna 2007).

In all the royal masks from Calakmul, jade was chosen as the principal material. Different tonalities of jade were selected for each one of the pieces. One might suppose that this precious mineral was imported from the Motagua River region, the only geological source known to date (for further information on the Motagua sources, see: Bishop 2001; Digby 1972; Foshag and Robert 1955; Harlow 1993; Keverne 1995; Lange 1993). In each case, the size of the tesserae that formed the mosaic is different and the range of colour selected is also characteristic of each mask. The artists who created four of the masks carved bevelled edges on each tessera which allowed for almost perfect assembly of the entire puzzle. In only one case (the mask from Tomb 3, Structure XV, inv. no. 10-566426) tessellae were used, in other words tiny pieces of jade to form the mosaic. In all the masks, a pigmented lime paste was employed to fill in the gaps between the tesserae and to unify the surface. The mask found in the tomb of Yuknoom Yich' Aak K'ak (Jaguar Paw Smoke, Tomb 4, Structure sub-IIB), who was ruler at Calakmul between AD 686 and 695, is particularly interesting because it has an edge decorated with an inscription that mentions the name of his predecessor, Yuknom the Great, as the owner of the mosaic mask (García Vierna 2007; Martin and Grube 2000: 109).

Just as we saw for the masks from Tikal, Palenque and Oxkintok, the new mounting undertaken by INAH restorers has shown that four of the examples from Calakmul display elaborate headdresses, diadems, earflares and pectorals that complemented the royal iconography and embellished the face of the ruler portrayed. The intentional handling of colour is evident in the mask from the reign of K'an in which a variety of hues was used to mark the different sections of the face, but above all to emphasise the elements of adornment: the diadem in the first case and the headdress and pectoral in the second.

The example of Dzibanché, Quintana Roo, where five mosaic masks were found (Nalda 2004), is quite different. Four masks were associated with burials and the last one was from a cache. A characteristic trait of masks from this zone is that jade was not used to make them. On the contrary, the artists selected a mineral with beautiful green tonalities: chrysoprase, a variety of green quartz was identified with XRD (Sánchez and Robles 2007: 18). It is likely that the inhabitants of this Maya centre, which was under the influence of Calakmul, did not have access to jade that came from the Motagua River zone and therefore they chose another type of greenstone. In Dzibanché, four of the masks that were made in the Late Classic have irises made of grey obsidian, the sclera is made of shell, including *P. mazatlanica* and a species of the *Unio* genus

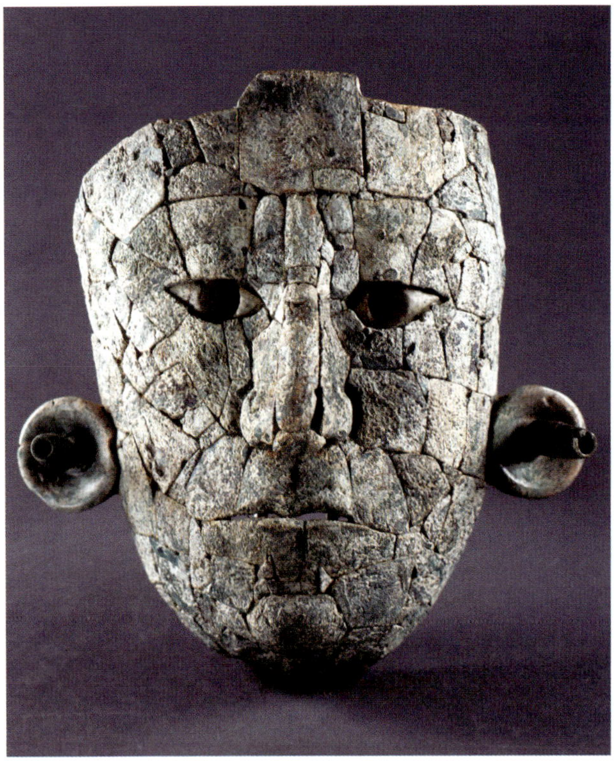

Figure 10. The Red Queen from Palenque, Chiapas. Funerary mask made with malachite tesserae over wood. Museo de Sitio, Palenque Archaeological Site. Photograph courtesy of CONACULTA-INAH.

and the teeth of *Turbinella* (Martínez del Campo Lanz 2007b). The fourth mask, from the Early Classic and found by Enrique Nalda (2004) in Tomb 1 in the Kinichná group at Dzibanché, is special because the sclera was carved in milky-coloured rock crystal and the iris from jade. To date, this is the only Maya mask that is known with this type of appliqué.

Although the Mayas chose different classes of minerals to create the tesserae of their masks, all the stones used were of tonalities that ranged from very light green to intense green. *Ya'x* is the Maya word for both green and blue. Green and blue are the colours of the centre of the cosmos, the point of convergence, the axis mundi. The significance of green is linked to vegetation, water as the vital liquid, maize, fertility, centrality, and as Karl Taube (1996, 2001) has discussed, vital essence, breath and rulership. Therefore it was synonymous with status, wealth, eternity, precious qualities, life and its continuity, and its use was an elite prerogative and restricted to whoever held power.

Kingdom of *Baakal* (Palenque, Chiapas)

From the 1950s to the present, eight greenstone mosaic pieces have been excavated at Palenque. Five of them come from royal jade belts (recovered from the Temple of the Inscriptions, Temple XVIIIa and the Olvidado) and one from a sceptre (Temple XIII-sub). Here I will concentrate on the description of two exceptional funerary masks from the Late Classic period. The first was excavated in Temple XIII-sub and is known as the mask of the Red Queen, and the second is that found by Alberto Ruz in the Temple of the Inscriptions.

In 1994, a new tomb was discovered at Palenque (González Cruz 2004, 2011). Unlike other sepulchres of the city, this time it lacked inscriptions identifying the individual interred. Anthropometric studies determined that the bone remains belonged to a woman (Tiesler et al. 2004). She was accompanied by a rich offering – necklaces, armlets, a mask from a royal belt and a diadem, all made of jade. In addition, the tributes included a beautiful unique mask crafted with tesserae made of malachite. This copper mineral had formerly only been identified in pigments used at Palenque and Bonampak for mural painting. A mine that exploited copper in the nineteenth century was known in the zone of Teapa in the northern part of the state of Chiapas (González Cruz 2008). The mineral could well have been brought from there during the Classic period as it is a zone easily accessed via the region's river network (González Cruz 2011). To date, the mask of the Red Queen is the only one that has been recovered in a woman's burial (Fig. 10).

The tomb in the Temple of the Inscriptions was discovered in November 1952 by archaeologist Alberto Ruz. There he found a spectacular burial in the interior of a massive stone sarcophagus and a beautiful jade mask that had become completely disarticulated (Ruz Lhuillier 1953, 1973). In September 2001, we had the opportunity to study and restore the mask of the most celebrated ruler of Palenque (Filloy Nadal 2010; Filloy Nadal and Martínez del Campo 2002).

Pakal's mask is composed of two pieces of grey obsidian, four pieces of seashell and 340 pieces of jade in six tonalities (Filloy Nadal and Martínez del Campo Lanz 2010: 114), including jadeite, ureyite and albite (these minerals were identified both by XRD, by Sánchez Hernández et al. 2010, and laser ablation-inductively coupled plasma-mass spectrometry (LA-ICP-MS) by Neff et al. 2010) – probably obtained from sources of the Motagua River. As for the mask's eyes, the grey obsidian used for the iris came from Guatemala, probably from El Chayal (Pastrana 2010). To make the pupils, a small black circle was painted with black pigment on the back of the obsidian iris. In turn, to make the sclera, the artisan used shell of the *P. mazatlanica* species (Velázquez et al. 2010b).

Through meticulous analysis of the pieces of jade mosaic, we were able to see that they were worked first by means of cutting and shaping. Once the piece was cut, the back was polished, while the front was burnished to an even finer lustre. In the technological study conducted on some of the tesserae (replication polymers/SEM, as discussed above), by comparing the traces detected in the archaeological pieces with those produced in the laboratory, it was noted that the pieces were probably shaped on limestone slabs, polished with another piece of jadeite, and burnished with leather. This technical sequence is common in the Maya area (Filloy 2011b; Filloy Nadal and Melgar Tisoc 2009; Kovacevich 2007, 2010) and the traces are very different from those obtained by polishing with sand and burnishing with leather. In summary, the technical sequence to produce the greenstone pieces in Janaab' Pakal's mask could have been: (a) shaping on limestone grinding stones; (b) cutting and incising with obsidian blades; (c) perforating with awls or pointed flint chips or blades, (d) polishing the surface with jadeite nodules, and (e) burnishing with pieces of leather.

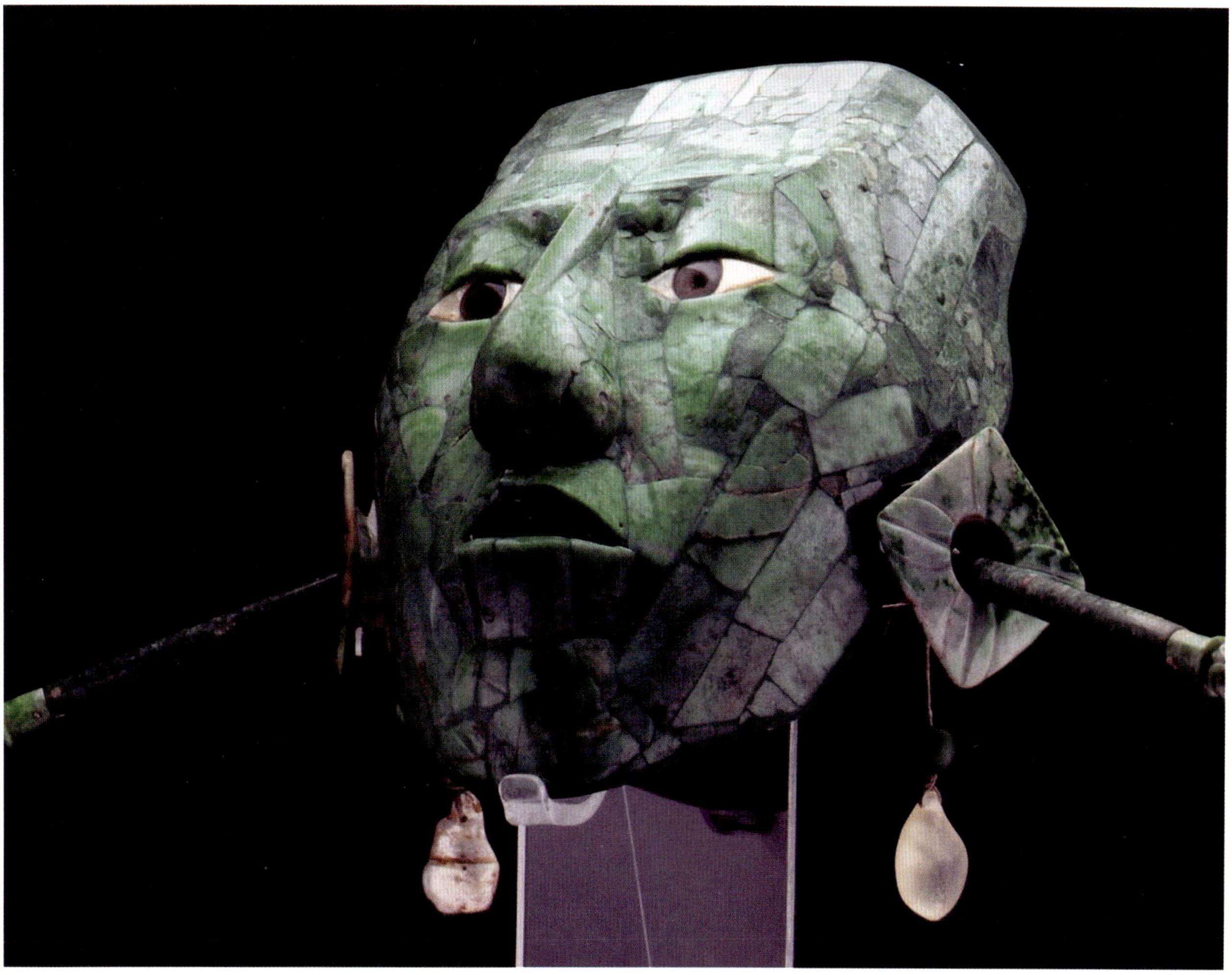

Figure 11. K'inich Janaab' Pakal from Palenque. Funerary mask made of jadeite, mother-of-pearl and obsidian tesserae. National Museum of Anthropology. Courtesy of CNME-CONACULTA-INAH. (Photo: Ignacio Guevara.)

Among the 340 pieces of jade, no two are completely alike but all have slanted edges for precise assembly; some pieces still preserve traces of stucco and cinnabar on the back (identified by XRF), and a large number of jade pieces are reused fragments of jade ornaments. There are also fragments of pendants, beads and ear plugs. Other fragments have carving on the back of human faces, geometric motifs or glyphs. Also, some pieces display old perforations that the creators of the mask covered with small jade plugs sealed with a resinous substance from trees of the Burseraceae family, a beeswax and lime paste (Maynard and Berdan 2010). In addition, two large pieces had prehispanic restorations.

Once all the pieces of the mosaic were finished, the artists assembled the puzzle on a rigid backing, probably originally made of wood, sticking the bits of jade onto the surface with a layer of lime paste, perfectly adjusting the joins to create relief. Once the jade pieces were in position, they accentuated Pakal's features by highly polishing the mask. After that, they applied a fine milk of lime (a suspension in water of fine calcium hydroxide particles) to reinforce the mosaic and to fill in the tiny gaps. Thanks to the recent discovery of other Maya jade masks, we have been able to see that lime paste used between the joins of the jade fragments was tinted to match the colour of the stone and was even polished to obtain a sheen resembling that of jade (García Vierna 2007).

The use of a wood support and lime paste as an adhesive made it possible to fit the pieces together exactly. During the process of remounting Pakal's mask we realised that there was a logical sequence for the placement of the tesserae. The mask had to be assembled by beginning with the nose, the eyes and the mouth. It continued on the cheekbones, the cheeks and chin; then the forehead and diadem were assembled. The final step was to place the ear ornaments and the edge. We observed that the mask was divided into six different shades of green and noted that there is a kind of bilateral symmetry between the pieces of mosaic because they were more or less similar on both sides of the face, both in terms of position as well as in form. Placing the largest pieces on the outer part of the eyes determines the inclination of the iris and simulates the internal squint, one of Pakal's specific features (Filloy 2011a,b). After the restoration, we were certain that the harmonious face of this mask was anatomically conceived prior to its manufacture and was carved with surprising realism. It is without doubt a faithful portrait of the most celebrated king of Palenque (Filloy Nadal and Martínez del Campo Lanz 2004) (Fig. 11).

When Alberto Ruz raised the lid of the sarcophagus in the Temple of the Inscriptions he could see that all the inner walls were painted with a bright red colour, which he identified as cinnabar (Ruz Lhuillier 1973: 56). In the photographs of the discovery, it is evident that both the bones and the associated objects also bore remains of this mineral. In fact, as we mentioned earlier, when we analysed the different artefacts (including the back of the mask's greenstone tesserae) the remains of cinnabar (which was identified by XRF, Ruvalcaba 2010) could be seen with the naked eye.

Conclusions

In the course of this brief survey, we have attempted to highlight the mastery and skill of the technique used by artists from Teotihuacan and the Maya zone to produce sublime stone mosaics. The green minerals employed in both regions to make the tesserae are of considerable hardness. Using tools of andesite and flint in the case of central Mexico and tools of limestone and obsidian in the Maya zone, artists managed to produce mosaics of great artistry. We might wonder whether the Mesoamerican mosaic tradition was inherited by the Mexica and Mixtec heirs to these earlier cultures, whose master artists achieved the refinement of pieces that are today safeguarded in the world's museums.

References

Aguirre, P. (1922) 'Informe del Ayudante Porfirio Aguirre', *Boletín del Museo Nacional de Arqueología, Historia y Etnografía* **I**: 34–35.

Bishop, R. (2001) 'Jade', in S.T. Evans and D.L. Webster (eds), *The Archaeology of Ancient Mexico and Central America*. New York: Garland Publishing Co., 381–385.

Carrasco, R. (2004) 'Maya architecture and funerary rites', *Arqueología Mexicana*, Edición especial **16**: 78–79.

Castillo Ledón, L. (1922) 'Máscara con mosaico de turquesa. Dictámenes periciales', *Boletín del Museo Nacional de Arqueología, Historia y Etnografía* **I**: 33.

Cummings, B. (1922) 'Dictamen', *Boletín del Museo Nacional de Arqueología, Historia y Etnografía* **I**: 38–39.

de la Fuente, B. (2003a) 'Peldaños en la conciencia. Rostros en la plástica prehispánica', in B. de la Fuente and V. Hernández Díaz (eds), *El arte, la historia y el hombre. Arte prehispánico de México: estudios y ensayos* (Obras de Beatriz de la Fuente, t.1). Mexico City: El Colegio Nacional, 191–359.

de la Fuente, B. (2003b) 'Introducción a la escultura precolombina en Mesoamérica', in B. de la Fuente, L. Staines Cicero and M.T. Uriarte (eds), *La escultura prehispánica de Mesoamérica*. Mexico City/Milan: CONACULTA/Jaca Book, 11–57.

de la Fuente, B. (2003c) 'Lo humano y lo digno', in B. de la Fuente and V. Hernández Díaz (eds), *La escultura del México Antiguo* (Obras de Beatriz de la Fuente, t.2). Mexico City: El Colegio Nacional, 221–231.

Devendahl, K. (2008) *Calakmul in Sight: History and Archaeology of an Ancient Maya City*. Yucatan, Mexico: Unas Letras, Industria Editorial.

Digby, A. (1972) *Maya Jades*. London: The Trustees of the British Museum.

Domínguez Carrasco, M.R. and Folan, W.J. (1999) 'Hilado, confección y lapidación: los quehaceres cotidianos de los artesanos de Calakmul, Campeche, México', in J.P. Laporte and H.L. Escobedo (eds), *XII Simposio de Investigaciones Arqueológicas en Guatemala, 1998*. Guatemala City: Museo Nacional de Arqueología y Etnología, 628–646.

Estrada, G. (1922) 'Dictamen', *Boletín del Museo Nacional de Arqueología, Historia y Etnografía* **I**: 40–42.

Filloy Nadal, L. (2004) 'Se restaura una máscara de jade y un pendiente zoomorfo de concha de Oxkintok, Yucatán', *Arqueología Mexicana* **XI**(65): 10–11.

Filloy Nadal, L. (ed.) (2010) *La máscara funeraria de K'inich Janaab Pakal*. Mexico City: Instituto Nacional de Antropología e Historia (INAH).

Filloy Nadal, L. (2011a) 'Imágenes convencionales o retratos. Los rostros de Pakal II de Palenque', in M.C. Valverde and V. Solanilla Demestres (eds), *Las imágenes precolombinas, reflejo de Saberes*. México: Instituto de Investigaciones Filológicas/ Universidad Nacional Autónoma de México (UNAM), 161–192.

Filloy Nadal, L. (2011b) 'Costume et insignes d'un roi maya de Palenque : K'inich Janaab Pakal. Vie et mort d'un souverain maya du Classique Récent', PhD thesis, Université de Paris I-Pantheon-Sorbonne, Paris (unpublished).

Filloy Nadal, L. and Martínez del Campo Lanz, S. (2002) 'El nuevo rostro de Pacal', *Arqueología Mexicana* **IX**(51): 12.

Filloy Nadal, L. and Martínez del Campo Lanz, S. (2004) 'La restauración de la máscara funeraria de Pacal: esencia e imagen del soberano', in R. Cobos (ed.), *Culto funerario en la sociedad maya, Memoria de la Cuarta Mesa Redonda de Palenque*. Mexico City: Instituto Nacional de Antropología e Historia (INAH), 81–98.

Filloy Nadal, L. and Martínez del Campo Lanz, S. (2010) 'El rostro eterno de K'inich Janaab' Pakal. La máscara funeraria', in L. Filloy Nadal (ed.), *Misterios de un rostro maya. La máscara funeraria de K'inich Janaab Pakal*. Mexico City: Instituto Nacional de Antropología e Historia (INAH), 109–130.

Filloy Nadal, L. and Melgar Tisoc, E.R. (2009) 'Manufacture technique of a figurine made of a greenstone mosaic from the Pyramid of the Moon, Teotihuacan, Mexico', paper presented at *XVIII International Materials Research Congress*, Cancún, Quintana Roo, México, 18 September 2009 (unpublished).

Filloy Nadal, L., Gumí, M.E. and Watanabe, Y. (2006) 'La restauración de una figura antropomorfa teotihuacana de mosaico de serpentina', in S. Sugiyama and L. López Luján (eds), *Sacrificios de consagración en la Pirámide de la Luna*. Mexico City: Instituto Nacional de Antropología e Historia (INAH)/CONACULTA / Museo del Templo Mayor/Arizona State University, 61–79.

Foshag, W.F. and Robert, L. (1955) 'Jadeite from Manzanal, Guatemala', *American Antiquity* **21**: 81–83.

García Vierna, V.A. (2007) 'Imágenes materializadas: las máscaras de jadeíta como retrato y representación en el Clásico Maya', MD thesis, Mexico City, Escuela Nacional de Antropología e Historia (unpublished).

González, C.J. and Olmedo, B. (1986) 'El trabajo de la turquesa', in L. Manzanilla (ed.), *Unidades Habitacionales Mesoamericanas y sus Áreas de Actividad*. Mexico City: Universidad Nacional Autónoma de México (UNAM), 103–114.

González Cruz, A. (2004) 'The Red Queen's masks: symbols of power', *Arqueología Mexicana*, Edición especial **16**: 80–81.

González Cruz, A. (2008) 'La máscara facial de la Reina Roja. Su origen y posible significado', paper presented at *VI Mesa Redonda de Palenque: Arqueología, Imagen y Texto*, Palenque, Chiapas, México, November 2008 (unpublished).

González Cruz, A. (2011) *La reina roja: una tumba real de Palenque*. Mexico: Turner/ Instituto Nacional de Antropología e Historia (INAH).

Harbottle, G. and Weigand, P.C. (1992) 'Turquoise in pre-Columbian America', *Scientific American* **266**(2): 78–85.

Harlow, G.E. (1993) 'Middle American jade: geologic and petrologic perspectives on variability and source', in F.W. Lange (ed.), *Precolombian Jade: New Geological and Cultural Interpretations*. Salt Lake City: Utah University Press, 9–29.

Harrison, P.D. (2000) *The Lords of Tikal: Rulers of an Ancient Maya City*. London: Thames and Hudson.

Instituto Nacional de Antropología e Historia (INAH) (2009) *Teotihuacan Ciudad de los Dioses*. Mexico City: CONACULTA/ INAH.

Izeki, M. (2008) *Conceptualization of 'Xihuitl': History, Environment and Cultural Dynamics in Postclassic Mexica Cognition*. BAR International Series 1863. Oxford: BAR.

Keen, M.A. (1971) *Sea Shells of Tropical West America*. Stanford, CA: Stanford University Press.

Keverne, R. (ed.) (1995) *Jade*. London: Lorenz Books.

Kidder, A.V. (1947) *The Artifacts of Uaxactun, Guatemala*. Washington, DC: Carnegie Institution of Washington (Publication no. 576).

Kovacevich, B. (2007) 'Ritual, crafting, and agency at the Classic Maya kingdom of Cancuen', in E.C. Wells and K.L. Davis Salazar (eds), *Mesoamerican Ritual Economy: Archaeological and Ethnological Perspectives*. Boulder, CO: University Press of Colorado, 67–114.

Kovacevich, B. (2010) 'Crafting classic Maya jades', in E. Melgar Tísoc, R. Solís Ciriaco and E. González Licón (eds), *Producción de bienes de prestigio ornamentales y votivos de la América antigua*. Doral, FL: Syllaba Press, 149–158.

Lange, F.W. (ed.) (1993) *Precolombian Jade: New Geological and Cultural Interpretations*. Salt Lake City: Utah University Press.

López Luján, L. (1989) *La recuperación mexica del pasado teotihuacano* (Colección Divulgación). Mexico City: Instituto Nacional de Antropología e Historia (INAH)/GV Editores/ Asociación de Amigos del Templo Mayor.

Martin, S. and Grube, N. (2000) *Chronicle of Maya Kings and Queens: Deciphering the Dynasties of the Ancient Maya*. London: Thames and Hudson.

Martínez del Campo Lanz, S. (2004) 'Restauración en el MNA. Máscara funeraria de Calakmul, Campeche', *Arqueología Mexicana* **XII**(67): 10.

Martínez del Campo Lanz, S. (2007a) 'Los mosaicos ceremoniales de la Estructura III de Calakmul. Representaciones de la divinidad', *Arqueología Mexicana* **XVI**(88): 16–18.

Martínez del Campo Lanz, S. (2007b) 'Dos máscaras de Dzibanché, Quintana Roo', *Arqueología Mexicana* **XIV**(84): 16–19.

Maynard, D. and Berdan, F. (2010) 'El adhesivo y el material de reparación de la máscara', in L. Filloy Nadal (ed.), *Misterios de un rostro maya. La máscara funeraria de K'inich Janaab Pakal*. Mexico City: Instituto Nacional de Antropología e Historia (INAH), 155–164.

Mirambell, L.E. (1968) *Técnicas lapidarias prehispánicas*. Mexico City: Instituto Nacional de Antropología e Historia (INAH).

Nalda, E. (2004) 'The jade masks of Dzibanché, Quintana Roo', *Arqueología Mexicana*, Edición especial **16**: 81–82.

Neff, H., Kovacevich, B. and Bishop, R.L. (2010) 'Caracterización de los compuestos de la jadeíta mesoamericana: breve revisión a partir de los resultados obtenidos durante el estudio de la máscara de K'inich Janaab' Pakal', in L. Filloy Nadal (ed.), *Misterios de un rostro maya. La máscara funeraria de K'inich Janaab Pakal*. Mexico City: Instituto Nacional de Antropología e Historia (INAH), 131–138.

Níven, W. (1922) 'Dictamen', *Boletín del Museo Nacional de Arqueología, Historia y Etnografía* **I**: 39–40.

Noriega, J.M. (1922) 'Dictamen', *Boletín del Museo Nacional de Arqueología, Historia y Etnografía* **I**: 41–42.

Palomero Sánchez, M.A. (1989) 'Informe sobre los objetos de Oxkintok. Temporada 1988', in M. Rivera Dorado (ed.), *Oxkintok* 2. Madrid: Misión Arqueológica de España en México, 93–107.

Pastrana, A. (2010) 'La obsidiana de los ojos', in L. Filloy Nadal (ed.), *Misterios de un rostro maya. La máscara funeraria de K'inich Janaab Pakal*. Mexico City: Instituto Nacional de Antropología e Historia (INAH), 151–154.

Quintanar Isaías, A. (2005) *Identificación preliminar de una muestra de madera proveniente del Entierro 6 de la Pirámide de la Luna*. Mexico City: Departamento de Biología, Universidad Autónoma Metropolitana-Iztapalapa (unpublished).

Rochette, E.T. (2009) 'Jade in full: prehispanic domestic production of wealth goods in the middle Motagua Valley, Guatemala', in K.G. Hirth (ed.), *Archaeological Papers of the American Anthropological Association* **19**(1): 205–224.

Ruvalcaba, Sil, J.L. (2010) 'Informe de los análisis practicados *in situ* al ajuar del Templo de las Inscripciones (cerámica, pigmentos y lítica) y a los cinturones del Templo Olvidado y del Templo XVII-sub de Palenque (13 al 22 de diciembre de 2010)', IFF/UNAM. Mexico City: Archivo del Laboratorio de Conservación del Museo Nacional de Antropología, INAH (unpublished).

Ruvalcaba Sil, J.L., Filloy Nadal, L., Vaggi, M., Tapia Gálvez, L.H. and Sánchez Becerra, R. (2010) 'Estudio no destructivo in situ de la Máscara de Malinaltepec', in S. Martínez del Campo Lanz (ed.), *La Máscara de Malinaltepec*. Mexico City: Instituto Nacional de Antropología e Historia (INAH), 153–168.

Ruz Lhuillier, A. (1953) 'The mystery of the Temple of the Inscriptions, Palenque', *Archaeology* **6**: 3–11.

Ruz Lhuillier, A. (1973) *El Templo de las Inscripciones* (Colección científica, arqueología 7). Mexico City: Instituto Nacional de Antropología e Historia (INAH).

Sahagún, Fray Bernardino de (2000) *Historia general de las cosas de Nueva España*, A. López Austin and J. García Quintana (eds), 3 vols. Mexico City: CONACULTA.

Sánchez, R. and Robles, J. (2003) 'Difractometría en muestras de teselas representativas del mosaico de la máscara de Oxkintok, Yucatán (10-290383). Informe preliminary'. Mexico City: Instituto Nacional de Antropología e Historia (INAH) (unpublished).

Sánchez, R. and Robles, J. (2005) 'Petrografía y mineralogía de los componentes líticos de una figura antropomorfa de mosaico procedente de la Pirámide de la Luna, Zona Arqueológica de Teotihuacan. Informe preliminar'. Mexico City: Instituto Nacional de Antropología e Historia (INAH) (unpublished).

Sánchez, R. and Robles, J. (2007) 'Dos máscaras de Dzibanché, Quintana Roo', *Arqueología Mexicana* **XIV**(84): 18.

Sánchez Hernández, R. and Robles Camacho, R. (2002) 'Amazonita en la Máscara de Malinaltepec, Guerrero', *Arqueología Mexicana* **X**(56): 7.

Sánchez Hernández, R. and Robles Camacho, R. (2010) 'Mineralogía, petrografía y características lapidarias de la Máscara de Malinaltepec', in S. Martínez del Campo Lanz (ed.), *La Máscara de Malinaltepec*. Mexico City: Instituto Nacional de Antropología e Historia (INAH), 125–152.

Sánchez Hernández, R., Robles Camacho, R. and Reyes Salas, M. (2002) 'Uso alternativo de amazonita y turquesa en objetos mesoamericanos', *GEOS, Unión de Geofísica Mexicana A.C., Boletín Informativo* **22**: 287–288.

Sánchez Hernández, R., Robles Camacho, J. and Reyes Salas, M. (2010) 'La mineralogía de la piedra verde', in L. Filloy Nadal (ed.), *Misterios de un rostro maya. La máscara funeraria de K'inich Janaab Pakal*. Mexico City: Instituto Nacional de Antropología e Historia (INAH), 139–144.

Schmidt, P. (2004) 'The masks of Oxkintok, Yucatán', *Arqueología Mexicana*, Edición especial **16**: 82–83.

Schmitter-Villada, E. and Martín del Campo de Schmitter, R. (1980) *Glosario de Especies Minerales*. Mexico City: Instituto de Geología/ Universidad Nacional Autónoma de México (UNAM).

Smith, A.L. and Kidder, A.V. (1951) *Excavations at Nebaj, Guatemala*. Washington, DC: Carnegie Institution (Publication 594).

Sugiyama, S. and Cabrera Castro, R. (2007) 'The Moon Pyramid Project and the Teotihuacan state politics: a brief summary of the 1998–2004 excavations', *Ancient Mesoamerica* **18**: 109–125.

Sugiyama, S. and López Luján, L. (2007) 'Dedicatory burial/offering complexes at the Moon Pyramid, Teotihuacan. Preliminary report of 1998–2004 explorations', *Ancient Mesoamerica* **18**: 127–146.

Taube, K. (1996) 'The Olmec maize god: the face of corn in formative Mesoamerica', *RES: Anthropology and Aesthetics* **29–30**: 39–81.

Taube, K. (2001) 'Maize: iconography and cosmological significance', in D. Carrasco (ed.), *The Oxford Encyclopedia of Mesoamerican Cultures*. Oxford: Oxford University Press, 150–152.

Tiesler, V., Cucina, A. and Romano A. (2004) 'Vida y muerte del personaje del Templo XIII-sub, Palenque, Chiapas. Una mirada bioarqueológica', in R. Cobos (ed.), *Culto funerario en la sociedad maya, Memoria de la Cuarta Mesa Redonda de Palenque*. Mexico City: Instituto Nacional de Antropología e Historia (INAH), 455–482.

Turner, M. (1992) 'Style in lapidary technology: identifying the Teotihuacan lapidary industry', in J.C. Berlo (ed.), *Art, Ideology, and the City of Teotihuacan*. Washington, DC: Dumbarton Oaks Research Library and Collections, 89–112.

USA Geological Survey (2002) 'Feldspar', http://minerals.usgs.gov/minerals/pubs/commodity/gemstones/sp14-95/feldspar.html (accessed 20 November 2009).
Velázquez Castro, A., Zúñiga Arellano, B. and Valentín Maldonado, N. (2003) 'Informe de las especies malacológicas presentes en la máscara de piedra verde de Oxkintok, Yucatán (10-290383)'. México: Proyecto técnica de manufactura de los objetos de concha del México prehispánico/ Instituto Nacional de Antropología e Historia (INAH) (unpublished).
Velázquez Castro, A., Zúñiga Arellano, B. and Valentín Maldonado, N. (2004) 'Informe de las especies malacológicas presentes en el pendiente de Oxkintok, Yucatán (10-517327)'. Mexico City: Proyecto técnica de manufactura de los objetos de concha del México prehispánico/ Instituto Nacional de Antropología e Historia (INAH) (unpublished).
Velázquez Castro, A., Zúñiga Arellano, B., Valentín Maldonado, N., Melgar Tísoc, E.R. and Solís Ciriaco, R.B. (2010a) 'Análisis tecnológico de la máscara y el collar de Malinaltepec, Guerrero', in S. Martínez del Campo Lanz (ed.), *La Máscara de Malinaltepec*. Mexico City: Instituto Nacional de Antropología e Historia (INAH), 169–188.
Velázquez Castro, A., Zúñiga Arellano, B. and Valentín Maldonado, N. (2010b) 'La madreperla de los ojos', in L. Filloy Nadal (ed.), *Misterios de un rostro maya. La máscara funeraria de K'inich Janaab Pakal*. Mexico City: Instituto Nacional de Antropología e Historia (INAH), 145–150.
Velázquez Valadez, R. (n.d.) 'Informe preliminar de la Tumba 8, estructura CA-14 4-sub del grupo arquitectónico Ah Canul'. Mexico City: Centro INAH Yucatán (unpublished).
Waitz, P. (1922) 'Dictamen', *Boletín del Museo Nacional de Arqueología, Historia y Etnografía* **I**: 45–48.
Weigand, P.C. (1968) 'The mines and mining techniques of the Chalchihuites culture', *American Antiquity* **33**: 45–61.
Weigand, P. and Weigand, A. (2001) 'A macroeconomic study of the relationships between the ancient cultures of the American Southwest and Mesoamerica', in V. Fields, V. Zamudio-Taylor and E. Gamboa Carrera (eds), *The Road to Aztlan*. Los Angeles: Los Angeles County Museum of Art, 184–195.
Weigand, P.C., Harbottle, G. and Sayre, E.V. (1977) 'Turquoise sources and source analysis: Mesoamerica and the Southwestern U.S.A.', in T.K. Earle, and J.E. Ericson (eds), *Exchange Systems in Prehistory*. New York: Academic Press, 15–34.

Author

Laura Filloy Nadal, Museo Nacional de Antropología, Laboratorio de Conservación, Mexico City, Mexico

Cracking the code of pre-Columbian turquoise trade networks and procurement strategies

Sharon Hull and Mostafa Fayek

Introduction

The procurement and trade of turquoise was as important to the inhabitants of the pre-Columbian societies of the Valley of Mexico and the southwestern United States as silver was to the colonial societies of New Spain. Although over one million pieces of turquoise have been recovered from archaeological sites throughout the Greater Southwest and Mesoamerica (Harbottle and Weigand 1992; Weigand et al. 1977), the spatial and temporal patterns of turquoise trade and procurement strategies remain elusive.

Turquoise ($A_{0-1}B_6(PO_4)_4(OH)_8{\bullet}4H_2O$) is a hydrous copper aluminum phosphate mineral with copper (Cu^{2+}) or ferrous iron (Fe^{2+}) as the most common substitutions for the A site and aluminium (Al^{3+}) and ferric iron (Fe^{3+}) for the B site. Calcium (Ca^{2+}) or zinc (Zn^{2+}) can occupy the A site in some of the more rare members of the turquoise family (Foord and Taggart 1998). For decades archaeologists have sought to chemically 'fingerprint' turquoise, thus allowing them to link specific artefacts to their provenance regions. This would allow them to reconstruct turquoise trade networks and investigate turquoise procurement strategies. These studies have met with only limited success due to the intrinsic limitations of trace and rare earth element or chemical analysis of such a complex mineral that can range in colour and thus vary chemically within a single sample or turquoise deposit (Fig. 1). To overcome these limitations we developed a technique that utilises the stable isotope of hydrogen ($^{2}H/^{1}H$ or D/H) and copper ($^{65}Cu/^{63}Cu$) to identify the provenance regions of turquoise artefacts, thus allowing us to link turquoise artefacts to prehistoric turquoise

Table 1. δD and $\delta^{65}Cu$ values of turquoise provenance regions and artefacts.

Location	δD	2σ	$\delta^{65}Cu$	2σ	Provenance Region	Associated Dates
Turquoise Provenance Regions						
Sleeping Beauty, AZ	−76	7	7.3	0.5		
Morenci, AZ	−114	5	1.6	0.5		
Halloran Springs, CA	−98	8	6.6	0.8		
Crescent Peak, NV	−87	8	3.9	0.6		
Royston, NV	−116	7	3.6	0.6		
Number Eight, NV	−106	8	−2.3	0.6		
Old Hachita, NM	−107	4	2.3	0.5		
Castillian, NM	−95	8	1.1	0.5		
Turquoise Artefacts Chaco Canyon						
29SJ1360 – trash midden TT1	−117	4	6.0	0.4	Unknown	AD 700–820
29SJ1659 – pithouse Y	−75	6	6.9	0.4	Sleeping Beauty, AZ	AD 550–700
29SJ423 – pithouse A	−124	7	2.3	0.4	Morenci, AZ	AD 1020–1120
29SJ625 – room E	−110	8	4.6	0.4	Royston, NV	AD 920–1020
29SJ627 – kiva G	−113	4	3.9	0.8	Royston, NV	AD 1000–1040/50
29SJ627 – room 10, floor 2	−121	6	3.5	0.4	Royston, NV	AD 900–1000
29SJ627 – room 10, level 2	−125	5	5.2	0.5	Unknown	AD 1000–1040/50
29SJ628 – pithouse E	−111	6	3.1	0.4	Royston, NV	AD 600–820
29SJ628 – pithouse A	−112	5	5.7	0.5	Unknown	AD 600–820
29SJ629 – trash midden 59	−113	5	2.8	0.5	Old Hachita, NM	AD 820–920
29SJ633 – room 7, layer 8	−114	5	6.3	0.4	Unknown	AD 1100
29SJ633 – room 8, layer 2	−106	4	3.5	0.4	Royston, NV	AD 1200
Guadalupe Community						
ENM 848	−116	6	0.8	0.5	Morenci, AZ	
ENM 848	−99	6	−1.7	0.6	Number Eight, NV	
ENM 848	−86	6	3.0	0.4	Crescent Peak, NV	
ENM 848	−109	4	3.8	0.4	Royston, NV	
ENM 848	−105	4	5.8	0.5	Halloran Springs, CA	
ENM 848	−111	4	3.1	0.4	Royston, NV	
ENM 848	−111	4	8.1	0.4	Unknown	

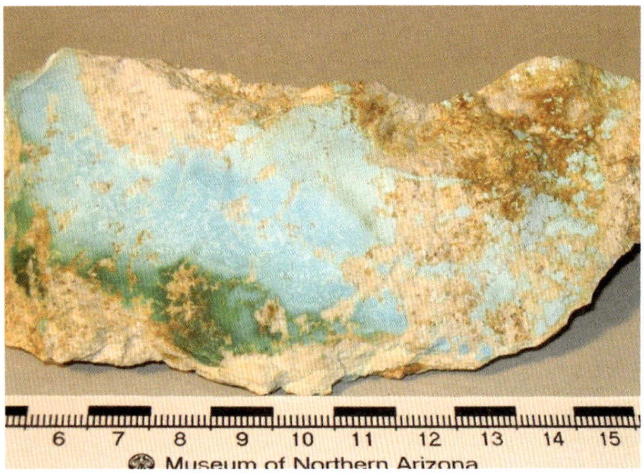

Figure 1. Turquoise sample from Nevada showing the range in colour in a single vein.

deposits hundreds, and in some instances, thousands of kilometres from the archaeological sites where the turquoise artefacts were recovered (Hull et al. 2008).

This technique is successful for several reasons. First, the hydrogen and copper isotopic signatures of turquoise are dictated by the geography and the geology of the turquoise deposits. Secondly, we use the *in situ* microanalytical capabilities of the secondary ion mass spectrometer (SIMS) to measure the isotopic ratios of the hydrogen and copper in turquoise. In this study, *in situ* is defined as a single analysis on solid material at the micrometer scale; not to be confused with the archaeological term *in situ*. SIMS is capable of performing accurate (e.g. ‰ to sub-‰) *in situ* isotopic measurements as small as 50 μm across and 2 μm deep on solid samples, allowing us to avoid inclusions of other minerals in the turquoise samples that could cause scatter in the data if bulk analytical techniques are

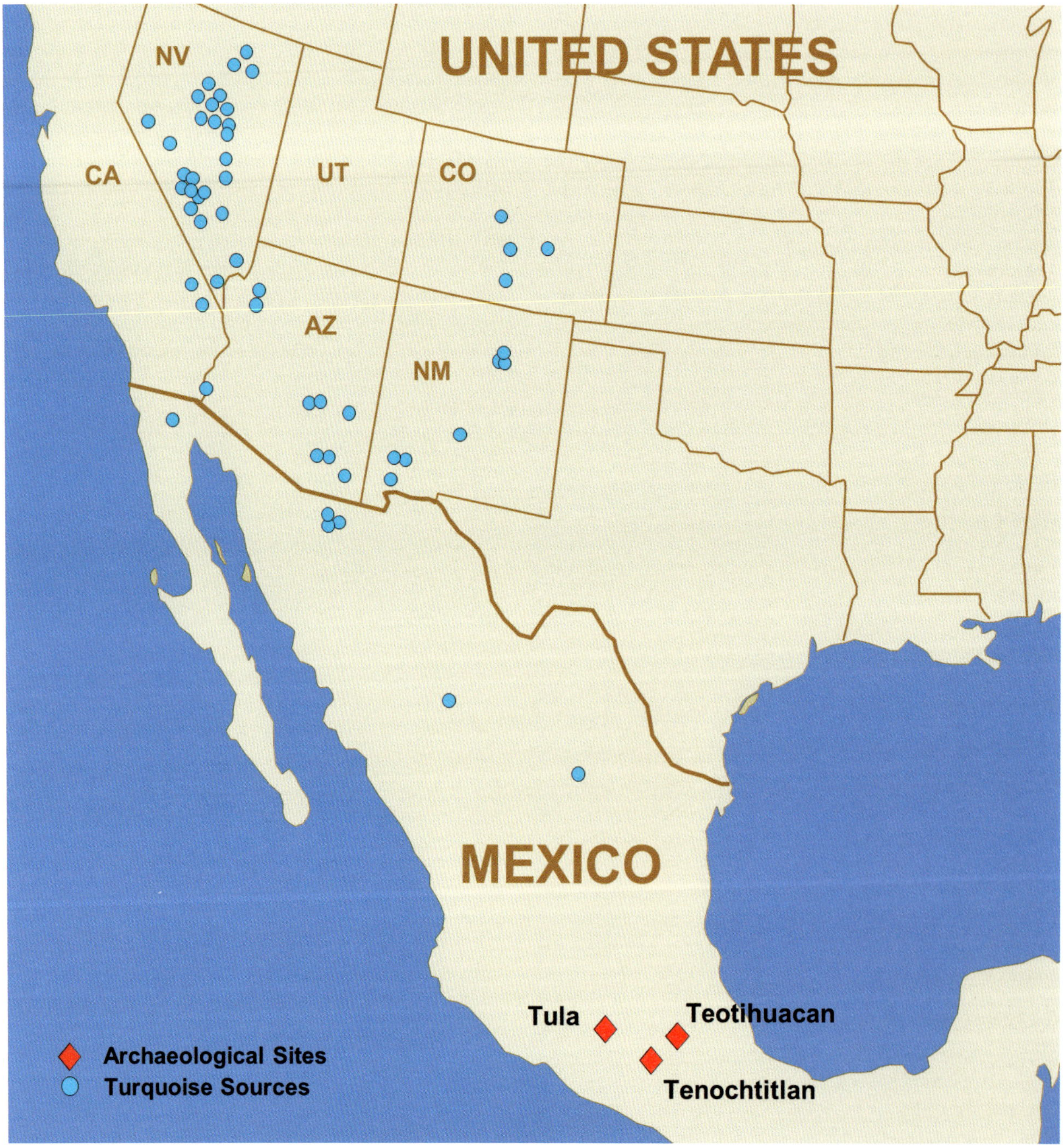

Figure 2. Turquoise deposits and Mesoamerican archaeological sites (modified from Weigand and Harbottle 1993).

employed on powdered samples. The small sample size required for analysis is particularly advantageous for analysing precious archaeological artefacts. Preparation for our technique includes polishing a flat surface of the turquoise artefact and several small (less than 50 μm) analysis pits that can only be seen under a high-powered microscope. Thus the technique is relatively non-destructive and turquoise artefacts can be returned to their original collections without affecting their physical appearance.

With this new technique we were able to isotopically characterise turquoise provenance regions and we are currently in the process of developing a database that contains the hydrogen and copper isotope ratios of analysed turquoise provenance regions and artefacts, thus allowing us to investigate pre-Columbian turquoise procurement strategies and exchange networks. For example, in a preliminary study we analysed 19 turquoise artefacts from the Guadalupe community and several archaeological sites in Chaco Canyon (see Table 1). Our results improved our knowledge of the Chacoan exchange and interaction, and the role of the inhabitants of several small sites in Chaco Canyon and Guadalupe Ruin in turquoise procurement and production.

Turquoise: a significant resource in the Greater Southwest and Mesoamerica

The Greater Southwest is an area that encompasses the southwestern United States and northern Mexico that includes the culture areas of the Hohokam, Mogollon, and the Ancestral Puebloans, formally known as the Anasazi. Shell objects from coastal shorelines and turquoise from throughout the western United States have been commonly recovered in archaeological sites throughout these culture areas (e.g. Di Peso 1974; Fish 1989; Harbottle and Weigand 1992; Pepper 1920). Other exotic items that have been recovered in these archaeological sites include macaws and copper bells that most likely originated from the complex societies far to the south in the Valley of Mexico. It appears that trade goods were being transported in both directions. Large quantities of turquoise, presumably from the western United States, have been recovered throughout Mesoamerican archaeological sites thousands of kilometres from where most of the known turquoise deposits are located (Fig. 2; Harbottle and Weigand 1992). Some turquoise artefacts have been recovered at Teotihuacan, an early regional centre in the Valley of Mexico, suggesting that turquoise was obtained either directly or through exchange between Mesoamericans and the inhabitants of the Greater Southwest for over a thousand years (Weigand et al. 1977).

Pre-contact trade and the extent of diffusional influence or even migration of Mesoamericans into the Greater Southwest culture area are highly controversial. The current literature is rather polarised on this issue. On one end of the spectrum Judge (1989) labels as 'Mexicanists' those whose paradigm is based on a very large macro-regional scale and argues that the Greater Southwestern cultures were directly influenced by the more highly complex societies in Mesoamerica. The influence of the large complex centres in Mesoamerica extended far beyond the Valley of Mexico for the purpose of exploiting the natural resources (e.g. turquoise) in the northern frontiers (e.g. Di Peso 1968, 1974; Kelley and Kelley 1975; Lister 1978; Rafferty 1990; Washburn 1980; Weigand et al. 1977; Weigand and Harbottle 1993; Weigand 1994). At the other end of the spectrum are the 'Indigenists' with a far more focused micro-regional perspective arguing that the development of complexity in the Greater Southwestern culture areas was independent of outside influences and there was little if any contact with Mesoamerica. Exotic trade commodities are considered too few and not important to the development of these communities (e.g. Cordell and Plog 1979; Judge 1989; Martin and Plog 1973; Vivian 1970). Although there is much debate over influence and contact between the Greater Southwest and Mesoamerica it is clear that turquoise was an important resource to the inhabitants of both regions.

Considering the amount of turquoise recovered, the hardships of prehistoric turquoise extraction, and the distance between the turquoise deposits and the sponsoring communities, turquoise was clearly a significant status symbol and an important commodity of pre-Columbian trade structures. Commodities that are transported through long-distance trade structures are generally considered exotic or significant items. Turquoise was a chief component in the validation of prestige and status within these pre-contact social organisations. Even as core regional centres changed through time (e.g. Teotihuacan, Tula and Tenochtitlan) there was a continual quest for specific exotic items, such as turquoise.

Many trade and exchange models have been developed to interpret the interregional economic relationships between Mesoamerica and the Greater Southwest and between the cultural groups within the Greater Southwest (e.g. world systems model, Wallerstein 1974; pochteca-like trader model, Frisbie 1978; prestige goods model, Nelson 1986). To test and add validity to these models and fully understand pre-contact turquoise procurement strategies and trade networks, a robust and reliable technique that identifies the provenance regions of recovered turquoise artefacts is required.

Development of the technique

The first attempt to identify the provenance regions of turquoise artefacts was to compare the colour and other visual similarities between the artefacts and turquoise source samples (Mathien 1981). Unfortunately, the colour and other visual similarities are subjective and range within a single turquoise deposit. As a result, researchers attempted to develop a technique to identify turquoise provenance regions by using trace and rare earth element patterns similar to the methods used to identify the provenance regions of obsidian artefacts (e.g. Shackley 2005). Once a similar trace element profile was established between a source and an artefact, the researcher had evidence that could support relationships and connections between the known provenance area and the unknown artefact sample. The assumption here is that the chemical variation within discrete compositional groups is less than the variation between compositional groups, also known as the provenance postulate (Weigand et al. 1977). The provenance postulate is the underlying basis for provenance-based studies of archaeological materials and allows researchers to examine relationships and connections between the known provenance area and the unknown artefact sample.

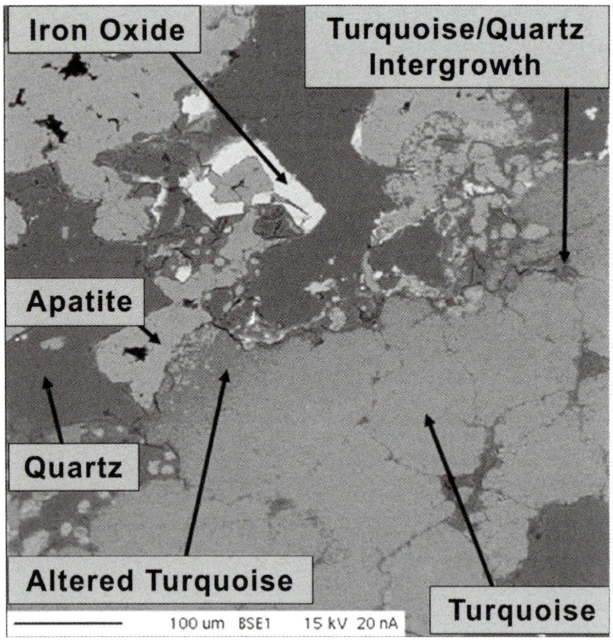

Figure 3. Backscatter electron image of a sample from the Castillian Mine in the Cerrillos Hills Mining District, New Mexico, displaying several minerals at the micrometer scale.

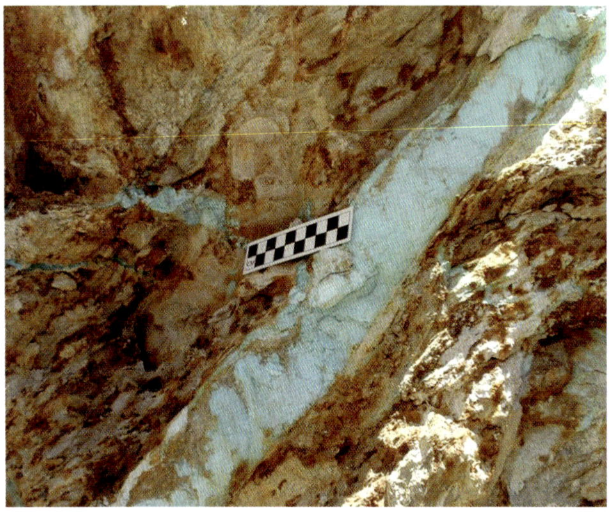

Figure 4. Turquoise in the fractures of copper porphyry at Mineral Park, Arizona.

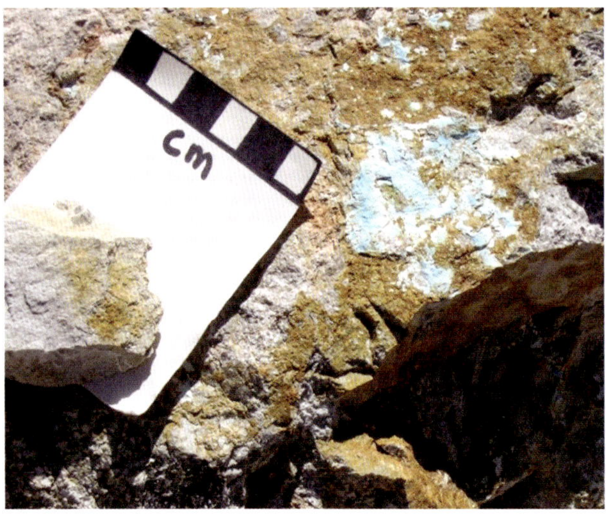

Figure 5. Turquoise weathering to clay minerals at Tiffany Mine, Cerrillos Hills Mining District, New Mexico.

Studies attempting to characterise trace and rare earth element compositions for turquoise provenance regions span four decades and an array of different instrumentation including X-ray fluorescence (XRF) (Judd 1954; Mathien and Olinger 1992; Ronzio and Salmon 1967), neutron activation analysis (NAA) (Bishop 1979; Harbottle and Weigand 1992; Sigleo 1975; Weigand and Harbottle 1993; Weigand et al. 1977), electron microprobe analyses (EMPA) (Ruppert 1982, 1983), atomic emission spectroscopy (AES) (Sigleo 1970), and proton induced X-ray emission (PIXE) (Kim et al. 2003). However, the success of these studies was limited due to overlap in the trace and rare earth element concentrations of the turquoise provenance regions.

We defined four major limitations with previous studies: (1) the complex mineralogy of turquoise affects the trace element composition within a single locale; (2) other blue-green minerals were mistaken for turquoise; (3) the geology and formation processes of turquoise deposits are similar between provenance regions; and (4) the weathering of turquoise can cause variation in trace and rare earth element concentrations. The first limitation is the complicated mineralogy of turquoise that can be determined chemically, which can cause variation in trace and rare earth element concentrations in a single deposit. The turquoise group consists of at least six end members including aheylite ($Fe^{2+}Al_6(PO_4)_4(OH)_8{\cdot}4(H_2O)$), chalcosiderite ($CuFe^{3+}{}_6(PO_4)_4(OH)_8{\cdot}4(H_2O)$), faustite ($ZnAl_6(PO_4)_4(OH)_8{\cdot}4H_2O$), planerite ($Al_6(PO_4)_2(PO_3OH)_2(OH)_8{\cdot}4(H_2O)$), turquoise ($(CuAl_6(PO_4)_4(OH)_8{\cdot}4(H_2O)$), and an unnamed iron rich end member ($Fe^{2+}Fe^{3+}{}_6(PO_4)_4(OH)_8{\cdot}4H_2O$) (Cid-Dresdner 1965; Foord and Taggart 1998). These end members have different chemical compositions but the same mineral structure. In addition to the complex mineralogy of turquoise that can be determined chemically, a study by Hull et al. (2008) showed that turquoise samples can consist of numerous mineral inclusions at the micrometer scale (Fig. 3).

The second limitation is the misidentification of the mineral turquoise. Other blue-green minerals, such as azurite ($Cu_3(OH)_2(CO_3)_2$), malachite ($Cu_2(OH)_2CO_3$), and chrysocolla ($(Cu, Al)_2H_2Si_2O_5(OH)_4{\cdot}nH_2O$) have been used by Native Americans in the same context as turquoise (Weigand et al. 1997). Kim et al. (2003) showed that many artefacts in collections were misidentified as turquoise by modern researchers. Trace element analysis of other blue-green minerals that may be mistaken for turquoise would result in large variation in trace element compositions and confusion when attempting to identify the provenance regions of 'turquoise' artefacts.

The third limitation is the geology of turquoise deposits. Turquoise forms in the fractures of igneous, sedimentary and metamorphic rocks throughout the western United States, drawing its elemental constituents from the surrounding host rock (Fig. 4; Morrissey 1968; Northrop 1975; Pogue 1974). Turquoise is a secondary mineral formed by the alteration of pre-existing rock. The majority of turquoise deposits are associated with copper porphyry intrusive bodies including many of the key economic copper deposits in the American Southwest and the Great Basin region of Nevada (Anthony et al. 1977; Arrowsmith 1974; Morrissey 1968; Sigleo 1970). In the western United States, turquoise deposits are generally hosted by similar rock types (e.g. apatite-rich monzonites/

granites), which have relatively similar trace and rare earth element concentrations (Longstaffe et al. 1982). Therefore, it is not surprising that trace and rare earth element studies were unsuccessful at distinguishing between turquoise provenance regions.

The fourth limitation is the susceptibility of turquoise to weathering, which can cause variation in trace and rare earth element contents. Highly altered and fractured copper porphyries that experienced erosion and exposure to long periods of supergene enrichment were an important dynamic in setting the discrete physical conditions for turquoise to form. Therefore, turquoise is stable in very specific geological environments and conditions and will weather or alter to other minerals when removed from these conditions (Abdu et al. 2011; Bergen et al. 2007; Hull et al. 2005). Extended exposure of turquoise to surface conditions will eventually leach important elements (e.g. phosphate) from turquoise, altering it to clay minerals (Fig. 5). This alteration causes extensive chemical and mineralogical changes (Abdu et al. 2011; Bergen et al. 2007; Hull et al. 2005) that result in variation in the trace and rare earth element concentrations of turquoise.

The trace and rare earth element concentrations in turquoise are largely governed by the geology of turquoise deposits, the chemistry of the supergene fluids present during the mineralisation of turquoise, and the weathering of turquoise. Therefore, it is clear why trace element analysis of turquoise has not satisfactorily differentiated between all of the turquoise provenance regions. As a result, several researchers are now working with isotope ratios to fingerprint turquoise artefacts and their provenance regions (Fayek et al. 2002; Hull et al. 2008; Thibodeau et al. 2007, 2008; Young et al. 1994).

Hydrogen and copper isotopes

Identifying the provenance regions of minerals using the isotopic ratios of hydrogen and oxygen (H_2O) is an established technique in geochemistry (Sheppard 1986). Isotope fractionation of hydrogen in meteoric water (i.e. rain) is affected by latitude (distance from the equator), longitude (distance inland from the ocean), and altitude, where the heavier deuterium is depleted as rain clouds move towards the poles, further inland, and higher in altitude (Albarède 2003). For example, ocean water contains both deuterium and hydrogen. As the evaporation process begins, the water vapour will contain more hydrogen than deuterium relative to sea water. The heavier isotopes do not partition into the vapour phase as readily as the lighter isotopes. In this respect, the heavier isotopes will partition into the liquid phase (rainwater), which causes rainfall to have different a hydrogen isotopic composition in different geographical locations (Hoefs 2004). Rainwater will have less of the heavier isotopes as the clouds move from the equator to the poles and from the ocean inward into the continent. Consequently, the hydrogen and oxygen isotopic composition of meteoric water varies as a function of geography.

Turquoise is a supergene mineral that forms from meteoric water. The geologic origin of turquoise has long been debated. Some researchers (e.g. Clarke and Diller 1887; Johnson 1903; Pogue 1974; Weber 1979) suggest that turquoise is a hydrothermal mineral, precipitating from hot fluids or vapours that flowed upward through fractures, whereas others believe it forms as a result of supergene processes where the host rock is weathered by descending rainwater through fractures and faults (Bergen et al. 2007; Leuth 1998; Paige 1912; Sterrett 1909). Based on our observations, as well as other studies (e.g. Leuth 1998; Paige 1912), the supergene model best explains the features of most turquoise deposits. The meteoric water is bound in the turquoise mineral structure. Therefore, the hydrogen isotopic composition of minerals, such as turquoise, which precipitate from rainwater, will also reflect these temporal and geographic variations in their hydrogen isotopic composition (Craig 1961; Dansgaard 1964; Epstein and Mayeda 1953; Friedman 1953; Friedman et al. 1964; Sheppard 1986; Yurtsever and Gat 1981). The hydrogen isotopic composition of turquoise can be particularly diagnostic because once turquoise forms, it holds onto its water, making it very difficult to alter the hydrogen isotopic composition of turquoise (Anderson et al. 1962; Manly 1950).

A survey of published data on the hydrogen isotopic composition of meteoric fluids (rainwater) from the turquoise source localities shows that they are distinct and have remained constant over the past 100 million years (Sheppard 1986; Taylor 1979). A preliminary study using gas source mass spectrometry showed variation in the hydrogen and oxygen isotopic compositions of several turquoise deposits located in Arizona, Colorado, New Mexico, and Nevada (Fayek et al. 2002).

Transition metal stable isotopes (e.g. iron, copper, zinc and chromium) offer a second potential avenue for isotopic fingerprinting of turquoise. Natural variations in transition metal stable isotopes are now known to occur in a wide variety of minerals and these variations may be produced by either abiotic or biotic processes (Bullen et al. 2001). Copper has two stable isotopes: ^{63}Cu and ^{65}Cu. Although copper isotope geochemistry is still in its formative stage, a relatively abundant data set exists for the isotopic compositions of copper minerals that shows that natural variations in δ^{65}Cu values vary by 34‰ (Klein et al. 2010; Larson et al. 2003; Mason et al. 1997; Mathur et al. 2009a,b; Palacios et al. 2011). Therefore, variations in the copper isotope ratios of turquoise samples are large enough, when used in conjunction with hydrogen, to be a successful discriminator. The fractionation of hydrogen and copper isotopes are driven by different geological processes, thus potentially displaying more variation between turquoise source regions than using the isotopic compositions of hydrogen and oxygen, which are governed by similar geological processes.

Methods

For turquoise provenance studies, SIMS has shown to be the least destructive technique for measuring isotope ratios. SIMS (Fig. 6) allows analyses of solid samples (i.e. the ability to focus the primary ion beam on clear turquoise grains) with a spatial resolution on the scale of a few micrometres (Fig. 7), simple sample preparation, and short analysis times. For hydrogen isotope analysis in turquoise, analysis times are less than 12 minutes (Liu et al. 2010). The sample preparation and SIMS analysis is relatively non-destructive to the sample, which is

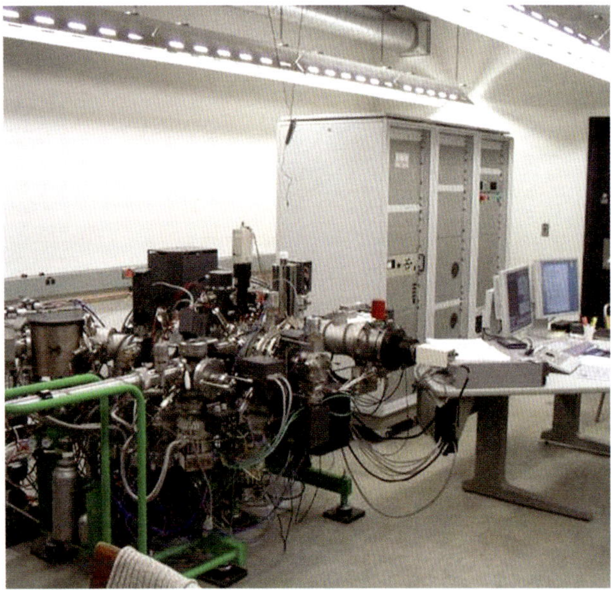

Figure 6. Secondary ion mass spectrometer (SIMS) at the University of Manitoba, Winnipeg, Manitoba.

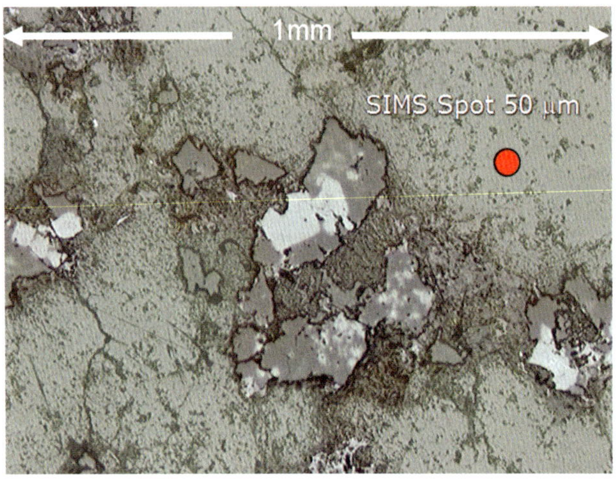

Figure 7. SIMS analysis spot of a sample from the Castillian Mine, Cerrillos Hills Mining District, New Mexico.

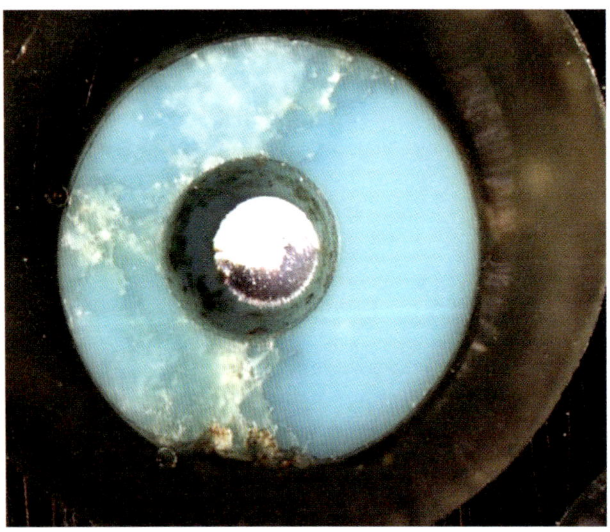

Figure 8. A turquoise bead analysed by SIMS. Alteration occurs along the fractures and mineral inclusions appear as yellow patches along the bottom portion of the bead. The clear blue area (on the right) was analysed by SIMS and is devoid of alteration and inclusions. Note that the analysis pits are not visible.

important when analysing archaeological artefacts (Fig. 8). Once the artefacts have been analysed they can be returned to museum or other archaeological collections (Hull et al. 2008).

For SIMS analyses, turquoise samples are cut into 1 mm to 1 cm pieces and prepared in epoxy mounts. Large artefacts (<2 cm, e.g. pendants) are mounted in phenol rings, and turquoise source samples and small artefacts (<6 mm, e.g. beads) are mounted in drilled-out 25 mm diameter aluminium mounts using Buehler 'Epoxide' epoxy resin. Mounts are polished using various grit (600–2400) SiC sandpaper and 1–15 μm diamond polishing compounds. As mentioned before, polishing is the most destructive part of the procedure and care is taken with the positioning of the artefact in the epoxy so that any areas that are relatively flat and polished are exposed, lessening the impact of the polishing. The polished mounts are washed with a dilute soap solution and rinsed in deionised water and then placed in an oven at 60 °C to remove absorbed water. Reflected-light photomaps of each mount are made to identify the regions devoid of inclusions and alteration. A ~200 Å thick Au coat is sputter-deposited on the sample mount surfaces to ensure a surface conductivity of 5–10 ohms/cm.

Mounts are placed in stainless steel sample holders and the entire assembly is placed in the SIMS and held at high vacuum for a minimum of eight hours prior to the start of analysis. During the measurement of isotopic ratios by SIMS, a beam of primary ions (a few μm in diameter) is focused on the solid sample surface, thus obtaining a localised spot analysis. Atoms, ions and molecules are removed by the primary beam, a phenomena referred to as sputtering. The ions are extracted, focused and accelerated by a 'secondary' ion beam through a slit and into a mass spectrometer. During the SIMS analysis, an intrinsic mass dependent bias is introduced, which is referred to as instrumental mass fractionation (IMF) and typically favours the low mass isotope. The greatest contributor to the IMF is the ionisation process dependent upon sample characteristics such as chemical composition. This is referred to as compositionally dependent fractionation or 'matrix effects' (Riciputi et al. 1998). Therefore, accurate isotopic SIMS analysis requires correction for IMF by standardising the IMF using mineral standards that are compositionally similar to the unknown samples. SIMS results from the standard are compared to its accepted isotopic composition in order to compute a correction factor that is applied to the unknown samples measured during the same analysis session (Riciputi et al. 1998). To calibrate the mass bias, standards with known iron contents and $\delta^{65}Cu$ and δD isotopic values are analysed during the same analytical session as the samples.

The D/H and $^{65}Cu/^{63}Cu$ isotopic measurements are performed with a Cameca IMS-7F ion microprobe installed in the University of Manitoba in 2006. Positive secondary ions are produced by an O^- bean with impact energy of 22.5 keV. The samples and standards are analysed using 40 nA, 12.5 kv O^- primary beam focused on a ~50 μm spot. The largest contrast (400 μm) and field (1800 μm) apertures, in conjunction with 150 μm image field and an energy bandpass of ±25 eV, are used to maximise sensitivity. The secondary column high voltage is set to 10 kV. For D/H isotopic measurements, the secondary ion mass spectrometer is operated at a mass resolution of ~800 to separate H^{2+} from D^+. Interferences for $^{65}Cu/^{63}Cu$

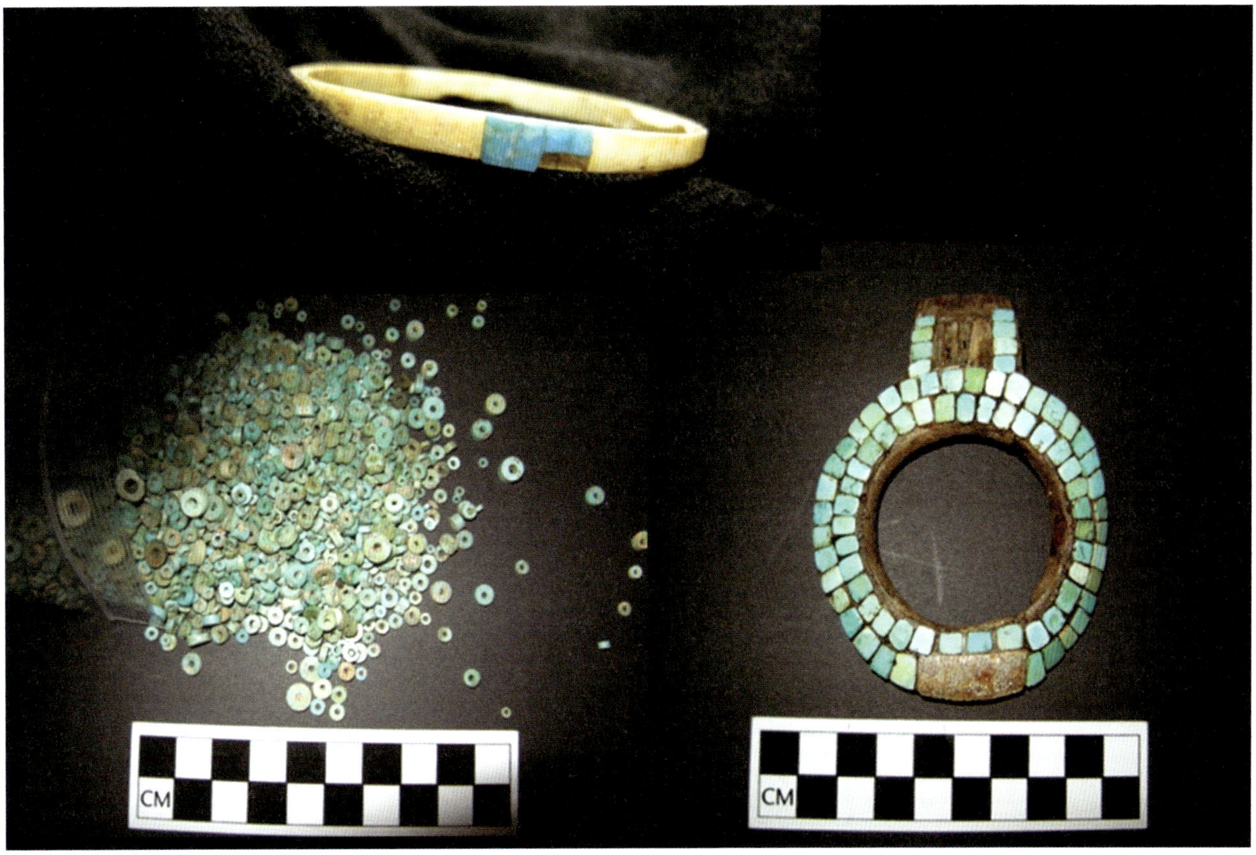

Figure 9. Turquoise artefacts recovered from Chacoan archaeological sites. These artefacts are from the collection at the American Museum of Natural History, New York.

isotopic measurements are resolved by offsetting the sample high voltage by –50 V (e.g. 9950 V), while maintaining the electrostatic analyser in the secondary column at 10 kV.

Hydrogen and copper isotopic compositions are reported as delta (δ) values in units of per mil (parts per thousand) relative to the standard, Vienna Standard Mean Ocean Water (V-SMOW) for hydrogen and NIST976 for copper such that:

$$\delta_A=[(R_A-R_{stn})/\ R_{stn}] \times 10^3$$

where R_A and R_{stnd} are the absolute ratios of $^2H/^1H$ (D/H; D=deuterium) or $^{65}Cu/^{63}Cu$ in the sample (turquoise) and the standard (V-SMOW or NIST976), respectively. The absolute $^2H/^1H$ ratio of V-SMOW is 155.76×10^{-6} (Hagemann et al. 1970) and the absolute $^{65}Cu/^{63}Cu$ ratio of NIST976 is defined as 4.4563×10^{-1}.

A preliminary study in the American Southwest

Large amounts of turquoise are commonly recovered in large pre-contact regional centres (e.g. Chaco Canyon) throughout the southwestern United States. For example, over 200,000 pieces of turquoise (Fig. 9) have been recovered from excavations of sites throughout Chaco Wash (Fig. 10; Harbottle and Weigand 1992; Pepper 1920). Considering the amount of turquoise recovered in the canyon and surrounding Chacoan communities, Judge (1976, 1989) argued that turquoise became so important in the exchange system that it may have developed into currency.

Chaco Canyon's nearest turquoise provenance region is over 200 km away in the Cerrillos Hills Mining District, near Santa Fe, New Mexico (Mathien 1999), where there is evidence of both ancient and modern exploitation of turquoise deposits. This area has been documented and studied more than most pre-Columbian turquoise mines. The extensive prehistoric mining in the Cerrillos Hills suggests that it was a significant source of turquoise in the Southwestern exchange system with at least ten prehistoric mining pits displaying intense removal of turquoise (Fig. 11; Mathien 2001).

Some researchers advocate that turquoise was collected from the Cerrillos Hills Mining District and then taken to Chaco Canyon where it was worked and then redistributed. The Guadalupe community, a Chacoan outlier, may have been an important location for the transport of turquoise from the Cerrillos Hills Mining District to Chaco Canyon and may have served as a link for the establishment of control over the Cerrillos Hills resource area. Judge (1989) suggested that the turquoise recovered in excavations at Guadalupe was intended for Chaco Canyon, the proposed dominant producer of finished turquoise items. Judge implies that there is not enough evidence to support the hypothesis that turquoise ornaments were worked and prepared at Guadalupe. However, Durand and Durand (2000) argue that the inhabitants of the Guadalupe community were active participants in the production of turquoise items. Almost 200 pieces of turquoise were recovered in a Pueblo II/III site (ENM 848) that is part of the Guadalupe community. The recovered pieces were small fragments, suggesting that the inhabitants of Guadalupe were working and producing turquoise items (Durand and Durand

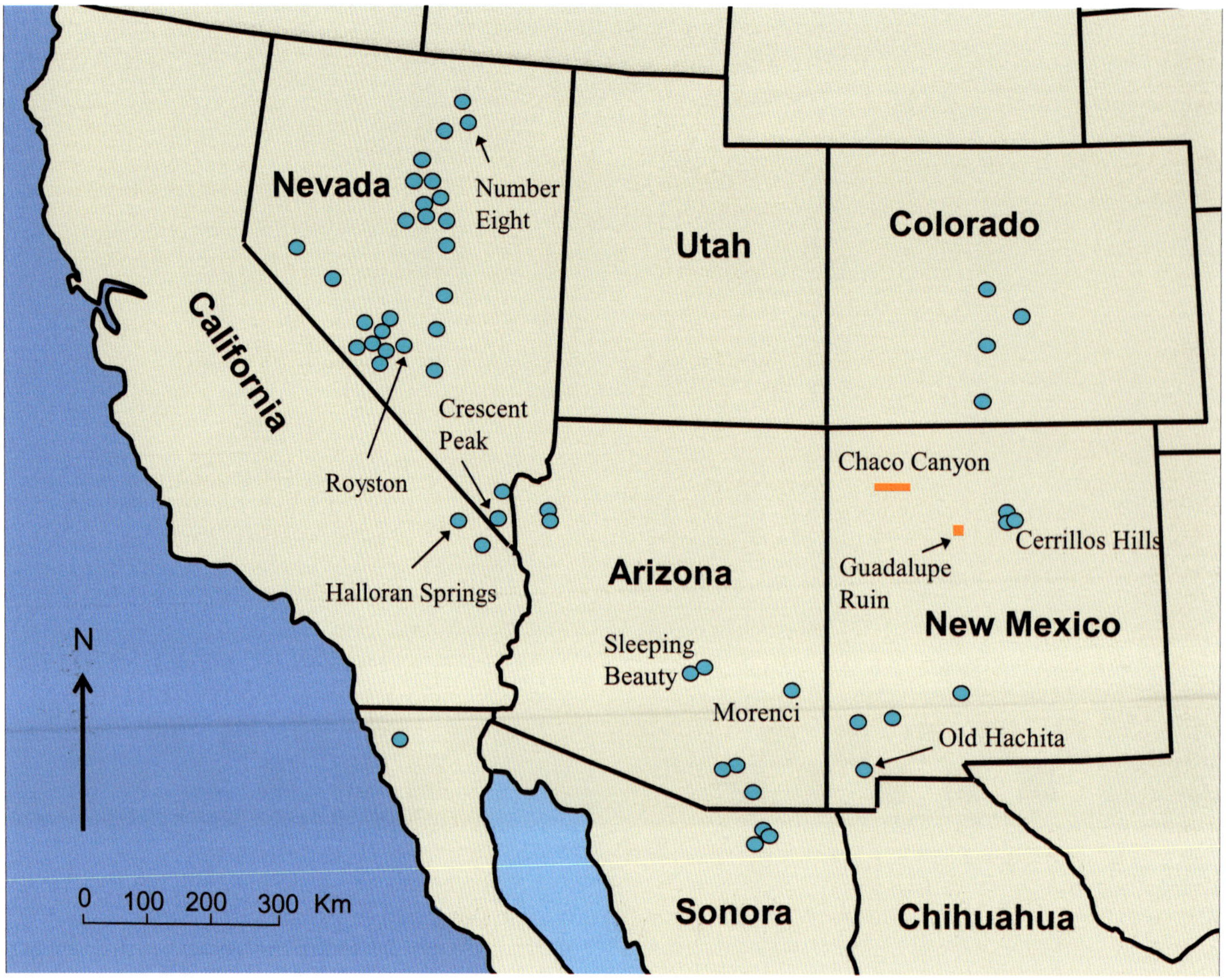

Figure 10. Map of archaeological sites and turquoise provenance regions used in the preliminary study.

2002) rather than just involved in the storage and movement of unmodified turquoise (Judge 1989).

We measured the isotope ratios of 19 artefacts to: (1) investigate the extent of turquoise procurement in the Chacoan world and (2) test Durand and Durand's (2000) hypothesis that the inhabitants of Guadalupe were active participants in the production of turquoise items. Twelve of these artefacts were from eight different archaeological sites in Chaco Canyon and seven were recovered from site ENM 848, part of the Guadalupe community (Table 1). Because of the proximity of Cerrillos Hills to Chaco Canyon and evidence of the pre-contact extraction of turquoise, it would be logical to make the assumption that the Chacoans obtained most of their turquoise from this provenance area. If the involvement in turquoise procurement or exchange of the inhabitants of the Guadalupe community was exclusively for the purpose of Chaco Canyon's control and supply of turquoise from Cerrillos Hills, then all of the turquoise recovered from the Guadalupe community should show that it was obtained from Cerrillos Hills. However, if the turquoise from the Guadalupe community came from several provenance regions, the data would support the argument that the inhabitants of the Guadalupe community were indeed active participants in the production of turquoise items and may have acted independently of Chaco Canyon in turquoise procurement strategies and exchange networks.

Our results show that the isotopic compositions of the Chacoan artefacts fell into the distribution patterns of several turquoise provenance regions including Sleeping Beauty and Morenci in Arizona, Royston, Nevada and Old Hachita, New Mexico (Fig. 12; Table 1) suggesting that the inhabitants of Chaco Canyon obtained their turquoise from several provenance regions and from many directions. These data suggest that turquoise exchange networks may have extended far into Arizona and Nevada or that the Chacoans travelled far from their sponsoring communities to obtain this valuable resource (Fig. 10). The turquoise artefacts that fell into the Sleeping Beauty, Royston and Old Hachita distribution patterns are associated with dates from AD 500 to AD 820 (Table 1), suggesting that the Chacoans were obtaining their turquoise from the far west (Arizona and Nevada) and from the south (southwestern New Mexico) long before Chaco Canyon truly flourished as a regional centre.

Turquoise artefacts recovered from the Guadalupe community plotted into the distribution patterns of Morenci, Arizona, Crescent Peak, Royston, Number Eight in Nevada and Halloran Springs, California (Fig. 12; Table 1). These data suggest that the inhabitants of the Guadalupe community were obtaining their turquoise from several provenance regions and were active participants in the procurement and production of turquoise (Fig. 10).

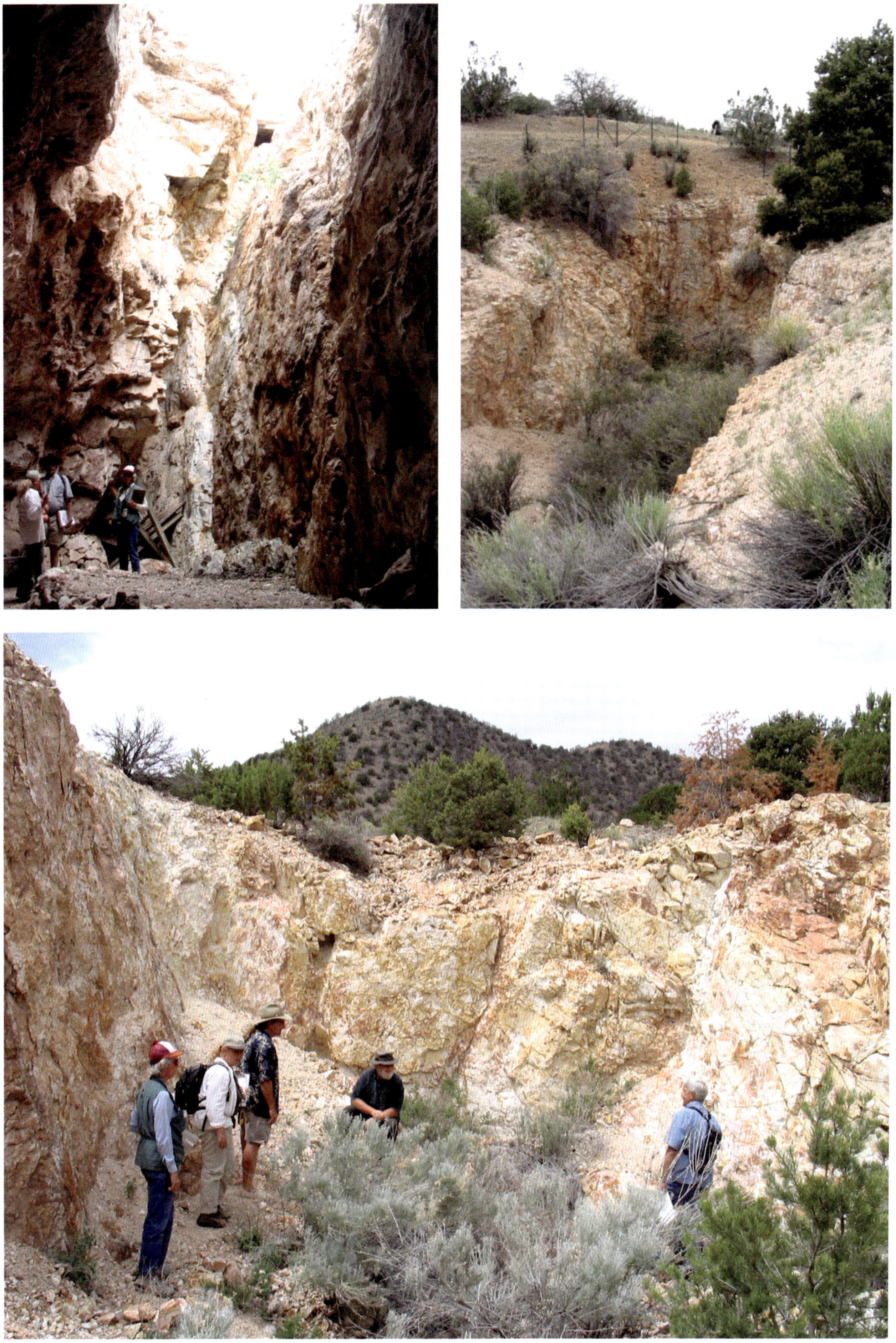

Figure 11. Turquoise deposits at the Cerrillos Hills Mining District, New Mexico.

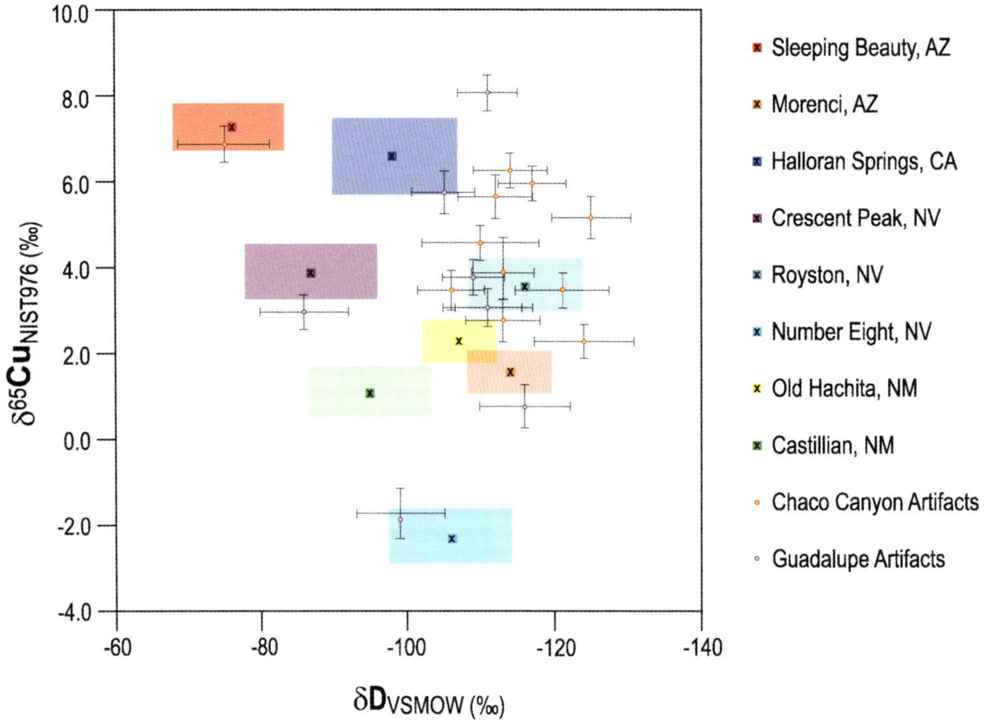

Figure 12. A plot of δ^{65}Cu and δD values of turquoise provenance regions and artefacts from communities in Chaco Canyon and Guadalupe Ruin. Error bars and shaded areas are 2σ.

Our study shows that turquoise procurement from the provenance region of the Cerrillos Hills Mining District was notably absent (Fig. 12; Table 1). These results raise questions as to why the inhabitants of these 'small sites' in Chaco Canyon or Guadalupe Ruin did not obtain their turquoise from the provenance region that was closest to their proximity and whether the inhabitants of smaller sites (not great house communities) had access to the mines in Cerrillos Hills. It is also important to consider the small sample size of the turquoise artefacts from each archaeological site. Five of the artefacts recovered from Chaco Canyon and the Guadalupe community did not plot into the distribution patterns of turquoise provenance regions that were characterised in this pilot study. The isotope compositions of these artefacts may represent turquoise provenance regions that have not been analysed.

Conclusions

Isotope research for archaeological provenance studies is somewhat new and is proving to be a good method for some of the more problematic minerals such as turquoise. The use of *in situ* stable isotopic analysis of hydrogen and copper by SIMS has been successful in defining turquoise provenance regions. The advantages of this technique are: (1) isotope ratios of turquoise are geographically distinct thus overcome the limitations of trace and rare earth element analyses; (2) *in situ* microanalysis avoids inclusions of other minerals, which could profoundly affect the trace and rare earth element and isotopic composition of turquoise; (3) except for minor polishing of the surface of artefacts and 50 µm pits that are created during analysis, turquoise artefacts are relatively undamaged; and (4) the analyses are rapid and inexpensive. The major disadvantage of this technique is that altered turquoise artefacts cannot be reliably analysed and characterised using H and Cu isotopes because alteration of turquoise affects the hydrogen and copper isotopic composition.

We developed a preliminary database that contains the isotope ratios of several known turquoise provenance regions. As we continue to analyse more turquoise artefacts and provenance samples, their isotope signatures will be added to this database and more turquoise procurement and exchange patterns will emerge. This research will expand to include turquoise artefacts from Mesoamerican archaeological sites, enhancing our knowledge of the extent of exchange and interaction between the cultural groups that so greatly desired this blue-green mineral.

Acknowledgements

We thank Wendy Bustard (Chaco Culture National Historical Park Museum), Kathy Roler Durand (ENMU), Douglas Magnus (Castillian Mine), Dean and Donna Otteson (Royston Mine), Bill Crank (Crescent Peak), Virgil Leuth (NM Tech) Phil Weigand (Museum of Northern Arizona), and Robert Reynolds, Robert Hilburn, Ed Nazelrod, and Jim Sherer (Halloran Springs) for the samples used in this study. This work was funded by Canada Research Chairs in Environmental and Isotope Geochemistry, a Major Facilities Access Grant, Research Tools and Equipment, and Discovery Grant from the Natural Sciences and Engineering Research Council of Canada, a Canada Foundation for Innovation Grant, and by the National Science Foundation under Grant No. 0609638 to MF. Any opinions, findings, and conclusions or recommendations expressed in this material are those of the author(s) and do not necessarily reflect the views of the National Science Foundation. This contribution greatly benefited from the comments by two anonymous reviewers. We would also like to thank the scientific editors of this volume, Caroline Cartwright and Rebecca Stacey.

References

Abdu, Y., Hull, S., Fayek, M. and Hawthorn, F. (2011) 'The turquoise-chalcosiderite $Cu(Al,Fe^{3+})_6(PO_4)_4(OH)_8{\bullet}4H_2O$ solid solution series: a Mössbauer spectroscopy, XRD, EMPA and FTIR study', *American Mineralogist* **96**: 1433–1442.

Albarède, F. (2003) *Geochemistry: An Introduction*. Cambridge: Cambridge University Press.

Anderson, W.R., Stringham, B. and Welan, J.A. (1962) 'Secondary phosphate from Bingham, Utah', *American Mineralogist* **47**: 1303–1309.

Anthony, J.W., Williams, S.A. and Bideaux, R.A. (1977) *Mineralogy of Arizona*. Tucson: University of Arizona Press.

Arrowsmith, R. (1974) 'Introduction', in J. Pogue, *The Turquois: A Study of its History, Mineralogy, Geology, Ethnology, Archaeology, Mythology, Folklore, and Technology*. Glorieta, NM: The Rio Grande Press [originally published 1915, *Memoirs of the National Academy of Sciences*, Vol. XII, Part II, Washington, DC, National Academy of Sciences].

Bergen, L.L., Fayek, M. and Hull, S. (2007) 'The formation and alteration of turquoise deposits in the American Southwest', poster presented at the Geological Society of America 2007, Denver, Colorado.

Bishop, R.L. (1979) *Neutron Activation Analysis of Chacoan Turquoise: An Informal Report to the Division of Chaco Research, Southwest Cultural Resources Center, National Parks Services*. Albuquerque: University of New Mexico.

Bullen, T.D., White, A.F., Childes, C.W., Vivit, D.V. and Schulz, M.S. (2001) 'Demonstration of significant abiotic iron isotope fractionation in nature', *Geology* **29**: 699–702.

Cid-Dresdner, H. (1965) 'Determination and refinement of the crystal structure of turquoise, $CuAl_6(PO_4)_4(OH)_8{\bullet}4H_2O$', *Zeitschrift für Kristallographie Bd.* **121**, S.: 87–113.

Clarke, F.W. and Diller, J.S. (1887) 'Turquoise from New Mexico', *United States Geological Survey Bulletin* **42**: 39–44.

Cordell, L.S. and Plog, F. (1979) 'Escaping the confines of normative thought: a reevaluation of Puebloan prehistory', *American Antiquity* **44**(3): 405–429.

Craig, H. (1961) 'Isotopic variations in meteoric waters', *Science* **133**: 1702–1703.

Dansgaard, W. (1964) 'Stable isotopes in precipitation', *Tellus* **16**: 436–468.

Di Peso, C.C. (1968) 'Casas Grandes and the Gran Chichimeca', *El Palacio* **75**: 47–61.

Di Peso, C.C. (1974) *Casas Grandes: A Fallen Trading Center of the Gran Chichimeca, Vols 1–3*. The Amerind Foundation Publication No. 9. Flagstaff, AZ: Northland Press.

Durand, S.R. and Durand, K.R. (2000) 'Notes from the edge settlement pattern changes at the Guadalupe community', in J. Kantner and N.M. Mahoney (eds), *Great House Communities across the Chacoan Landscape*. Tucson: University of Arizona Press, 101–109.

Durand, K.R. and Durand, S.R. (2002) 'A closer look at the Guadalupe Ruin Great House Community', poster presented at the 67th Annual Meeting of the Society for American Archaeology, Denver, Colorado.

Epstein, S. and Mayeda, T. (1953) 'Variations of ^{18}O content of waters from natural sources', *Geochimica et Cosmochimica Acta* **4**: 213–224.

Fayek, M., Riciputi, L.R., Milford, H. and Mathien, F.J. (2002) 'Sourcing turquoise using oxygen and hydrogen isotopes', paper presented at the Geological Society of America 2002, Denver, Colorado.

Fish, P.R. (1989) 'The Hohokam: 1,000 years of prehistory in the Sonoran Desert', in L.S. Cordell and G.J. Gumerman (eds), *Dynamics of Southwest Prehistory*. Washington, DC: Smithsonian Institution Press, 19–63.

Foord, E.E. and Taggart, Jr., J.E. (1998) 'A reexamination of the turquoise group: the mineral aheylite, planerite (redefined), turquoise and coeruleolacite', *Mineralogical Magazine* **62**: 93–111.

Friedman, I. (1953) 'Deuterium content of natural water and other substances', *Geochimica et Cosmochimica Acta* **4**: 89–103.

Friedman, I., Redfield, A.C., Schoen, B. and Harris, J. (1964) 'The variation in deuterium content of natural waters in the hydrogenic cycle', *Reviews of Geophysics* **2**: 177–224.

Frisbie, T.R. (1978) 'High status burials in the Greater Southwest: an interpretative synthesis', in C.L. Riley and B.C. Hendrick (eds), *Across the Chichimec Sea: Papers in Honor of J. Charles Kelley*. Carbondale: Southern Illinois University Press, 202–227.

Hagemann, R., Nief, G. and Roth, E. (1970) 'Absolute isotopic scale for deuterium analysis of natural waters: absolute D/H ratio of SMOW', *Tellus* **22**: 712–715.

Harbottle, G. and Weigand, P.C. (1992) 'Turquoise in pre-Columbian America', *Scientific American* **266**(2): 78–85.

Hoefs, J. (2004) *Stable Isotope Geochemistry*, 5th edn. Berlin: Springer.

Hull, S., Fayek, M., Anovitz, L.M., Mathien, F.J. and Shelley, P. (2005) 'The effects of alteration on sourcing archaeological turquoise', paper presented at the Geological Society of America 2005, Salt Lake City, Utah, Abstract 37-7.

Hull, S., Fayek, M., Mathien, F.J., Shelley, P. and Durand, K.R. (2008) 'A new approach to determining the geological provenance of turquoise artifacts using hydrogen and copper stable isotopes', *Journal of Archaeological Science* **35**: 1355–1369.

Johnson, D.W. (1903) 'The geology of the Cerrillos Hills, New Mexico', *Columbia University School of Mines Quarterly* **24**: 493–499.

Judd, N.M. (1954) *The Material Culture of Pueblo Bonito*. Smithsonian Miscellaneous Collections 124. Washington, DC: Smithsonian Institution.

Judge, W.J. (1976) *The Development of a Complex Cultural Ecosystem in the Chaco Basin, New Mexico*, Contribution No. 21 of the Chaco Center, National Park Service. Albuquerque: University of New Mexico.

Judge, W.J. (1989) 'Chaco Canyon-San Juan Basin', in L.S. Cordell and G.J. Gumerman (eds), *Dynamics of Southwest Prehistory*. Washington, DC: Smithsonian Institution Press, 209–261.

Kelley, J.C. and Kelley, E.A. (1975) 'An alternative hypothesis for the explanation of Anasazi culture history', in T.R. Frisbie (ed.), *Collected Papers in Honor of Florence Hawley Ellis. Papers of the Archaeological Society of New Mexico No. 2*. Norman, OK: Hooper Publishing, 178–223.

Kim, J., Simon, A.W., Ripoche, V., Mayer, J.W. and Wilkens, B. (2003) 'Proton-induced X-ray emission analysis of turquoise artefacts from Salado Platform Mound sites in the Tonto Basin of central Arizona', *Measurement Science and Technology* **14**: 1579–1589.

Klein, S., Brey, P.B., Durali-Müller, S. and Lahaye, Y. (2010) 'Characterisation of the raw metal sources used for the production of copper and copper-based objects with copper isotopes', *Archaeological and Anthropological Science* **2**: 45–56.

Larson, P.B., Maher, K., Ramos, F.C., Chang, Z., Gaspar, M. and Meinert, L.D. (2003) 'Copper isotope ratios in magmatic and hydrothermal ore-forming environments', *Chemical Geology* **201**: 337–350.

Lister, R.H. (1978) 'Mesoamerican influence at Chaco Canyon, New Mexico', in C.L. Riley and B.C. Hedrick (eds), *Across the Chichimec Sea: Papers in Honor of J. Charles Kelley*. Carbondale: Southern Illinois University Press, 233–241.

Liu, R., Hull, S. and Fayek, M. (2010) 'A new approach to measuring D/H ratios with the Cameca IMS-7F', *Journal of Surface and Interface Analysis* (www.interscience.wiley.com) DOI 10.1002/sia.3467.

Longstaffe, F.J., McNutt, R.H. and Crocket, J.H. (1982) 'Rare-earth element modeling of Archean meta-igneous and igneous rocks, Lake Despair Area, Northwestern Ontario', *Precambrian Research* **17**: 275–296.

Lueth, V.W. (1998) 'Two diverse origins of turquoise at the Orogrande Mining District, Otero County, New Mexico', *New Mexico Geology* **20**: 64–65.
Manley, Jr., R.L. (1950) 'Differential thermal analysis of certain phosphates', *American Mineralogist* **35**: 108–115.
Martin, P.S. and Plog, F. (1973) *The Archaeology of Arizona: A Study of One Southwest Region.* New York: Doubleday Natural History Press.
Mason, T.F., Weiss, D.J., Chapman, J.B., Wilkinson, J.J., Tessalina, S.G., Spiro, B., Horstwood, M.S.A., Spratt, J. and Coles, B.J. (1997) 'Zn and Cu isotopic variability in the Alexandrinka Volcanic-hosted Massive Sulphide (VHMS) ore deposit, Urals, Russia', *Chemical Geology* **221**: 170–187.
Mathien, F.J. (1981) 'Neutron activation of turquoise artifacts from Chaco Canyon, New Mexico', *Current Anthropology* **22**: 293–294.
Mathien, F.J. (1999) 'Identifying sources of prehistoric turquoise in North America: problems and implications for interpreting social organization', *Beads. Journal of the Society of Bead Researchers* **12–13**: 17–37.
Mathien, F.J. (2001) 'The organization of turquoise production and consumption by the prehistoric Chacoans', *American Antiquity* **66**(1): 103–118.
Mathien, F.J. and Olinger, B. (1992) 'An experiment with x-ray fluorescence to determine trace element variability in turquoise composition', in M.S. Duran and D.T. Kirpatrick (eds), *Archaeology, Art, and Anthropology: Papers in Honor of J.J. Brody.* Albuquerque: The Archaeological Society of New Mexico, 123–134.
Mathur, R., Titley, S., Barra, F., Brantley, S., Wilson, M., Phillips, A., Munizaga, F., Maksaev, V., Vervoort, J. and Hart, G. (2009a) 'Exploration potential of Cu isotope fractionation in porphyry copper deposits', *Journal of Geochemical Exploration* **102**: 1–6.
Mathur, R., Titley, S., Hart, G., Wilson, M., Davignon, M. and Zlatos, C. (2009b) 'The history of the United States cent revealed through copper isotope fractionation', *Journal of Archaeological Science* **36**: 430–433.
Morrissey, F. (1968) *Turquoise Deposits of Nevada, Report 17.* Reno: Mackay School of Mines, University of Nevada.
Nelson, R.S. (1986) 'Pochtecas and prestige: Mesoamerican artifacts in Hohokam sites', in F.J. Mathien and R.H. McGuire (eds), *Ripples in the Chichimec Sea: New Considerations of Southwestern-Mesoamerican Interaction.* Carbondale: Southern Illinois University Press, 153–182.
Northrop, S.A. (1975) *Turquoise and Spanish Mines in New Mexico.* Albuquerque: University of New Mexico Press.
Paige, S. (1912) 'The origin of turquoise in the Burro Mountains, New Mexico', *Economic Geology* **7**: 382–392.
Palacios, C., Rouxel, O., Reich, M., Cameron, E.M. and Leybourne, M.I. (2011) 'Pleistocene recycling of copper at a porphyry system, Atacama Desert, Chile: Cu isotope evidence', *Mineralium Deposita* **46**: 1–7.
Pepper, G.H. (1920) *Pueblo Bonito: Anthropological Papers of the American Museum of Natural History 27.* New York: Museum of Natural History.
Pogue, J. (1974) *The Turquois: A Study of Its History, Mineralogy, Geology, Ethnology, Archaeology, Mythology, Folklore, and Technology.* Glorieta, NM: The Rio Grande Press [originally published, 1915, *Memoirs of the National Academy of Sciences*, Vol. XII, Part II, Washington, DC, National Academy of Sciences].
Rafferty, K. (1990) 'The Virgin Anasazi and the pan-southwestern trade system, A.D. 900–1150', *The Kiva* **56**: 3–24.
Riciputi, L.R., Paterson, B.A. and Ripperdan, R.L. (1998) 'Measurement of light stable isotope ratios by SIMS: matrix effects for oxygen, carbon, and sulfur isotopes in minerals', *International Journal of Mass Spectrometry* **178**: 81–112.
Ronzio, A.R. and Salmon, M.L. (1967) 'Relation between source and composition of turquoise', *Journal of the Colorado-Wyoming Academy of Science* **5**: 30–31.
Ruppert, H. (1982) 'Zur verbreitung and herkunft von Türkis und Sodalith in präkolumbischen Kulturen der Kordilleren', *Baessler-Archiv* **30**: 69–124.
Ruppert, H. (1983) 'Geochemische Untersuchungen an Türkis und Sodalith aus Lagerstätten und präkolumbischen Kulturen der Kordilleren', *Berliner Beiträge zur Archäometrie* **8**: 101–210.
Shackley, M.S. (2005) *Obsidian: Geology and Archaeology in the North American Southwest.* Tucson: University of Arizona Press.
Sheppard, S.M.F. (1986) 'Characterization and isotopic variations in natural waters', in J.W. Valley, H.P. Taylor and J.R. O'Neil (eds), 'Stable isotopes in high temperature geological processes', *Reviews in Mineralogy* **16**: 165–183.
Sigleo, A.M.C. (1970) 'Trace-element geochemistry of Southwestern turquoise', MA thesis, University of New Mexico (unpublished).
Sigleo, A.M.C. (1975) 'Turquoise mine and artifact correlation for Snaketown Site, Arizona', *Science* **189**: 459–460.
Sterrett, D.B. (1909) 'Gems and precious stones', *United States Geological Survey Mineral Resources* **2**: 846.
Taylor, H.P. (1979) 'Oxygen and hydrogen isotopic relations in hydrothermal mineral deposits', in H.L. Barnes (ed.), *Geochemistry of Ore Deposits*, 2nd edn. New York: J. Wiley and Sons, 236–277.
Thibodeau, A.M., Ruiz, J. and Chesley, J.T. (2007) 'Lead and strontium isotopes as tracers of turquoise', *SAS Bulletin* **30**: 10–13.
Thibodeau, A.M., Ruiz, J., Chesley, J.T. and Killick, D.J. (2008) 'An expanded lead and strontium isotopic database for turquoise deposits in the Greater American Southwest: implications for provenance studies of archaeological turquoise', poster presented at the *37th International Symposium on Archaeometry*, Siena, Italy.
Vivian, R.G. (1970) 'An inquiry into prehistoric social organization in Chaco Canyon, New Mexico', in W.A. Longacre (ed.), *Reconstructing Prehistoric Pueblo Societies.* Santa Fe: School of American Research and Albuquerque: University of New Mexico Press, 59–83.
Wallerstein, I. (1974) *The Modern World System: Studies in Political Discontinuity.* New York: Academic Press.
Washburn, D.K. (1980) 'The Mexican connection: cylinder jars from the Valley of Oaxaca', *Transactions of the Illinois State Academy of Science* **72**(4): 70–85.
Weber, R.H. (1979) 'Turquoise in New Mexico', *New Mexico Geology* **28**: 39–40.
Weigand, P.C. (1994) 'Observations on ancient mining within the northwestern regions of the Mesoamerican civilization, with emphasis on turquoise', in A.K. Craig and R.C. West (eds), 'In quest of mineral wealth: Aboriginal and colonial mining and metallurgy in Spanish America', *Geoscience and Man* **33**: 21–35.
Weigand, P.C. and Harbottle, G. (1993) 'The role of turquoises in the ancient Mesoamerican trade structure', in J.E. Ericson and T.G. Baugh (eds), *The American Southwest and Mesoamerica: Systems of Prehistoric Exchange.* New York: Plenum Press, 159–177.
Weigand, P.C., Harbottle, G. and Sayre, E.V. (1977) 'Turquoise sources and source analysis: Mesoamerica and the southwestern U.S.A.', in T.K. Earle and J.E. Ericson (eds), *Exchange Systems in Prehistory.* New York: Academic Press, 15–34.
Young, S.M.M., Phillips, Jr., D.A. and Mathien, F.J. (1994) 'Lead isotope analysis of turquoise in the southwestern U.S.A. and Mesoamerica', paper presented at the *29th International Symposium on Archaeometry* 1994, Ankara, Turkey.
Yurtsever, Y. and Gat, J.R. (1981) 'Atmospheric waters', in J.R. Gat and R. Gonfiantini (eds), *Stable Isotope Hydrology: Deuterium and Oxygen-18 in the Water Cycle.* Technical Report Series #210. Vienna: International Atomic Energy Agency, 103–142.

Authors

- Sharon Hull, Department of Geological Sciences, University of Manitoba, Winnipeg, Canada
- Mostafa Fayek, Department of Geological Sciences, University of Manitoba, Winnipeg, Canada

Conservation of a turquoise mosaic disk from Tula, Mexico

Patricia Meehan and Valerie Magar

Introduction

Turquoise mosaics from ancient Mexico are rare and extraordinary objects, which often fascinate, not only because of their design and colour, but also on account of the workmanship and dexterity exhibited by the use of perfectly fitting small tesserae in shaped patterns. The most well-known examples of these turquoise mosaics can be found in European museums, and are thought to be objects made in the fifteenth or early sixteenth century, shipped to Europe during the early contact period with the New World. Intact mosaics from burial contexts are much rarer as their fragile materials are extremely sensitive to the changing conditions in the ground. All objects found in Mexican collections, including ceremonial disks, have come from archaeological sites.

One of these turquoise mosaic disks was found in 1992 inside an offering at the Palacio Quemado, in the archaeological site of Tula in central Mexico (Fig. 1). The disk was severely degraded, both in the physical state of its materials and in their coherence as an object. Two years of intensive research and conservation resulted in a hypothetical reconstruction, based on (at that time) an innovative methodology (Magar and Meehan 1995), and a challenge that allowed a divergence from what would normally have been declared unrestorable. The strategy covered various aspects, including the understanding of the values and significance of this artefact, while at the same time gaining enough information to evaluate the feasibility of a conservation process and disseminating the information obtained through research.

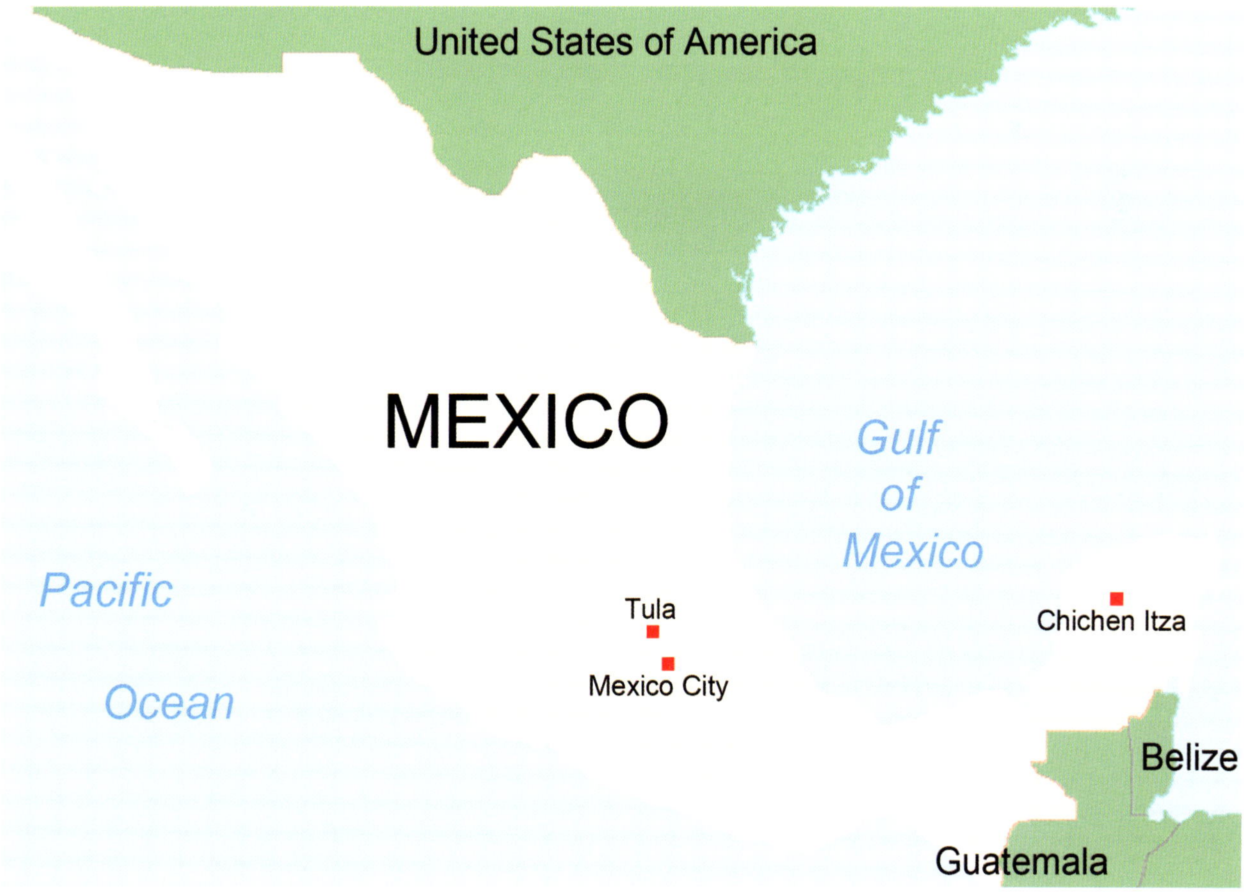

Figure 1. Location of Tula (image: V. Magar).

In order to achieve this, the strategy essentially comprised three main steps: (1) research, (2) decision-making process and (3) conservation – restoration process.

Research

The first step was a thorough research, focusing on two main areas:

- Historic-symbolic aspects: to define the values and significance of the disk, according to its use, meaning, iconography, archaeological context and comparison to other similar artefacts. This included bibliographical research and observation of the technological and design evidence from the disk and its context.
- Materials: this area of research focused on the physical properties of the various materials comprising the disk, with the aim of discovering technological data (constituent materials and technology), and defining the decay mechanisms and any distribution patterns within the remains of the disk. It was carried out through micro-excavation and the merging of the different materials into a single temporary support, their identification by laboratory analysis, a detailed observation of original materials and their interconnection, and the disk's relationship with the archaeological context.

These two research objectives were undertaken simultaneously and they constantly complemented each other.

Decision-making process

This step consisted of cross-checking the information gained during the previous phase (significance, values and data), along with discussions and feedback from colleagues, including conservators, archaeologists, geologists and other specialists, in order to reach the decision to highlight the significance of the offering.

Conservation – restoration process

The third and final step consisted of a series of actions to recover the disk's materials and present them in a way in which its main values were best portrayed.

Background

The ancient city of Tula, located in the central high plains of Mexico, was the capital of the Toltec state between the ninth and eleventh century (Fig. 2). Its political power included the valleys of Mexico, Puebla and Tlaxcala, and its influence reached the Maya area, as evidenced by a number of similar architectural, artistic and symbolic features. According to Aztec legends, the Toltecs were the inventors and creators of turquoise mosaics, a craftsmanship they would have passed on to other Mesoamerican civilisations.

The first important archaeological work in Tula was carried out in the 1940s and 1950s by Jorge Acosta (Acosta 1960). More recent important archaeological research, between

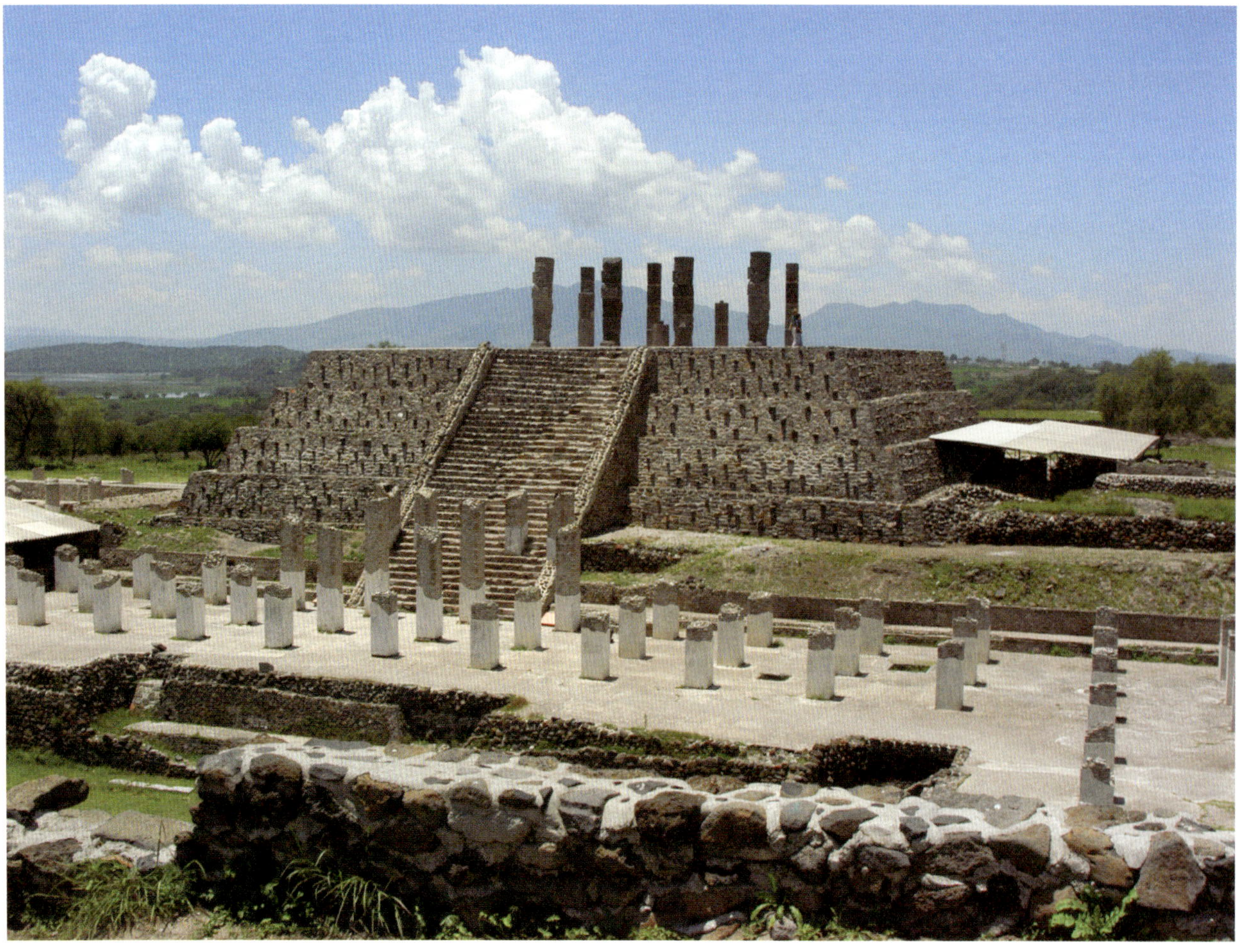

Figure 2. View of Tula (image: V. Magar).

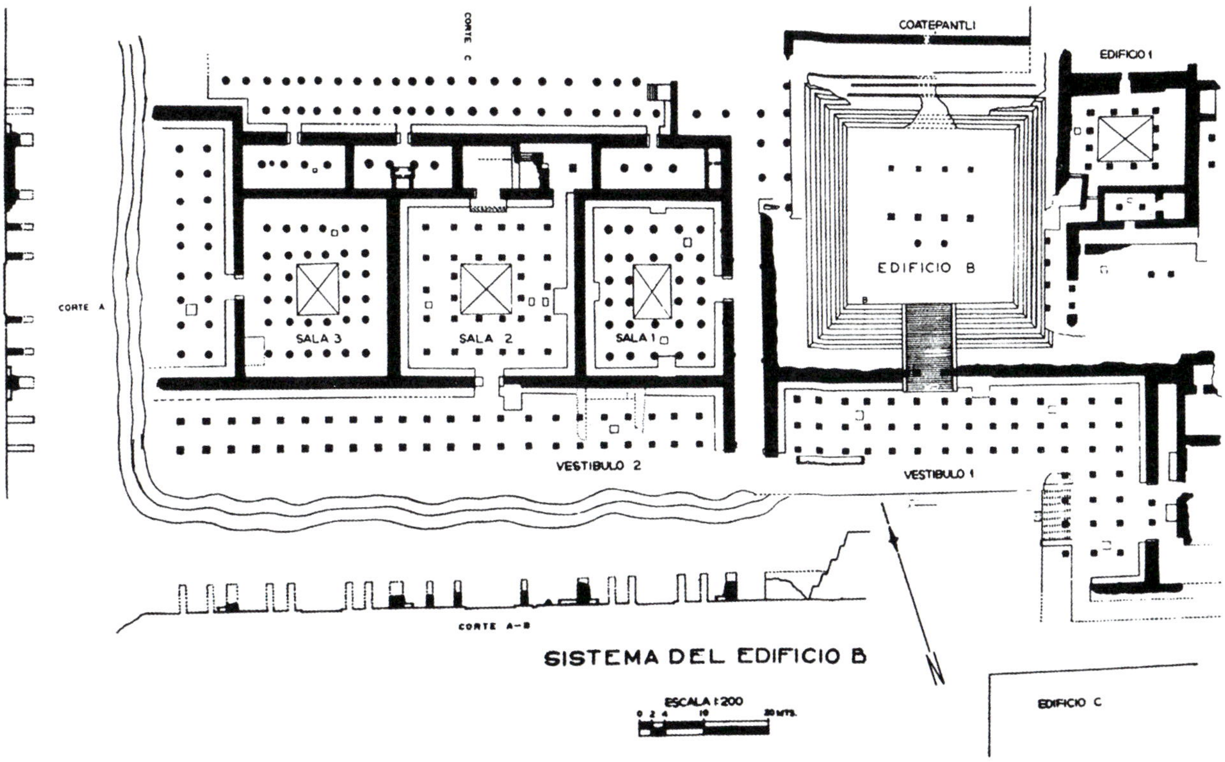

Figure 3. Plan of the Palacio Quemado (image: V. Magar and P. Meehan, redrawn from INAH 1967).

the 1970s and 1990s, was carried out by Robert Cobean and Guadalupe Mastache.

The mosaic disk was found in 1992 during the 'Maintenance, Conservation and Research Project of the Archaeological Site of Tula', directed by Dr Cobean. Within this project, the Palacio Quemado, a ceremonial building, was explored in order to continue Acosta's initial research regarding construction techniques, and also to locate and revitalise original drainage systems (Fig. 3).

The drainage system of rooms 1 and 3 had been located by Acosta, but room 2 offered no evidence of this. The archaeological campaign of 1992 was therefore focused on this central room. A large irregular cavity in the northern side of the courtyard, dating from Acosta's excavations, showed the stratigraphy of the fills of the original floors (Fig. 4). In 1992, during the excavation, an offering (No. 1) was found in the fourth layer at the centre of the room, under a circular lid.

When the lid was lifted, the offering was partially covered with soil therefore it was difficult to identify the turquoise mosaic. Additionally, over it lay shell plaques, jade pendants (over the sandstone disk), shell beads (placed in the north), a gorgonian collar leaf (in the east), and the remains of a marine animal (in the west), all of which formed part of the offering (Fig. 5).

These associated materials were lifted after preliminary documentation and the archaeologists then rapidly realised that a mosaic artefact lay between all those materials. The support was completely decayed so the mosaic elements were loose and were easily moved as cleaning progressed. A sandstone disk, located in the centre of the offering, was retrieved separately, as well as a large number of turquoise tesserae.

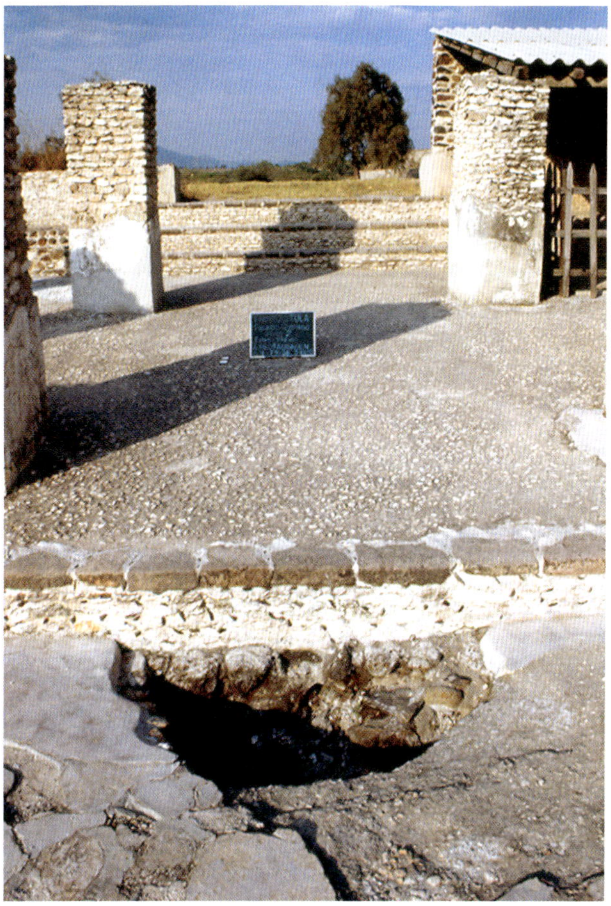

Figure 4. View of the cavity in the northern side of the courtyard, room 2, Palacio Quemado (image: Humberto Illera, Maintenance, Conservation and Research Project of the Archaeological Site of Tula).

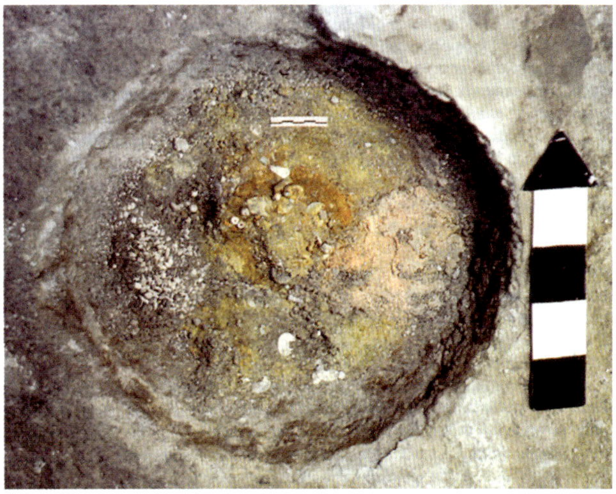

Figure 5. View of the offering *in situ* (image: Humberto Illera, Maintenance, Conservation and Research Project of the Archaeological Site of Tula).

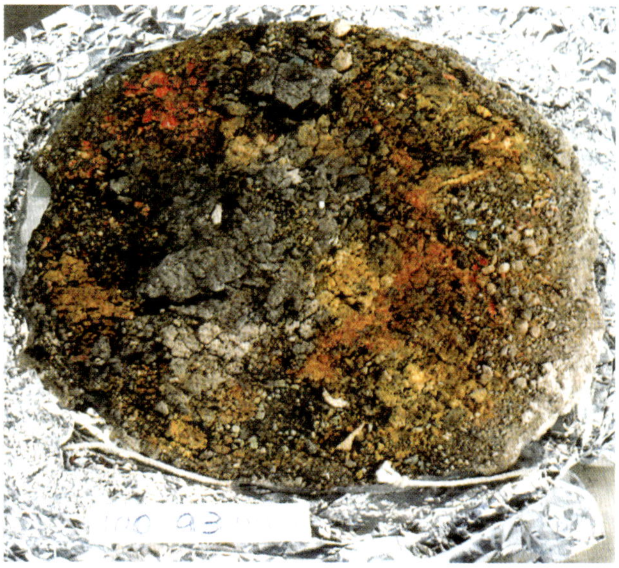

Figure 7. Detail of one of the paraffin wax blocks as it reached the conservation laboratories (image: V. Magar).

Once archaeologists could see the mosaic had a design (very likely a disk) and that all previous actions taken had altered its coherence, they proceeded to extract the mosaic artefact by pouring molten paraffin wax on it in order to lift it all *in situ*. This method was partially successful in immobilising the mosaic elements, but not all of them were retrieved in the first attempt, so it had to be repeated twice in order to retrieve all the materials (Fig. 6).

Subsequently the elements comprising the offering were taken to the conservation laboratories of the Instituto Nacional de Antropología e Historia (INAH) in Mexico City. At the time, the offering included the materials embedded in the three paraffin wax blocks (Fig. 7), and a vast quantity of separate materials, for some of which the overall original location was known, including turquoise tesserae, the central sandstone disk, samples of different materials for analysis and other associated elements.

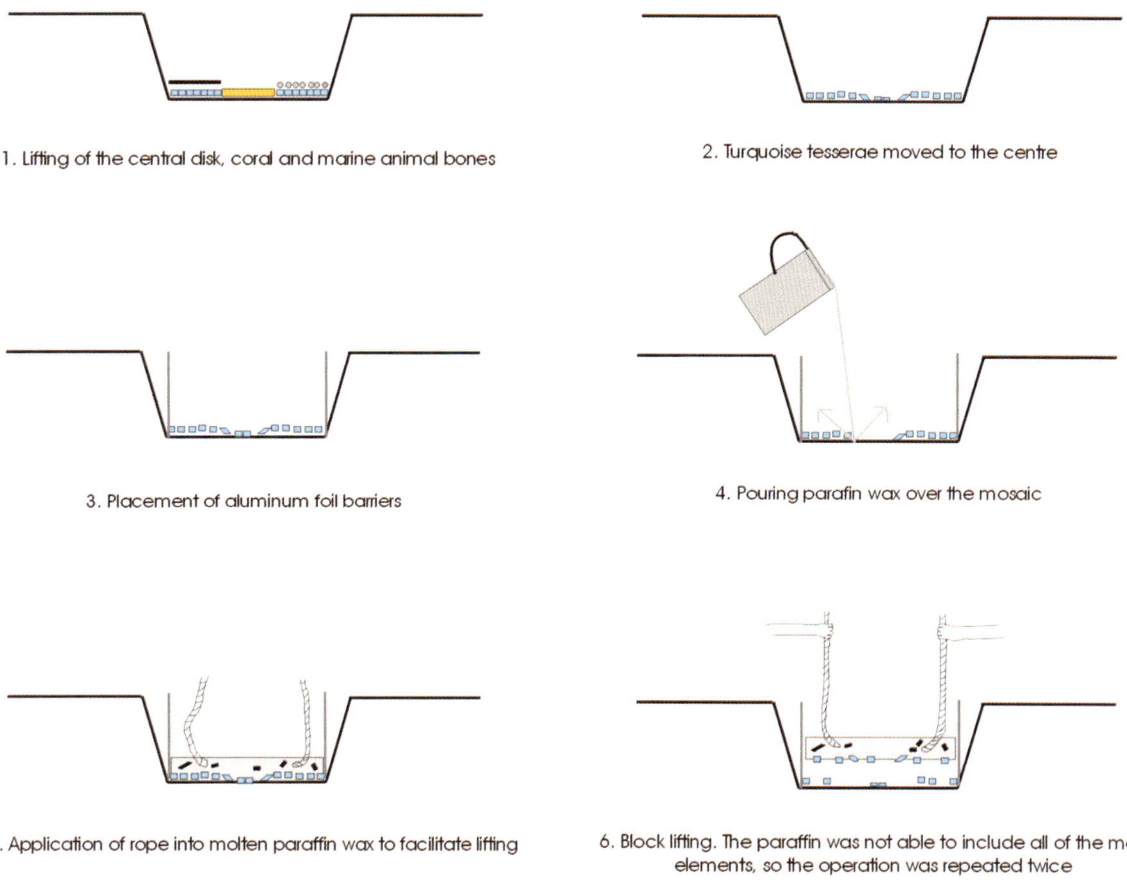

Figure 6. Lifting procedure of the turquoise mosaic (image: P. Meehan and V. Magar).

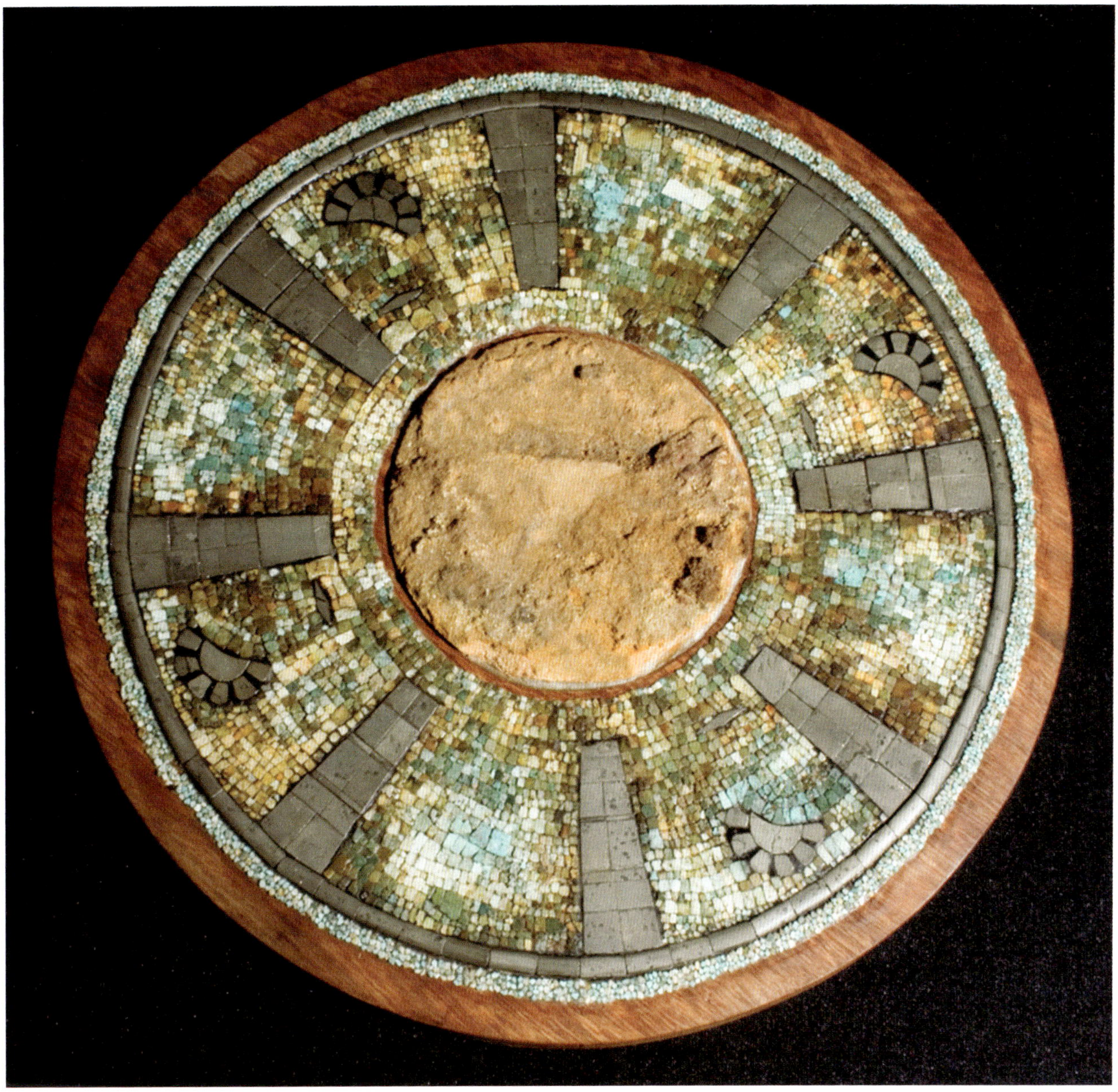

Figure 8. View of the disk (image: A.J. Ruigómez, Museo Nacional de Antropología).

Prior to describing the scope of this research and conservation project, we will briefly describe the mosaic disk as it presumably was, for ease of reference in subsequent descriptions of the text (Fig. 8).

The disk had a diameter of 31 cm and was supported by a wooden backing, except for its central part. The central area was occupied by a smaller disk made of fine sandstone, decorated with pyrite plaques. Surrounding this central disk, on the wooden support, was a series of concentric bands, largely composed of turquoise tesserae. The inner band (1.5 cm wide) was formed by very small, rounded turquoise tesserae. The next band was wider (ca. 6.5 cm) and was divided into eight sections in the form of a trapezoid. Two different designs were repeated in four alternate sections. The sections oriented to the north, south, east and west on the offering were composed only of turquoise tesserae shaped as trapezoids and arranged in lines, with larger tesserae in the centre. The northwest, northeast, southeast and southwest sections represented a synthesised serpent made of pyrite plaques, turquoise tesserae and shell, with a background of smaller turquoise tesserae arranged in lines. The sections were divided by smaller trapezoidal bands made of pyrite plaques. Following this main band towards the exterior, there was a fine line of pyrite plaques of approximately 0.5 cm wide, in cabochon form. Finally, the external part of the disk consisted of a line of very small turquoise tesserae of irregular forms. In other examples of similar disks and iconographical representations (for example in some disks from Chichen Itza), this line is sometimes divided in lobes whose number may vary from 14–18 to 24. In this case there was no evidence of the presence of such lobes.

Research and planning of the conservation of the mosaic

Several turquoise mosaic disks have been found in archaeological contexts in Mexico, many of which have been lost during their lifting. At the time of our research, only six disks found in archaeological contexts had previously been restored. The vast majority could be found in museum deposits (including mosaics found in Tula, Templo Mayor, and Chichen Itza), and

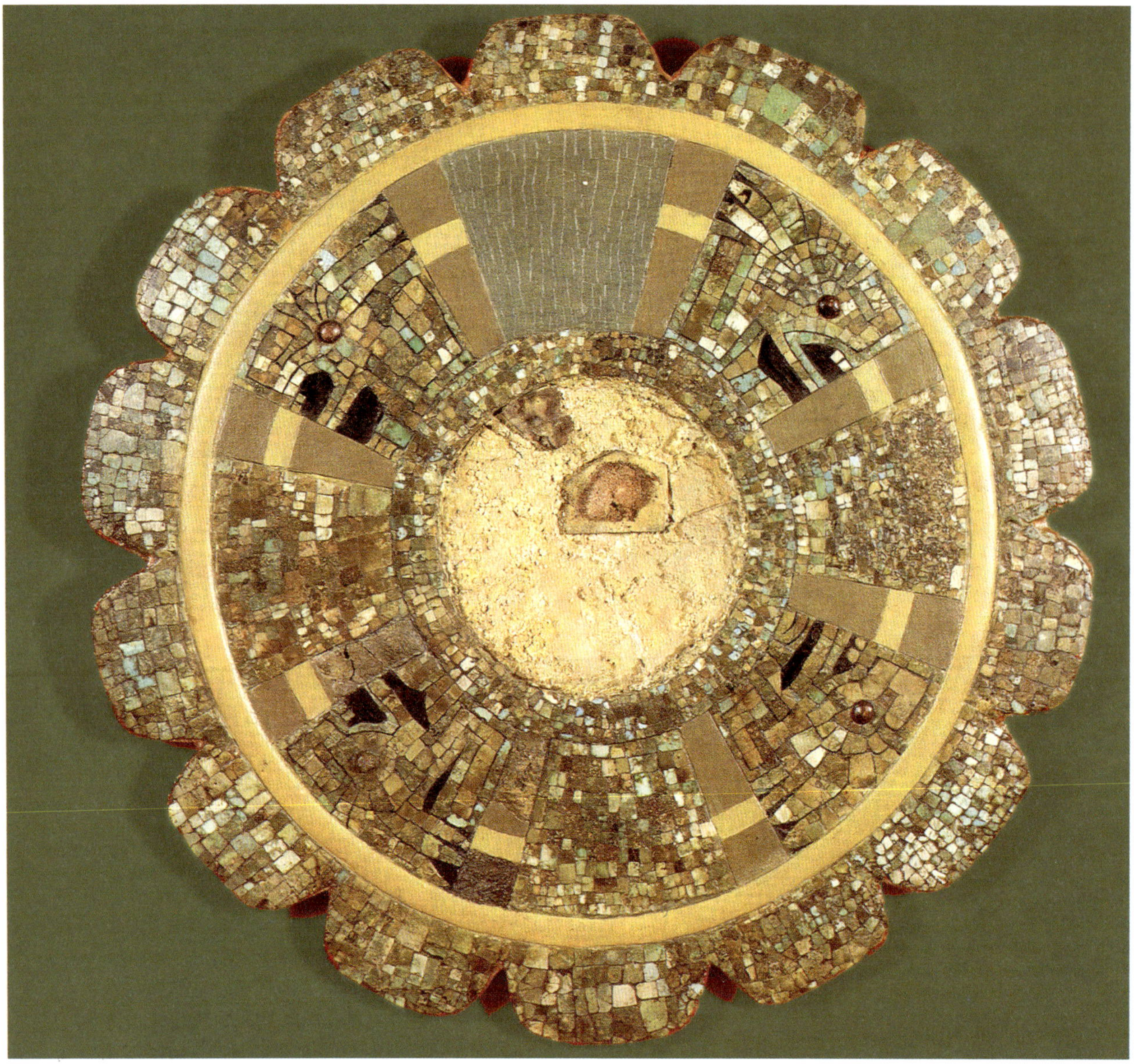

Figure 9. Disk from Chichen Itza (image: Prem and Dyckerhoff 1986: 366).

only a few were exhibited in museums. No mosaic disk from Tula had ever been restored in the past. Therefore, exploring the possibility of restoring the disk from Offering No. 1 had a particular relevance.

However, from the existing disk mosaics, only one could be used as secure reference for comparative analysis and for a hypothetical reconstruction of Tula's mosaic disk (Fig. 9). Discovered by Earl Morris in 1931 in the Temple of the Warriors in Chichen Itza (Morris et al. 1931), the restoration was done at the time by Schoichi Ichikawa at the Museum of Natural History, New York (Ichikawa 1932). His complete reports, the careful lifting and the fact that most of the mosaic elements remained adhered together conserving the original design, made it the most reliable external source of information for the conservation of the disk from Tula – given their striking similarity.

Historic and symbolic values

The turquoise mosaic disk from Tula is highly significant for learning more about the Toltec civilisation and the confusion existing between the Toltecayotl legends regarding the essence and legacy of the Toltecs or 'Toltecayotl' described by the Aztecs and written in colonial chronicles, and the actual archaeological remains found at the archaeological site. Toltecayotl are identified in three levels: one as inventors of the arts, techniques and crafts (Sahagún 1956/1975: X, XXIX, 595–598). The second level refers to them as conquerors and creators of a tributary system with great material power, and lastly as artists and craftsmen linked with Cholula, the metropolis that inherited part of the legacy from Teotihuacan (Noguez 1995: 197; Vaillant 1941/1985: 69).

Regarding the disk's context, an aspect highlighting its significance was the fact that it was found in the Palacio Quemado, identified as a religious and administrative building situated in the main ceremonial Plaza of Tula. Even more significant was its location within the Palacio Quemado in the centre of the building, a place of great importance in prehispanic cosmovision.

The disk was part of an offering and can therefore be considered as a way of communicating with the gods by means of a ritual, i.e. a symbolic individual or collective act

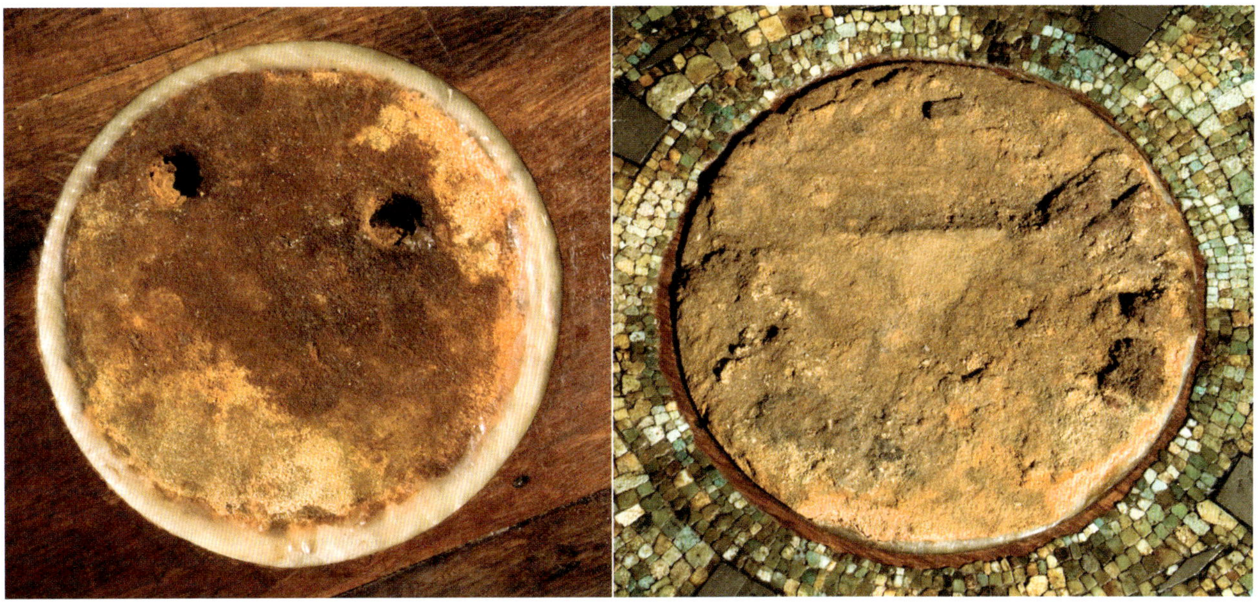

Figure 10. View of the front and back sides of the central sandstone disk (image: A.J. Ruigómez, Museo Nacional de Antropología).

that responded to certain rules. Thus its integral elements plus its location were the means to guarantee this communication (López Luján 1993: 53). As these ritual acts were repetitive, similar disks have been found in other archaeological contexts. Its design can also be seen in the monumental sculptures of Tula called Atlantes, in the slabs of the benches decorated with bas-reliefs of the Palacio Quemado and in the representations of warriors in Chichen Itza and Tula.

Many authors state that these disks represent Xiuhcoatl, the 'turquoise serpent' (Di Peso 1974, II: 497, 557; Gutiérrez Solana 1978: 12; Acosta 1961: 224; Toscano 1944/1970: 180). This meaning refers to the fire serpent, one of the manifestations of Xiuhtecuhtli, the god of fire. Other authors consider that the turquoise serpent is linked to Tezcatlipoca, the war god (Di Peso 1974, II: 557). This would coincide with the use of the disk by a warrior and with the function that, archaeologically and historically, has been given to the Palacio Quemado, a meeting place of the warrior-priests. This point is reinforced by the warriors carved in the benches decorated with bas-reliefs found in this building.

Other interpretations link Xiuhcoatl with Quetzalcoatl, the feathered serpent (Di Peso 1974, II: 556–557; Beyer 1965a: 229–239; Gutiérrez Solana 1978: 12) and Acosta (1960: 69) mentions the existence of reliefs depicting this deity together with Tlaloc in the friezes of room 2 of the Palacio Quemado.

Both interpretations seem to have common aspects: the presence of a serpent, either fire or feathered, and the blue colour. Even if they represent apparently opposing topics, one war and the other fertility, blue colour is logical for both; in the first as the literal meaning of Xiuhcoatl and, in the second, due to its more obvious relationship to water and fertility.

During our research we were inclined to relate the mosaic disk from Offering No. 1 to Xiuhcoatl for two reasons. First, the rear of the disk was painted red (cinnabar). Red colour was associated with war, blood and fire (Beyer 1965b: 482–487). The second reason refers to its function. Disks are represented as part of warrior array, as can be seen in the Atlantes and the pillars of the Tlahuiscalpantecuhtli Temple (Structure B in Fig. 3) in Tula, and in the Temple of the Warriors in Chichen Itza. In the latter, the disks worn by warriors are painted in blue colour. All representations show warriors wearing them in the lower back area.

Due to its constituent materials we consider Tula's mosaic disk to be a war shield that was probably used for a ritual dedicated to Xiuhtecuhtli (Toscano 1970:179), after which it was given in offering. Its location in the centre of the Palacio Quemado gives it great importance, as it is the axis mundi in prehispanic cosmovision. From its physical location to the smallest detail in its manufacture, its uniqueness and value are evident.

The disk and the offering's elements were composed of a variety of materials, most of which were likely to have come from remote provenances. If so, this would confirm the existence of long-distance trade or tributary routes attributed to the Toltecs by various sources (Noguez 1995: 197, 207; Vaillant 1941/1985: 69; Harbottle and Weigand 1992). The repetition of part or all of the design in artefacts found elsewhere suggests there was an effective communication system between distant cultures. This is obvious between Chichen Itza and Tula; stone and pyrite disks (such as the central one in Offering No.1) have been found in both sites, as well as in other less related sites such as in Querétaro, Durango and Zacatecas. Similar designs made on copper plates were found in Paquimé (Chihuahua).

The origin of the materials and the work required in their construction are the result of a great effort. The quality of the turquoise mosaic disk attests to the skilfulness and craftsmanship of those who made it. If the Toltecs constructed it, it could be argued that this art gained importance and skill in Tula, even if they were not the inventors as they were thought to be.

The turquoise mosaic disk from Tula provided important information on the Toltec civilisation, concerning which there has been much confusion in the legends regarding their magnificence described by Aztecs and written in colonial chronicles, and the actual archaeological remains found in Tula.

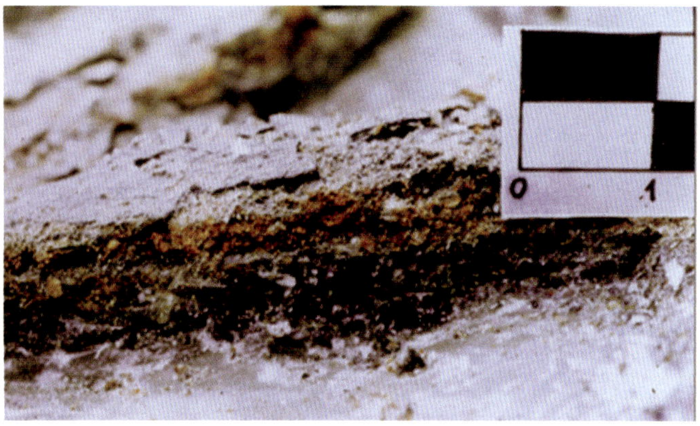

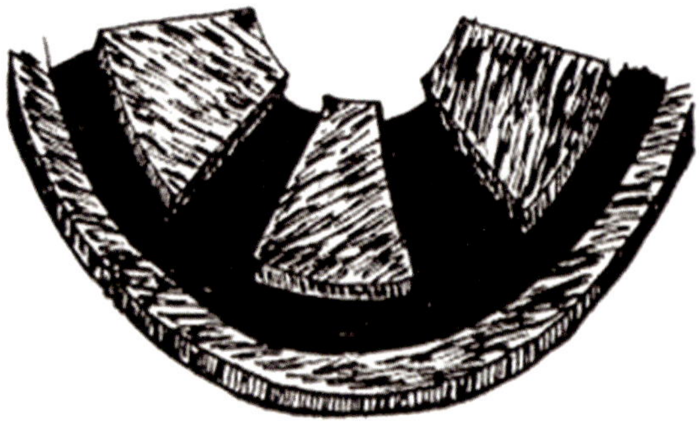

Figure 11. Detail of the pyrite elements in the paraffin wax block (left), which would be placed on lower areas of the wooden support (right) (images: V. Magar and P. Meehan).

Fabric and physical properties

When considering the technology involved, given the complexity of the mosaic design, it seems likely that a preliminary drawing or sketch must have been used to guide the work. This was probably done on paper in a process similar to that described by Sahagún for the fabrication of quill mosaics (Sahagún 1956/1975: IX, IV, 531).

Central disk

The first step was quite possibly to carve the central sandstone disk, as it would dictate the overall size of the disk. Made of soft sandstone, it has a slightly irregular circular form. On the rear of the disk, the edges are bevelled for their assemblage with the wooden backing supporting the mosaic (Fig. 10). In the upper part of the rear of this central disk two perforations were made, with a thin channel on the front side. These were possibly made so that a cord could go around them for fastening the whole assemblage to the warrior's back, without distorting the front surface. The front of the central disk was covered with yellow iron and sulphide oxides, resulting from the degradation of pyrite plaques. Originally these were cut in rectangular or irregular geometrical shapes in order to cover the entire surface, thus forming a mirror.

Wooden support

Wooden remains were found at the bottom of the offering, as well as traces on the black adhesive used to glue the pyrite plaques onto the mosaic. However, due to its severe decay, at the time of our research it was not possible to identify the type of wood used.

We believe that the wooden support was adapted in form to accommodate the central sandstone disk. Once the round shape was cut, the centre was perforated with sharp instruments, based on the shape and size of the central sandstone disk. Later, incisions were made in the areas where pyrite sections and decorations were placed. These levels reflect a practical necessity: the pyrite plaques are thicker than the turquoise tesserae and are adhered with a thicker adhesive, and they had to be placed on a lower level on the support to result in the same level on the mosaic surface. This difference in levels was visible in the materials contained in the paraffin wax blocks (Fig. 11).

The use of different levels on the wooden support is also found in other mosaics from Mesoamerica, created, however, with a different purpose; in those instances used to create patterns in the various features of the design, creating small 3D reliefs (McEwan et al. 2006: 65).

Turquoise tesserae

The tesserae were identified through physical analyses (including the measurement of hardness, density, resistance to abrasion and ignition) and X-ray diffraction (XRD), undertaken at the Geology Institute of the Universidad Nacional Autónoma de México (UNAM). The results indicated the presence of turquoise, with traces of silica in its original composition, as well as a number of surface additions composed of amorphous elements. Preliminary results of current research carried out by Emiliano Melgar, Reyna Solís and José Luis Ruvalcaba from the Templo Mayor Museum Laboratories has revealed additional information regarding technological evidence. Optical and scanning electron microscopy analyses provided evidence of cutting and polishing techniques used for the turquoise tesserae, indicating that they were abraded with sandstone and polished with leather. Additional marks also indicated the use of obsidian blades for the cuts (Melgar 2009). During the last decade there have been significant advances in analytical techniques, allowing a deeper knowledge of the composition of materials, notably turquoise, and providing greater insight into provenance and trade routes (see articles by Cartwright et al., Stacey et al., Hull and Fayek, and Thibodeau et al., in this vol., pp. 9–14, 55–64, 29–40 and 65–74 respectively). These new techniques open up the potential for an entirely new set of questions on the Tula disk, which will be explored in the near future.

Most tesserae are perfectly cut to fill every space with specific shapes to ensure an exact fit in the main part of the mosaic. The value of turquoise as a material is evident in the irregular tesserae in the border of the disk that used even the extremities of turquoise nodules extracted from the mines. Most of the tesserae, especially the trapezoidal ones, have bevelled edges for a perfect bond between them and effective adhesion to their support.

Turquoise adhesive

Turquoise tesserae were adhered by means of a whitish adhesive of which small remains were found and analysed. This adhesive was composed of resin, wax and gypsum, as revealed through direct chemical analyses with reagents, in addition to Fourier transform infrared spectroscopy (FTIR) analysis undertaken at the Faculty of Chemistry (Department of Analytical Chemistry and Foods, UNAM). Although at this stage of the research it was not possible to define the species, resins have been reported as adhesives in other mosaics from Mesoamerica. The use of single as well as mixed resins (pine and copal) has been characterised for adhesives on the mosaics in the collections of the British Museum (Stacey et al. 2006). Beeswax mixed with resin is reported for the mosaic mask depicting two entwined serpents; here it was not used as an adhesive, but rather as a paste to mould small decorative reliefs (McEwan et al. 2006: 50). A mixture of resins and waxes is reported for the Coixtlahuaca mask, although no specifications are provided as to how the analyses were made (Montero 1968: 102).

It is worth noting that the adhesive used in the Tula disk was mixed with a filler, which could not only have made the adhesive more plastic, but may also have served the purpose of filling in spaces between tesserae and allowing their levelling on the surface. It also provided a white background, something not seen, for example, in the Aztec/Mixtec mosaics in the British Museum's collections.

Pyrite plaques

Pyrite plaques were carved in different shapes: trapezoidal, half circle and rectangular with a rounded surface. The sizes and shapes depended on the design, making use of the forms of the matrix. The trapezoidal pyrite plaques are also bevelled but with less of an angle than on the turquoise tesserae. They were also polished on both sides, but due to their decay all metallic lustre has been lost (Fig. 12).

Pyrite adhesive

The pyrite fragments were adhered to the support by a black adhesive identified as tar through physical analysis (ignition) and FTIR analyses. Traces of the plaques on adhesive remains indicate that the tar was in a viscous state at the time they were placed, most likely by previous heating.

Red paint

The rear was decorated in red using cinnabar applied over a primer made of similar materials as the white adhesive. This layer must have been the last to be applied, possibly due to its toxicity. Evidence of this was shown on traces found over some turquoise tesserae, probably as a result of contamination during its application.

Red pigment (cinnabar and red ochre) was used on the rear of both turquoise mosaic masks from the British Museum collections (McEwan et al. 2006: 47, 50). However the difference is that no preparation layer to underlie the red is reported for these mosaics, as can be seen on the Tula disk.

Alteration and decay mechanisms

Defining the alteration mechanisms of the offering was essential in order to have a better understanding of the disk. An analysis of the construction system of the floor in room 2 of the Palacio Quemado showed that it was carefully planned to protect the offering. However, this highly stable system was

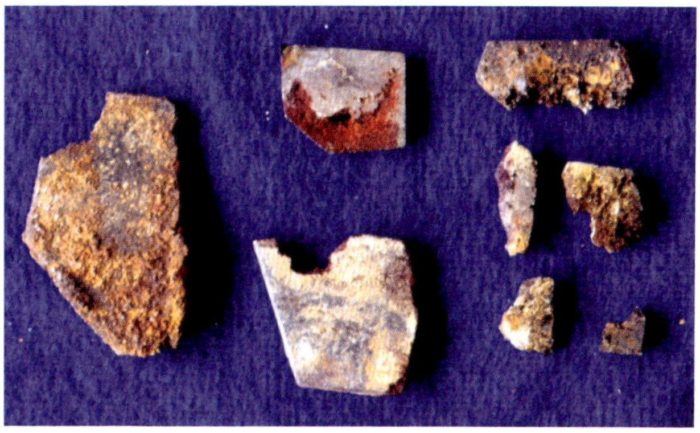

Figure 12. View of decayed pyrite plaques from the disk (images: V. Magar and P. Meehan).

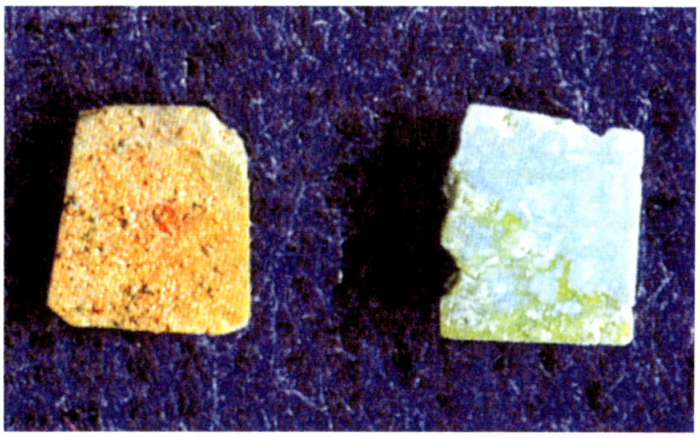

Figure 13. View of turquoise tesserae with material deposited on the surface, altering the blue colouration (images: V. Magar and P. Meehan).

disrupted by a cavity on the floor that remained in the north side following the 1950s excavations (Fig. 4), which allowed the access of water and organic matter for nearly 40 years. This drastic change in the burial conditions and microclimate of the offering led to numerous chemical reactions. The most sensitive materials degraded almost completely. Pyrite, which is mainly iron sulphide, decayed into iron oxides, iron sulphates and sulphuric acid. The liberated iron combined with organic matter formed stable compounds, which can be seen on the turquoise surfaces.

The alteration of turquoise was defined in different levels according to their proximity to pyrite within the mosaic. This selective alteration was an important tool for the reconstruction of the disk. The iron crust was found in the surface of all turquoise, but increased in thickness and in red-brown colour in the tesserae that were closer to pyrite elements (Fig. 13).

The decay of the wooden support was the first main cause of the disk's loss of coherence, which was later exacerbated by the excavation interventions. These can be summarised in a lack of anticipation and of conservation specialists working *in situ* to control conditions and implement more successful lifting methods. The paraffin wax method was successful up to a certain point in immobilising the mosaic but it did not permit the recovery of most organic materials therein.

Decision-making process

After assessing all the information previously gained through research, some considerations were taken into account in order to decide the future of the mosaic disk.

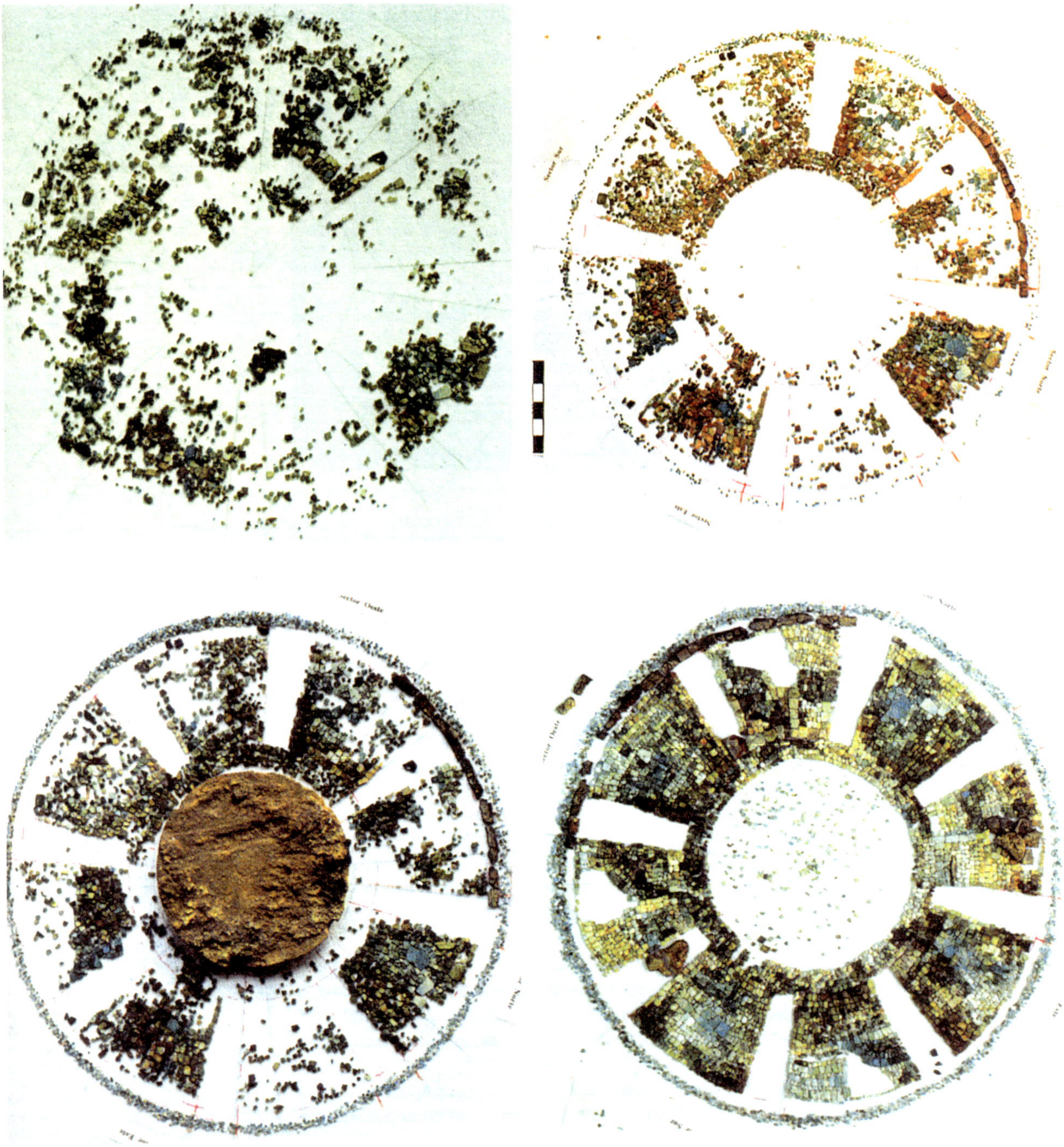

Figure 14. View of the various stages of placement of the materials on a provisional support (images: V. Magar and P. Meehan).

This included its historic and symbolic significance, the scarcity of information available on turquoise mosaics from Tula and the opportunity to disseminate its technological values.

The options were:

1. To keep all materials in bags and boxes and publish an article with the results of the research carried out. This would mean another mosaic disk remaining in museum storage.
2. To integrate the disk's elements from the paraffin wax blocks and glue them to a support as they were found *in situ*. This would have meant keeping the disk in a chaotic, incomprehensible state that was essentially a result of recent interventions.
3. To do a hypothetical reconstruction that would help us understand these materials better, and which could be used as didactic material to disseminate its aesthetic, historic, symbolic and technological values.

Following discussions by a multidisciplinary group, the third option was chosen. This option could not be called a restoration; it had to be clear that it was a hypothetical reconstruction so that the general public would know this and researchers would study the mosaic disk with due caution.

The support for this decision included the observation of distribution patterns of tesserae in the paraffin wax blocks; even if they had moved, they still conserved certain coherence in shape, colour and the places they occupied. Other

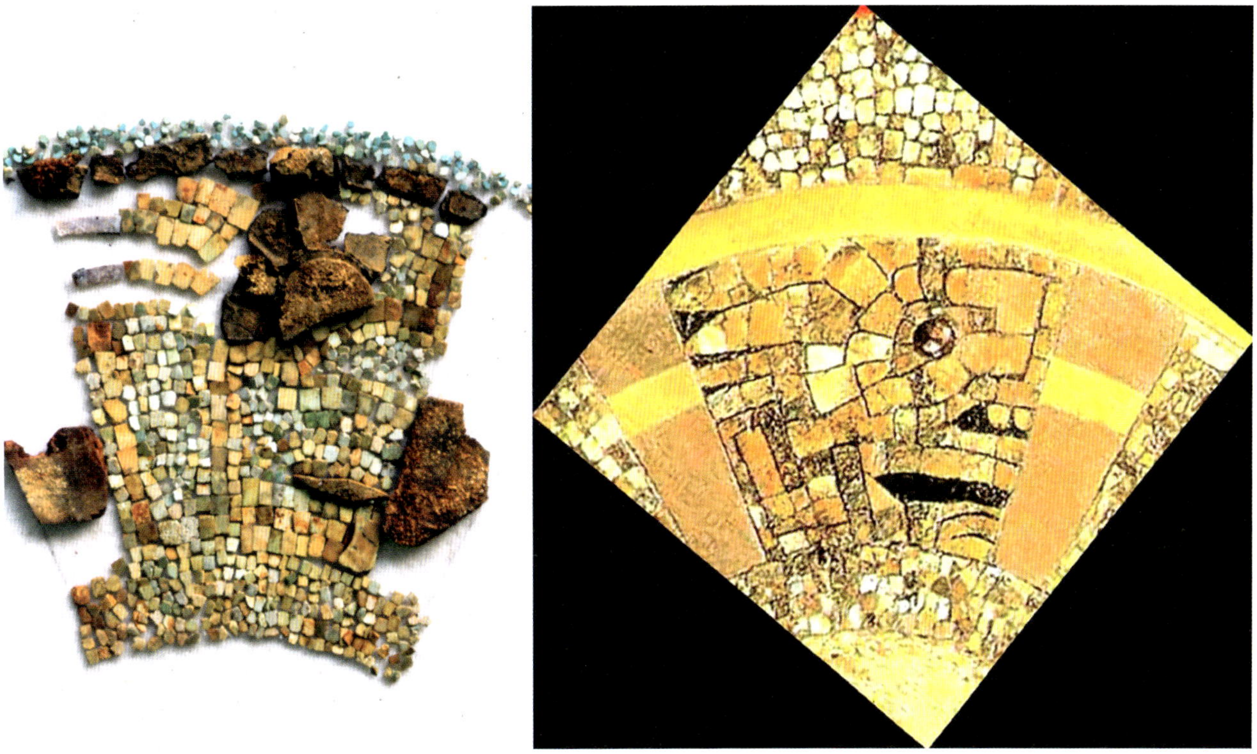

Figure 15. Trial design of the serpent (left) and detail from the disk of Chichen Itza (right) (images: V. Magar and P. Meehan; Prem and Dyckerhoff 1986: 366).

important elements included the information gained through recording, the selective alteration of the turquoise tesserae, the repetitive designs in other disks, and the fact that previously excavated mosaics from Tula were essentially found unrestored in museum deposits.

Conservation and hypothetical reconstruction

The reconstruction included a series of progressive steps, the first of which was a preparatory stage and included:

- the observation and classification of materials into groups;
- sifting of the contextual soil to gather all the offering's materials;
- cleaning by extracting soil, gravel and plaster from the paraffin blocks to be able to observe and record the mosaic elements;
- graphic and photographic recording that was carried out continuously throughout the project;
- X-radiography of the three paraffin wax blocks to enable better observation of the materials and their distribution.

In the second stage, the paraffin wax blocks were divided into eight sections using the orientation marked by archaeologists. This division into sections was based on observations of the distribution of elements within the blocks. All the recoverable materials were extracted from the paraffin wax blocks, starting with the turquoise tesserae. They were cleaned of paraffin wax and placed on a provisional support, respecting their exact location, but inverting them as the materials were extracted from the bottom (Fig. 14).

The third stage involved the search for the design, and consisted of finding distribution patterns based on colour, size and shape of the turquoise tesserae. This was done with a combination of direct observation, enhancing digital images with computer programs, and also comparing patterns with other disks. Through these enhanced images, and after long months of close observation of the materials, we were actually able to see clear distinctions and arrangements of the tesserae in specific areas of the mosaic.

Once the patterns had been recognised, the next step was to rearrange the mosaic elements. A second provisional support was made with corrected measurements and all tesserae were placed on it. The tesserae that had been displaced during the lifting process were moved to logical places within the sections and bands (for example, the ones that fell into division sectors and the central disk area, once it was removed). The loose tesserae were also taken from the bags and integrated into their original sectors.

In the next stage, a trial of the reconstruction of the serpent sectors was carried out using the loose materials, based on the disk's indicative elements and on the disk from Chichen Itza. Many coinciding elements between them were detected. The result was satisfactory and illustrative (Fig. 15). It was therefore possible to proceed to a final distribution of tesserae in the provisional support, starting with the serpent design sectors, followed by turquoise ones, and lastly the outer line of small irregular tesserae.

It was then necessary to undertake the construction of the new, definitive support. This was done using a tropical aged wood, which was hard and resistant to biological attack. It was cut slightly larger than the original disk to provide protection during handling. The support was perforated in the centre and incisions were made in the areas where the pyrite plaques were to be glued (Fig. 16).

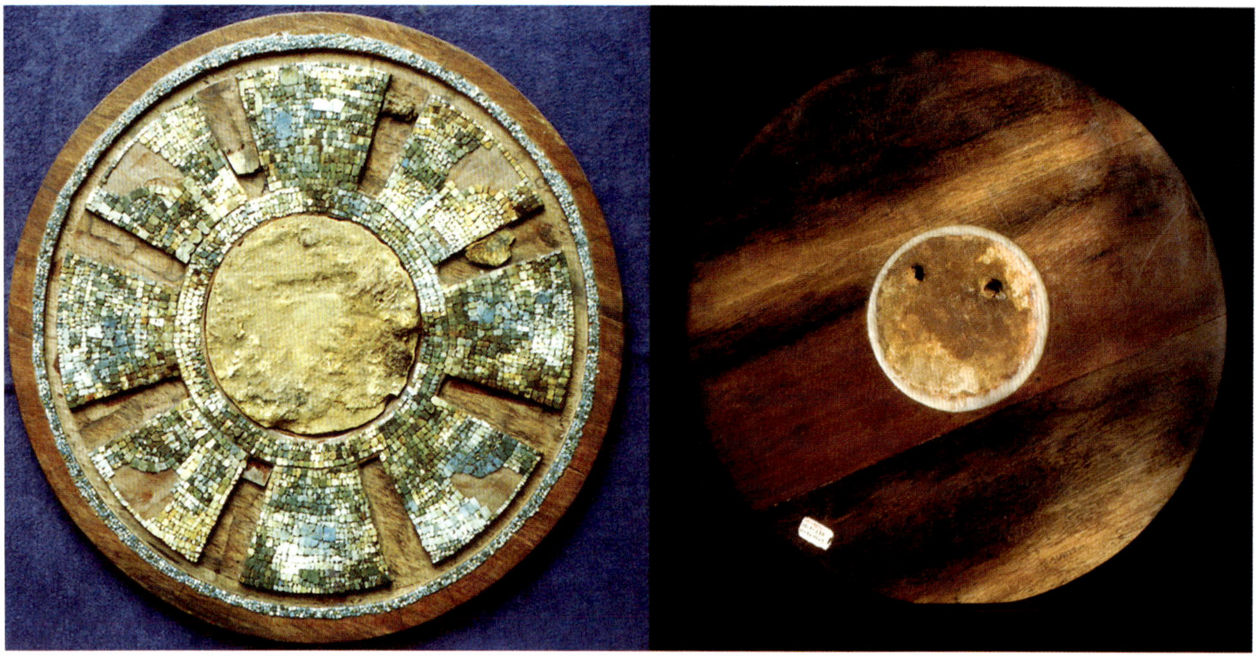

Figure 16. View of the front (left) and back (right) sides of the new wooden support (images: V. Magar and P. Meehan; A.J. Ruigómez, Museo Nacional de Antropología).

Moving the tesserae to the new support was done by facing the whole mosaic on the provisional support, using a non-woven Japanese paper and a water-soluble adhesive. This permitted a single move to the new support. The tesserae were glued using a similar adhesive to the original one, composed of beeswax, carnauba wax and dammar resin. New pyrite plaques were cut and adhered to the support with tar. Finally, the central sandstone disk was fixed to the wooden support by the bevelled edges with the same adhesive as that used for the turquoise tesserae (Fig. 17).

Conclusions

The hypothetical reconstruction of the Palacio Quemado's turquoise mosaic disk was concluded in early 1995, more than 15 years ago. After this was accomplished, the decision was made that it should become part of the collection of the Museo Nacional de Antropología (MNA) in Mexico City. It was part of the permanent collection for a few years and has also been displayed in various temporary exhibits in Mexico and abroad. But the fact that it is a hypothetical reconstruction has never been mentioned, regardless of our recommendations at the end of the process.

The reconstruction was based on solid research, but we were of course concerned about the compromised authenticity of the turquoise mosaic disk. Significant information was gained through this research on historic, symbolic, technological and decay mechanisms, and was reflected as much as possible in the reconstruction. The main aim in carrying out this reconstruction was to be able to use the materials to try to replicate the construction techniques, and to have one first example of a mosaic from Tula, even if it was the result of an invasive treatment. More importantly, the research also provided information that should allow improved planning and research for future excavations, including better retrieval techniques. The physical integrity of the disk was recovered to a certain extent, but obvious limits in using this material should always be clearly evidenced.

Another major drawback in the presentation of the mosaic since it was reinstated as part of the collection of the MNA is that there have never been references made to the other materials of the offering, which may be highly relevant to the meaning of the disk in its broader context.

In recent years, the mosaic disk was moved to a vault in the MNA, where fragile and valuable objects are kept. As a result of its travels, the disk suffered some damage, and received conservation treatments from MNA staff. There are apparently two main issues with the stability of the mosaic: on one hand, the turquoise tesserae that were recovered from the bags (i.e. the first ones that were retrieved by the archaeologists as they were lifting the separate materials) seem to be suffering a loss of material in the form of surface powdering. They appear to be in a much more fragile condition than the tesserae that were embedded in the paraffin wax. A more significant matter of concern is the stability of the remaining pyrite plaques on the central disk, which were left untreated, and are extremely susceptible to changes in relative humidity. The other fragile element is the new wooden support, which, in spite of being made of an aged tropical wood, developed a small crack on one side. The latter has been treated, but the decay of the pyrite plaques could still be active. We can presume that its appearance in exhibitions may be diminished as a result.

In an ideal situation, an intervention such as the one carried out on this disk should not be required, as maintenance of buildings and excavation, and lifting methods improve. However, this project has proved to be extremely valuable in providing a deeper understanding of mosaic manufacture in Tula and of turquoise mosaic disks in general. With the development of analytical techniques in the last two decades, a new phase of research has been launched that should soon provide more detailed information.

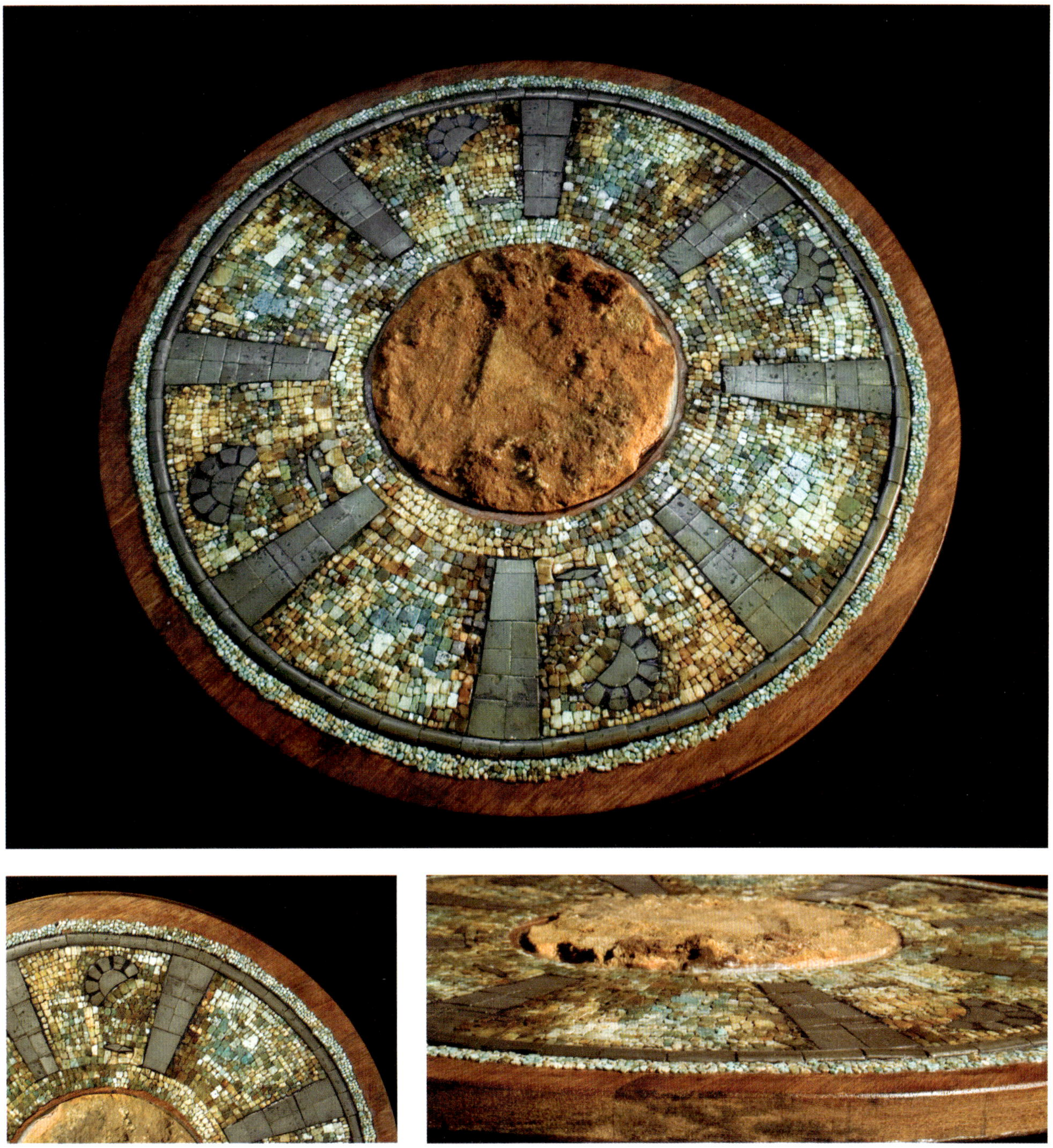

Figure 17. Final view of the disk and details (images: A.J. Ruigómez, Museo Nacional de Antropología).

Acknowledgements

We would like to thank Dr Robert Cobean, Director of the Tula Archaeological Project, and Luisa María Mainou, our advisor during this conservation and research project, for their support and guidance.

References

Acosta, J. (1960) 'Las exploraciones arqueológicas en Tula, Hidalgo durante la XI Temporada, 1955', *Anales del INAH. Tomo* XI, *1957–58*, No 40 Colección INAH-SEP. Mexico City: Instituto Nacional de Antropología e Historia (INAH)/Secretaría de Educación Pública, 39–74.

Acosta, J. (1961) 'La indumentaria en las cariátides de Tula', in *Homenaje a Pablo Martínez del Río en el XXV Aniversario de la edición de Los Orígenes Americanos*. Mexico City: Instituto Nacional de Antropología e Historia (INAH), 221–228.

Beyer, H. (1965a) 'El llamado "Calendario Azteca". Descripción e interpretación de Cuauhxicalli de la Casa de las Aguas', in *El México Antiguo, Tomo X*. Mexico City: Sociedad Alemana Mexicanista, Instituto Cultural Mexicano Alemán Alejandro de Humboldt, 134–256.

Beyer, H. (1965b) 'El color rojo en el simbolismo de los antiguos mexicanos', in *El México Antiguo, Tomo X*. Mexico City: Sociedad Alemana Mexicanista, Instituto Cultural Mexicano Alemán Alejandro de Humboldt, 482–487.

Di Peso, C.C. (1974) *Casas Grandes: A Fallen Trading Center of the Gran Chichimeca, Vols 1–3*. The Amerind Foundation Publication No. 9. Flagstaff, AZ: Northland Press.

Gutiérrez Solana, N. (1978) 'Xiuhcóatl tallada en piedra del Museum of Mankind; Londres', *Anales del Instituto de Investigaciones Estéticas* **48**. Mexico City: Universidad Nacional Autónoma de México (UNAM), 5–14.

Harbottle, G. and Weigand, P.C. (1992) 'Turquoise in Pre-Columbian America', *Scientific American* **266**(2): 78–85.

Ichikawa, S. (1932) 'La restauración del mosaic de turquesas', *Boletín del Museo Nacional de Arqueología, Historia y Etnografía* **3**, Quinta Epoca, Número 2: 7–8.

INAH (1967) *Tula, Guía Oficial*. México: Instituto Nacional de Antropología e Historia (INAH).

López Luján, L. (1993) *Las Ofrendas de Templo Mayor de Tenochtitlan, México*. Mexico City: Instituto Nacional de Antropología e Historia (INAH).

Magar, V. and Meehan P. (1995) 'Investigación para la interpretación y la conservación de un disco de mosaico de turquesa', BA dissertation, Escuela Nacional de Conservación, Restauración y Museografía, Instituto Nacional de Antropología e Historia (INAH), Mexico City (unpublished).

McEwan, C., Middleton, A., Cartwright, C. R. and Stacey, R. (2006) *Turquoise Mosaics from Mexico*. London: The British Museum Press.

Melgar, E. (2009) Museo de Templo Mayor, personal communication (November 2009).

Montero, S.A. (1968) 'Restoration of turquoise masks from Coixtlahuaca and Zaachila, Oaxaca', *Studies in Conservation* **13**: 102–104.

Morris, E., Charlot, J. and Morris A.A. (1931) *The Temple of the Warriors at Chichen Itza in Yucatan*, 2 vols. Washington, DC: Carnegie Institution of Washington.

Noguéz, X. (1995) 'La zona del Altiplano central en el Posclásico: La etapa Tolteca', in L. Manzanilla and L. López Luján (eds), *Historia Antigua de México, Volumen III: El horizonte Posclásico y algunos aspectos intelectuales de las culturas mesoamericanas*. Mexico City: Editorial Porrua/Instituto Nacional de Antropología e Historia (INAH)/ Universidad Nacional Autónoma de México (UNAM), 189–224.

Prem, J.H. and Dyckerhoff, U. (1986) *El antiguo México*. Mexico City: Plaza & Janés.

Sahagún, B. (1956/1975) *Historia general de las cosas de Nueva España*. Anotaciones y apéndices Ángel María Garibay. Mexico City: Editorial Porrúa.

Stacey, R.J., Cartwright, C.R. and McEwan, C. (2006) 'Chemical characterisation of ancient Mesoamerican "copal" resins: preliminary results', *Archaeometry* **48**: 323–340.

Toscano S. (1944/1970) *Arte precolombino de México y América Central*. Mexico City: Universidad Nacional Autónoma de México (UNAM).

Vaillant G. (1941/1985) *La civilización Azteca. Origen, grandeza y decadencia*, S.B. Vaillant (ed.). Mexico City: Fondo de Cultura Económica.

Authors

- Patricia Meehan, Coordinación Nacional de Conservación del Patrimonio Cultural, Instituto Nacional de Antropología e Historia, Mexico City, Mexico
- Valerie Magar, Coordinación Nacional de Conservación del Patrimonio Cultural, Instituto Nacional de Antropología e Historia, Mexico City, Mexico

An integrated approach to understanding the selection and fate of turquoise on Mexican mosaics in the British Museum

Rebecca Stacey, Caroline Cartwright, Giovanni Verri and Colin McEwan

Background

Within the British Museum's extensive Mesoamerican collections are nine turquoise mosaics from Mexico. These complex composite objects are formed from mosaic worked over carved wood, or, in one case, a human skull. They include a variety of forms such as masks, a knife, a shield and various animals. Their precise provenance remains obscure (see Caygill, in this vol., pp. 183–196) but they are thought to be the products of Mixtec craftsmanship and to date to the early sixteenth century. Although they are typically referred to as 'turquoise mosaics', many other materials were selected for the construction of these objects. In addition to turquoise, minerals such as lignite, malachite and pyrite are used in the mosaic work, as well as various species of shell (see Cartwright et al., in this vol., pp. 9–14). There are also painted and gilded surfaces, organic components such as wood, agave fibres, deerskin and the resin adhesives that hold all of these materials together.

The nine mosaics have been the subject of a long-term research project at the British Museum involving a multidisciplinary team of scientists (see Cartwright 2009; Cartwright and Meeks 2007; McEwan et al. 2006; McEwan et al. forthcoming; Stacey et al. 2006; Stacey and Cartwright 2009 for details). The removal of the mosaics from their permanent gallery display in preparation for the British Museum's *Moctezuma* exhibition in 2009 offered an opportunity for a fresh campaign of scientific work to address previously unresolved questions.

This paper draws on both newly collected and previously published data to consider the turquoise of the mosaics in terms of its distribution on the objects, the way that it was used, the impact of subsequent alterations and the implications for ongoing scientific study of turquoise used in mosaic work.

Experimental

The choice of analytical techniques used for the examination of the turquoise (and other minerals) on the objects was dictated principally by the requirement for non-invasive analysis. The goals of the analyses were: the identification of the minerals present, examination of their variability (in terms of colour, quality and composition), the mapping of their distribution on the objects and the assessment of the extent of recycling and later repairs. Source characterisation (provenance) analysis was not attempted (see later discussion). Non-invasive sampling locations on the objects were chosen to gain an overview of the range of material represented, although sometimes the selection of sampling spots for analysis was limited by the ability to orientate the objects so as to place specific tesserae into the path of the analytical beam.

Optical microscopy

Examinations were carried out using a Leica stereomicroscope at magnifications ranging from ×8 to ×80.

Raman spectroscopy

Raman spectroscopy was carried out using a Jobin Yvon LabRam Infinity spectrometer with green (532 nm) and near infrared (785 nm) lasers with maximum powers of 2.4 and 4 mW at the sample respectively, a liquid nitrogen cooled CCD detector and an Olympus microscope system. This allowed specific areas of individual tesserae to be targeted for analysis, with a sample spot size in the order of a few micrometres, depending on the power of objective lens used. Spectra were collected for between 20 and 100 seconds, with at least five repeat measurements being used to produce each spectrum. The resultant spectra were identified by comparison with the British Museum's in-house database.

X-ray fluorescence (XRF)

The XRF analyses were carried out using a Bruker Artax spectrometer on uncleaned surfaces; therefore they are only qualitative as the analyses may be affected to some extent by the presence of surface materials (such as dust, dirt, coatings or conservation treatments). The spectrometer, fitted with a molybdenum X-ray tube, was operated at 50 kV and 0.8 mA. Spectra were collected for between 100 and 300 seconds. The area of analysis was ca. 0.65 mm in diameter and helium was used to flush the area in order to improve the detection of light elements.

Ultraviolet-induced luminescence (UIL) imaging

Images were taken using a Canon 40D camera body modified by removing the IR-blocking filter. For UIL imaging the excitation was provided by two Wood radiation sources

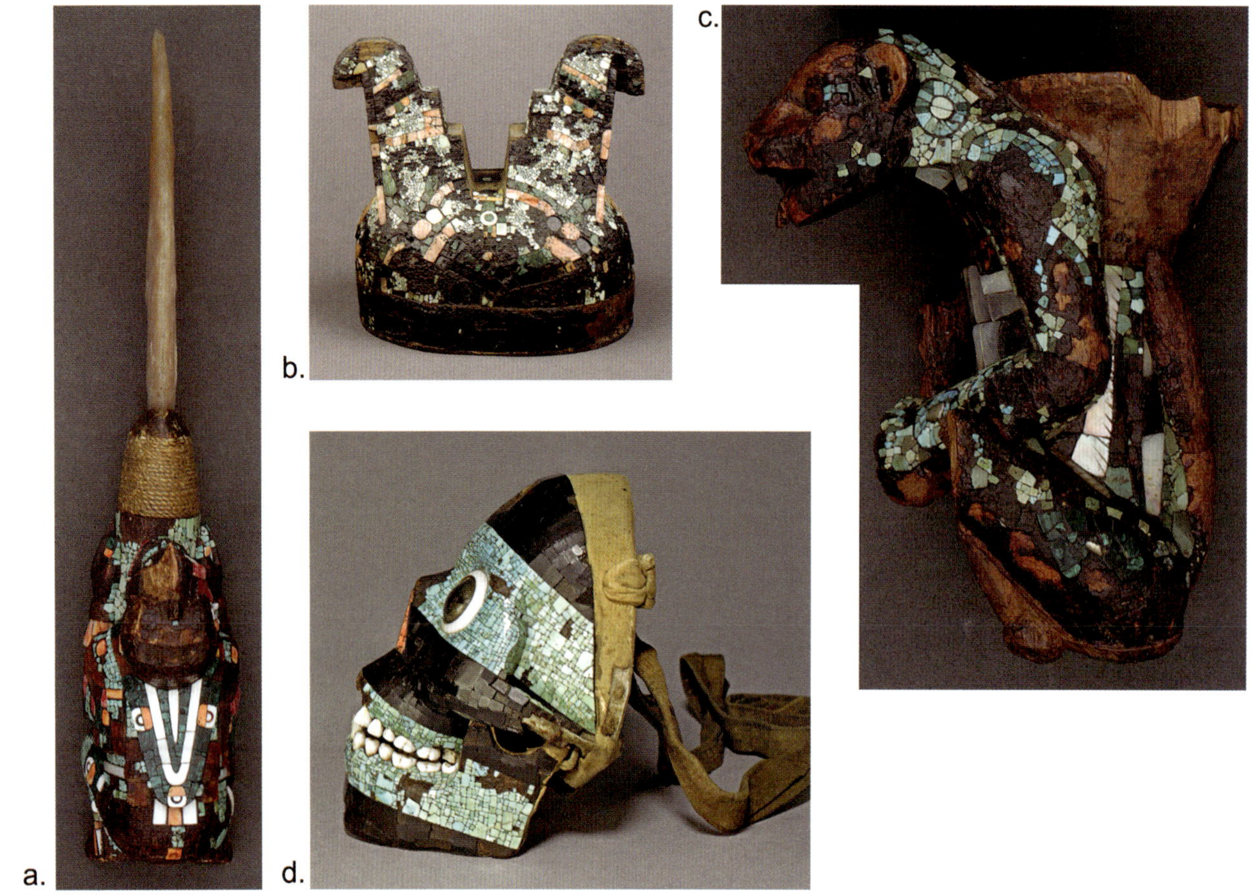

Figure 1. Mexican mosaics from the collections of the British Museum displaying prominent use of non-turquoise mosaic materials: (a) knife handle (Am, St.399) viewed from above; (b) helmet (Am, +.6382); (c) jaguar (Am, +.165); (d) skull (Am, St.401) (© Trustees of the British Museum).

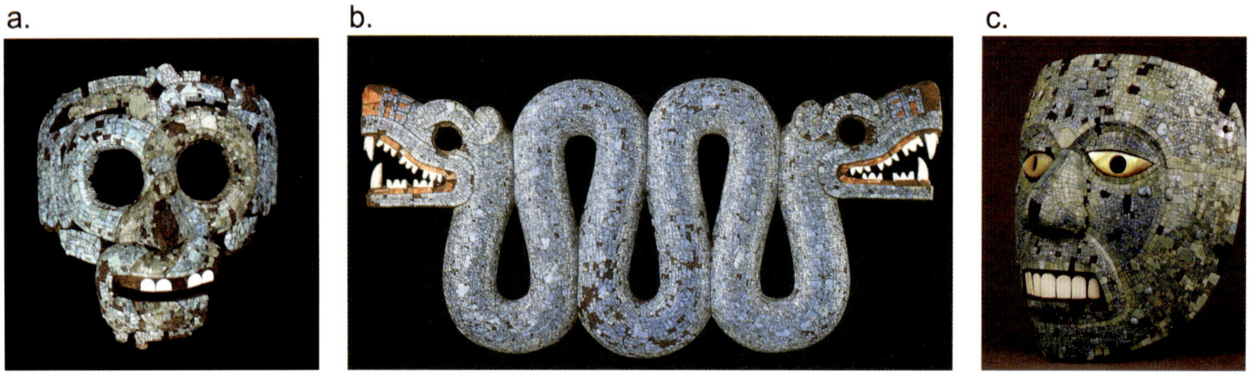

Figure 2. Mexican mosaics from the collections of the British Museum showing predominant use of turquoise in the mosaic work: (a) Tlaloc mask (Am, 1987,Q.3); (b) double-headed serpent (Am, 1894,-.634); (c) 'warty' mask (Am, St.400) (© Trustees of the British Museum).

(365 nm) filtered with a Schott DUG11 interference bandpass filter (280–400 nm) and the camera was fitted with a Schott KV418 cut-on filter (50% transmission at ca.418 nm) and an IDAS-UIBAR bandpass filter (400–700 nm) (Verri et al. 2008).

Results and discussion

Turquoise on the mosaics

Although the mosaics are usually described as 'turquoise', the extent of the use of this mineral in the mosaic work varies considerably. In some of the mosaics in the British Museum collections non-turquoise materials are used extensively and, in places, dominate the overall design (Fig. 1). The costume of the eagle warrior knife handle (Am, St.399), for example, is worked predominantly in malachite with white conch and red *Spondylus princeps* shell while the complex designs on the side of the helmet (Am, +.6382) exhibit a great variety of colour, again using different colours of shell and malachite. The jaguar (Am, +.165) shows particularly extensive use of mother-of pearl shell and also pyrite within the mosaic. The prominent use of black lignite enhances the visual impact in the banded design of the skull (Am, St.401). In contrast, other mosaics in the British Museum collections (the masks (Am, 1987,Q.3 and Am, St.400) and the double-headed serpent (Am, 1894,-.634)) display almost exclusive use of turquoise, with other materials reserved for features such as eyes and teeth or for limited areas of contrast decoration (Fig. 2). The

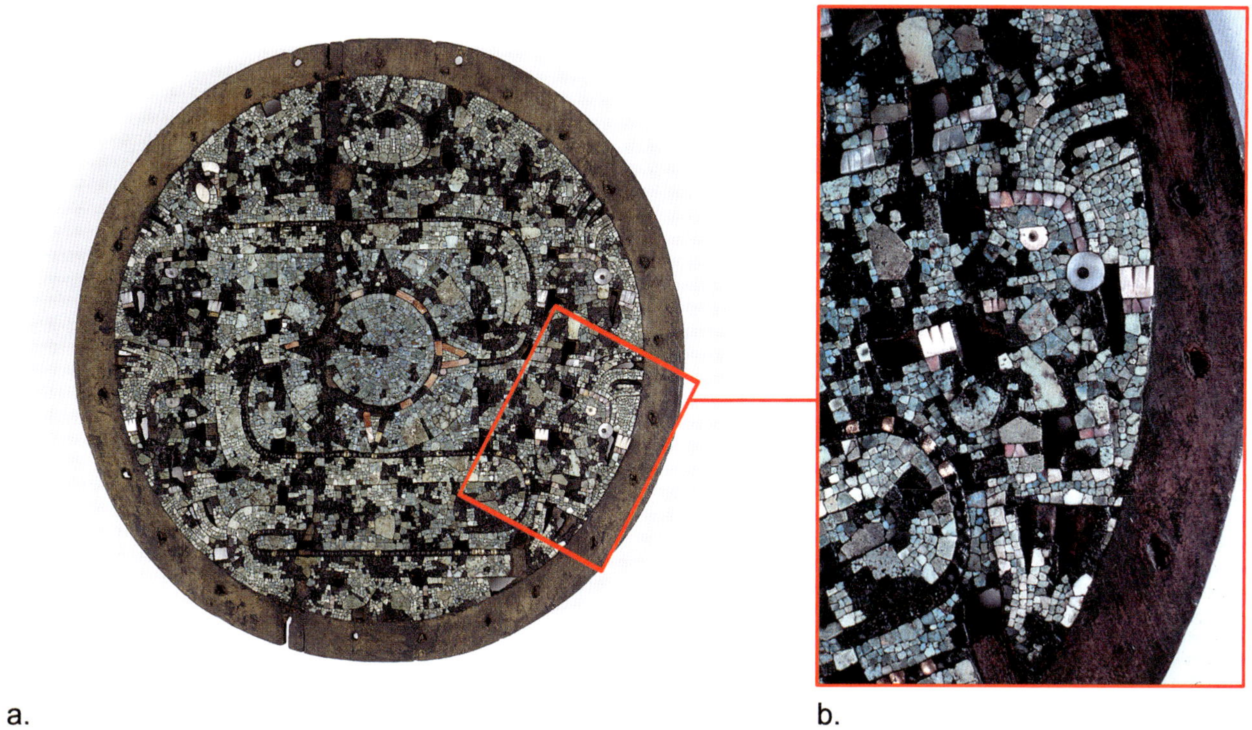

Figure 3. Mosaic shield from the British Museum collections: (a) the front of the shield (Am, St.397a) showing the complex design; (b) detail of the sky bearer (see Velázquez et al., in this vol., pp. 75–87 for comparison) (© Trustees of the British Museum).

term 'turquoise mosaic' ought to be qualified as it risks arbitrarily promoting the importance of this material above others represented. While the cultural significance of turquoise in ancient Mexico is undeniable (Weigand and Weigand 2001), the role played by other materials used in the mosaic work should not be overlooked. The prestige associated with the mosaic materials may be linked to their scarcity (for example the difficulty in harvesting the deep water *S. princeps*), specific properties (the mirrored finish of pyrite) or their association with ritual activities (resins used as adhesives on the mosaics were also valued as incense and offerings) (López Luján 1994). It is vital, therefore, to interpret the turquoise in the context of its use as one component among many in these complex composite objects.

It is interesting that none of the nine mosaics curated by the British Museum exhibits the kind of iconographic detail in the mosaic work that is seen on some mosaics in other collections. Examples of the latter include shields held by the Naturhistorisches Museum in Vienna (see Feest, in this vol., pp. 103–116), the Smithsonian National Museum of the American Indian, the turquoise disc excavated at the Templo Mayor in Tenochtitlan (see Velázquez et al., in this vol., pp. 75–87) and the Cueva Cheve tablet (Steele and Snavely 1997). All have complex figurative scenes for which engraved and shaped tesserae have been used to create images in detailed relief. The imagery on the shield in the British Museum collections (Am, St.397a), although complex, is rather different as the images are created entirely through the tessellation of the mosaic (Fig. 3). Some of the shell tesserae are engraved and drilled but the overall concept in creating these detailed images is very different to that seen, for example, on the Smithsonian shield. This is also true of the complex designs on the helmet in the British Museum collections (Am, +.6382) although these are more difficult to interpret due to extensive tessera loss. The incidence among the known mosaics of these contrasting approaches to achieving iconographic content and detail merits further research as it may have significance for understanding the relationship of the objects in terms of artists/workshops, time periods or cultural attribution.

The natural variability in the colour of turquoise offers a wide-ranging palette of blue-green hues ranging from pale apple green to deep blue, and this colour range is fully exploited in the mosaic work, with clear selection of particular colours for visual effect. The Tlaloc mask (Am, 1987,Q.3) provides perhaps the most striking example: the design is based on a pair of intertwining serpents, one worked in green turquoise shades and the other in blue (Fig. 2a). Different turquoise shades are also used to create gradients of colour change. For example, on the skull (Am, St.401) there is a marked gradation in colours from bright blue hues at the front of the face to much greener shades at the sides (Fig. 1d). A similar effect can be observed on the 'warty' mask (Am, St.400) where intense blue colours are used on the cheeks and forehead, diffusing out to lighter and greener shades towards the edges of the face (Fig. 2c). The variety in the colour of the turquoise used extends beyond blue-green hues to include turquoise tesserae of yellow and brown as well as some very deep shades of blue (Fig. 4a and b). The texture and quality of the stone is also highly variable; individual tesserae often exhibit a high degree of heterogeneity and some have a very dull weathered appearance (Fig. 4c and d). These observations emphasise the importance of both colour and texture of materials in respect of their selection for artistic effect.

Variability and mapping

One of the fundamental aims of this current phase of research on the mosaics at the British Museum is to map the use of different materials on the objects in order to understand

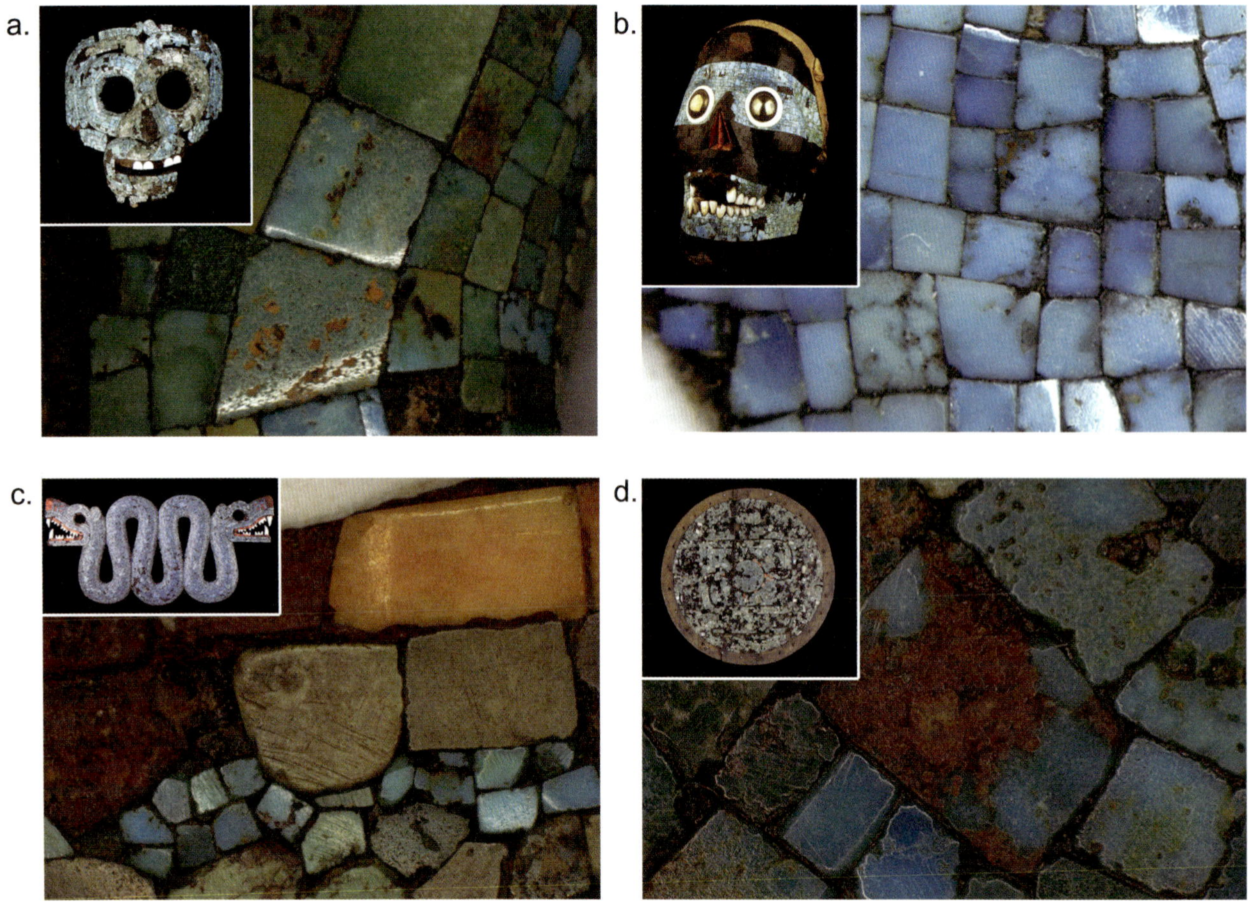

Figure 4. (a) Yellow, brown and green tesserae on the Tlaloc mask (Am, 1987,Q.3); (b) deep blue turquoise used on the skull (Am, St.401); (c) dull grey weathered turquoise on the reverse of the double-headed serpent (Am, 1894,-.634); (d) inhomogeneous and irregular turquoise tesserae on the shield (© Trustees of the British Museum).

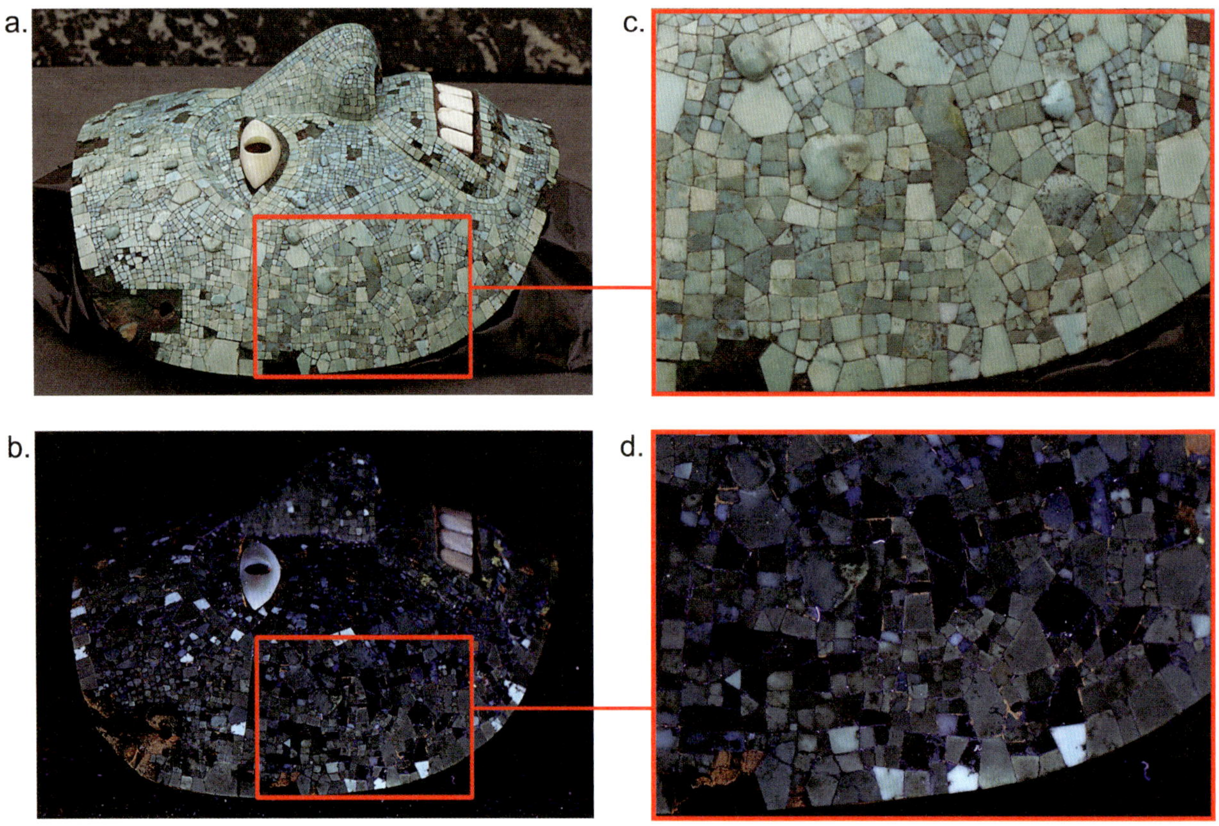

Figure 5. (a) Visible and (b) UV-induced luminescence images of the 'warty' mask (Am, St.400) with close-ups (c) and (d) showing the high variable luminescence behaviour of the turquoise (© Trustees of the British Museum).

more fully which material properties drive the selection and use of particular materials. Imaging technologies are useful techniques as they provide spatial information. UV-induced luminescence (UIL) imaging was used to differentiate tesserae of similar colour that could not otherwise be distinguished with the naked eye (Gaft et al. 2005). The approach has proved helpful in detecting the use of different minerals among the turquoise in other studies (Ruvalcaba Sil 2008). However, due to the extreme variability in the luminescence behaviour of the turquoise on these objects, it was not possible to draw conclusions about mineral origins from these data. Figure 5 shows the UIL image obtained for the 'warty' mask (Am, St.400); the wide range of luminescence response for the turquoise is immediately obvious. Comparison with the visible image clearly shows that there is no correlation between the visible colour of the turquoise and the UIL behaviour. XRF analysis, undertaken to support the mineral characterisation by Raman spectroscopy, indicated that the elemental composition of the turquoise is also extremely variable; see Figure 6 which shows the variation in relative abundances of iron, zinc and arsenic. At this stage of the research, the origin of the luminescence observed exciting in the UV range is not understood. The luminescence could be caused by major (detected by XRF), minor or even trace elements (not detected by XRF). Similarly, the presence of ions such as iron could cause quenching phenomena, introducing a further element of variability. Nevertheless, UIL imaging provides a tool for the rapid discrimination of phenomena that informed a non-invasive sampling strategy for *in-situ* analysis by XRF and Raman spectroscopy.

Non-turquoise greenstones

Malachite could usually be readily identified by optical microscopy, but the high variation in colour and texture meant that the turquoise tesserae were not always so easily characterised. Raman spectroscopy and XRF analyses enabled confirmation of turquoise where this was in question and allowed non-turquoise stones to be distinguished. A number of different non-turquoise stones were noted using these techniques, including dark green-brown tesserae on the shield (Am, St.397a) and the helmet (Am, +.6382), and some very bright, almost iridescent, leaf green tesserae on the jaguar (Am, +.165) (Fig. 7). Interferences from surface coatings and/or past restoration treatments have hampered interpretation of data for these stones so that, to date, definitive characterisation has not been possible. Ongoing study of later alteration to the mosaics will assist in the fuller interpretation of these sporadic non-turquoise tesserae in the longer term. It is notable that these stones are used indiscriminately among the turquoise even though they are not particularly turquoise in colour. This contrasts with the way that malachite is used. Although malachite is used extensively in the mosaic work on the mosaics in the British Museum collections, it is typically applied to discrete parts of the designs to produce contrasting areas of darker green, such as the serpents depicted on the helmet, the eye surrounds on the small animal head (Am, St.400a) and for circle motifs on the jaguar (Fig. 8). So far, only a few examples of indiscriminate use of malachite in turquoise areas have been noted, despite its extensive use elsewhere on the objects. This suggests that malachite was largely defined and used as a distinct entity by the mosaic artists and applied accordingly.

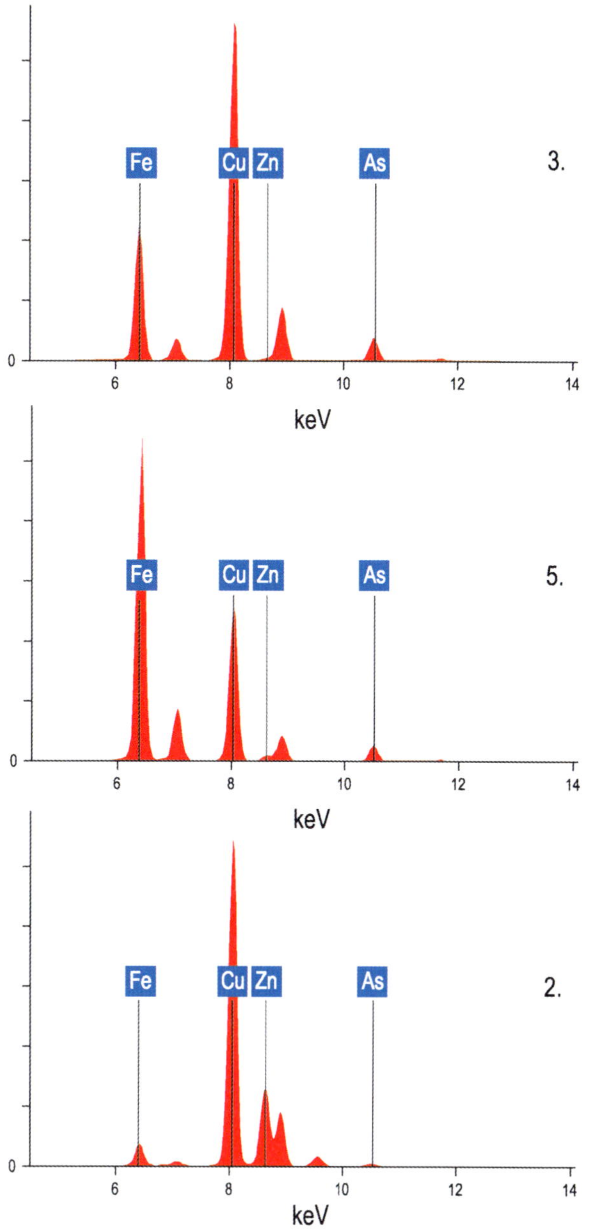

Figure 6. XRF spectra showing variability in the relative abundance of major elements detected in three tessera from the double-headed serpent (Am, 1894,-.634) (© Trustees of the British Museum).

The wide variation in the colour of turquoise, coupled with the apparent willingness of the artists to substitute or intermingle non-turquoise tesserae, suggests that a more fluid approach was adopted in the selection and application of these materials.

Turquoise provenancing: challenges and questions

The origin of turquoise in ancient Mexico has long been linked to the Cerrillos Hills Mining District of New Mexico and it is thought that turquoise mined, and quite possibly worked, at this northern frontier found its way south into Mexico via trade and tribute networks. The results of provenance analyses seem to support this hypothesis (Harbottle and Weigand 1992; Mathien 1981; Young et al. 1994) although there are also possible sources in Mexico (see Berdan, in this vol., pp. 91–102) and it remains unclear to what extent these were exploited. Developing methodologies for the analytical provenance of

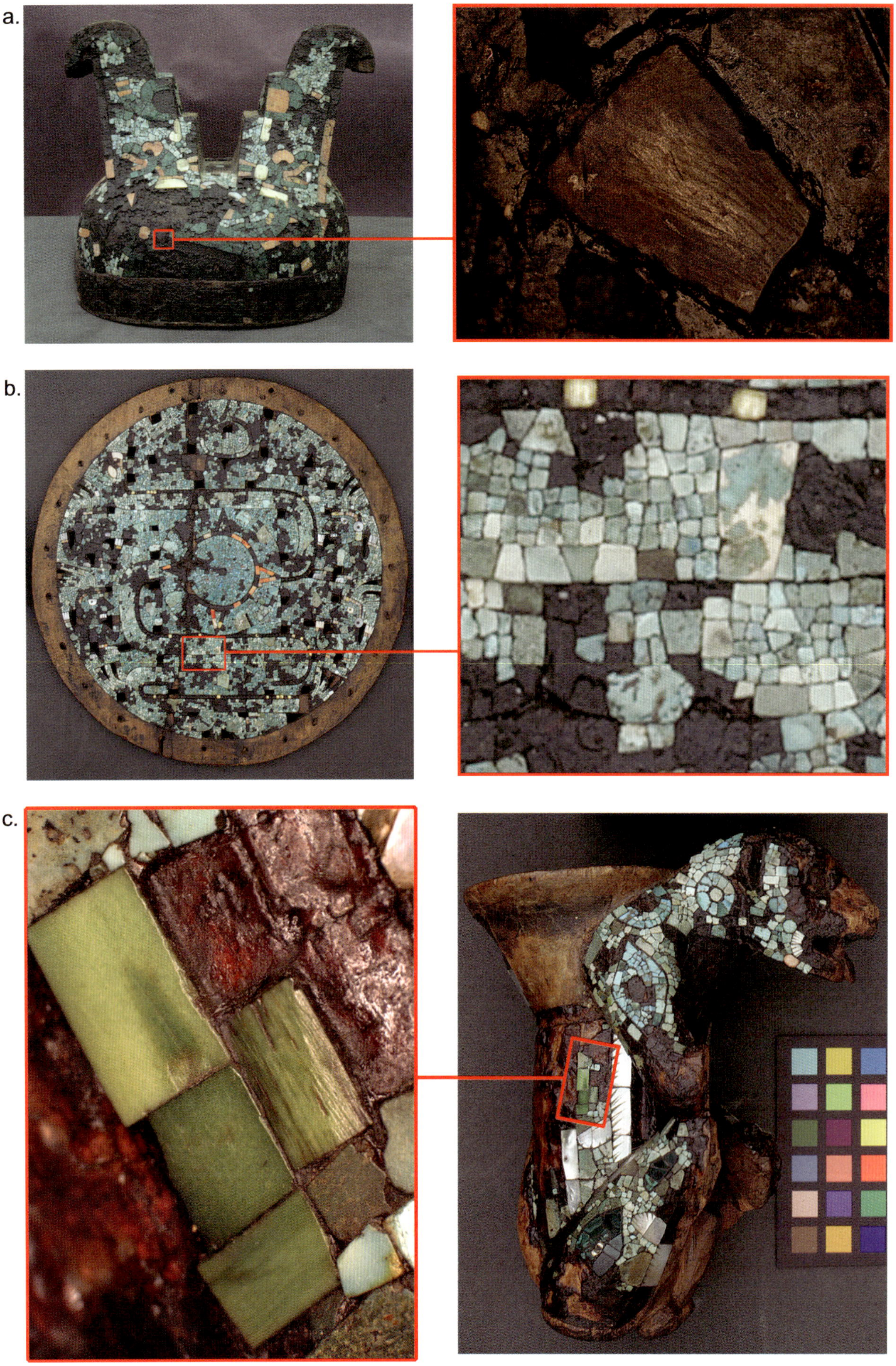

Figure 7. Non-turquoise tesserae: (a) green-brown tessera on the helmet (Am, +.6382) and (b) on the shield (Am, St.397a); (c) green tesserae on the jaguar (Am, +.165) (© Trustees of the British Museum).

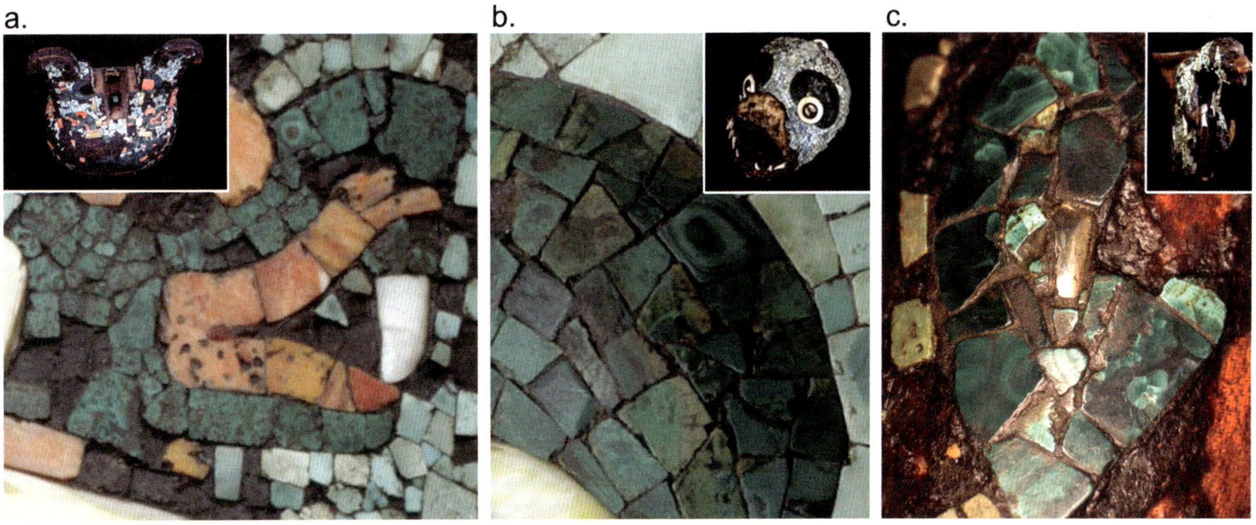

Figure 8. Details of mosaic design using malachite: (a) serpent head from the helmet; (b) eye surround on the animal head; (c) bordering a pyrite tessera on the jaguar (Am, +.165) (© Trustees of the British Museum).

turquoise (see Hull and Fayek, and Thibodeau et al., in this vol., pp. 29–40 and 65–74 respectively) offer considerable scope for further investigation of this question. The time is ripe, therefore, for a serious consideration of the potential for the application of such techniques to turquoise mosaic objects.

To date, no work to provenance the turquoise on the British Museum mosaics has been undertaken and the current research programme has instead focused on characterisation of the range of minerals present, their distribution, variability and selective use. A number of factors has hindered the application of provenance techniques to these objects in the past and these obstacles still merit careful consideration. First is the sheer number of tesserae on the objects (a factor that has also limited the amount of analysis undertaken to mineralogically characterise individual tesserae). There are, for example, some 15,000 turquoise tesserae on the double-headed serpent. Even the helmet, which has suffered major losses of tesserae and displays very extensive use of other materials, incorporates in excess of 2,000 pieces of turquoise. Analysis of all tesserae on these objects is clearly not feasible, thus a representative and meaningful sampling strategy is essential if any useful interpretation is to arise from such work.

Any consideration of sampling must take into account the implications of reuse of tesserae. Close examination of the objects reveals numerous engraved pieces used among the original mosaic work (Fig. 9a and b), and many of these are worked to the same design as tesserae that are used in detailed images on other objects, such as the Smithsonian shield described earlier. These pieces almost certainly represent reuse of tesserae from earlier dismantled or broken objects. There may, of course, be many more reused tesserae that are indistinguishable because they are not engraved. Clearly, therefore, within the original mosaic work there is evidence to indicate that the tesserae are most likely to have a mixture of origins. The Codex Mendoza indicates that both loose tesserae and fully finished objects were given in tribute (Berdan and Anawalt 1992) and it seems likely that the supply of tesserae available to ancient lapidaries was a mixed pool from all sources incorporating both freshly prepared and recycled material, some of it potentially in circulation for long periods.

There is also considerable evidence for later reworking, perhaps occurring after the arrival of the mosaics in Europe. Figure 9 (c and d) shows some details from the animal head that displays extensive and obvious areas of reworking where haphazard, widely spaced, relaid tesserae contrast with the neat coherence of the original mosaic work. These areas of reworking show a variety of different-shaped tesserae as well as engraved pieces, which may have been used in the original mosaic but could equally have been salvaged from other objects to carry out these repairs. Not all areas of reworking are as evident from the mosaic work and the UIL imaging data provide vital evidence for identifying areas of reworking (Fig. 10). Sometimes these alterations can be found in areas of apparently pristine mosaic work. An example is the 'warty' mask, which is one of the best-preserved mosaics with high-quality surfaces and limited areas of tessera loss. A key feature of this mask is the use of many cabochon turquoises that give the face a 'warty' appearance. These cabochon tesserae are also the most vulnerable to loss, and re-adhered examples are readily identified in the UV images (Fig. 10). Unfortunately it is not possible to determine if these are original cabochons that have been re-adhered, or replacements that have been added over the years. Some of the latter may not even be New World turquoise.

The challenges for conducting provenance studies on these mosaics are thus very clear:

1. Devising an appropriate sampling strategy for the very large number of tesserae.
2. The need for a non-destructive analytical method that can be applied *in situ*.
3. The need to take account of recycling of turquoise tesserae.
4. The need to consider later reworking in terms both of reuse of turquoise and of possible introduction of Old World turquoise.

In the face of these challenges it is important to evaluate the likely outcomes of provenance analysis and the potential of such work to enhance the understanding of these objects.

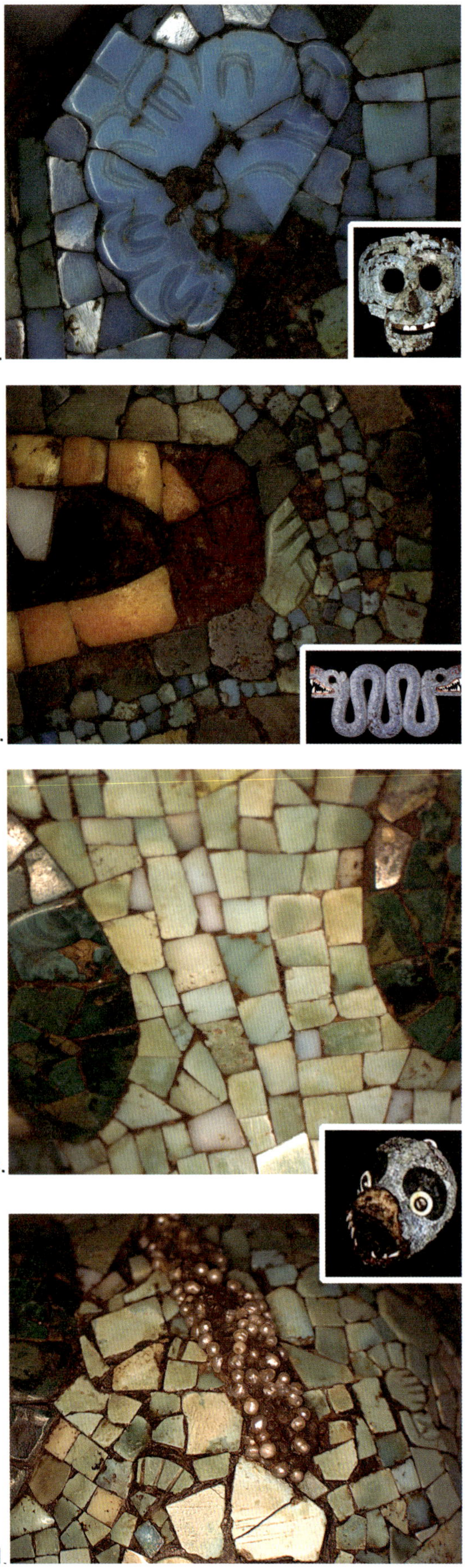

a. b. c. d.

Figure 9. Engraved tesserae used in original mosaic work on: (a) the Tlaloc mask (Am, 1987,Q.3); (b) the reverse of the double-headed serpent (Am, 1894,-.634). Details from the animal head showing: (c) original mosaic work; (d) areas of later reworking (© Trustees of the British Museum).

Most of the work that has been undertaken on archaeological turquoise sourcing has largely focused on material from the southwestern USA (see Hull and Fayek, in this vol., pp. 29–40 and Thibodeau et al., in this vol., pp. 65–74 and references therein) so provenance studies have enormous potential to inform about the origin of turquoise in ancient Mexico and to test the predominant theory that most of it was brought in from the north. The mosaics are far from ideal material to form the basis for such a study since they are themselves unprovenanced. Many excavated finds of turquoise in the form of loose tesserae and beads offer candidate material that bypasses the problem of non-invasive analysis and moreover would allow the examination of the distribution of turquoise with relation to source. Mosaic turquoise might assist in this evaluation if specific or exclusive sources were used for mosaic work, although given what is known from the documentary sources about the movement of turquoise, coupled with the evidence for recycling seen on these objects, this seems unlikely.

The most important questions that provenance studies could address must therefore be what analysis of the turquoise on these objects would reveal about the mosaics themselves. Again the movement and recycling of turquoise set the parameters for interpretation of any such data, and thus the source of the turquoise is unlikely to be related to the location of manufacture. Given a sufficiently representative number of analyses per object, such a study might provide useful insights into the range of source material available at a given, albeit unknown, point in time when the objects were manufactured, although not necessarily sources in active exploitation because of recirculation of turquoise through recycling. The relationships of the mosaics to one another and the interpretation of later alterations is perhaps the route of enquiry that has the most to offer and, in fact, this perspective is an active area of ongoing research with respect to the adhesives used both for the original manufacture and later modification and repair of the mosaics in the British Museum collections.

Conclusions

Turquoise is an important component of the nine Mexican mosaics in the British Museum's collections. It contributes alongside a range of other prestige materials to the manufacture and significance of these complex composite objects and it is important that it is interpreted and understood in this context.

Detailed examination and analysis has highlighted the variety (in terms of colour, quality and composition) of turquoise used. The high elemental variability in the turquoise limits the scope for the use of UV-induced luminescence imaging as a rapid method for distinguishing sporadic non-turquoise tesserae, but the technique has proved to be a powerful tool for detecting areas of later reworking.

Recycling of tesserae, both at the time of manufacture and in later reworking, presents a particular challenge for future provenance studies of turquoise on these objects. The work reported here highlights the vital need to establish a detailed understanding of turquoise use and reuse on individual mosaics if provenance studies are to yield informative results.

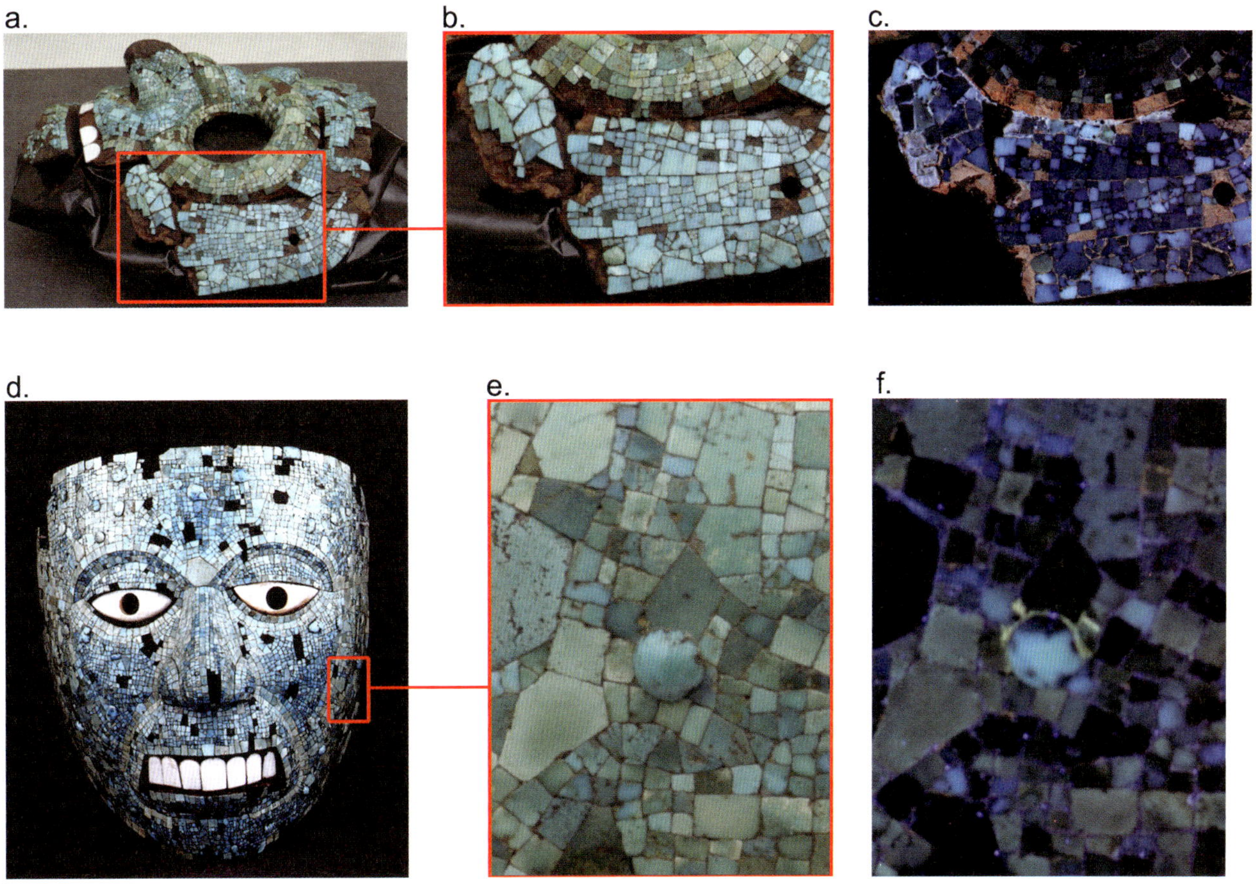

Figure 10. (a) Tlaloc mask (Am, 1987,Q.3) with (b) detail visible and (c) UV-induced luminescence images showing areas of repair. Original adhesive (pine resin) appears pink in the UV-induced luminescence image whereas adhesives used in later repairs appear blue. (d) 'Warty' mask (Am, St.400) with (e) detail visible and (f) UV-induced luminescence images showing later yellow/green luminescence of adhesives used to re-adhere cabochon turquoises (© Trustees of the British Museum).

The research questions that provenance studies might address also require careful consideration.

The current phase of research on the mosaics at the British Museum is focused on mapping the materials present, in particular examining the adhesives to identify areas of change and alteration with a view to understanding relationships between mosaics that appear to have been subjected to similar campaigns of alteration or restoration. This detailed information will provide a sound foundation for considering how more extensive work on the turquoise might be undertaken and what such research can reasonably be expected to achieve.

Acknowledgements

We are grateful to Janet Ambers for assistance with Raman spectroscopy and to Antony Simpson for formatting the images for publication and also to two anonymous referees for their helpful comments. One of the authors (GV) was a Mellon postdoctoral research fellow from 2007 to 2010 and this study was partially supported by The Andrew W. Mellon Foundation during this period.

References

Berdan, F.F. and Anawalt, P.R. (1992) *The Codex Mendoza*. Berkeley, CA: University of Berkeley Press.

Cartwright, C.R. (2009) 'Scanning electron microscopy at the British Museum and the Aztec turquoise mosaics from Mexico', *Hitachi Electron Microscopy News* **3**: 38–42.

Cartwright, C.R. and Meeks, N.D. (2007) 'Aztec conch shell working: high-tech design', *British Museum Technical Research Bulletin* **1**: 35–42.

Gaft, M., Reisfeld, R. and Panczer, G. (2005) *Luminescence Spectroscopy of Minerals and Materials*. Berlin: Springer, 45–116.

Harbottle, G. and Weigand, P.C. (1992) 'Turquoise in Pre-Columbian America', *Scientific American* **266**(2): 78–85.

López Luján, L. (1994)*The Offerings of the Templo Mayor of Tenochtitlan*, B.R. Ortiz de Montellano and T. Ortiz de Montellano (tr.). Niwot, CO: University Press of Colorado.

Mathien, F. J. (1981) 'Neutron activation of turquoise artifacts from Chaco Canyon, New Mexico', *Current Anthropology* **XXII:** 293–294.

McEwan C., Middleton, A., Cartwright, C.R. and Stacey, R. (2006) *Turquoise Mosaics from Mexico*. London: British Museum Press.

McEwan C., Stacey, R.J. and Cartwright, C.R. (forthcoming) 'The "Tezcatlipoca" skull mosaic in the British Museum collections: new insights and questions of identity', in E. Baquedano (ed.), *Tezcatlipoca: Trickster and Supreme Aztec Deity*. Niwot, CO: University of Colorado Press, 119–127.

Ruvalcaba Sil, J.L. (2008) Personal communication, December 2008.

Stacey, R.J. and Cartwright, C.R. (2009) 'Construction, modification and repair of Mexican mosaics: evidence from the adhesives', in J. Ambers, D. Saunders and L. Harrison (eds), *Holding it all Together: Ancient and Modern Approaches to Joining, Repair and Consolidation*. London: Archetype Publications, 61–67.

Stacey, R.J., Cartwright, C.R. and McEwan, C. (2006) 'Chemical characterisation of ancient Mesoamerican "copal" resins: preliminary results', *Archaeometry* **48**: 323–340.

Steele, J.F. and Snavely, R. (1997) 'Cueva Cheve Tablet', *Journal of Cave and Karst Studies* **59**: 26–32.

Verri, G., Comelli, D., Cather, S., Saunders, D. and Pique, F. (2008) 'Post-capture data analysis as an aid to the interpretation of ultraviolet-induced fluorescence images', in D.G. Stork and J. Coddington (eds), *Proceedings of SPIE, Computer Image Analysis in the Study of Art, Vol. 6810*. Bellingham: IS&T/SPIE, pp. 681002.1–681002.12.

Weigand, P. and Weigand, A. (2001) 'A macroeconomic study of the relationships between the ancient cultures of the American Southwest and Mesoamerica', in V. Fields, V. Zamudio-Taylor and E. Gamboa Carrera (eds), *The Road to Aztlan*. Los Angeles: Los Angeles County Museum of Art, 184–195.

Young, S.M.M., Phillips, D.A. and Mathien, F.J. (1994) 'Lead isotope analysis of turquoise sources in the southwestern U.S.A. and Mesoamerica: a preliminary report', in S. Demirci, A.M. Özer and G.D. Summers (eds), *Archaeometry '94: Proceedings of the 29th International Symposium on Archaeometry*. Ankara: Tubitak, 147–150.

Authors

- Rebecca Stacey, Department of Conservation and Scientific Research, The British Museum, London, UK
- Caroline Cartwright, Department of Conservation and Scientific Research, The British Museum, London, UK
- Giovanni Verri, Department of Conservation and Scientific Research, The British Museum, London, UK
- Colin McEwan, Department of Africa, Oceania and the Americas, The British Museum, London, UK

An alternative approach to the prehispanic turquoise trade

Alyson M. Thibodeau, John T. Chesley, Joaquin Ruiz, David J. Killick and Arthur Vokes

Introduction

The major turquoise deposits of North America are concentrated within the southwestern United States across Arizona, New Mexico, Colorado, Nevada and southern California, and in northern Mexico in the state of Sonora (Fig. 1). Turquoise is a blue-green hydrated copper aluminium phosphate ($CuAl_6(PO_4)_4(OH)_8{\bullet}4H_2O$) that was both highly valued and widely exchanged among many prehispanic societies in the American Southwest and Mexico. In the Southwest, turquoise was fashioned into a wide variety of forms, including beads and pendants, as well as nose and lip plugs. It was also used to construct elaborate inlays, as a part of intricate mosaic designs, and carved into zoomorphic forms (Snow 1973; Jernigan 1978). Turquoise appears most frequently in the archaeological record of the Southwest after ~ AD 900 (Snow 1973; Windes 1992), and large quantities have long been associated with great house and small house sites in Chaco Canyon, New Mexico (Windes 1992; Mathien 2001). Turquoise was also used in Mesoamerica, thousands of kilometres away

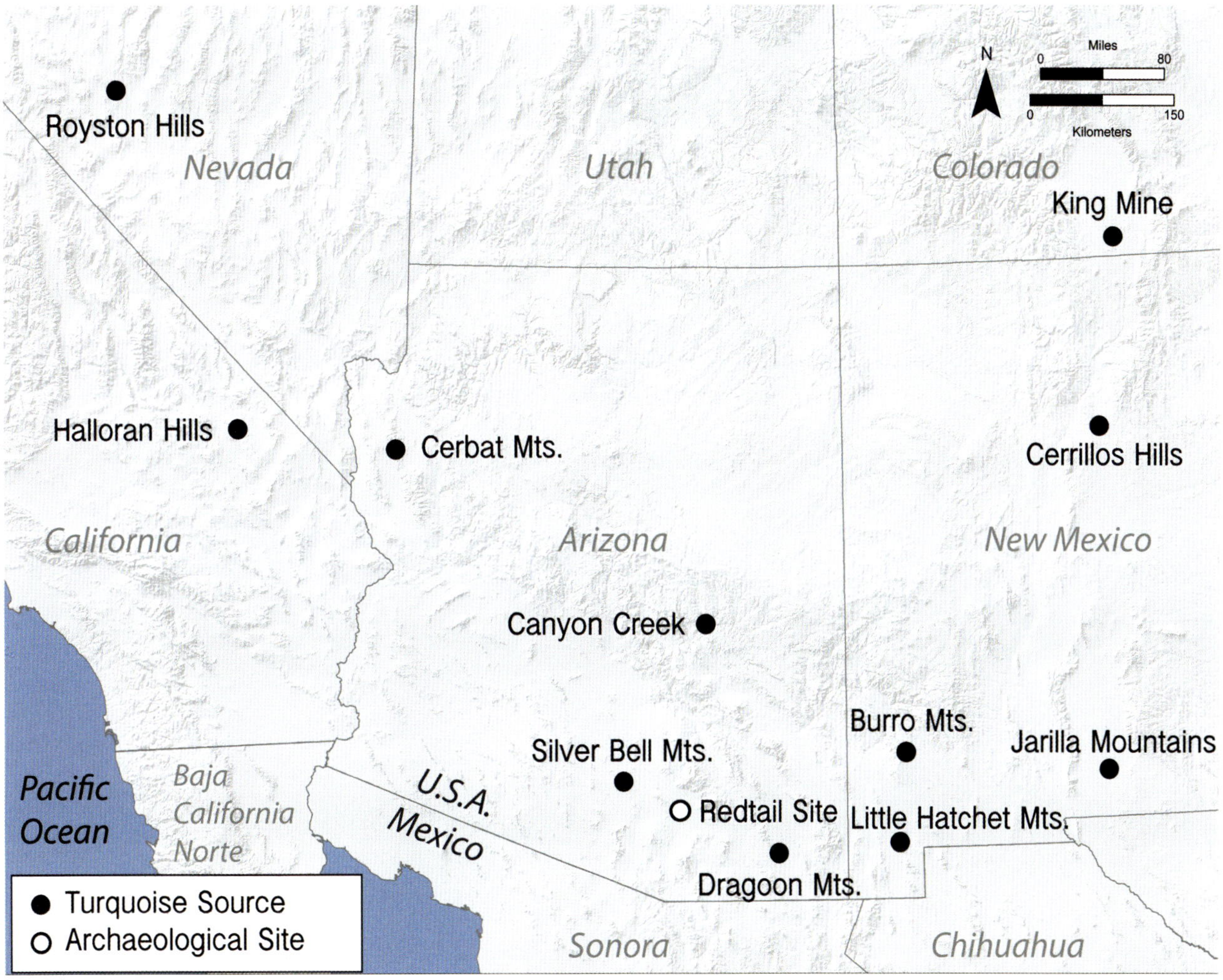

Figure 1. Locations of turquoise sources and archaeological sites from the American Southwest that are discussed in the text. Many additional sources (not shown here) can be found across central and northern Nevada (Morrissey 1968), and in Colorado (Modreski and Murphy 2002), as well as in Arizona, New Mexico and Sonora. All sources shown have evidence of prehispanic mining.

from the well-known mines scattered across the present-day American Southwest and northern Mexico. In Mesoamerica, blue-green stones, particularly jade, were highly valued and widely circulated for thousands of years prior to the arrival of the Spanish. Although turquoise is reported to have been found in isolated instances in Mesoamerica as early as the Formative Period (e.g. Vaillant 1935), its most widespread use was during the Postclassic (AD 900–1521), when it was employed by the Mixteca and the Mexica to create elaborate mosaics and other ceremonial and status objects (e.g. McEwan et al. 2006; Saville 1922). Because there is little evidence of turquoise mining in Mesoamerica, many southwestern and Mesoamerican archaeologists believe that turquoise may have been acquired through trade with the American Southwest, an idea that has sparked an enduring debate (e.g. Kelley and Kelley 1975; Weigand et al. 1977; Lister 1978; McGuire 1980; Frisbie 1983; McGuire 1993; Wilcox 1986; Doyel 1991; Weigand and Harbottle 1993; Harbottle and Weigand 1992; Weigand 1994).

Since the 1970s, both archaeologists and physical scientists have sought to reconstruct the prehispanic turquoise trade by developing methods to identify the geologic provenance of turquoise recovered from the archaeological record or associated with ancient objects. Because of its long history of use in much of the prehispanic American Southwest and Mexico, turquoise presents a significant opportunity to address questions of resource procurement and exchange across a wide range of societies and at an array of spatial and temporal scales. However, turquoise has proved to be a difficult mineral to trace and efforts to do so have encountered a variety of the archaeological, geological, and chemical complexities associated with the problem. Lead and strontium isotopes may offer new answers to longstanding questions about the prehispanic turquoise trade. In the present study, lead and strontium isotopic measurements are used to link turquoise from an archaeological site in the Tucson Basin to prehispanic turquoise mines in the Silver Bell Mountains of southern Arizona. The positive results of this case study illustrate the potential of these two geochemical tracers to overcome the problems identified by previous researchers, and provide a tool that can link turquoise from site to source.

Prehispanic turquoise mines in the American Southwest and Mexico

Since the mid-nineteenth century, ideas about prehispanic turquoise mining and exchange have abounded in both the scientific and popular imagination. Geologist William Blake was the first to formally comment on the grand scale of the prehispanic turquoise mines found near Cerrillos, New Mexico. In an article for the *American Journal of Science*, Blake (1858: 227) states, 'On reaching the locality I was struck with astonishment at the extent of the excavation ... [i]t is an immense pit ... and tens of thousands of tons of rock have been broken out.' Blake was describing the great pit on the western side of Mount Chalchihuitl in the Cerrillos Hills Mining District, now the most famous locale associated with prehispanic turquoise mining in North America. Situated some 20 km southwest of present-day Santa Fe, New Mexico, the district has a history of mining that spans the past 1,000 years or more (Mathien 1998). Mined initially for turquoise and later for lead ores by Ancestral Puebloan groups (Schroeder 1979; Warren and Weber 1979; Bice et al. 2003), the Cerrillos Hills also contain ores of copper, zinc, gold and silver that were later worked by both Spanish and American miners (Disbrow and Stoll 1957; Mathien 1998). In 1915, Joseph Pogue, a geologist with the National Academy of Sciences, published an exhaustive monograph on turquoise, which included a compilation of the known turquoise deposits across the world (Pogue 1974). Despite his knowledge of other ancient turquoise mines in the American Southwest, the immensity of the prehispanic workings in the Cerrillos Hills Mining District led him to declare, 'this deposit must have supplied much of the turquoise which was so widely used in pre-Spanish times, not only in the Southwest, but in Mexico as well, for no other sources are now known at all adequate to account for the quantities employed' (1974: 52).

In the decades that followed Blake's original description of the Cerrillos Hills mines, many prehispanic turquoise workings from across the American Southwest were rediscovered and described by prospectors, journalists and geologists (Bennett 1970). In New Mexico, Snow (1891) reported on the ancient workings of the Burro Mountains, and Hidden (1893) reported the presence of prehistoric mines near the town of Hatchita (in the Little Hatchet Mountains), as well as in the Jarrilla Mountains (Fig.1). The discovery of hundreds of additional prehispanic turquoise workings broadened perspectives on the scope and scale of turquoise mining in the prehispanic American Southwest, leaving many with the impression that 'very few turquoise mines have ever been found that did not show some evidence of having been previously mined by Indians' (Bennett 1970: 35).

Consequently, it has been speculated that many of the major deposits of turquoise in the American Southwest were the source of turquoise used by Mesoamerican groups in prehistory. On 3 June 1895, the *Tombstone Prospector* commented on the prehistoric turquoise mines in the Dragoon Mountains of southeastern Arizona, stating, 'Even centuries ago the value of this stone was apparent among the extinct races, as there is evidence of extensive workings at Turquois by the Aztecs who are supposed to have inhabited this entire section of the country after this precious mineral.' On 18 March 1898, a headline in the *San Francisco Call* announced, 'Long Lost Mines of Precious Gems are Found Again, Located in the Remotest Wilds of San Bernardino Country and Marked by Strange Hieroglyphics' regarding the turquoise mines of the Halloran Hills in California's eastern Mojave Desert. The accompanying text, which introduces an article by explorer Gustav Eisen (1898), states, '[t]here have been found the mines of the ancients, long lost to knowledge and so the secret source from which the Aztecs drew their supply of turquoise is no longer a secret locked in the mists of the past.' Although there is no evidence that any Mesoamerican group ever mined turquoise in the American Southwest, these early ideas about the source of the turquoise used in Mesoamerica still reverberate in present-day archaeological discussions of long-distance trade.

In spite of the likely importance of turquoise in local, regional and long-distance exchange networks, there has been little direct study of prehispanic mines in the American

Southwest or Mexico. Formal archaeological investigation of the tools, ceramics, shelters or workshops found at or near known mining sites is rare (for examples of such studies see: Leonard and Drover 1980; Warren and Mathien 1985; Wiseman and Darling 1986). A number of factors have converged to make the archaeological study of turquoise mines difficult, and in some cases, impossible. Because turquoise in the American Southwest often occurs near economically significant deposits of copper, molybdenum, lead and silver, historic and modern mining practices have encroached upon, or sometimes completely obliterated sites of ancient mining. In New Mexico, ancient workings in the Burro Mountains have been destroyed by the expansion of open-pit copper mines. Ancient turquoise mining sites on Ithaca Peak in the Cerbat Mountains and in the Silver Bell Mining District (both in Arizona) have met a similar fate. Even in the famous Cerrillos Hills Mining District, which contains some of the best-preserved and studied prehispanic mines in North America, centuries of mining and artefact collection had already compromised the integrity of the archaeological remains before archaeological surveys began (Warren and Mathien 1985). Additionally, much of the literature surrounding ancient turquoise mines is vague, cannot be verified, and does not provide specific location information. As Bennett (1970: 35) notes, '[a] singular lack of imagination was used in naming the hills or ridges or mountains on which the mines were situated. Turquoise Ridges; Turquoise Mountains; Turquoise Hills; and Castle Dome Ridges, Mountains, and Hills are scattered all over several states.'

Understanding the exchange of turquoise between American southwestern and Mesoamerican societies becomes even more difficult in light of the paucity of information available regarding turquoise sources closer to major population centres of Mesoamerica. For example, hypotheses regarding the trade of turquoise from the American Southwest to Mesoamerica imply that major turquoise deposits did not exist in Mesoamerica, or if they did, were either not known to or not mined by Mesoamerican groups. To accurately evaluate such hypotheses it is necessary to consider the evidence for turquoise deposits closer to the core areas of Mesoamerican civilisations that utilised the stone. The geologic literature documents turquoise in several localities in Baja California Norte, in association with areas of major copper mineralisation in Sonora, and at several localities in Zacatecas (Panczer 1987: 382–383). Weigand (1994: 31) reports evidence of prehistoric workings in at least five areas in Sonora and two areas in Zacatecas, as well as in Chihuahua, Coahuila, and San Luís Potosí. In addition to mines in Sonora and Zacatecas, Di Peso (1974: 277) provides references attesting to the existence of turquoise sources in Coahuila and Chihuahua and to deposits of chrysocolla and turquoise in Jalisco. He also indicates that there are small surface outcrops of turquoise from Guerrero and Aguascalientes (Di Peso 1974: 207). However, the references do not disclose their sources of information about these occurrences. Furthermore, there have also been reports of turquoise from Veracruz, Tehuantepec (Michoacan), and Chiapas (Mathien 1981a: 178, citing Sigleo). However, the lack of available information about the locations of these turquoise sources, the common confusion of turquoise with other blue-green minerals, and conflicting information between geological and archaeological reports makes it difficult to verify many of these claims, particularly for the reported mines in central and southern Mexico. Sinkankas (1976: 143–144) makes the illuminating observation that in the Concepción del Oro area of Zacatecas, the mineral prosopite ($CaAl_2(F,OH)_8$) is often mistaken for turquoise.

Past approaches to tracing turquoise

To better understand ancient turquoise mining and trade, many researchers have utilised geochemical methods to investigate the source of turquoise artefacts. The majority of these studies approach the problem from a compositional perspective and use trace and major element concentrations to draw links between turquoise objects and known mines (Sigleo 1975; Weigand et al. 1977; Mathien 1981b; Harbottle and Weigand 1992; Mathien and Olinger 1992; Weigand 1994; Kim et al. 2003). Early applications of trace and major element measurements were partially driven by the ready availability and low cost of analytical methods such as neutron activation analysis (NAA) and X-ray fluorescence (XRF), which allowed large numbers of samples to be analysed. Although these efforts to address questions of turquoise provenance have made important contributions to the topic, published data from compositional studies have not been able to define unique or characteristic signatures for the many mines in the American Southwest. However, new approaches to turquoise provenance are now being explored, including the use of copper and hydrogen isotopes (Hull et al. 2008) and lead and strontium isotopes (Thibodeau et al. 2007b; this study).

The geology and mineralogy of turquoise

The difficulty of using elemental concentrations to trace turquoise can be explained by the mineralogical characteristics of turquoise and the circumstances of its formation. Turquoise comes in a variety of colours, from the familiar 'robin's egg' blue, to apple green. It was once commonly believed that visual inspection was sufficient to identify the source of a piece of turquoise. Although some claim that high-grade turquoise may have distinctive types of matrix that could help narrow the field of possible sources (Arrowsmith 1974), appearance cannot be reliably applied to determine geologic provenance, particularly of archaeological samples. In his description of turquoise at Mount Chalchihuitl, Blake (1858: 229) observed, 'various shades of apple-green and pea-green, passing into bluish green ... [s]ome fragments having a blue color were found.' He went on to note:

> On breaking open one of the dark green fragments a small cavity was found at the center ... the color gradually shading into white. There are not any distinct layers ... but the color gradually diminishes from the surface to the center. A variation in the different amount of coloring matter in different specimens according to the circumstances of formation is thus indicated, it is seen that the composition of the mineral cannot be regarded as constant.

These early observations accurately note a common characteristic of turquoise: its variable composition, even on the scale of a single nodule or vein. Ideally, turquoise has the formula $CuAl_6(PO_4)_4(OH)_8{\bullet}4H_2O$ and crystallises with triclinic symmetry (Cid-Dresdner 1965; Foord and Taggart 1998). However, high purity crystalline turquoise is rare (Braithwaite 1981; Foord and Taggart, 1998; Barwood 2003). Instead it is usually cryptocrystalline or massive, incorporates many impurities, and may have a wide variety of compositional and visual characteristics.

Table 1. Formulas of turquoise group minerals (source: Foord and Taggart 1998).

Mineral	Formula
Turquoise	$CuAl_6(PO_4)_4(OH)_8{\cdot}4H_2O$
Chalcosiderite	$CuFe_6(PO_4)_4(OH)_8{\cdot}4H_2O$
Aheylite	$FeAl_6(PO_4)_4(OH)_8{\cdot}4H_2O$
Fausite	$ZnAl_6(PO_4)4(OH)_8{\cdot}4H_2O$
Planerite	$Al_6(PO_4)_2(PO_3OH)_2{\cdot}4H_2O$
Unnamed	$Fe^{2+}Fe^{3+}{}_6(PO_4)_4(OH)_8{\cdot}4H_2O$

Turquoise is classified by mineralogists as one of the six known minerals within the 'turquoise group' (Foord and Taggart 1998). The other five minerals, listed in Table 1, are similar to turquoise, and differ primarily due to substitution of other cations (positively charged ions) for Cu^{2+} or Al^{3+}. Many minerals within the turquoise group are known to form partial or complete solid solution series with one another, a phenomenon that can greatly impact the composition of a given sample. For example, an extensive solid solution series exists between the turquoise and chalcosiderite end members. Minerals that lie along this series may have various proportions of Al^{3+} to Fe^{3+}, which result in a wide range of blue to green colouration in samples of 'turquoise'. Even more notable is the complete solid solution series between turquoise and planerite, which is a light blue to blue-green copper-deficient analogue to turquoise. Foord and Taggart (1998) suggest that most samples described as 'turquoise' in the literature may be closer in composition to planerite. The recognition that turquoise, as popularly defined, is actually a group of minerals with widely variable compositions is consistent with the wide array of colours observed in 'turquoise' from the same deposit, outcrop or even hand sample. The prehispanic cultures of the American Southwest and Mexico embraced these variations and used 'turquoise' of all colours and quality. In many of the famous Mixtec pieces, the artists used differences in the colour of various mosaic pieces to achieve specific design features (McEwan et al. 2006). However, these same variations have hindered the use of elemental concentrations to fingerprint mines and identify the source of turquoise.

The variable nature of turquoise can be explained through consideration of the processes that create it. Turquoise forms in the weathering environment, via the interaction of meteoric waters with copper, phosphorous, and aluminium-bearing rocks. Major deposits of turquoise are confined to arid or semi-arid regions worldwide, and are known in Egypt, Iran, China, Afghanistan and the USA. In parts of the American Southwest and northern Mexico, turquoise is often found in the oxidised portions of copper porphyry deposits that are especially common across Arizona, New Mexico, and Sonora. In the American Southwest, it is commonly found at shallow depths (30 m or less) and precipitates as veins or nodules in highly weathered and fractured felsic igneous rocks, especially monzonites, latites and trachytes. In many cases, these host rocks are inferred to have supplied the elemental and molecular constituents necessary to form turquoise; copper from disseminated copper-bearing minerals (e.g. chalcopyrite), phosphorous from apatite, and aluminium from feldspars, clays (especially kaolinite) or other aluminium-rich phases (Paige 1912; Pogue 1974; Pearl 1941).

However, turquoise is not limited to forming from the weathering of igneous rocks. Turquoise deposits on the Sinai peninsula in Egypt are found in sedimentary rocks where copper and phosphorous-charged ground waters interacted with aluminium-rich concretions made of alunite and gibbsite (Hussein et al. 1994–1995). In the American Southwest, turquoise can be found within weathered aluminium-bearing sedimentary rocks in many locales, including the Canyon Creek mine (Moore 1968; Welch and Triadan 1991), the Dragoon Mountains (Crawford and Johnson 1937), the Jarrilla Mountains (Crook 2002), and at numerous deposits in central and northern Nevada (Morrissey 1968). The source of copper and phosphorous in these deposits may lie in adjacent igneous rocks (which sometimes also contain turquoise), or within the sedimentary rocks that host the turquoise. A detailed study examining the genesis of turquoise mineralisation in the Jarilla Mountains reveals that turquoise hosted by shale deposits derived phosphate from layers of shale that are rich in apatite and xenotime (Crook 2002). Notably, turquoise in the Jarrilla Mountains can also be found in weathered quartz-monzonites, and, from a geochemical perspective, it is likely that turquoise within the area has two distinct geological origins (Lueth 1998).

The weathering processes that create turquoise can produce localised and uneven effects, and thus create heterogeneities in the chemical composition of turquoise across a deposit, or within a single nodule or vein. Additionally, a chunk of turquoise, whether hit off an outcrop or made into a pendant, often contains other mineral phases. These impurities have also played a role in confounding efforts to trace archaeological turquoise. Minerals commonly associated with turquoise include (but are not limited to) quartz, pyrite, limonite, kaolinite, alunite, sericite and jarosite. The common appearance of these minerals near or within turquoise deposits is unsurprising, since most can be associated with turquoise formation. For example, pyrite is an iron-sulphide mineral that, when oxidised, acidifies meteoric water and makes it possible to leach copper and phosphate from country rock. The frequent association of iron oxides with turquoise (e.g. limonite) is testament to the role of pyrite oxidation and decomposition in the processes that form turquoise (e.g. see Paige 1912). As previously mentioned, minerals such as kaolinite, alunite and sericite are often implicated as the source of aluminium in turquoise. In several deposits that have been studied in detail, jarosite is interpreted to have formed from solutions that originally precipitated turquoise, but became depleted in phosphorous and copper (Paige 1912; Lueth 1998).

Lead and strontium isotopes as tracers of turquoise

Why lead and strontium?

High-precision lead and strontium isotope geochemistry offers a new opportunity to reconstruct turquoise procurement and exchange. Lead and strontium occur as minor or trace elements in most rock-forming minerals, and can be found in measurable concentrations in turquoise. Both elements have four stable isotopes, but the relative abundance of these isotopes is widely variable in nature. Variation within these two isotopic systems is created by the radioactive decay of long-lived parent isotopes of rubidium (for strontium) and uranium and thorium (for lead).

Lead and strontium isotopic ratios have an established track record as powerful tracers and can be precisely and routinely measured using multi-collector-inductively coupled plasma-mass spectrometry (MC-ICP-MS), and thermal ionisation mass spectrometry (TIMS). In the geosciences, isotopic ratios of lead and strontium are commonly used to track the sources of pollutants and environmental contaminants (e.g. Bolhofer and Rosman 2001), to determine the source of metals found in ore deposits (e.g. Bouse et al. 1999), and to gain insight into weathering processes and water–rock interaction (e.g. Probst et al. 2000; Négrel et al. 2010). They are also used as tracers of archaeological materials, and can be employed to link metals and minerals from the archaeological record to specific ore deposits (e.g. Gale and Stos-Gale 1982; Farquhar et al. 1995; Stos-Gale et al. 1997; Thibodeau et al. 2007c), and applied to teeth and bone to provide insight into human movement and migrations (e.g. Müller 2003, Price et al. 2002). Discussions of lead and strontium isotope geochemistry, and a review of their applications to both geological and archaeological questions can be found elsewhere (Faure 1986; Brothwell and Pollard 2005; Faure and Mensing 2005).

Aside from their myriad established applications, lead and strontium isotopes may also prove to be effective tracers of turquoise. Lead isotopes were first applied to turquoise by Young et al. (1996) but they were unable to use their results to clearly differentiate among several mines in the American Southwest. However, these conclusions can now be re-evaluated with high-precision techniques that offer far greater resolution. For example, a pilot study that applied high precision lead and strontium isotopic measurements to 17 turquoise fragments from five locales found that samples from each deposit had distinct isotopic signatures (Thibodeau et al. 2007b). Unlike other tracers that have been applied to turquoise, lead and strontium isotopes are not significantly fractionated or partitioned by weathering processes, making them appropriate tracers for a mineral that forms in the weathering environment and may be altered after deposition. Furthermore, the processes that cause lead and strontium to vary in their isotopic composition are fundamentally different from those that produce variation in elemental concentrations, or those that alter the isotopic ratios of either hydrogen or copper.

Lead and strontium incorporated into turquoise should derive from the rocks or minerals that interacted with the turquoise-precipitating waters. Thus, assuming no significant in-growth of radiogenic lead and strontium after deposition, the isotopic composition of turquoise should broadly reflect the geologic context of specific deposits, and the circumstances under which the turquoise formed. As described above, turquoise can form as a weathering product of the same rocks that host the mineralisation, and many of the minerals associated with turquoise are related to the processes that form it. Under these conditions, lead and strontium within turquoise may be derived from its immediate environment, and thus, any major impurities would be expected to have a similar isotopic signature.

Turquoise from the Redtail site: a case study

The potential advantages of applying lead and strontium isotopes to questions of turquoise provenance can be demonstrated through a case study. The goal of the analyses described here is to identify the source of turquoise recovered from the Redtail site (AZ AA:12:149), a Hohokam village located in the northern Tucson Basin and occupied between ca. AD 700 and AD 950 (Fig. 1; Bernard-Shaw 1989; Klucas 2007). Over 3,000 fragments of turquoise have been recovered from the Redtail site and over 60 percent of this material has been classified as unworked turquoise, turquoise debitage or manufacturing debris. The quantity, form and distribution of turquoise across the site has been interpreted as evidence that the inhabitants of Redtail were involved in the acquisition and processing of turquoise during the Cañada del Oro Phase, in the early Hohokam Colonial period (AD 760–820) (Bernard-Shaw and Hohmann 1989). Thus, Redtail may have been part of an early exchange network involving turquoise in the region (Bernard-Shaw and Hohmann 1989).

The closest documented source of turquoise to the Redtail site can be found on the periphery of the Silver Bell Mountains, which lie about 56 km to the northwest of the site (Fig. 1). In addition to being the closest known source of turquoise to Redtail, the Silver Bell Mountains contain a prehispanic turquoise mining and habitation site that also dates to the Cañada del Oro Phase (Slawson 1997). This site, called Scorpion Village (AA:10:22), lies close to outcrops of turquoise, and thousands of fragments of the mineral were recovered from pit houses at the site, along with processing tools such as drills, anvils and slab abraders (Slawson 1997). Notably, detailed documentation of these mines and their associated archaeological sites postdates many previous studies of turquoise exchange and provenance. The Silver Bell Mountains are not mentioned as a prominent source in any of the previous literature examining the turquoise trade (e.g. see Weigand and Harbottle 1993). This disparity highlights the problems that can arise from having an incomplete knowledge of prehispanic turquoise mining across the American Southwest and Mexico.

Because of the spatial proximity and temporal overlap of these two sites, Vokes and Gregory (2008) suggest that turquoise found at the Redtail site could have been mined in the Silver Bell Mountains. To test this idea, we apply lead and strontium isotopic measurements to turquoise samples recovered from the Redtail site and Scorpion Village. This comparison is based on the assumption that the collection of turquoise excavated from Scorpion Village derives from the prehispanic mines of the Silver Bell Mountains. If turquoise recovered at the Redtail site was mined from this area, the lead and strontium isotopic ratios of samples collected from both sites should be similar.

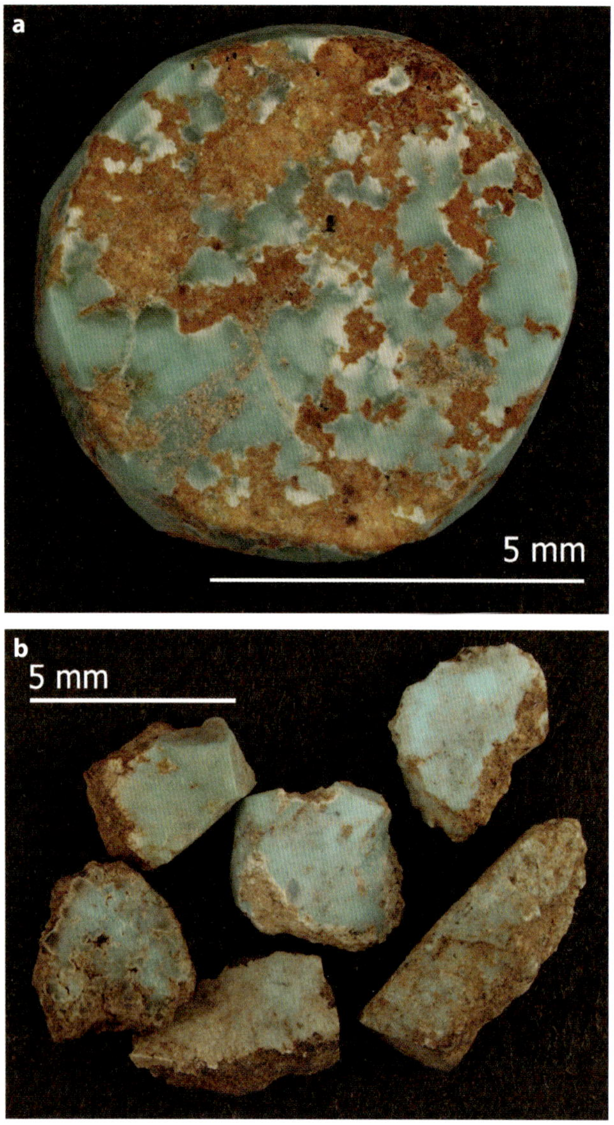

Figure 2. Examples of turquoise objects investigated for this study: (a) turquoise artefact from Scorpion Village that shows variable colouration and the close association of turquoise with alunite; (b) turquoise fragments from the Redtail site showing the presence of considerable host rock attached to the turquoise.

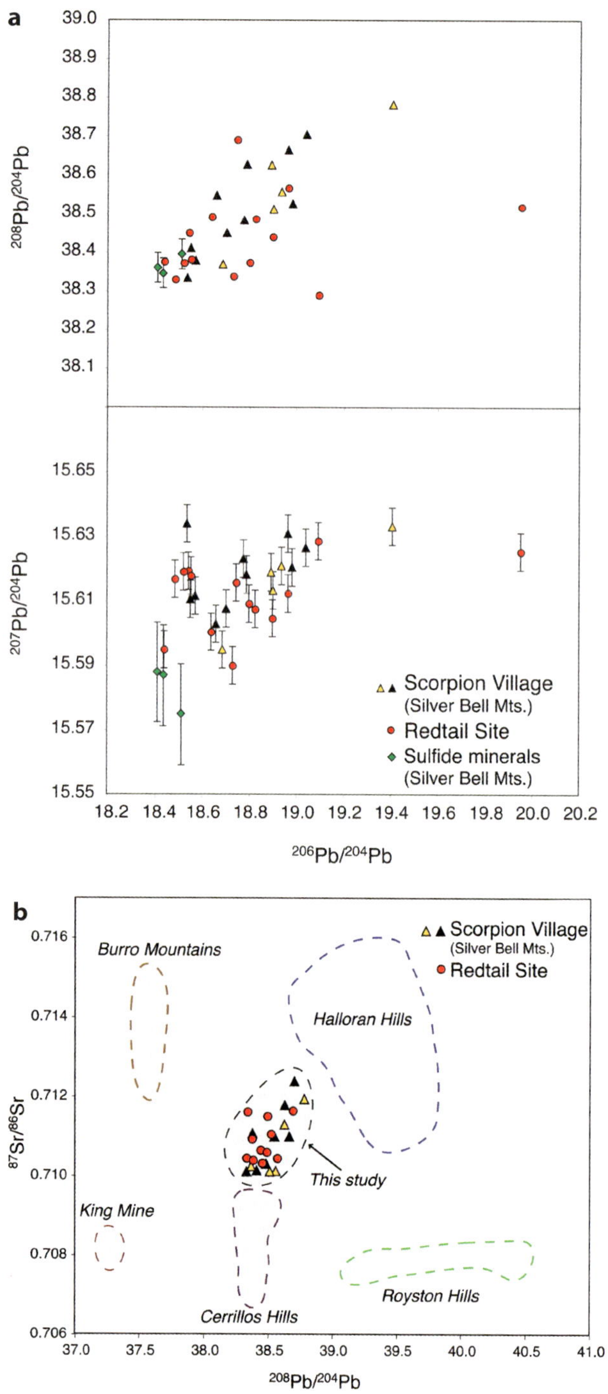

Figure 3. Plots of lead and strontium isotopic ratios of turquoise objects from the Redtail site and Scorpion Village (data from Table 2). Data plotted for Scorpion Village include samples that underwent total and partial dissolution (see 'Sample selection and methods' section for discussion): shaded triangles represent samples that were dissolved in hydrochloric acid and open triangles represent those that were dissolved in a combination of hydrofluoric, nitric and hydrochloric acids. (a) Plot comparing the lead isotopic compositions ($^{206}Pb/^{204}Pb$ vs. $^{207}Pb/^{204}Pb$ and $^{206}Pb/^{204}Pb$ vs. $^{208}Pb/^{204}Pb$) of turquoise from the Redtail site to those of turquoise from Scorpion Village. Also shown are lead isotopic ratios reported for sulphide minerals (galena and molybdenite) from the Silver Bell Mining District (Bouse et al. 1999). Except where shown, symbols are larger than 2s external errors associated with the measurements. (b) Multi-isotope plot ($^{87}Sr/^{86}Sr$ vs. $^{208}Pb/^{204}Pb$) comparing isotopic ratios of turquoise from the Redtail site to isotopic ratios of turquoise from Scorpion Village. These measurements are contrasted with shaded regions that represent the isotopic signatures of turquoise from five other turquoise mining areas in the American Southwest. These geological data are part of a larger isotopic investigation of turquoise (being prepared by Thibodeau and colleagues) that is not part of this study. Locations of the turquoise sources are shown in Figure 1. All data points and shaded regions are larger than the 2s external errors associated with measurements.

Sample selection and methods

To make the comparison, lead and strontium isotopic measurements were applied to 10 samples of turquoise from Scorpion Village and 17 samples of turquoise from the Redtail site. The lead and strontium isotopic analyses applied in this study are bulk analyses, which require the dissolution of a small amount of turquoise powder (typically between 25 and 100 mg). Bulk analyses have been critiqued because they limit the ability to control for impurities (Hull et al. 2008). Thus, to address this critique and better assess the effect of impurities, the samples selected for study included raw and worked turquoise objects that varied visibly in quality, purity, and colour (Fig. 2). Minimal effort was made to remove impurities prior to analysis.

To confirm that these objects contain turquoise (and not another, similar-looking blue-green mineral), a subset of the samples was analysed with scanning electron microscopy using an energy-dispersive X-ray attachment (SEM-EDX). This technique produces semi-quantitative data on the concentrations of major elements. All samples tested contained copper,

Table 2. Lead and strontium isotopic ratios of turquoise samples from the Redtail site and Scorpion Village.

Lab ID	FN#	Feature	Dissolution method	$^{87}Sr/^{86}Sr$	% std err	$^{208}Pb/^{206}Pb$	$^{207}Pb/^{206}Pb$	$^{206}Pb/^{204}Pb$	$^{207}Pb/^{204}Pb$	$^{208}Pb/^{204}Pb$
Scorpion Village (AZ:AA:10:22)										
SB1	1166	32	HCl	0.71013	0.0009	2.0704	0.8415	18.550	15.611	38.410
SB2	1957	56	HCl	0.71106	0.0016	2.0664	0.8406	18.572	15.612	38.378
SB3	1967	53	HCl	0.71029	0.0031	2.0497	0.8321	18.775	15.623	38.483
SB4a	2122	56	HCl	0.71237	0.001	2.0326	0.8207	19.041	15.627	38.705
SB5a	2178	52	HCl	0.71010	0.001	2.0679	0.8434	18.536	15.634	38.333
SB6a	2178	52	HCl	No data		2.0559	0.8345	18.702	15.608	38.450
SB7a	2188	56	HCl	0.71098	0.001	2.0388	0.8242	18.965	15.631	38.664
SB8	2074	53	HCl	0.71098	0.001	2.0659	0.8362	18.660	15.603	38.547
SB9	2074	53	HCl	0.71177	0.001	2.0561	0.8314	18.787	15.618	38.627
SB10a	615	2	HCl	No data		2.0297	0.8229	18.983	15.620	38.526
SB4b	2122	56	total	0.71192	0.0011	1.9983	0.8056	19.407	15.633	38.782
SB5b	2178	52	total	0.71011	0.001	2.0360	0.8248	18.936	15.621	38.556
SB6b	2178	52	total	0.71023	0.001	2.0533	0.8346	18.686	15.595	38.369
SB7b	2188	56	total	0.71128	0.001	2.0444	0.8267	18.892	15.619	38.626
SB10b	615	2	total	0.71010	0.0009	2.0376	0.8261	18.901	15.613	38.512
Redtail Site (AZ:AA:12:149)										
RT1	189	4	HCl	0.71045	0.0028	2.0735	0.8449	18.485	15.617	38.330
RT2	189	4	HCl	0.71065	0.0011	2.0340	0.8257	18.898	15.605	38.439
RT3	4309	94	HCl	0.71032	0.0008	2.0738	0.8424	18.543	15.619	38.450
RT4	4309	94	HCl	0.71106	0.0010	1.9306	0.7832	19.952	15.625	38.518
RT5	4313	94	HCl	0.71093	0.0011	2.0719	0.8433	18.521	15.619	38.372
RT6	4313	94	HCl	0.71039	0.0010	2.0686	0.8417	18.552	15.618	38.381
RT7	4313	94	HCl	0.71044	0.0013	2.0338	0.8233	18.965	15.612	38.567
RT8	4313	94	HCl	0.71161	0.0010	2.0469	0.8324	18.730	15.590	38.338
RT9	4144	94	HCl	No data		2.0053	0.8185	19.094	15.629	38.289
RT10	4144	94	HCl	0.71060	0.0010	2.0446	0.8291	18.823	15.608	38.486
RT11	223	4	HCl	No data		2.0813	0.8458	18.438	15.595	38.375
RT12	223	4	HCl	0.71150	0.0012	2.0652	0.8370	18.639	15.600	38.491
RT13	432	4	HCl	No data		2.0413	0.8303	18.798	15.609	38.373
RT14	1770	18	HCl	0.71164	0.0010	2.0641	0.8331	18.744	15.616	38.691

aluminium and phosphorous, consistent with turquoise. The data collected via SEM-EDX also revealed the presence of other elements within the turquoise, especially silicon (Si) and calcium (Ca), further demonstrating the heterogeneous nature of the turquoise mineralisation.

All samples were powdered using a clean mortar and pestle and dissolved using twice-distilled ultra-pure hydrochloric acid (HCl). HCl was chosen because it dissolves turquoise, but will not digest many of the other mineral phases that may be present, especially silicates. Thus, the use of HCl to dissolve samples decreases the influence of non-turquoise minerals on the isotopic signature. However, to gather additional data to assess whether or not the impurities associated with the samples would alter the lead and strontium isotopic signature, we re-sampled five objects from Scorpion Village and dissolved the powders using a combination of hydrofluoric (HF), nitric (HNO_3), and hydrochloric acids (see Table 2). While HCl should primarily dissolve the turquoise, the use of HF + HNO_3 + HCl will fully digest all major mineral phases present. The details of the procedures used to prepare all the samples and to measure the isotopic ratios are reported in the technical appendix.

Results and interpretations

The lead and strontium isotopic ratios of all objects in this study are listed in Table 2 and presented in Figure 3. To visually display the results, the data are plotted according to the ratios $^{87}Sr/^{86}Sr$, $^{206}Pb/^{204}Pb$, $^{207}Pb/^{204}Pb$, and $^{208}Pb/^{204}Pb$. As predicted, the measured lead and strontium isotopic ratios on samples from both Scorpion Village and the Redtail site are very similar to one another. Although not all combinations of measured isotope ratios are plotted due to space considerations, any chosen pair of ratios will produce similar results. Thus, turquoise found at the Redtail site has lead and strontium isotopic compositions that are consistent with an origin in the Silver Bell Mountains.

Additionally, these data can be interpreted to support the use of lead and strontium isotopes to fingerprint turquoise and to suggest that bulk analyses do not pose a significant problem when using these tracers. The samples reported here, especially from Scorpion Village, have a wide range of visual characteristics, including colour, hardness and the relative proportion of matrix minerals and host rock associated with the turquoise. Despite these differences among samples, the isotopic composition of the turquoise (or more accurately, turquoise-bearing) samples from both the Redtail site and Scorpion Village vary within roughly the same parameters. Even when five objects were re-sampled and completely dissolved for analysis, these values closely corresponded with the values measured from only dissolving the sample in HCl (for example, compare SB4a to SB4b in Table 2).

Although we only measure archaeological samples in this study, it is intriguing to note that published lead isotopic measurements made on galena and molybdenite from the Silver Bell Mountains closely align with values we report for turquoise from Scorpion Village and the Redtail site (Bouse et al. 1999; Fig. 3a). This observation further supports the conclusion that turquoise at Scorpion Village and the Redtail site come from the Silver Bell Mountains, as it is expected that the turquoise would broadly reflect the isotopic characteristics of the metal-rich geologic environment in which it formed.

The agreement between the archaeological evidence and the geochemical data supports the conclusion that turquoise at Redtail was likely to have been mined in the Silver Bell Mountains and demonstrates that lead and strontium isotopes can be effective tracers of turquoise. Although destructive, and therefore not appropriate for all archaeological objects, this method can be applied to a wide range of turquoise mineralisation, allowing for the analysis of weathered, poor quality or matrix-rich turquoise. The use of bulk analyses also confers the advantage of measuring isotopic ratios with high precision and accuracy via well-established and routine methods of analysis. This ability to collect high precision data is likely to be a significant advantage to differentiating between the numerous turquoise deposits across the American Southwest.

Conclusions and future directions

Measurements of turquoise collected from mines around the American Southwest have been shown to have distinct ranges of lead and strontium isotopic ratios, but a detailed discussion of these data is beyond the scope of this paper (Thibodeau et al. 2007a, 2008, 2011). However, to offer a necessary perspective on the data and conclusions drawn here, Figure 3b uses a multi-isotope plot to compare the lead and strontium isotopic ratios of samples of the turquoise reported in this study to ranges of ratios collected on turquoise samples from five mines in the American Southwest. This figure demonstrates that the range of values measured on samples from the Redtail site and Scorpion Village is relatively small when compared to the array of ratios measured from other mines and locales.

The application of lead and strontium isotopes to turquoise offers a new opportunity to address longstanding questions about the origins of turquoise artefacts. High precision measurements of these two tracers have the potential to overcome the problems that have negatively impacted past studies of turquoise provenance, and may be robust enough to identify the source of samples with a wide range of compositional characteristics and associated impurities. Sourcing a complex mineral such as turquoise will continue to provide challenges for researchers for decades to come, and the continued destruction of the prehispanic mining record in the American Southwest makes studies of turquoise procurement particularly urgent. Although many turquoise mines have been abandoned or destroyed, the turquoise objects created by prehispanic cultures endure, capturing the imagination of the public and the scientific establishment alike. Through careful and systematic scientific analysis, it should be possible to begin to understand these objects, and the people who created them, from a new, isotopic perspective.

Technical appendix: methodology

To prepare samples for analysis, each was powdered using a clean mortar and pestle and accurately weighed. Because fragments of turquoise from Redtail were small in size, in some cases, multiple pieces were combined to ensure enough material was present for isotopic analysis. Typical sample sizes ranged between 15 and 100 mg. All acid solutions used during sample preparation, separation and analysis were made from twice-distilled acid and ultra-pure Milli-Q water (Millipore, 18.2 MΩ/cm). Powders were transferred to clean Teflon® vials and dissolved on a hot plate at 125 °C using concentrated HCl, or in several steps using a combination of HF, HNO_3 and HCl. After dissolution, lead and strontium are separated from single samples using Sr-Spec resin (Eichrom, Darien IL). Lead isotope analysis is conducted on a GV-Instruments MC-ICP mass spectrometer at the University of Arizona. Procedures for lead isotope analysis are detailed in Thibodeau et al. (2007c). The maximum 2s external errors associated with the measurements are 0.035% for $^{206}Pb/^{204}Pb$, 0.035% for $^{207}Pb/^{204}Pb$, 0.037% for $^{208}Pb/^{204}Pb$, 0.021% for $^{208}Pb/^{206}Pb$, and 0.009% for $^{207}Pb/^{206}Pb$.

After separation, strontium samples are loaded onto degassed tantalum (Ta) filaments using phosphoric acid and tantalum gel to enhance ionisation. All samples were analysed in a VG Sector 54 multi-collector thermal ionisation mass spectrometer in dynamic collection mode at the University of Arizona. The $^{87}Sr/^{86}Sr$ ratios are corrected for mass fractionation of using $^{86}Sr/^{88}Sr = 0.1194$. The average $^{87}Sr/^{86}Sr$ value of the NBS-987 standard (all runs combined) is 0.71024 ± .000016 (n=14, 1s), based on 50–100 cycles of dynamic data collection. Total process blanks were prepared alongside all samples, and were an insignificant proportion of the total lead and strontium extracted from each sample.

Acknowledgements

All samples analysed in this study were curated at the Arizona State Museum in Tucson, Arizona. Mark Baker provided assistance in the laboratory. Lisa Molofsky and two anonymous reviewers read earlier versions of this paper, and greatly improved its final content.

References

Arrowsmith, R. (1974) 'Introduction', in J. Pogue, *The Turquois: A Study of its History, Mineralogy, Geology, Ethnology, Archaeology, Mythology, Folklore, and Technology*. Glorieta, NM: The Rio Grande Press [originally published 1915, *Memoirs of the National Academy of Sciences*, Vol. XII, Part II, Washington, DC, National Academy of Sciences].

Barwood, H. (2003) 'The Bishop copper prospect: near Lynch Station, Campbell County, Virginia (Famous Mineral Localities)', *The Mineralogical Record* **34**: 215–218.
Bennett, E.M. (1970) *Turquoise and the Indian*. Chicago: The Swallow Press, Inc.
Bernard-Shaw, M. (ed.) (1989) *Archaeological Investigations at the Redtail Site, AA:I2:149 (ASM), in the Northern Tucson Basin*. Tucson: Center for Desert Archaeology Technical Report 89-8.
Bernard-Shaw, M. and Hohmann, W.D. (1989) 'Turquoise artifacts and other minerals', in M. Bernard-Shaw (ed.), *Archaeological Investigations at the Redtail Site, AA:I2:149 (ASM), in the Northern Tucson Basin*. Tucson: Center for Desert Archaeology Technical Report 89-8, 151–157.
Bice, R.A., Davis, P.S. and Sundt, W.M. (2003) *Indian Mining of Lead for use in Rio Grande Glaze Paint: Report of the AS-5 Bethsheba Project near Cerrillos, New Mexico*. Albuquerque, NM: Albuquerque Archaeological Society.
Blake, W.P. (1858) 'The chalchihuitl of the ancient Mexicans: its locality and association, and its identity with turquoise', *American Journal of Science* **25**: 227–232.
Bolhofer, A. and Rosman, K.J.R. (2001) 'Isotopic source signatures for atmospheric lead: the Northern Hemisphere', *Geochemica and Cosmochimica Acta* **65**: 1727–1740.
Bouse, R.M., Ruiz, J., Titley, S.R., Tosdal, R.M. and Wooden, J.L. (1999) 'Lead isotope compositions of late Cretaceous and early Tertiary igneous rocks and sulfide minerals in Arizona: implications for the sources of plutons and metals in porphyry copper deposits', *Economic Geology* **94**: 211–244.
Braithwaite, R.S.W. (1981) 'Turquoise crystals from Britain and a review of related species', *The Mineralogical Record* **12**: 349–353.
Brothwell, D.R. and Pollard, A.M. (eds) (2005) *Handbook of Archaeological Sciences*. Chichester: Wiley & Sons.
Cid-Dresdner, H. (1965) 'Determination and refinement of the crystal structure of turquoise, $CuAl_6(PO_4)_4(OH)_8{\cdot}4H_2O$', *Zeitschrift für Kristallographie Bd.* **121**, S.: 87–113.
Crawford, W.P. and Johnson, F. (1937) 'Turquoise deposits of Courtland, Arizona', *Economic Geology* **32**: 511–523.
Crook, J. (2002) 'The Origin of Turquoise at the Iron Mask Claim, Orogrande Mining District, Otero County, New Mexico', MS thesis, New Mexico Institute of Mining and Technology (unpublished).
Di Peso, C.C. (1974) *Casas Grandes: A Fallen Trading Center of the Gran Chichimeca, Vols 1–3*. The Amerind Foundation Publication No. 9. Flagstaff, AZ: Northland Press.
Disbrow, A.E. and Stoll, W.C. (1957) *Geology of the Cerrillos Area, Santa Fe County, New Mexico, Bulletin 48*. Socorro, NM: New Mexico Bureau of Mines and Mineral Resources.
Doyel, D.E. (1991) 'Hohokam exchange and interaction', in P.L. Crown and W.J. Judge (eds), *Chaco and the Hohokam*. Santa Fe, NM: School of American Research Press, 225–252.
Eisen, G. (1898) 'The prehistoric turquoise mines of California and the ancient Indian workers', *San Francisco Call*, 18 March 1898.
Farquhar, R.M., Walthall, J.A. and Hancock, R.G.V. (1995) '18th century lead smelting in central North America: evidence from lead isotope and INAA measurements', *Journal of Archaeological Science* **22**: 639–648.
Faure, G. (1986) *Principles of Isotope Geology*. New York: Wiley.
Faure, G. and Mensing, T.M. (2005) *Isotopes: Principles and Applications*. Hoboken, NJ: John Wiley & Sons, Inc.
Foord, E.E. and Taggart, J.E. (1998) 'A reexamination of the turquoise group: the minerals aheylite, planerite (redefined), turquoise and coeruleolactite', *Mineralogical Magazine* **62**: 93–111.
Frisbie, T.R. (1983) 'Anasazi-Mesoamerican relationships: from the bowels of the earth and beyond', in J.E. Smith (ed.), *Proceedings of the 1981 Anasazi Symposium*. Mesa Verde, CO: Mesa Verde Museum Association, 215–227.
Gale, N.H. and Stos-Gale, Z.A. (1982) 'Bronze Age copper sources in the Mediterranean: a new approach', *Science* **216**: 11–19.
Harbottle, G. and Weigand, P.C. (1992) 'Turquoise in pre-Columbian America', *Scientific American* **266**(2): 78–85.
Hidden, W.E. (1893) 'Two new localities for turquoise', *American Journal of Science* **46**: 400–402.
Hull, S., Fayek, M., Mathien, F.J., Shelley, P. and Durand, K.R. (2008) 'A new approach to determining the geological provenance of turquoise artifacts using hydrogen and copper stable isotopes', *Journal of Archaeological Science* **35**: 1355–1369.
Hussein, A.A.A., Eldougdoug, A., Takla, M.A. and El-Eraky, F.Y. (1994–1995) 'Mineralogy, chemistry, and origin of Egyptian turquoise, west central Sinai, Egypt', *Annals of the Geological Survey of Egypt* **20**: 361–379.
Jernigan, E.W. (1978) *Jewelry of the Prehistoric Southwest*. Albuquerque: University of New Mexico Press.
Kelley, J.C. and Kelley, E.A. (1975) 'An alternative hypothesis for the explanation of Anasazi culture history', in T.R. Frisbie (ed.), *Collected Papers in Honor of Florence Hawley Ellis. Papers of the Archaeological Society of New Mexico No. 2*. Norman, OK: Hooper Publishing, 178–223.
Kim, J., Simon, A.W., Ripoche, V., Mayer, J.W. and Wilkens, B. (2003) 'Proton-induced x-ray emission analysis of turquoise artefacts from Salado Platform Mound sites in the Tonto Basin of central Arizona', *Measurement Science and Technology* **14**: 1579–1589.
Klucas, E. (2007) 'Recent research at the Redtail site', *Archaeology Southwest* **21**: 6.
Leonard, N.N. and Drover, C.E. (1980) 'Prehistoric turquoise mining in the Halloran Springs district, San Bernardino County, California', *Journal of California and Great Basin Anthropology* **2**: 245–256.
Lister, R.H. (1978) 'Mesoamerican influence at Chaco Canyon, New Mexico', in C.L. Riley and B.C. Hedrick (eds), *Across the Chichimec Sea: Papers in Honor of J. Charles Kelley*. Carbondale: Southern Illinois University Press, 233–241.
Lueth, V.W. (1998) 'Two diverse origins of turquoise at the Orogrande Mining District, Otero County, New Mexico', *New Mexico Geology* **20**: 64–65.
Mathien, F.J. (1981a) *Economic Exchange Systems in the San Juan Basin*. Albuquerque: Department of Anthropology, University of New Mexico.
Mathien, F.J. (1981b) 'Neutron activation of turquoise artifacts from Chaco Canyon, New Mexico', *Current Anthropology* **22**: 293–294.
Mathien, F.J. (1998) 'Tri-cultural use of the Cerrillos Mines', *Cultural Resource Management* **7**: 49–52.
Mathien, F.J. (2001) 'The organization of turquoise production and consumption by the prehistoric Chacoans', *American Antiquity* **66**: 103–118.
Mathien, F.J. and Olinger, B. (1992) 'An experiment with x-ray fluorescence to determine trace element variability in turquoise composition', in M.S. Duran and D.T. Kirpatrick (eds), *Archaeology, Art, and Anthropology: Papers in Honor of J.J. Brody*. Albuquerque: The Archaeological Society of New Mexico, 123–134.
McEwan, C., Middleton, A., Cartwright, C.R. and Stacey, R. (2006) *Turquoise Mosaics from Mexico*. London: British Museum Press.
McGuire, R.H. (1980) 'The Mesoamerican connection in the Southwest', *The Kiva* **46**: 3–38.
McGuire, R.H. (1993) 'The structure and organization of Hohokam exchange', in J. Ericson and T.G. Baugh (eds), *The American Southwest and Mesoamerica: Systems of Prehistoric Exchange*. New York: Plenum Press, 95–199.
Modreski, P.J. and Murphy, J.A. (2002) 'Colorado turquoise', *New Mexico Geology* **24**: 133.
Moore, R.T. (1968) *Mineral Deposits of the Fort Apache Indian Reservation, Arizona*. Arizona Bureau of Mines Bulletin 177. Tucson: University of Arizona.
Morrissey, F.R. (1968) *Turquoise Deposits of Nevada. Report 17, Nevada Bureau of Mines*. Reno: McKay School of Mines, University of Nevada.
Müller, W., Fricke, H., Halliday, A.N., McCulloch, M.T. and Wartho, J. (2003) 'Origin and migration of the Alpine Iceman', *Science* **302**: 862–866.

Négrel, P., Romain, M., Roy, S., Guerrot, C. and Pauwels, H. (2010) 'Lead isotopes in groundwater as an indicator of water–rock interaction (Masheshwaram catchment, Andhra Pradesh, India)', *Chemical Geology* **274**: 136–148.

Paige, S. (1912) 'The origin of turquoise in the Burro Mountains, New Mexico', *Economic Geology* **7**: 382–392.

Panczer, W.D. (1987) *Minerals of Mexico*. New York: Van Nostrand Reinhold Company.

Pearl, R.M. (1941) 'Turquoise deposits of Colorado', *Economic Geology* **35**: 335–344.

Pogue, J. (1974) *The Turquois: A Study of Its History, Mineralogy, Geology, Ethnology, Archaeology, Mythology, Folklore, and Technology*. Glorieta, NM: The Rio Grande Press [originally published, 1915, *Memoirs of the National Academy of Sciences*, Vol. XII, Part II, Washington, DC, National Academy of Sciences].

Price, T.D., Burton, J.H. and Bentley, R.A. (2002) 'The characterization of biologically available strontium isotope ratios for the study of prehistoric migration', *Archaeometry* **44**: 117–135.

Probst, A., El Gh'mari, A., Aubert, D., Fritz, B. and McNutt, R. (2000) 'Strontium as a tracer of weathering processes in a silicate catchment polluted by acid atmospheric inputs, Strengbach, France', *Chemical Geology* **170**: 203–219.

Saville, M. (1922) *Turquoise Mosaic Art in Ancient Mexico*. New York: Museum of the American Indian, Heye Foundation.

Schroeder, A.H. (1979) 'The Cerrillos mining area', in R.V. Ingersoll (ed.), *Archaeology and History of Santa Fe Country, Special Publication No. 8*. Socorro, NM: New Mexico Geological Society, 13–16.

Sigleo, A.M.C. (1975) 'Turquoise mine and artifact correlation for Snaketown Site, Arizona', *Science* **189**: 459–460.

Sinkankas, J. (1976) *Gemstones of North America, Volume II*. New York: Van Nostrand Reinhold Company.

Slawson, L.V. (ed.) (1997) *Turquoise Traders of the Silver Bells: Settlement, Subsistence, and Exchange on the Periphery of the Northern Tucson Basin*. Tucson, AZ: Aztlan Archaeology, Inc.

Snow, C.H. (1891) 'Turquois in southwestern New Mexico', *American Journal of Science* **21**: 511–512.

Snow, D.H. (1973) 'Prehistoric southwestern turquoise industry', *El Palacio* **79**: 17–32.

Stos-Gale, Z.A., Maliotis, G., Gale, N.H. and Annetts, N. (1997) 'Lead isotope characteristics of the Cyprus copper ore deposits applied to provenance studies of copper oxide ingots', *Archaeometry* **39**: 83–123.

Thibodeau, A.M., Ruiz, J. and Chesley, J.T. (2007a) 'Lead and strontium isotopes as geochemical tracers of turquoise', paper presented at the 2007 Geological Society of America Annual Meeting, Denver, Colorado.

Thibodeau, A.M., Chesley, J.T. and Ruiz, J. (2007b) 'Lead and strontium as tracers of turquoise', *SAS Bulletin* **30**: 10–14.

Thibodeau, A.M., Killick, D.J., Ruiz, J., Chesley, J.T., Deagan, K., Cruxent, J.M. and Lyman, W. (2007c) 'The strange case of the earliest silver extraction by European colonists in the New World', *Proceedings of the National Academy of Sciences* **104**: 3663–3666.

Thibodeau, A.M., Ruiz, J., Chesley, J.T. and Killick, D.J. (2008) 'An expanded lead and strontium isotopic database for turquoise deposits in the Greater American Southwest: implications for provenance studies of archaeological turquoise', poster presented at the *37th International Symposium on Archaeometry*, Siena, Italy.

Thibodeau, A.M., Chesley, J.T., Ruiz, J. and Killick, D.J. (2011) 'Determining the provenance of turquoise from Pueblo Bonito, Chaco Canyon, NM', poster presented at the 76th Annual Meeting of the Society for American Archaeology, Sacramento, CA.

Vaillant, G. (1935) *Excavations at El Arbollilo*. New York: Anthropological Papers of the American Museum of Natural History, Part 2, 136–279.

Vokes, A.W. and Gregory, D.A. (2008) 'Exchange networks for exotic goods in the Southwest and Zuni's place in them', in D.A. Gregory and D.R. Wilcox (eds), *Zuni Origins: Toward a New Synthesis of Southwestern Archaeology*. Tucson: University of Arizona Press, 318–357.

Warren, A.H. and Mathien, F.J. (1985) 'Prehistoric and historic turquoise mining in the Cerrillos district: time and place', in C.H. Lange (ed.), *Southwestern Culture History: Collected Papers in Honor of Albert H. Schroeder*. Papers of the Archaeological Society of New Mexico 10. Santa Fe: Archaeological Society of New Mexico and Ancient City Press, 93–128.

Warren, A.H. and Weber, R.H. (1979) 'Indian and Spanish mining in the Galisteo and Hagan basins', in R.V. Ingersoll (ed.), *Archaeology and History of Santa Fe Country, Special Publication No. 8*. Socorro, NM: New Mexico Geological Society, 7–11.

Weigand, P.C. (1994) 'Observations on ancient mining within northwestern regions of the Mesoamerica civilization with emphasis on turquoise', in A.K. Craig and R.C. West (eds), *In Quest of Mineral Wealth: Aboriginal and Colonial Mining and Metallurgy in Spanish America*. Baton Rouge, LA: Department of Geography and Anthropology, Louisiana State University, 21–33.

Weigand, P.C. and Harbottle, G. (1993) 'The role of turquoises in the ancient Mesoamerican trade structure', in J.E. Ericson and T.G. Baugh (eds), *The American Southwest and Mesoamerica: Systems of Prehistoric Exchange*. New York: Plenum Press, 159–177.

Weigand, P.C., Harbottle, G. and Sayre, E.V. (1977) 'Turquoise sources and source analysis: Mesoamerica and the southwestern U.S.A', in T.K. Earle and J.E. Ericson (eds), *Exchange Systems in Prehistory*. New York: Academic Press, 15–34.

Welch, J.R. and Triadan, D. (1991) 'The Canyon Creek turquoise mine, Arizona', *The Kiva* **56**: 145–64.

Wilcox, D. (1986) 'A historical analysis of the problem of southwestern Mesoamerican connections', in F.J. Mathien and R.H. McGuire (eds), *Ripples in the Chichimec Sea: New Considerations of Mesoamerican–Southwestern Interactions*. Carbondale: Southern Illinois University Press, 9–44.

Windes, T.C. (1992) 'Blue notes: the Chacoan turquoise industry in the San Juan Basin', in D.E. Doyel (ed.), *Anasazi Regional Organization and the Chaco System*. Anthropological Papers No. 5. Maxwell Museum of Anthropology. Albuquerque: University of New Mexico Press, 159–168.

Wiseman, R.N. and Darling, A.J. (1986) 'The bronze trail group: more evidence for a Cerrillos-Chaco turquoise connection', in A. Poore (ed.), *By Hands Unknown: Papers on Rock Art and Archaeology in Honor of James G. Bain*. Papers of the Archaeological Society of Mexico 12. Santa Fe: Ancient City Press, 115–143.

Young, S.M.M, Phillips Jr., D.A. and Mathien, F.J. (1996) 'Lead isotope analysis of turquoise sources in the southwestern USA and Mesoamerica: a preliminary report', in S. Demirci, A.M. Özer and G.D. Summers (eds), *Archaeometry 94: Proceedings of the 29th International Symposium*, Ankara, 9–14 May 1994. Ankara: TÜBİTAK, 147–150.

Authors

- Alyson M. Thibodeau, Department of Geosciences, University of Arizona, Tucson, USA
- John T. Chesley, Department of Geosciences, University of Arizona, Tucson, USA
- Joaquin Ruiz, Department of Geosciences, University of Arizona, Tucson, USA
- David J. Killick, School of Anthropology, University of Arizona, Tucson, USA
- Arthur Vokes, Arizona State Museum, University of Arizona, Tucson, USA

The turquoise disk from Offering 99 at the Templo Mayor in Tenochtitlan

Adrián Velázquez Castro, María Eugenia Marín Benito, Emiliano Ricardo Melgar Tísoc, Reyna Beatriz Solís Ciriaco and José Luis Ruvalcaba Sil

Excavation, analysis of provenance of raw materials and conservation

Discovery and extraction

Excavations conducted by the Urban Archaeology Program from 1994 to 2000 at 'La Casa de las Ajaracas' at the Templo Mayor in Tenochtitlan revealed four offerings buried by the Mexicas in front of the double stairway of their main temple. Offering 99 was located in front of the stairway that led to the temple of Huitzilopochtli, the Mexica patron god, almost next to the southern half of the structure dedicated to Tlaloc, the rain god, in fill that covered the second to last construction stage of the Aztec Great Temple (Epoch VI) (Figs 1 and 2). This phase may be chronologically situated from 1502 to 1521, corresponding to the final architectural phase of the Templo Mayor (Matos 1988: 176).

Offering 99 was deposited in a core of clay and sandy-lime soil covered by porous tezontle stone, occupying an area of approximately 1 m^2. At the deepest level, in the northwest quadrant, a large concentration of small pieces of a blue material was found (Carrillo n.d.) (Fig. 2b), which were delimited and covered with polyethylene and mud. These elements were held together solely by the soil between them, although a number of the pieces were more or less in their original position, while the other part was found dispersed around the area of the greatest concentration of fragments. Given the risk of losing even more of the original placement of the pieces, the decision was made to remove the entire block, together with the soil supporting it, for which metal sheets covered with two layers of cheesecloth were used (Fig. 3). A total of three blocks (northwest, east, and southwest) was extracted and kept moist by spraying distilled water in addition to the placing of thymol crystals to avoid the growth of microorganisms. At the moment of the excavation, the conditions of Offering 99

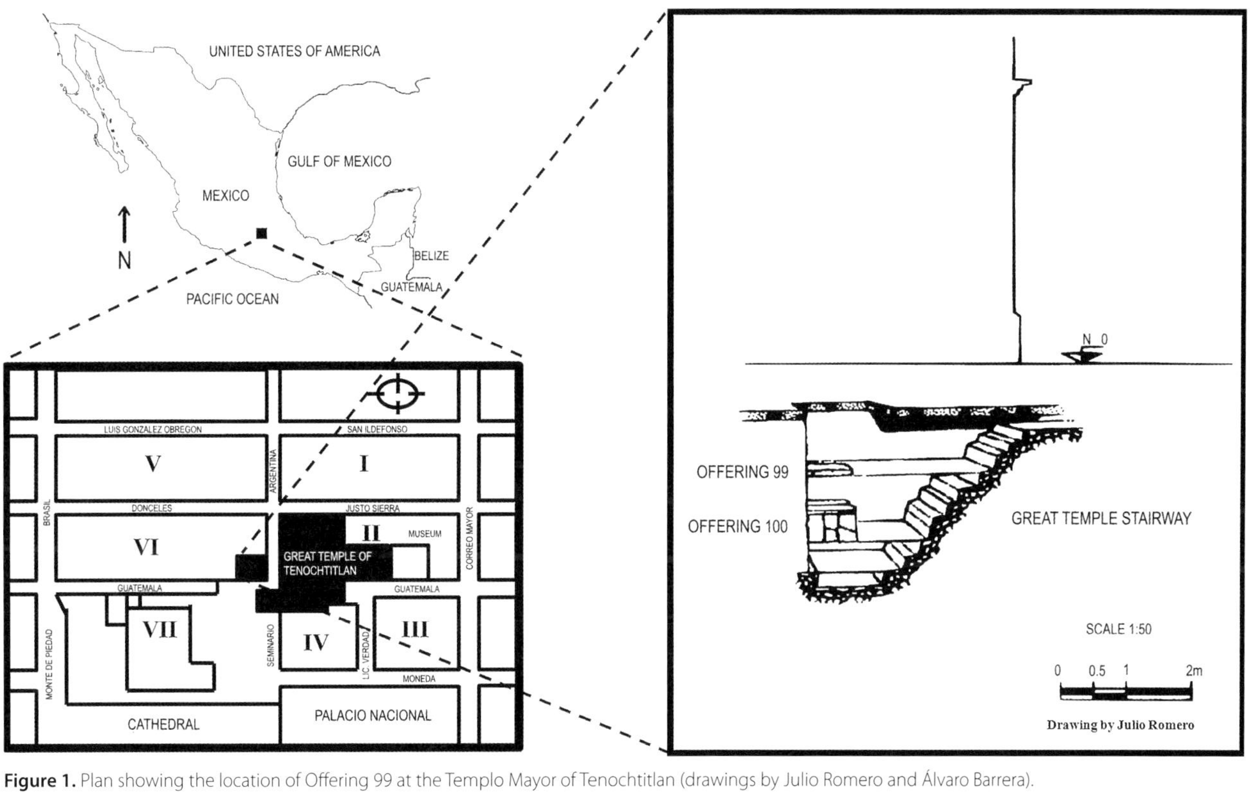

Figure 1. Plan showing the location of Offering 99 at the Templo Mayor of Tenochtitlan (drawings by Julio Romero and Álvaro Barrera).

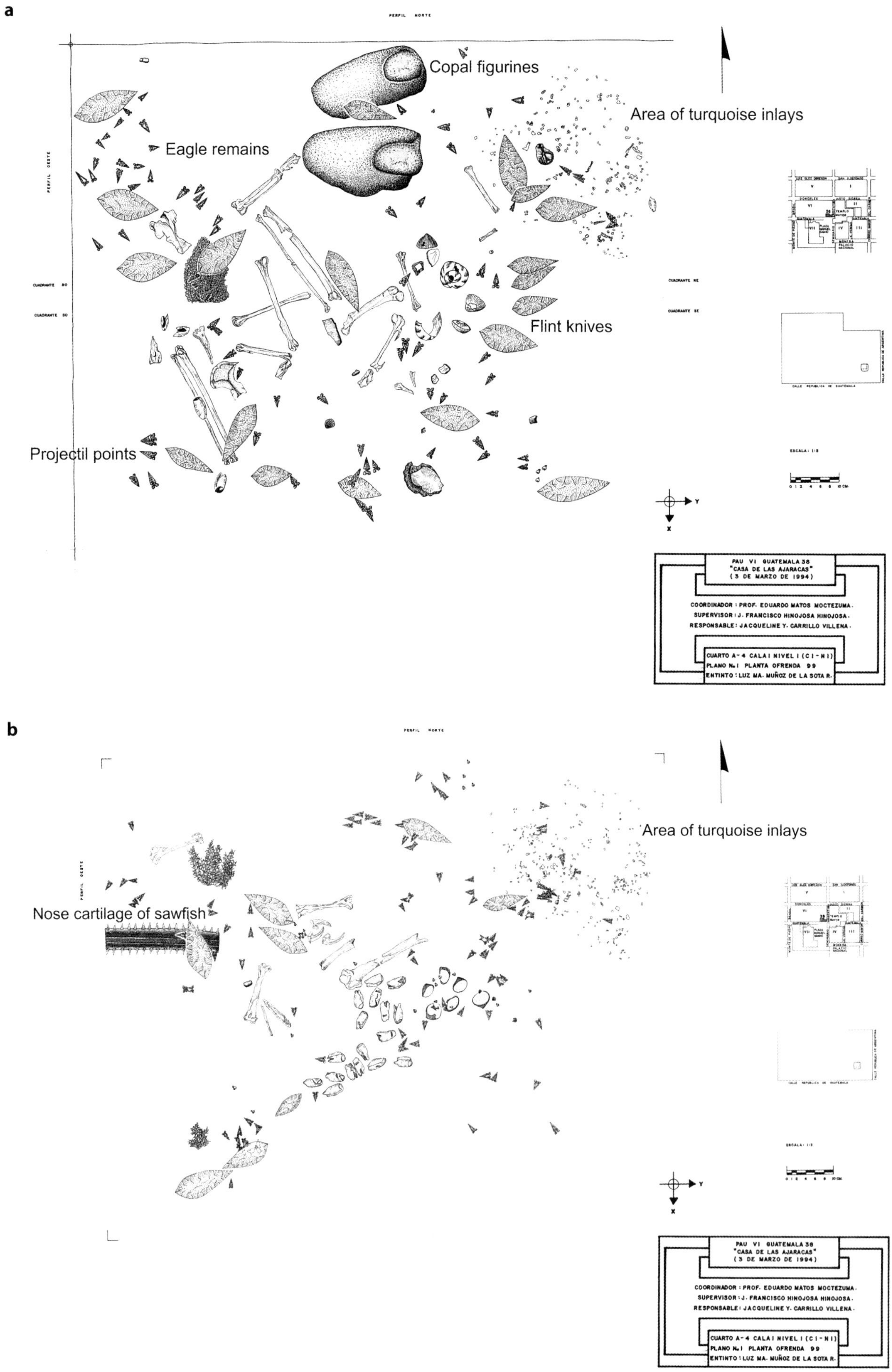

Figure 2. Plan of first (a) and second (b) levels of excavation of Offering 99 (Fondo Reservado del Museo del Templo Mayor).

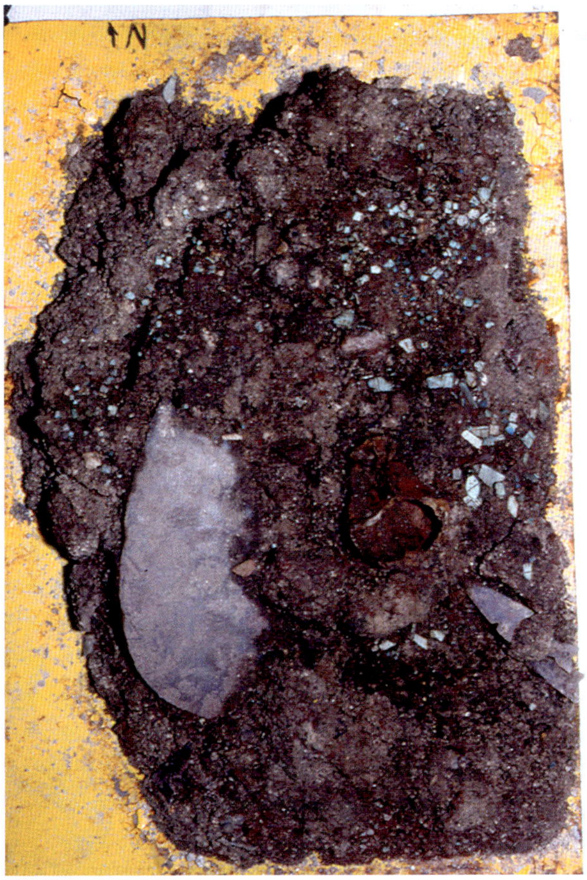

Figure 3. Block containing the tesserae.

were as follows: neutral soil (pH 7), moisture content 70%, which increased at night-time due to the absence of ventilation, and an average temperature of 13 °C.

Laboratory intervention

The blocks were moved to the restoration laboratory at the Museo del Templo Mayor, where micro-excavation was carried out in three layers, employing a grid of threads and dental tools. Throughout the process, the blocks were periodically sprayed with distilled water to stabilise the moisture level, and when work was completed for the day, the blocks were covered with a humidity chamber. Photographic and written records were made in addition to plans on millimetre graph paper and on acetate. The purpose of the latter drawings was to be able to superimpose the different levels of the excavation for a complete, clear vision of the complex.

The most generalised deterioration displayed by the pieces was as follows: dark spots, fading, fractures and lack of cohesion – the latter particularly in the case of white objects with a texture resembling chalk. As each piece was removed, it was cleaned with an alcohol-water mixture (1:1) applied with a paintbrush; in cases in which soil was encrusted on the surface, a bamboo swab was used. The most fragile pieces were consolidated through immersion in an acrylic copolymer at a low concentration (Paraloid B72 at 3% in toluol [toluene]-alcohol) and dried in a closed environment. Fractured pieces were adhered with a polyvinyl acetal (Mowithal B60H at 7.5% in ethyl alcohol and acetate 7:3). The cleaned and/or consolidated pieces were placed in a wooden box covered by a lid, lined with white flannel, and marked with a grid. A separate box for each level of the excavation made it possible to preserve the location of each piece at the level where it was found (Fig. 4). During the micro-excavation process various elements were recovered that were found on top of the area where the blue pieces were found. These included lithic elements such as knives and projectile points, copper bells, turtle carapaces, and bird and fish bones. Some of the blue fragments were found attached to these other objects, and even more were found after sieving the soil from the excavation.

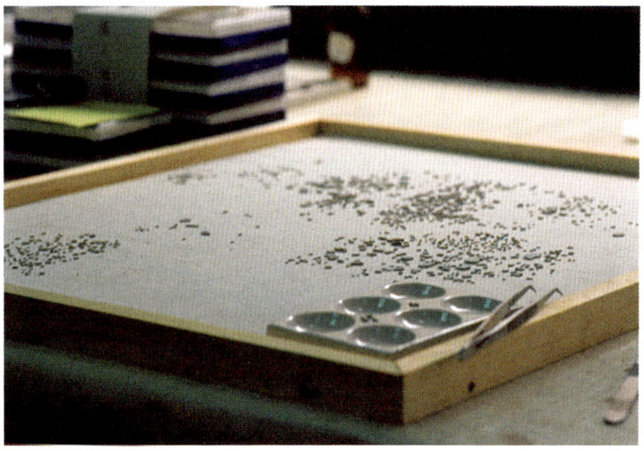

Figure 4. Pieces reassembled in wooden box.

Analysis of the composition and provenance of the tesserae

A total of 15,000 tiny tesserae of different shapes and thickness was found, ranging in tonality from white to blue and green. To determine if they were made of mineral turquoise (i.e. turquoise that can be determined chemically) from sources in the southwest of the United States or what one might call 'turquoise substitute' (or even 'cultural turquoise'), which includes blue-green minerals such as malachite or chrysocolla from deposits in Zacatecas (Weigand et al. 1977: 16; Weigand 1993: 315), several archaeometric tests were performed. In 2000, X-ray diffraction (XRD) was conducted on a sample of these pieces, which made it possible to confirm that the blue tesserae were made from mineral turquoise, perhaps from the American Southwest, but the green pieces might be malachite from Zacatecas, although there were no samples for comparison to allow us to reach a more definitive conclusion, and the white pieces were perhaps calcite (Neff 2000; Chiari 2000). Later, in 2004, new XRD analyses were performed on a group of 20 tesserae from the disk (Fig. 5a), as well as Rutherford backscattering spectrometry (RBS) (Fig. 5b) and particle induced X-ray emission (PIXE) (Fig. 5c and d). Consequently, it was possible to confirm that the blue pieces were mineral turquoise, while the green and white inlays were also identified as minerals of the turquoise family (Ruvalcaba et al. 2005: 105–109). Finally, in 2008, the most recent study of the composition of the disk's tesserae was conducted using X-ray fluorescence (XRF) *in situ* (Fig. 6), ultraviolet fluorescence (UVF) and infrared imaging (IR) (Fig. 7), comparing them with mineral turquoise samples from Arizona, New Mexico, and Nevada, as well as with 'turquoise substitute'(chrysocolla) from deposits at Chalchihuites in Zacatecas. Based on the UVF and IR analyses, a notable similarity was detected in the characteristics of the minerals of the tesserae of the entire disk.

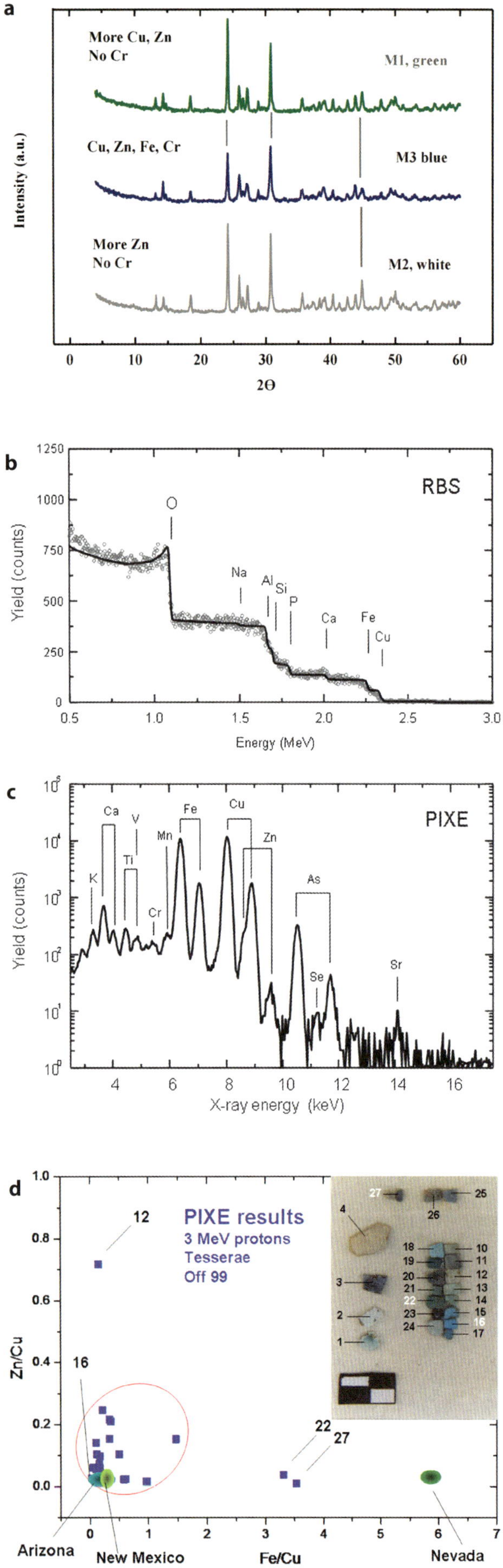

Figure 5. (a) XRD, (b) RBS, and (c–d) PIXE spectra.

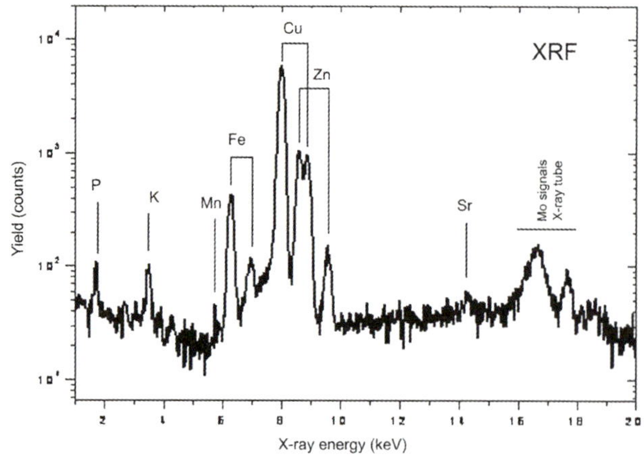

Figure 6. XRF spectrum.

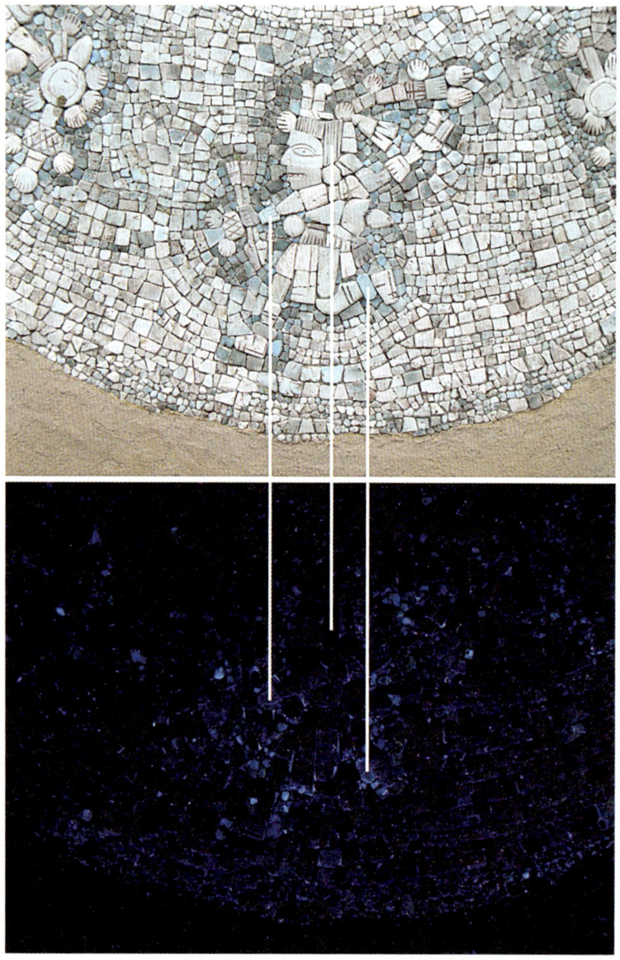

Figure 7. Detail of the disk in (a) visible light and (b) UV light.

Therefore, based on XRF measurements of more than 300 larger tesserae, it was possible to determine that the majority of them displays a spectrum similar to that of the mineral turquoise from Arizona and New Mexico, based on their copper, iron, zinc and arsenic content (Ruvalcaba et al. 2008). This implies that practically no 'turquoise substitute' was used in the manufacture of the disk and that the selection of materials was carefully made, which corresponded, in this case, to long-distance exchange routes. The type of materials from which the pieces were made, the different shapes that they displayed, and the fact that almost all of them had bevelled edges that allowed for a perfect fit, suggested that they were parts of a mosaic.

Assembling the piece

Through the superimposition of the drawings on the acetates from the micro-excavation, it was possible to determine that the piece had a roughly circular shape, measuring approximately 38 cm in diameter. Similarly, it was possible to note a concentration of dark pieces in the centre, followed by a lighter coloured band. In this last section, there were some pieces that stood out because of their larger size (up to 1 cm), as well as for their shape and their incised designs, which suggested that they were representations of human figures.

Based on this idea, the elements that were believed to form part of the figures were separated. They were moved to a backing covered with black cloth, where they were placed in their consecutive positions. Reassembling them was carried out through two complementary processes: (1) a survey of pre-Columbian figurative representations, as well as specialised iconographic studies, which helped in the identification of the figure attributes as well as the position and way they were often held; (2) spatial analysis of the pieces, whose position, proximity and ease of assembly with other elements made it possible to know how they originally fitted together. When the figures were completely defined, each of them was attached to rice paper with an emulsion of polyvinyl acetal (Fig. 8). It was only necessary to replace miniscule missing parts: the nasal septum of one figure and two sandals of two others (one from each figure), for which pieces recovered from the sifted soil were used or which formed part of the background of the disk.

An acrylic plaque was chosen since it is a stable, inert material, on which to place the mosaic. The idea was not to reconstruct the prehispanic technique given that the original backing and adhesive materials are unknown. The acrylic was covered with rice paper, stretched and attached to the back, on which were affixed the darker central tesserae, the band of lighter pieces, and the figures that had been assembled in their corresponding position and order. The rest of the pieces were placed around the latter, by quadrants, and without leaving any open spaces, bordering the figures with darker turquoise pieces. Later, the materials from the sieved soil (4%) were reintegrated. The same polyvinyl acetal used to join the fractured pieces was also used here as an adhesive. Finally, the rest of the paper was given an earthen finish.

Description of the piece

The final result of the reassembled piece was a disk composed of inlaid blue, green and white stone pieces, with an overall diameter of 28 cm (Fig. 9). In the middle there is a round area of dark pieces that measures 8.5 cm in diameter, bordered by a lighter band 1.5 cm in width. Finally, there is a wider band (8 cm in width), in which seven male human figures are represented. These figures appear in profile, three facing the west and four facing the east; the torso is shown frontally, with the arms to the sides of the body and the legs also in profile, slightly flexed, perhaps in a position of walking or running. We were able to recognise the following accoutrements (Fig. 10):

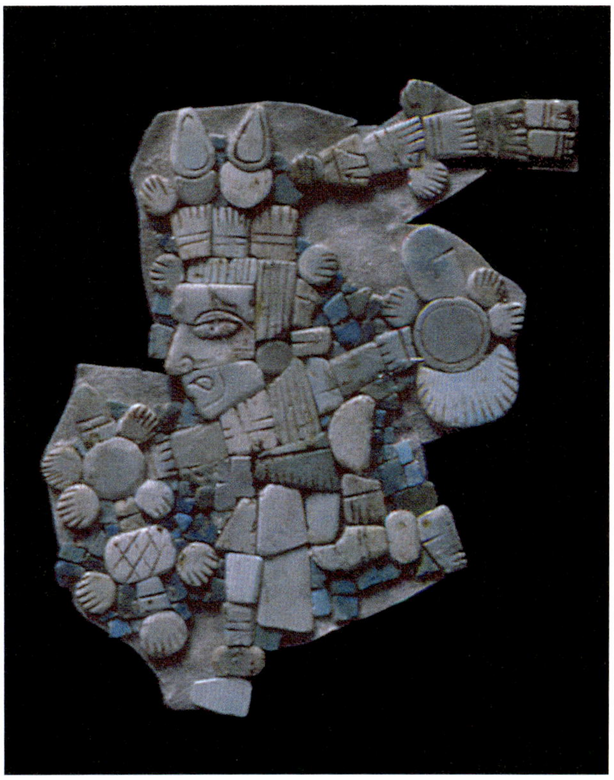

Figure 8. An assembled warrior.

- The figures wear feather headdresses from which hang long, slender plumes, some bearing curved elements and what seem to simulate small tubes. Above the forehead of two of the figures were pairs of crescent forms each with droplet-shaped motifs, with incisions displaying an identical silhouette in their interior (Fig. 10a and g). Instead of feathers, one of the figures has round and spindle-shaped pieces on his head (Fig. 10e), while another has two short bars whose upper ends are bifurcated in opposite directions (Fig. 10c).
- Five wear frontal bands, plain in two cases and the others are decorated in the middle with two parallel diagonal lines.
- All of them have short hair over the forehead and long hair at the back of the head.
- They have round earflares, which in two cases are perforated in the middle, and from which hang ornaments that range from one or several feathers to long elements resembling the crests.
- The faces of the figures are produced by means of two horizontal pieces that are joined where the nose begins. In five cases, the corners of the mouth are turned downward and the mouth is surrounded by semicircular lines. Similarly in the case of two of them the lips are straight and the teeth are visible (Fig. 10c and f). Five figures have open eyes and two half-closed. Only one figure has his face decorated with two parallel lines that go down the forehead, cross the eye, and break at the nose (Fig. 10e).
- They wear rectangular pectorals, five of which are decorated with pairs of parallel lines running approximately from the middle of their four corners towards

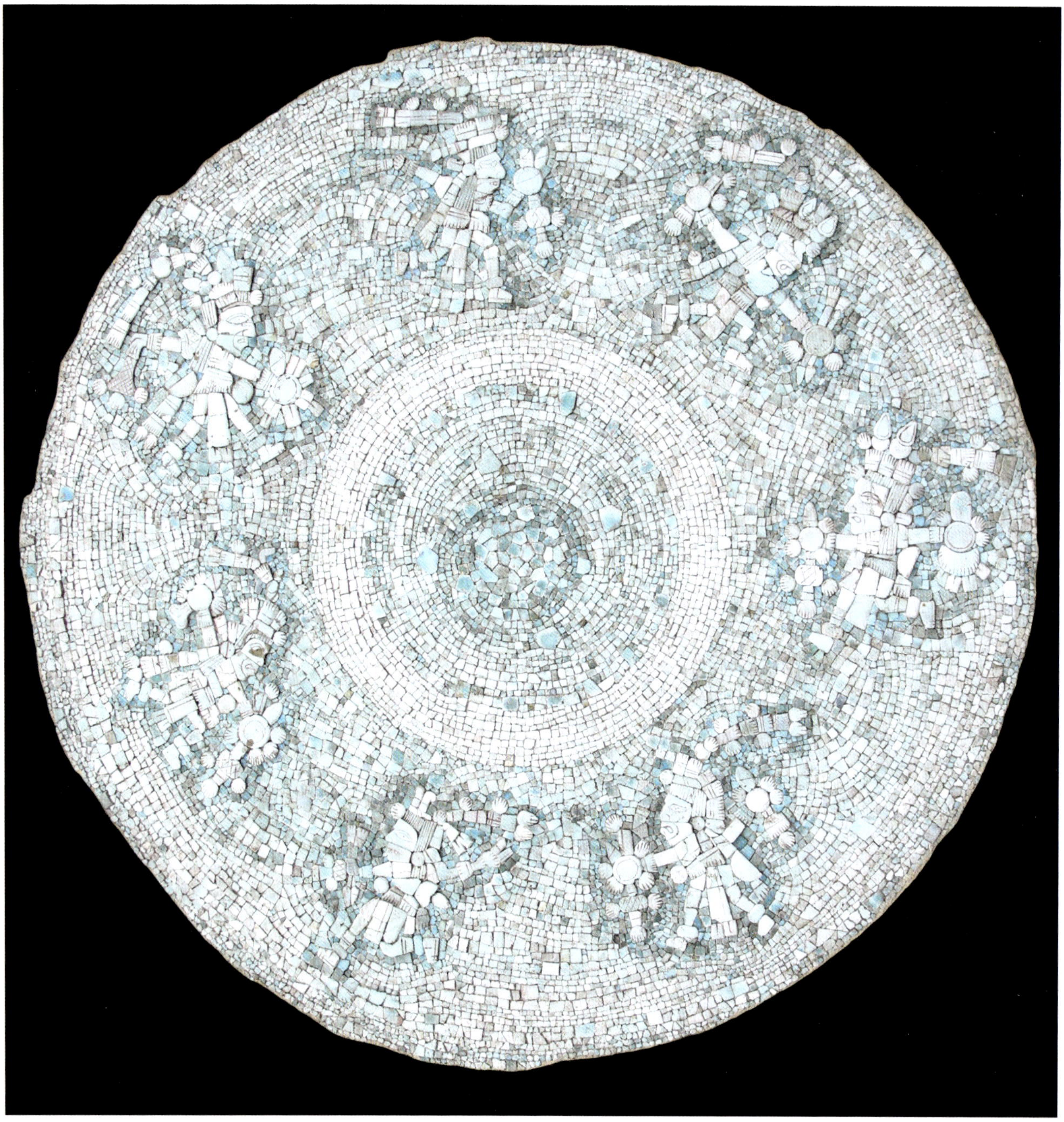

Figure 9. The complete reassembled disk from Offering 99.

the centre (Fig. 10a–e). Another of these ornaments has two curved parallel lines at one of its corners and on its edge there is a series of short incisions (Fig. 10f). On the remaining pectoral it is not possible to see any decoration (Fig. 10g).

- All the figures carry weapons in both hands, including round shields, arrows, net baskets, knives, spear throwers and banners.
- The rest of the attire is composed of a hipcloth, loincloth, wristlets, anklets and sandals. Four of them have a back disk with hanging strips.
- The weapons and the headdresses of the figures are profusely decorated with round feathers.

The disk from Templo Mayor Offering 99 is similar to other mosaic works that are held in museums outside of Mexico, such as the specimens in the British Museum in London, the Naturhistorisches Museum Wien in Vienna, and the Museum of the American Indian Heye Foundation (now the National Museum of the American Indian) in New York. It is believed that the piece in Vienna was among the treasures sent by Hernán Cortés to Europe, so it is likely it was contemporary with the Templo Mayor piece. Similarly, the one in New York was reportedly found in a cave in the Mixteca region (Saville 1922: 69, 71).

Tesserae manufacturing techniques

It has been suggested that the pieces of turquoise mosaic found at Mesoamerican sites were produced at workshops near the deposits of this mineral, such as those at Chaco Canyon,

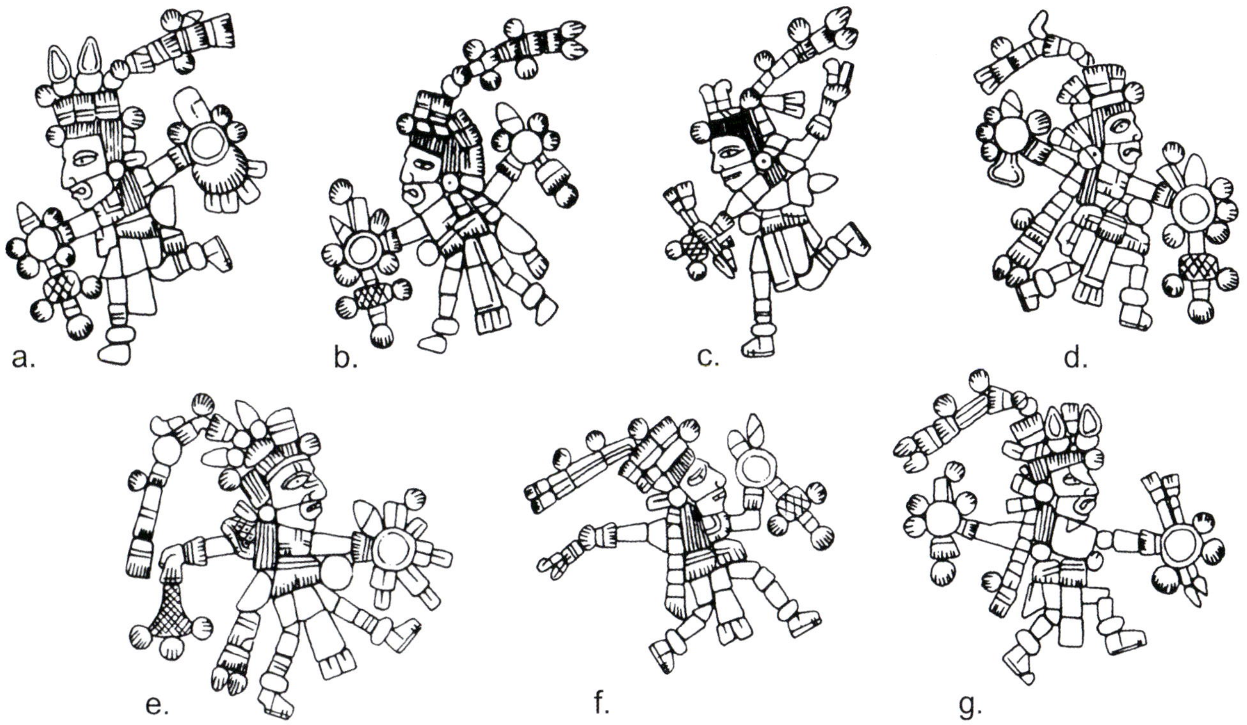

Figure 10. Drawing of the seven warriors (a–g) (drawings by Julio Romero).

New Mexico, and at Chalchihuites, Zacatecas (Weigand et al. 1977: 22; Weigand 1993: 336–337; Neitzel 1995: 403). To verify this, manufacturing traces on the surface, edges, and incised designs of a sample of 90 tesserae were analysed. The methodology followed Velázquez Castro (2007) and Melgar and Bautista (2006), according to which a series of observations was made with the naked eye, with a stereoscopic microscope at low magnifications (×10, ×30 and ×40) and with a scanning electron microscope (SEM) at ×100, ×300, ×600 and ×1000. The information obtained was compared with the database from the project 'La lapidaria del Templo Mayor: estilos y tradiciones tecnológicas' (Lapidary work at the Templo Mayor: styles and technological traditions), which includes an experimental archaeology program in which the modifications identified in archaeological materials have been reproduced, with the possible tools and processes used in the past (Fig. 11); these features have been characterised through the methods of observation mentioned above. The results obtained are given below.

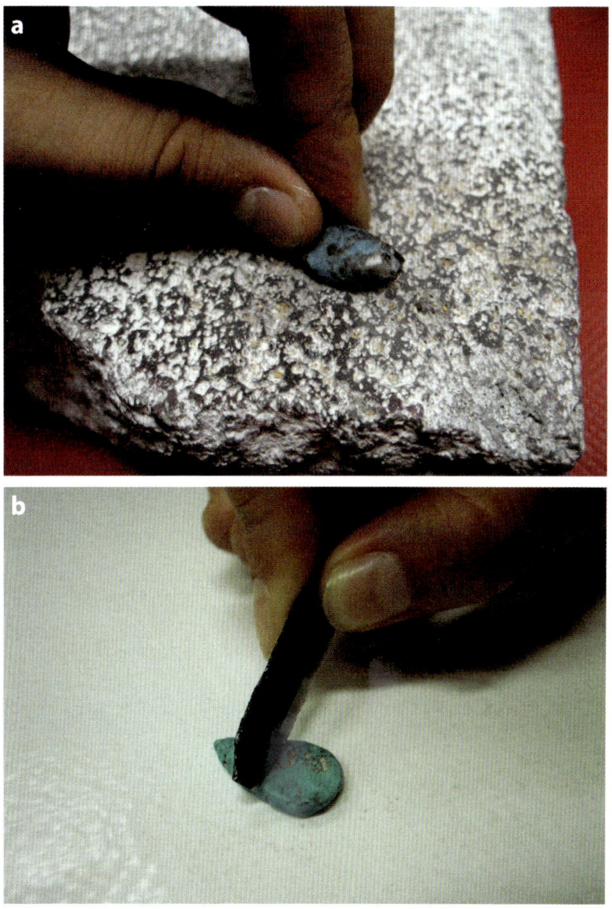

Figure 11. Experimental archaeology in turquoise: (a) abrasion with basalt and (b) cutting with obsidian blades.

Surfaces

Fifty samples were reviewed in which well-defined, straight scratches could be seen with the naked eye and at low magnifications (Fig. 12a and b). Observations with the SEM made it possible to observe straight bands measuring 100 μm in thickness, similar to those produced by abrasion with basalt rocks (Fig. 12c and d). There were no traces of polishing in some of the pieces; however, the fact that some of them have a lustre and that their traces of abrasion are softened suggest that these pieces were burnished with a soft material, perhaps leather or a textile.

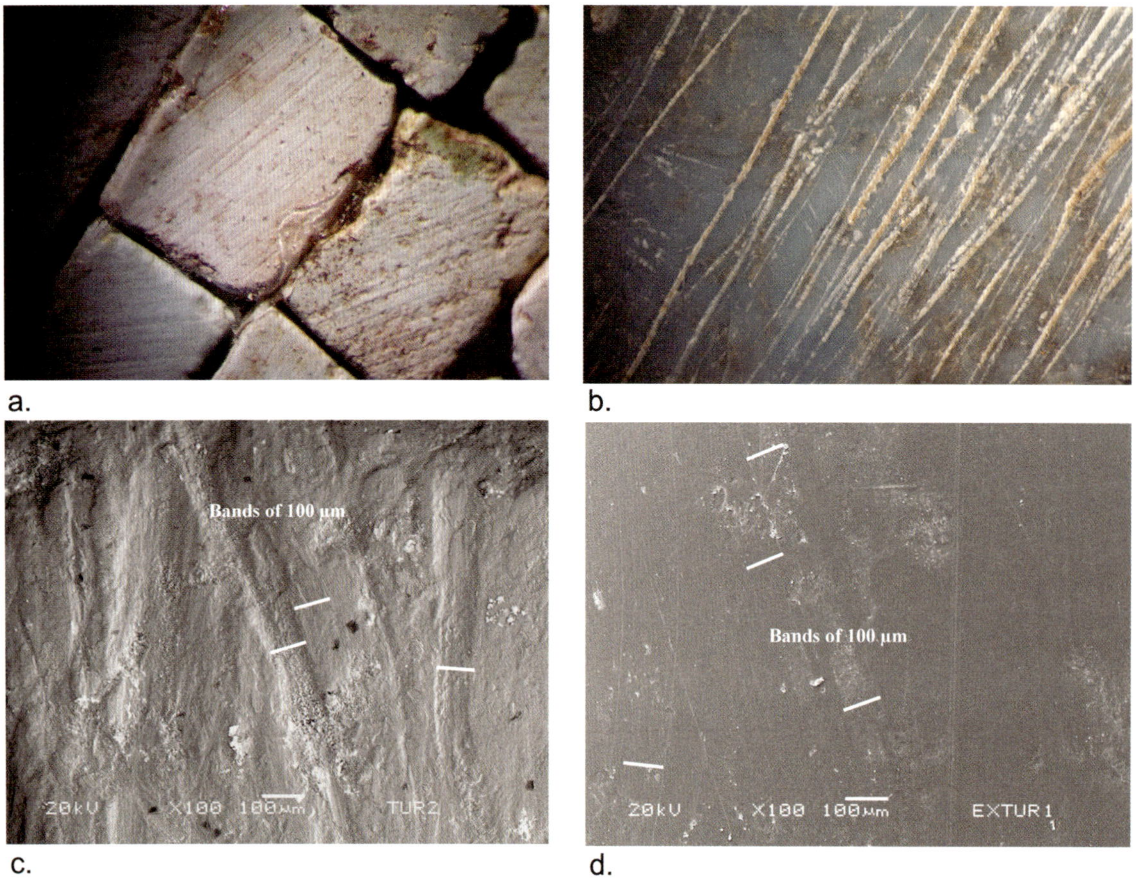

Figure 12. Analysis of abrasion: (a) archaeological surface and (b) experimental surface with stereomicroscope (×10); (c) archaeological surface and (d) experimental surface with SEM (×100).

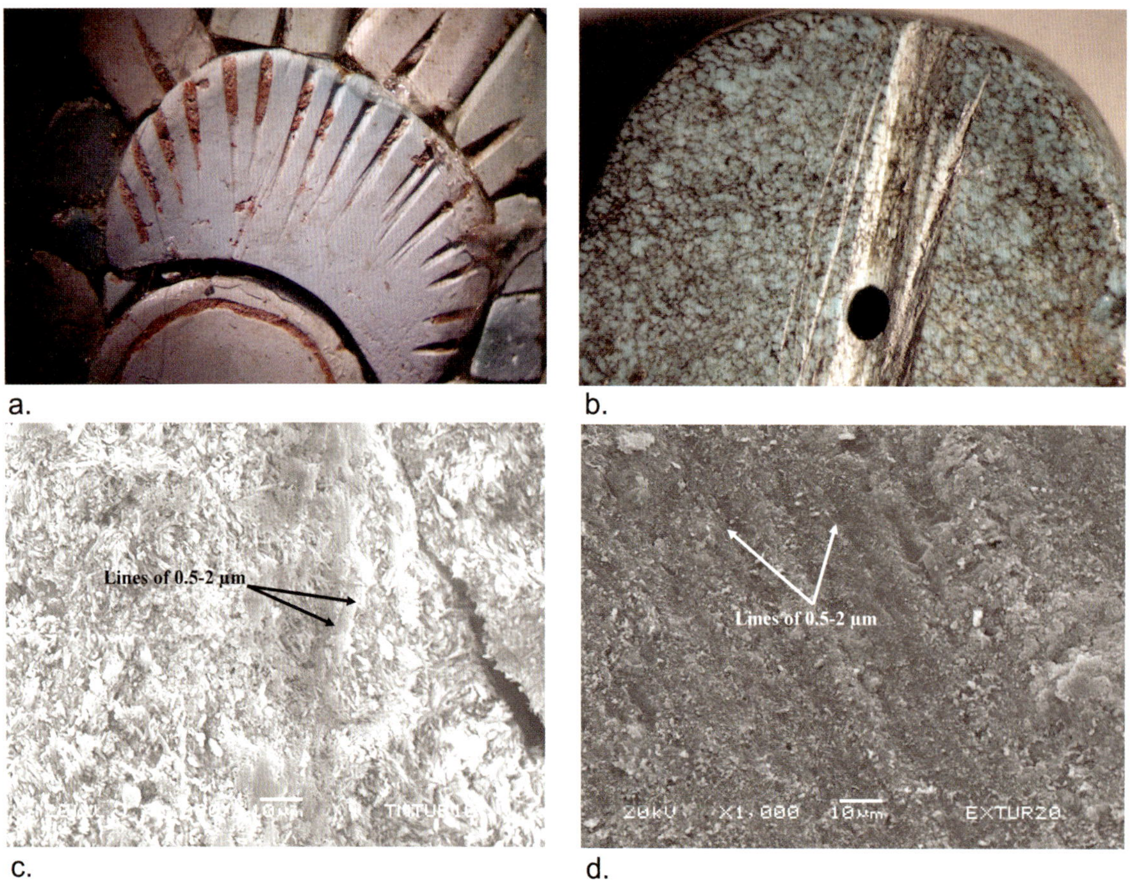

Figure 13. Analysis of cuts and incisions: (a) archaeological marks and (b) experimental marks with stereomicroscope (×10); (c) archaeological marks and (b) experimental marks with SEM (×1000).

Edges and incisions

In 40 elements a series of fine straight lines was visible with the naked eye and at low magnifications (Fig. 13a and b). In the SEM, fine sinuous lines could be seen between 0.5 and 2 µm wide; these traits coincide with those produced by the use of sharp obsidian tools to cut or produce incised designs (Fig. 13c and d).

The marks on the disk from Offering 99 do not coincide with those identified for the workshops at Chalchihuites and Pueblo Bonito. In the case of Chalchihuites, both the archaeological finds and the analysis of manufacturing marks attest to the use of rhyolite tools for abrasion (Fig. 14a) and flint for cutting (Fig. 14b) and polishing (Maldonado 2007). For Pueblo Bonito, the tools found at the site are sandstone abrasion tools (Fig. 14c) and obsidian and flint cutting tools (González and Olmedo 1986: 107–110; Neitzel 1995: 403–405; Peregrine 2001: 41–43; Mathien 2001: 105–110). The use of basalt for abrasion, obsidian for cutting, and the absence of polishing have been found in shell and lapidary pieces of clear Mexica manufacture at Tenochtitlan (Velázquez 2007; Melgar and Solís 2009).

Symbolism of the mosaic from Offering 99

To understand the symbolism of the mosaic disk from Templo Mayor Offering 99, it was necessary to unravel the associated meanings of each one of its constituent elements, as well as the contextual relationships within the deposit in which it was found. In the first case, it was necessary to identify the figures in the mosaic, for which we proceeded to search for the different elements of their insignia in pictographic and sculptural representations from pre-Columbian and early colonial times. In this way we hoped to find clues to their identity and character.

Character of the figures on the mosaic disk

One of the first elements that attracted our attention was the strong bellicose nature of the figures, given that all of them bear weapons and back disks, common in images of warriors from several parts of Mesoamerica. Similarly, the abundant, round feathers they display are those that were used to cover prisoners captured in battle and destined to die on the sacrificial stone. In some elements of the garb we were able to find references to specific deities. Therefore, the droplet-shaped designs on the headdress of two of the warriors are clearly the *oyohualli* worn by Tlahuizcalpantecuhtli (Códice Borbónico 1991: 9; Códice Vaticano A 1996: 21v) (Fig. 15a). The two vertical bars with bifurcated ends over the forehead of another one of the figures recall the motif that tops the diadem used by Huitzilopochtli in some images (Códice Borbónico 1991: 34 and 36) (Fig. 15b). The net basket is one of the frequent attributes of Mixcoatl (Códice Borbónico 1991: 33; Códice Vaticano A 1996: 49r; Códice Telleriano-Remensis 1964: VIII; Códice Magliabechi 1996: 42r) (Fig. 15c). The strange headdress composed of circles and spindle-shaped pieces perhaps refers to ears of corn and plant motifs characteristic of the maize god (Spranz 1993: 62) (Fig. 15d). The rectangular pectorals decorated with pairs of parallel lines perhaps represent the surface of the earth on which the four cardinal directions are marked, a motif known as the quincunx (Fig. 15e). Of the remaining insignia (bands over the forehead, earflares, hipcloth, loincloth and sandals), either we could not find clear parallels in representations or else they are worn by so many and such diverse deities that their presence was not a significant feature, at least in terms of the identity of the figures.

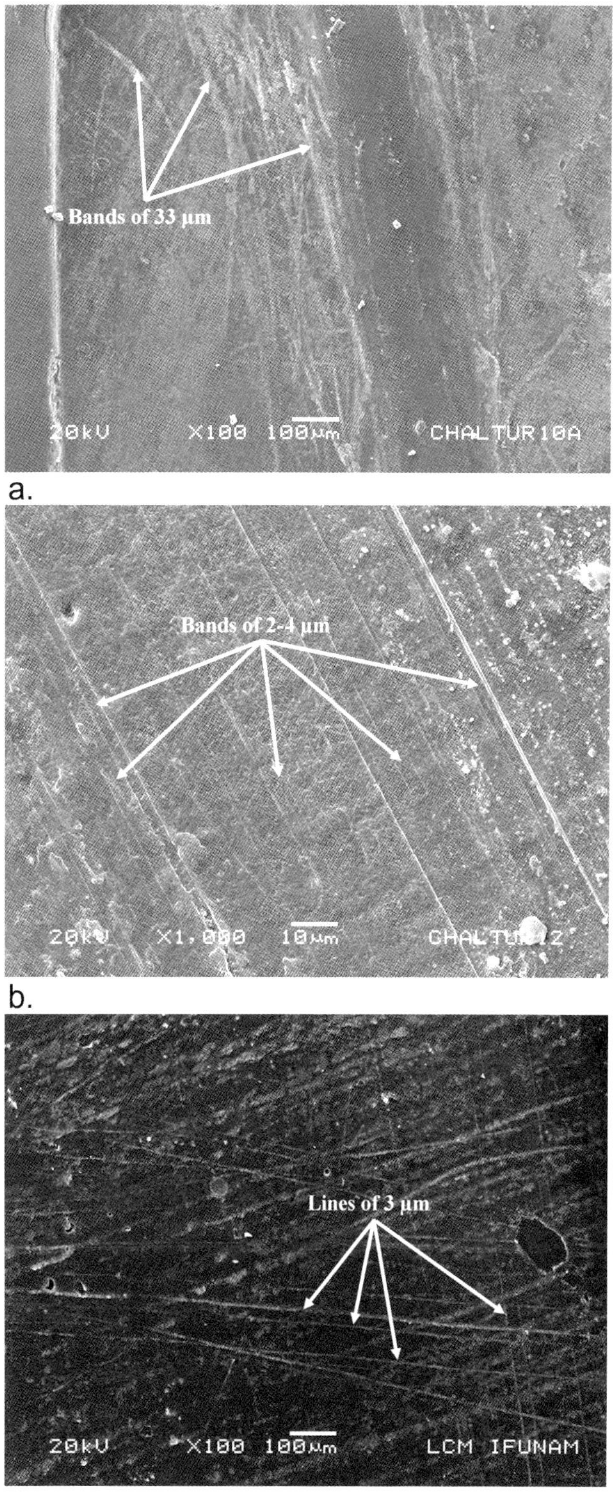

Figure 14. SEM analysis of archaeological pieces from workshops near sources: (a) surface and (b) edge from Chalchihuites (×100 and ×1000 respectively); (c) surface expected from Chaco Canyon (×100).

As for the face paint – which undoubtedly represents the incised designs on the faces of these figures – apart from not

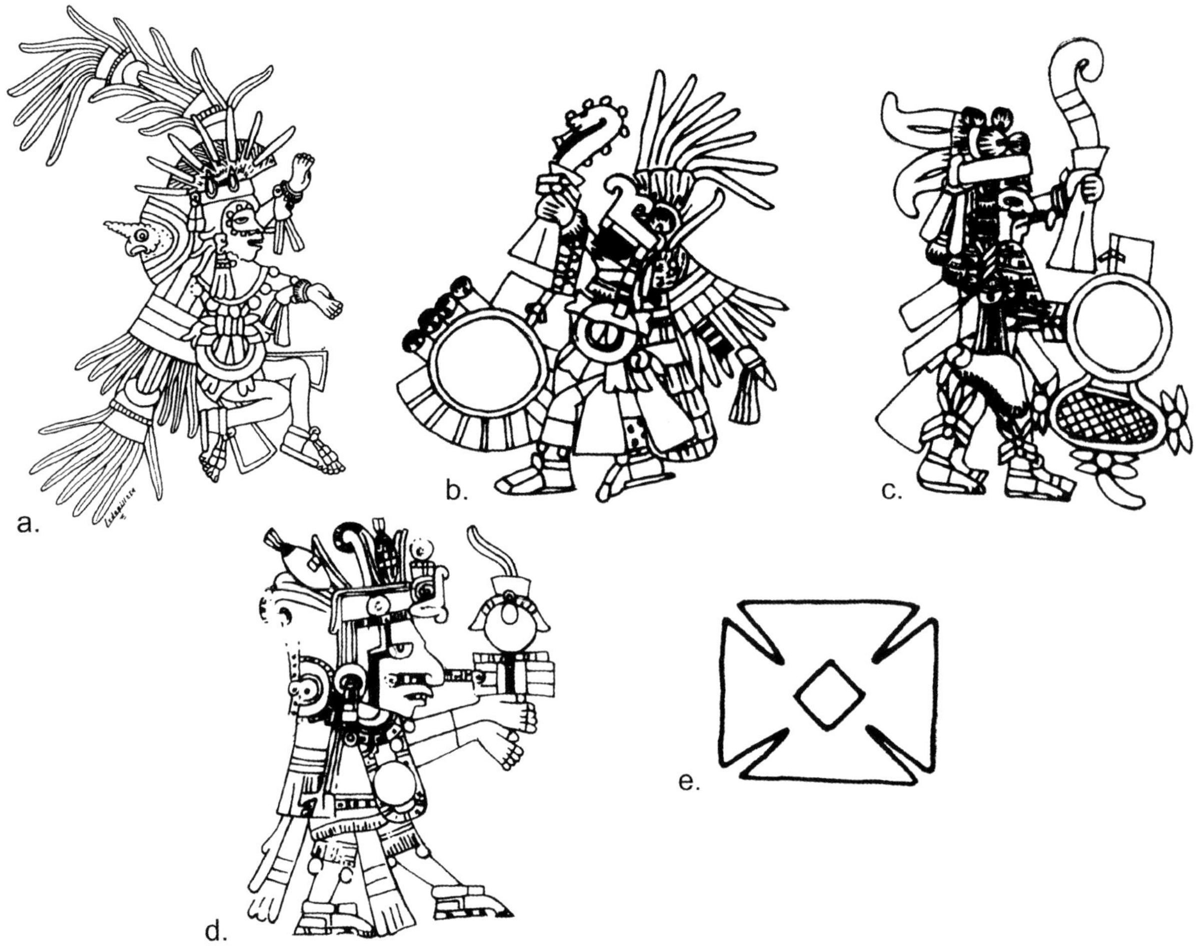

Figure 15. Images of deities and motifs from codices: (a) Tlahuizcalpantecuhtli (Códice Borbónico 1991: 9); (b) Huitzilopochtli (Códice Borbónico 1991: 34); (c) Mixcoatl (Códice Borbónico 1991: 33); (d) maize god (Spranz 1993: 14); (e) quincunx (Códice Borbónico 1991: 34) (drawings by Julio Romero and Manuel Urdapilleta).

knowing their colours, it is unclear whether or not the marks on the faces of two of the figures were made to simulate coloured areas. Similarly, it was not possible to identify the design that consisted of two parallel lines running down the forehead of one of the figures, crossing the eye and ending at the nose.

As for the headdresses with crests resembling those worn by the figures on the turquoise disk, these are not common in images from codices from the Basin of Mexico – where one of the few gods who uses them is Tlahuizcalpantecuhtli (Códice Borbónico 1991: 9) – instead they are worn by many and diverse deities in Mixteca-Puebla style documents (Chalchiuhtlicue, Chantico, Centeotl, Itztlacoliuhqui, Mixcoatl, Patecatl, Quetzalcoatl, Tezcatlipoca, Tlazolteotl, Tlaloc, Tlahuizcalpantecuhtli, Mictlantecuhtli, Tonacacihuatl, Tonatiuh, Huehuecoyotl, Xochipilli, Xochiquetzal, and Xolotl) (Spranz 1993: 35–36, 52, 63, 101, 121, 131, 149, 192, 214, 233, 253, 267, 311, 322, 342, 387, 408, 423).

Therefore, the attributes worn by the figures on the turquoise disk from Templo Mayor Offering 99 identify them as warriors and link them with three important gods of war: Huitzilopochtli, Mixcoatl and Tlahuizcalpantecuhtli. The first of these is none other than the Mexica war god and their patron deity; the second is mentioned in some sources as the creator of warfare (Historia de los mexicanos por sus pinturas 1985: 37); and the third is the warrior who shoots the Fifth Sun, after its creation, with an arrow to force it to move in the sky (Leyenda de los Soles 1992: 122).

Another element that these deities share in common is that they are stars. Huitzilopochtli is a Mexica solar god; Tlahuizcalpantecuthli is Venus as the Morning Star, responsible for preparing the way for the Sun at dawn; and Mixcoatl is described as a star (Códice Telleriano-Remensis 1964: VIII). Thus the turquoise disk bears three of the main star warriors of the Mexica cosmos and we believe it is highly likely that the remaining figures represent stars, the prototypes of men-of-arms whose important cosmic mission was to feed the Sun and the Earth with their blood and hearts. In this sense, it should not be forgotten that the *centzonmimixcohua* (the so-called 400 stars of the north) were created at the beginning of time to make war and to feed the Sun, and warriors who were immolated on the sacrificial stone were attired in their insignia. Given the failure of these deities, because they began to spend their time drinking pulque, seducing women and hunting animals, their five younger brothers – including Mixcoatl – defeated them in battle and sacrificed them (Anales de Cuauhtitlán 1992: 123). The fact that one of the figures wears a headdress that perhaps refers to that of the maize god is not strange in this context, given that in Mesoamerican thought the planning and germination of maize was equated with the journey of the stars through the interior of the earth (Sprajc 1996).

Relationship of the piece to other elements from Offering 99

Although it seems clear that the figures on the turquoise disk are star warriors from the Mexica cosmos, we believe that some of the attributes they carry are comprehensible only by analysing the locations and relations with other elements presented by the disk within the offering in which it was deposited. The disk was placed at the bottom of the oblation, together with burned calcareous materials – perhaps shells – on a layer of mortar that formed part of the construction fill of the Templo Mayor platform (Carrillo n.d.). This was followed by a layer of unmodified marine materials (mollusc shells, corals and sponges), a necklace of *Oliva* shell pendants, a large quantity of projectile points and six greenstone beads, three of which were found very close to the turquoise disk (Carrillo n.d.). On top of this were placed a sawfish and five turtle carapaces, of which the tail of the first and one of the second were found on top of the disk, which in turn was covered by round copper rattles. These were then covered with 36 flint knives, two human figures made of copal resin and projectile points (Carrillo n.d.; Valentín and Zúñiga 2007). On the uppermost superficial level of the offering were placed two golden eagles, as well as the skins of a roseate spoonbill and a hummingbird. These latter were found associated with the mosaic disk (Valentín and Zúñiga 2007). Finally, the elements of the offering were covered with red pigment (Carrillo n.d.). It is worth highlighting that the vast majority of elements deposited in Offering 99 are small, including the animals, which were juveniles (Carrillo n.d.; Valentín and Zúñiga 2007).

Research conducted to date on the meaning of Mexica offerings has demonstrated that some groups of gifts to the gods were placed in particular locations in offerings in order to recreate specific levels of the cosmos. Thus, the placement of a bed of sand at the bottom of deposits, along with greenstone beads, copper rattles and unmodified marine materials was a way of representing the Underworld, the lowest of the three levels in the Mexica conception of the universe; this was regarded as the interior of the earth, connected to underground water and seas (López Luján 1993: 240). On the other hand, the surface of the earth was recreated by the placement, on top of the aforementioned elements, of animal skins related to the *cipactli* (crocodiles, serpents, turtles and swordfish, among others), a telluric creature that floated on top of the ocean waters (López Luján 1993: 251–252).

Based on the above, we can infer that the turquoise disk from Offering 99 was located at the deepest part of the Underworld, because it was placed at the bottom of the deposit and covered by marine materials and animals that recreated the terrestrial level. The presence of star warriors depicted on this piece in the Underworld can be seen as an allusion to the cyclical journey that it was believed that celestial bodies made to the interior of the earth, during which they died and were reborn. Thus, the presence of the quincunx on the pectorals of some of the figures becomes meaningful because this motif has been identified as an element that marks the completion of temporal cycles (Klein 1976: 196). It is also interesting that the number of figures is seven, a number signifying good fortune that alluded to corn and food (Sahagún 1989: 102), because stars are celestial fires that periodically descend to the Underworld to fertilise the corn. Furthermore, the fact that the disk was crafted from turquoise mosaic is significant because it is a material intimately related to the Sun and to fire (Noguez 1973: 84; Izeki 2008: 28, 41–42). Also significant is that three greenstone beads were placed close to the piece, given that this number is directly associated with the fire god (López 1985: 269).

Two of the most important stars are rendered on the mosaic disk, namely the Sun and Venus. The first is personified as Huitzilopochtli, the regent of the fifth cosmic era, which is not unusual given the location of Offering 99 on the western side of the temple of the war god, in the direction of the universe where the Sun penetrated the kingdom of the dead. The presence of two figures with attributes of Tlahuizcalpantecuhtli perhaps shows the two forms of Venus, the Evening Star and the Morning Star, because this god was 'lord of the morning when the sun comes out and the same lord of that light when it wishes to become dark' (Códice Telleriano-Remensis 1964, 2ª: XIV). However, it strikes us as odd that the Evening Star is represented with an attribute such as the *oyohualli* related to the rebirth of stars and vegetation, which perhaps obeys the general meaning of fertility that the disk seems to convey.

As a result of the above, it is of supreme importance that the remains of a roseate spoonbill (*Ajaia ajaja*) were found on top of the turquoise disk (Valentín and Zúñiga 2007), an animal also known as *quecholli*, which is the name of the bird as well as a calendrical festival that was dedicated to Mixcoatl. Although the overall interpretation of Offering 99 goes beyond the limits of the present study, it seems to us that several of its elements are related to some of the rituals that were carried out during this *veintena* or the 20 days of this ritual period, as well as its general meaning.

Durán (1967: 281) translated the name of the 'month' of *quecholli* as 'sticks or arrows for throwing', which refers to the fact that during this festival large quantities of arrows were made, some of which were offered to Huitzilopochtli, while the others, small in size, were offered to warriors who had died in battle, together with warlike insignia including the paper banner that was carried tied to a dead hummingbird (Sahagún 1989: 157–158). In this connection, it is worth recalling that 226 miniature projectile points were found in the deposit containing the mosaic, as well as a hummingbird skin that was associated with the disk.

During this *veintena*, a ritual hunt was carried out in honour of Mixcoatl, the Chichimec hunting god (Sahagún 1989: 158), in which the participants dressed as the *mimixcohua* –with striped body paint, black paint around the eyes, with mouth outlined, and with feathers on the head – and those who managed to catch animals were singled out with honours similar to those of warriors who captured prisoners in battle (Durán 1967: 75). Living images in the form of sacrificial victims wearing the deity's attire were sacrificed in honour of Mixcoatl, pulque gods (Tlamatzincatl and Izquitecatl), of their respective spouses, who were earth goddesses, as well as captives who were brought with hands and feet bound, and hanging face down (Sahagún 1989: 158–159).

In connection with the symbolism of the mosaic disk, it is interesting to note the interpretation of the month of *quecholli* in the caption of plate VIII of the Códice Telleriano-Remensis: 'This is the feast of the descent of Miquitlatecutli and of Zontemoc and of the others so that they paint it with the adornments of war because he brought it to the world.

One should more correctly say the descent of the devils that were stars so now there are stars in the sky.'

According to a thorough analysis of the *veintena* festivals, Michel Graulich (1990: 358) reaches the conclusion that at *quecholli* the fertility of the earth by the stars was celebrated, specifically by the stars Mixcoatl, the old fire god who in mythological cycles from central Mexico played the role of the setting sun, which, when it penetrated the Underworld, produced its successor and the Morning Star; and the *mimixcohua*, who having failed in their mission to feed the Sun were converted into the first victims of sacrifice.

Conclusions

The materials used to produce the disk are mainly mineral turquoise (which can be determined chemically); the presence of 'turquoise substitute' (or 'cultural turquoise') is minimal. From this it may be inferred that there was a noteworthy selectivity in raw materials and a profound awareness of them, which made it possible at least to mix turquoise from Arizona and New Mexico with other blue-green minerals accessible from sources closer to home. Without doubt this is evidence of the high value of the piece under consideration. Currently it is unknown what exchange routes afforded Mexica society access to turquoise; it has been proposed that the tesserae and beads arrived already manufactured to west Mexico via the Pacific Coast and from there they reached the Central Highlands through trade routes or indirect tribute payments (Harbottle and Weigand 1992: 82). However, it is interesting to note that the manufacturing techniques identified on the tesserae do not coincide with tools known from workshops at Chaco Canyon; instead, there is a technological correlation with pieces made in Tenochtitlan. Furthermore, the imagery of the seven deities corresponds to Mesoamerican iconography and specifically with the Mexica, given that their patron god is depicted. All in all, evidence suggests that this piece was manufactured by Mexica lapidary workshops in Tenochtitlan.

According to our study, the turquoise disk from Templo Mayor Offering 99 represents the journey of stars through the Underworld, during which they fed the earth and completed their temporal cycles in which they died and were reborn. This interpretation is supported by the attributes of the mosaic figures, which identify them as star warriors, particularly the Sun and Venus; the location of the piece within the deposit is the conceptual location of the lowest level of the cosmos; the pectorals with the quincunx motif; and the possible relationship of the deposit with the *quecholli* festival, during which the mythical moment of the creation of the world and the descent of the stars from the sky, which had united opposite elements of the cosmos through war, was commemorated. In this regard, it seems significant that the piece was manufactured from turquoise, which is a material that occurs in sedimentary or metamorphic rocks near or at the contact with igneous intrusives, and that it was placed in an aquatic context. Both the possible presence of the maize god and the number of figures (seven) seem to be allusions to the fertile character of the journey of the stars to the interior of the earth.

Earlier, through the analysis of objects made from mollusc shells, the possibility had been proposed that some of the offerings deposited in the southern half of the platform of the Templo Mayor conveyed metaphors allusive to the death of the Sun, the journey of the Sun and Venus through the Underworld, and the rebirth of both – perhaps conflated with the maize god – as well as images of star warriors and warlike activity carried out in the lowest levels of the cosmos, conceived of as the starry night sky (Velázquez 2000). For us, the appearance of the mosaic disk contributes information that supports earlier hypotheses, because we can see more obviously the stars in the Underworld, perhaps at midnight, at the precise time of their death and resurrection. In our opinion, evidence of this interpretation is the presence of the *oyohualli* in the headdresses of the two figures that we have interpreted as manifestations of Venus, given that this attribute appears in two offerings from the eastern façade of the Templo Mayor in contexts that allude to the death of the Evening Star and the birth of the Morning Star (Velázquez 2000: 152–160). Therefore, while the ascent of the Sun toward the Zenith is represented on the floor and platform of the southern half of the platform of the Templo Mayor, beneath the nocturnal underground part of the star's cycle is captured.

References

Anales de Cuauhtitlán (1992) in Códice Chimalpopoca. Mexico City: Universidad Nacional Autónoma de México (UNAM).

Carrillo, J. (n.d.) 'Descripción de la cala I del rescate arqueológico en la Casa de las Ajaracas', Museo del Templo Mayor (unpublished report).

Carrillo, J. (n.d.) 'Reconstrucción del ritual de depositación de la ofrenda 99', Museo del Templo Mayor (unpublished report).

Chiari, G. (2000) Dipartimento di Scienze Mineralogiche e Petrologiche, Torino, Italy, personal communication (November 2000).

Códice Borbónico (1991) Mexico City: Fondo de Cultura Económica.

Códice Magliabecchiano (1996) Mexico City: Fondo de Cultura Económica/Akademische Druck und Verlagsanstalt.

Códice Telleriano-Remensis (1964) Mexico City: Secretaría de Hacienda y Crédito Público.

Códice Vaticano A (1996) Mexico City: Fondo de Cultura Económica/ Akademische Druck und Verlagsanstalt.

Durán, Fray D. de (1967) *Historia de las indias de Nueva España e islas de Tierra Firme.* Mexico City: Editorial Porrúa.

González, C.J. and Olmedo, B. (1986) 'El trabajo de la turquesa', in L. Manzanilla (ed.), *Unidades Habitacionales Mesoamericanas y sus Áreas de Actividad.* Mexico City: Universidad Nacional Autónoma de México (UNAM), 103–114.

Graulich, M. (1990) *Mitos y rituales del* México *antiguo.* Madrid: Ediciones Istmo.

Harbottle, G. and Weigand, P.C. (1992) 'Turquoise in pre-Columbian America', *Scientific American* **266**(2): 78–85.

Historia de los mexicanos por sus pinturas (1985) in *Teogonía e Historia de los mexicanos, tres opúsculos del siglo XVI.* Mexico City: Editorial Porrúa.

Izeki, M. (2008) *Conceptualization of 'Xihuitl': History, Environment and Cultural Dynamics in Postclassic Mexica Cognition.* BAR International Series 1863. Oxford: BAR.

Klein, C. (1976) *The Face of the Earth: Frontality in Two-Dimensional Mesoamerican Art.* New York and London: Garland Publishers Inc.

Leyenda de los Soles (1992) in Códice Chimalpopoca. Mexico City: Universidad Nacional Autónoma de México (UNAM).

López, A. (1985) 'El dios enmascarado del fuego', *Anales de antropología* **22**: 251–285.

López Luján, L. (1993) *Las Ofrendas de Templo Mayor de Tenochtitlan, México*. Mexico City: Instituto Nacional de Antropología e Historia (INAH).

Maldonado, A. (2007) 'Tecnologías de producción lapidaria en la región del río Suchil', Escuela Nacional de Antropología e Historia (unpublished report).

Marín, M. (n.d.) 'Restauración de la pieza de mosaico de turquesa de la ofrenda 99 del Templo Mayor', Museo del Templo Mayor (unpublished report).

Mathien, F. (2001) 'The organization of turquoise production and consumption by the prehistoric Chacoans', *American Antiquity* **66**: 103–118.

Matos, E. (1988) *The Great Temple of the Aztecs*. London: Thames and Hudson Ltd.

Melgar, E. and Bautista, P. (2006) 'Análisis de mosaicos de piedra verde incrustados en dos cráneos humanos a través de sus huellas de manufactura', in D. Mendoza, J. Arenas, V. Rodríguez and J. Ruvalcaba (eds), *La ciencia de materiales y su impacto en la arqueología*, vol. III. Mexico City: Academia Mexicana de Ciencia de Materiales, 161–175.

Melgar, E. and Solís, R. (2009) 'Caracterización de huellas de manufactura en objetos lapidarios de obsidiana del Templo Mayor de Tenochtitlan', *Arqueología* **42**: 118–134.

Neff, H. (2000) Research Reactor Center, University of Missouri, personal communication (27 July 2000).

Neitzel, J. (1995) 'Elite styles in hierarchically organized societies: the Chacoan regional system', in C. Carr and J. Neitzel (eds), *Style, Society, and Person: Archaeological and Ethnological Perspectives*. New York: Plenum Press, 393–417.

Noguez, X. (1973) 'La diadema de turquesa (xihuitzolli) y las alianzas de señoríos prehispánicos. Acercamiento iconográfico', *Mesa Redonda de la Sociedad Mexicana de Antropología* **XIII**: 83–94.

Peregrine, P. (2001) 'Matrilocality, corporate strategy and the organization of production in the Chacoan world', *American Antiquity* **66**: 36–46.

Ruvalcaba, J., Bucio, L., Marín, M. and Velázquez, A. (2005) 'Estudio por DRX y haces de iones de teselas de un disco de turquesas del Templo Mayor de Tenochtitlan', in D. Mendoza, J. Arenas and V. Rodríguez (eds), *La ciencia de materiales y su impacto en la arqueología*, vol. II. Mexico City: Academia Mexicana de Ciencia de Materiales, 95–111

Ruvalcaba, J., Melgar, E. and Solís, R. (2008) 'El disco de turquesa de la ofrenda 99 del Templo Mayor de Tenochtitlan', Museo del Templo Mayor (unpublished paper).

Sahagún, Fray B. de (1989) *Historia General de las cosas de Nueva España*. Mexico City: Alianza Editorial Mexicana and Consejo Nacional para la Cultura y las Artes.

Saville, M. (1922) *Turquoise Mosaic Art in Ancient Mexico*. New York: Museum of the American Indian Heye Foundation.

Sprajc, I. (1996) *Venus, lluvia y maíz*. Mexico City: Instituto Nacional de Antropología e Historia (INAH).

Spranz, B. (1993) *Los dioses en los códices mexicanos del grupo Borgia*. Mexico City: Fondo de Cultura Económica.

Valentín, N. and Zúñiga, B. (2007) 'Informe de la identificación de la fauna procedente de las ofrendas 99 y 100 del Templo Mayor de Tenochtitlan', Museo del Templo Mayor (unpublished report).

Velázquez, A. (2000) *El simbolismo de los objetos de concha de las ofrendas del Templo Mayor de Tenochtitlan*. Mexico City: Instituto Nacional de Antropología e Historia (INAH).

Velázquez Castro, A. (2007) *La producción especializada de los objetos de concha del Templo Mayor de Tenochtitlan*. Mexico City: Instituto Nacional de Antropología e Historia (INAH).

Weigand, P. (1993) *Evolución de una civilización prehispánica: arqueología de Jalisco, Nayarit y Zacatecas*. Zamora: El Colegio de Michoacán.

Weigand, P.C., Harbottle, G. and Sayre, E.V. (1977) 'Turquoise sources and source analysis: Mesoamerica and the southwestern U.S.A.', in T.K. Earle and J.E. Ericson (eds), *Exchange Systems in Prehistory*. New York: Academic Press, 15–34.

Authors

- Adrián Velázquez Castro, Museo del Templo Mayor del INAH, Mexico City, Mexico
- María Eugenia Marín Benito, Coordinación Nacional de Restauración del INAH, Mexico City, Mexico
- Emiliano Ricardo Melgar Tísoc, Museo del Templo Mayor del INAH, Mexico City, Mexico
- Reyna Beatriz Solís Ciriaco, Museo del Templo Mayor del INAH, Mexico City, Mexico
- José Luis Ruvalcaba Sil, Instituto de Física de la UNAM, Mexico City, Mexico

Culture and Collections

Turquoise in the Aztec imperial world

Frances Berdan

Introduction

This article focuses on the role of turquoise in the Aztec imperial world. It emphasises the practical side of turquoise in Aztec life, especially its place in Aztec economic, social and political life (in some measure of contrast with the articles in this volume contributed by Karl Taube and Patrick Johansson K., pp. 117–134 and 135–144 respectively, who address the more symbolic, abstract and metaphorical aspects of these treasured blue stones). Turquoise was a highly valued and symbolically laden material, reserved for use by the aristocracy, and this article addresses the manner in which the Aztec elite acquired and used both raw turquoise and turquoise objects. Worded somewhat differently, this article looks at questions of how turquoise served the needs of the state and members of the nobility, and how (and from where) turquoise was obtained by the imperial Aztecs.

We know about turquoise and related fine stones from extant surviving objects, and also from pictorial depictions and textual descriptions of these revered stones and artefacts in sixteenth-century codices and other written documents. This article relies most heavily on the historical sources, pictorial and textual, pre-Columbian and colonial. Most significantly, depictions of turquoise tribute were recorded in imperial tribute lists such as the *Matrícula de Tributos* (1982) and the Codex Mendoza (Berdan and Anawalt 1992), and illustrations of turquoise objects are found in numerous codices such as the *Primeros Memoriales* (Sahagún 1993), Codex Boturini (Johansson 2007), Codex Telleriano-Remensis (Quiñones Keber 1995), Códice Borbónico (Anders et al. 1991), and the Codex Magliabechiano (Boone 1983). Turquoise stones and objects were also described in several textual documents, including monumental colonial works by Bernardino de Sahagún (1950–1982), Francisco Hernández (1959) and Alvarado Tezozomoc (1975). While the information on turquoise gleaned from these several works is generally dispersed and somewhat scant in places, combined they reveal intriguing details of the use of this luxury in Aztec life.

The Aztec world

But first, a few words about the Aztecs and their imperial world. The Aztecs (or Mexica as they called themselves) were latecomers to central Mexico. They migrated to the Basin of Mexico from somewhere vaguely to the north (from currently unlocated Aztlan, Chicomoztoc and other stopping points) beginning in the twelfth century. Guided by their patron god Huitzilopochtli, they finally established their destined city of Tenochtitlan in AD 1325. The city began on a small, unpromising island in Lake Texcoco, one of a network of five linked lakes that dominated the Basin of Mexico at that time. Today, the ancient capital and its neighbouring cities are completely overlain by a sprawling Mexico City.

Over the next 200 years or so, Tenochtitlan was to become the greatest metropolis in all of Mesoamerica. It headed a political and military alliance with two other important cities, Texcoco to the east and Tlacopan to the west (called, not surprisingly, the Triple Alliance; see Fig. 1). This alliance embarked on a programme of military conquest and economic control, culminating in the most extensive empire in the history of Mesoamerica. The empire extended from some 80 km north of Tenochtitlan to the present-day Mexico–Guatemala border, and from the Pacific coast to the Gulf coast (Fig. 2).

While the Mexica spearheaded this extensive empire, a great deal of their cultural, social and political lives was inherited from earlier civilisations; they chose what they wished from the past (even ritually burying ancient artefacts), and placed their own stamp on already well-established institutions and traditions. Politically, they entered a world that was organised into a multitude of city-states, and they added their own, although their city far surpassed any other in size and splendour numbering 200,000–250,000 inhabitants by the time of the Spanish arrival. Economically, they developed a more commercialised economy than ever before, drawing on pre-existing ideas and patterns of specialised production, long-distance trade and markets. Socially, they were arranged

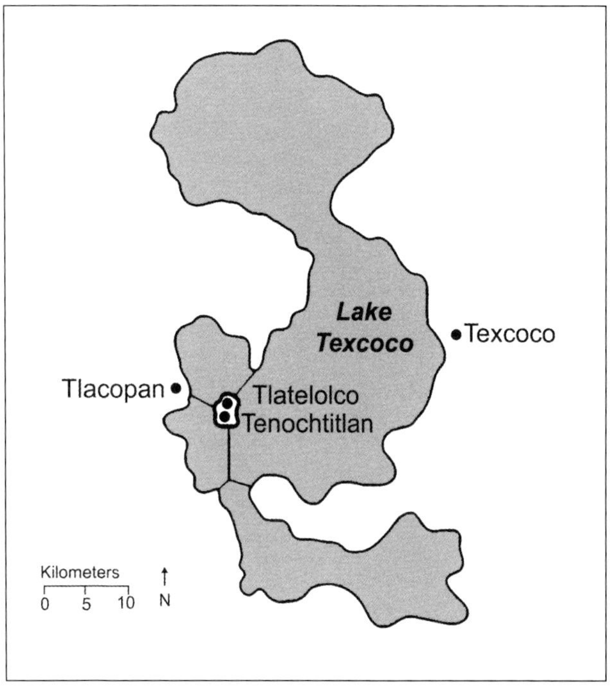

Figure 1. The Aztec Triple Alliance of Tenochtitlan, Texcoco and Tlacopan. Over time, Tenochtitlan became the dominant partner (map by Jennifer Berdan).

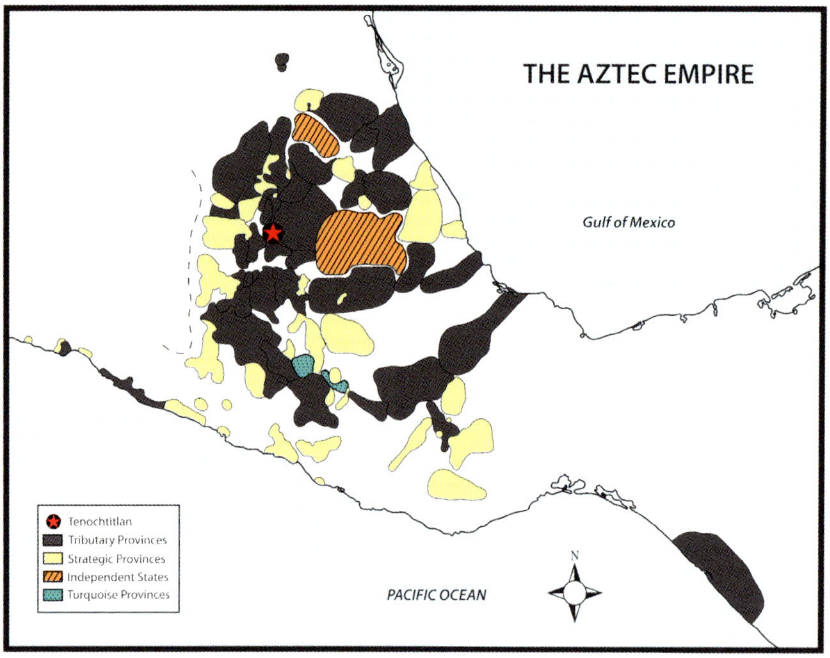

Figure 2. Geographical extent of the Aztec Empire, composed of tributary and strategic provinces (map by Jennifer Berdan).

into an intense, well-defined social hierarchy that visually and effectively separated nobles from commoners. Religiously, they adopted god after god in their encounters and conquests, adding their own paramount patron deity Huitzilopochtli to the mix. And they engaged in almost constant warfare and forged an empire of unsurpassed dimensions, commanding control over vast resources through incessant tribute demands. These were the hallmarks of Aztec life, in which turquoise played a significant role.

Turquoise used by the Aztecs

The ethnohistoric sources tell us, in Nahuatl, about several different types of turquoise stones that were used in Aztec or Mexica ornamentation. There was just plain *xihuitl*, common turquoise; *teoxihuitl*, fine turquoise; *matlalxihuitl*, blue turquoise; *uel popoca teuxiuitl*, very smoky fine turquoise; *xiuhtomolli*, turquoise that is half-round, half-flat; *xiuhtomoltetl*, a beautiful green and white stone; and *xixitl*, a soft low-grade turquoise that is flawed and spotted and may represent a misprint for *xihuitl*, which was considered less than perfect in any event (Sahagún 1950–1982, 11: 223, 224; 9: 82; 10: 60; Hernández, 1959, II: 411). The experienced lapidaries probably recognised many more distinctions, such as the *xiuhmatlaliztli* (turquoise-blue obsidian) which may have been turquoise or obsidian. *Tlapalteoxihuitl*, or fine red turquoise, is also somewhat problematical, considering the usual blue or blue-green tones of turquoise stones. These distinctions were based on colour, quality and shape. This is not surprising, in that these and other stones were prepared for use as showy, decorative objects. Of course, whether we would geologically consider all of these types categorically as turquoise is quite another matter. What is important here is what the Aztec or Mexica thought. And they considered that it was these blue and blue-green stones, called in some way *xihuitl* or turquoise, that were worthy of royal and noble attention.

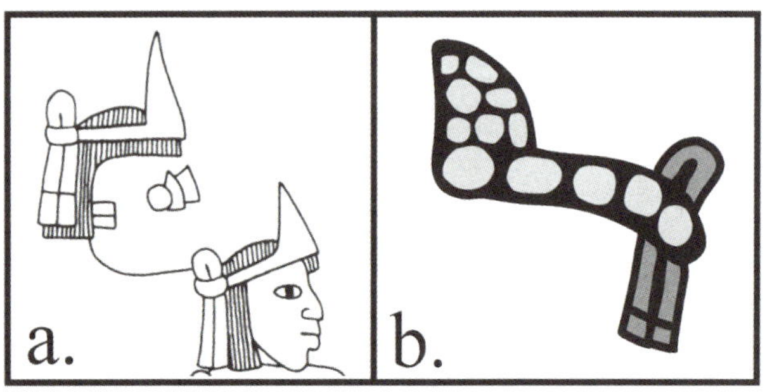

Figure 3. (a) The *xiuhuitzolli* or turquoise diadem of the Mexica ruler (from Berdan and Anawalt 1992, 4: f. 15v). (b) The turquoise diadem depicted as a mosaic (from Sahagún 1993: f. 51r; drawn by Jennifer Berdan).

Objects fashioned with turquoise were mosaics or beads, and included diadems, lip plugs, nose plugs, ear plugs, necklaces, pectorals, armbands, masks, helmets, shields and disks, mirrors, sacrificial knives, and possibly capes with turquoise pieces attached (Olko 2005: 218–223). All of these were of varying degrees of exquisiteness. The lengthy lists of loot obtained by Hernán Cortés and his company, and by prior and subsequent Spaniards, provide an enticing idea of what some of these now-lost objects were like. The inventories include various objects fashioned of turquoise, such as a turquoise necklace for a greyhound; a shield of blue stone mosaic with a gold rim; another similar shield but with a feathered rim; a large duck head of blue stone mosaic work; ear plugs of blue stones; shells with blue stones set around them; rings with turquoises; a cicada with three small turquoises set in the tail; earrings of gold and turquoises; masks of turquoises; a mitre of blue stone mosaic work; a mirror set in blue and red stone mosaic work; some leggings of blue stone mosaic work; ear ornaments to be set in the big head of a crocodile; head armours; and stone mosaic shields (Saville 1920: 8–104).

Worn or wielded by rulers, elites, priests, deities or deity impersonators, such objects were significant markers of

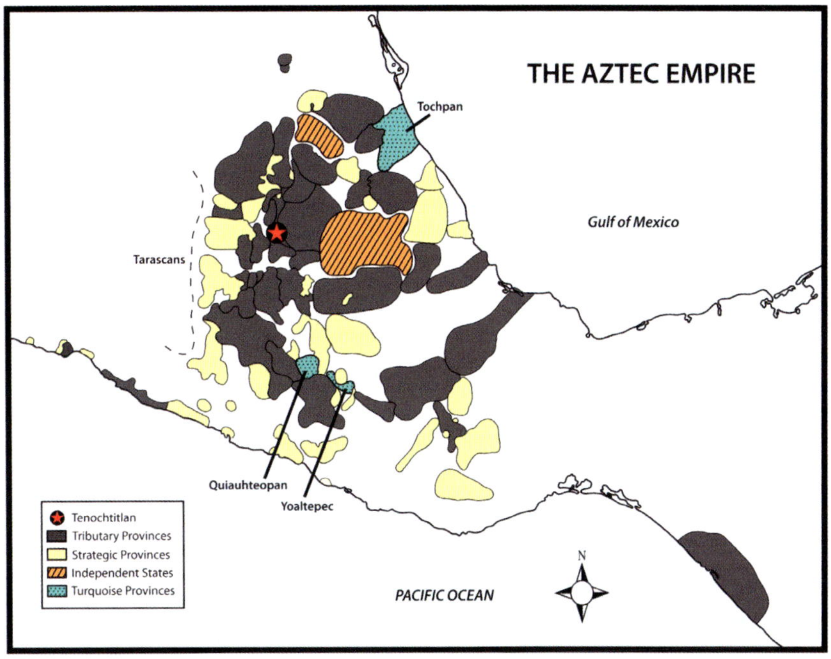

Figure 4. The turquoise-paying provinces of Tochpan, Quiauhteopan and Yoaltepec (map by Jennifer Berdan).

exalted social position and extraordinary, legitimate political supremacy. They were also presented, along with other symbols of luxury and power, as greeting gifts by people seeking favours or asylum (e.g. Bierhorst 1992: 74–75).

Turquoise, the deities and the nobility

The Mexica and their neighbours repeatedly demonstrated their dominance over human and material resources by extravagant displays on their persons, in their palaces, in sumptuous feasting, in the adoration of their gods and in flamboyant religious ceremonies. These displays required large quantities of exotic materials, the most notable of which were tropical feathers, gold, greenstones, and the focus of this article, turquoise. These materials, separately or combined, were fashioned into exquisite and symbolically laden objects destined for use by rulers and other elites, and for the adoration of the many deities.

The deities

As was surely fitting, turquoise objects comprised the adornment of several deities: Huitzilopochtli's image wore turquoise mosaic ear plugs (Sahagún 1950–1982, 12: 52; Sahagún 1993: f. 261r), Paynal had a turquoise nose rod and a turquoise mosaic shield (Sahagún 1950–1982, 1: 3, ill. 2), Cihuacoatl carried a 'turquoise [mosaic] weaving stick' (Sahagún 1950–1982, 1: 11) and Chalchiuhtlicue wore turquoise mosaic ear plugs (Sahagún 1950–1982, 1: 22). Cinteotl wore a turquoise mosaic mask (Sahagún 1950–1982, 9: 80) and carried a turquoise shield (Anders et al. 1991: 27). Xiuhtecuhtli's shield was encrusted with pieces of turquoise and mirror stone (Anders et al. 1991: 26, 28; Sahagún 1950–1982, 9: 30). Yacatecuhtli was adorned with a *xiuhtlalpilli* or turquoise/feather headpiece (Sahagún 1993: f. 262r). Images of Quetzalcoatl include a clearly depicted turquoise jewel in his headdress (Quiñones Keber 1995: ff. 8v, 10r; Anders et al. 1991: 3, 22, 26, 27); that same symbol appears in the headdress of Xolotl (Anders et al. 1991: 16). The Quetzalcoatl array bestowed upon Hernán Cortés consisted of a turquoise mosaic serpent mask, a turquoise mosaic back shield, and a turquoise spear thrower (Sahagún 1950–1982, 12: 11). Idols and priests were adorned with turquoise ear plugs, bracelets, lip pendants and masks during several of the flamboyant monthly ceremonies (Sahagún 1950–1982, 2: 66–166 passim; 12: 81). Turquoise-coloured clothing and accoutrements also commonly adorn participants in these same ceremonies (Sahagún 1993: ff. 250r–254r; Anders et al. 1991: passim). Ritual offerings around Tenochtitlan's Templo Mayor have yielded some turquoise-studded objects, most of which are spatially associated with Huitzilopochtli's temple (López Luján 1994; Izeki 2008). The presence of these turquoises, along with associated flints and skulls, suggests 'that turquoise symbolized the sun, heat, life, sacrifice, and blood' (Izeki 2008: 90).

The ruler

The ruler was a particularly ravenous consumer of turquoise finery. He danced with turquoise-studded leather wrist bands, turquoise-embedded collars, fine turquoise nose rods and fine turquoise and gold lip plugs (Sahagún 1950–1982, 8: 27–28). He wagered fine turquoise (*teuxiuitl*) and fine turquoise and greenstone bracelets in ball games and *patolli*, a game similar to parcheesi (Sahagún 1950–1982, 8: 29–30). When he went to war, his splendid martial attire consisted of costly war shirts, headdresses, back devices, shields, and 'a green stone necklace of round, large, green stone and fine turquoise [*teuxiuitl*] combined' (Sahagún 1950–1982, 8: 33). His persona as legitimate ruler was intimately linked to displays of turquoise, greenstones (*chalchihuitl*) and flowing tropical feathers. To assure appropriate supplies of these exquisite status symbols, the ruler maintained palace artisans, who included lapidaries (Sahagún 1950–1982, 8: 45). In this source, turquoise mosaic-makers are not specifically mentioned, although 'cutters of stones' and 'workers in greenstone mosaic' (*tlatecque,*

a

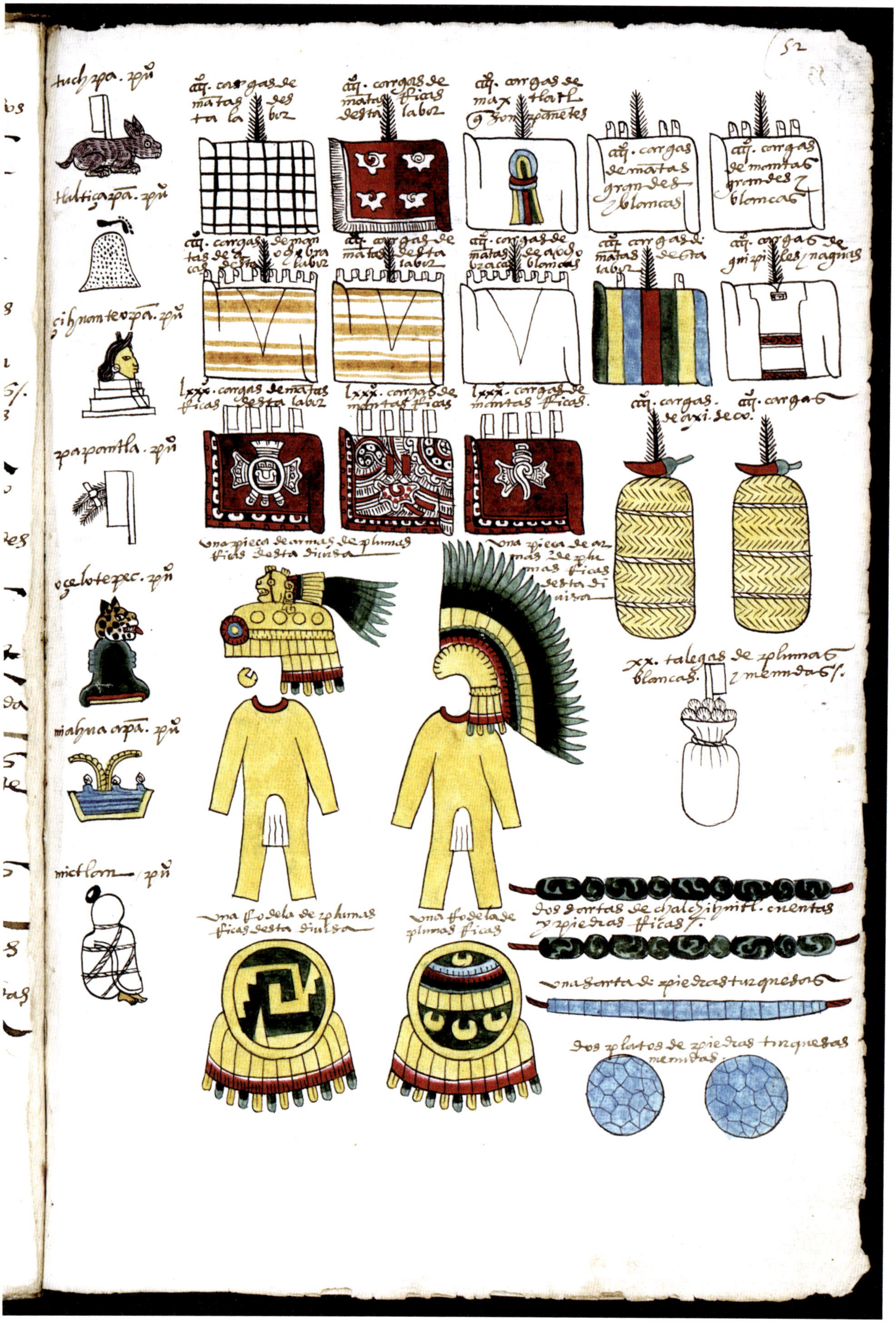

Figure 5. (a) Tribute from the province of Tochpan, on the Gulf coast. The turquoise and greenstones are on the lower right (Codex Mendoza, f. 52 r, Bodleian Library, Oxford). (b) Black and white image with translated captions (from Berdan and Anawalt 1992, 4: f. 52r).

b

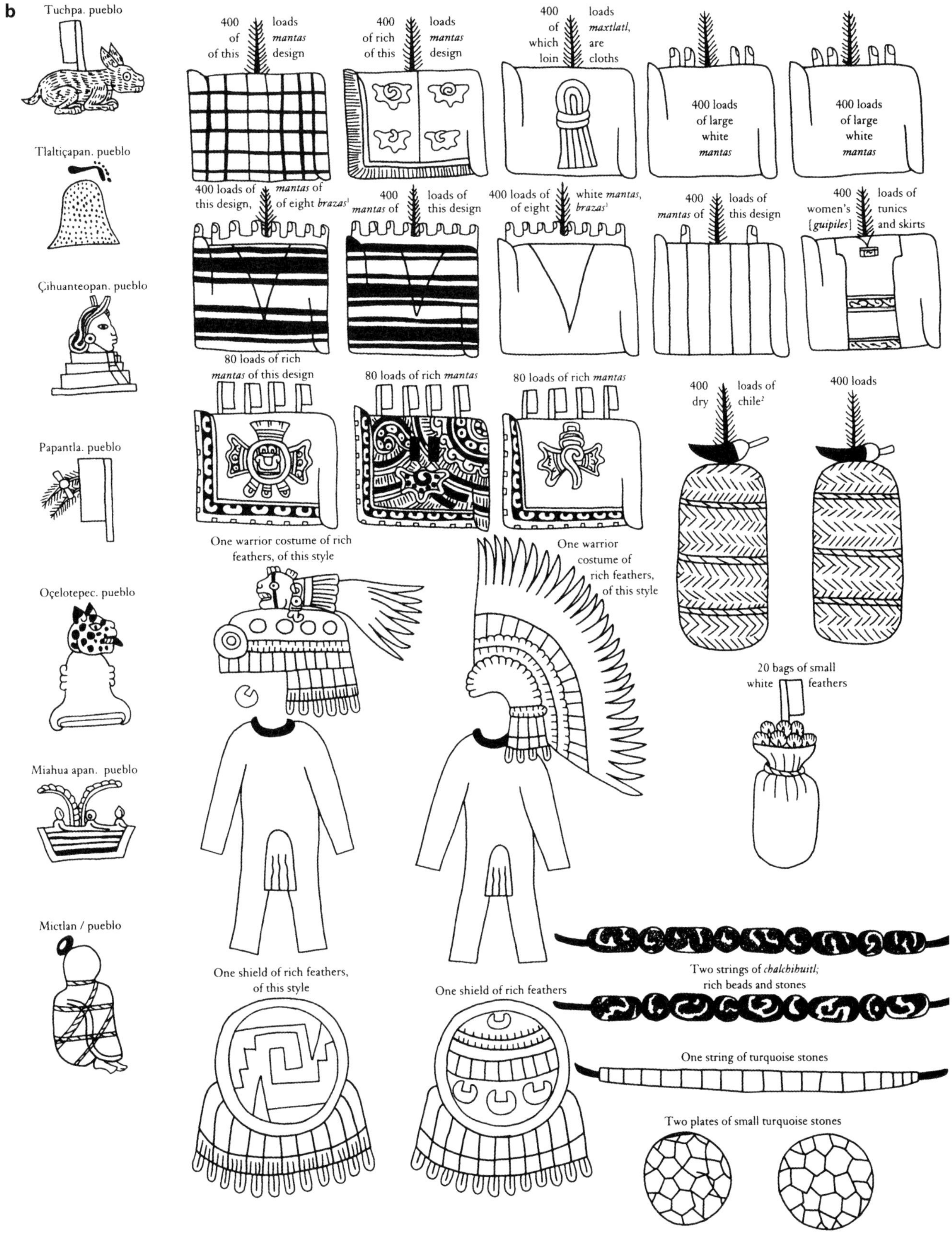

chalchiuhtlacuiloque) are described. Sahagún does not elaborate on turquoise mosaic-workers in his section on lapidaries either, although he does mention that these skilled artisans worked with round turquoise (*xiuhtomolli*) and fine turquoise (*teoxihuitl*) (Sahagún 1950–1982, 9: 82).

Especially meaningful was the aristocratic turquoise diadem (*xiuhuitzolli*), the quintessential symbol of high office. It is usually depicted as a solid blue object, although it is also shown as a mosaic (Sahagún 1993: ff. 51r, 51v), more likely its actual form (Fig. 3). According to Alvarado Tezozomoc

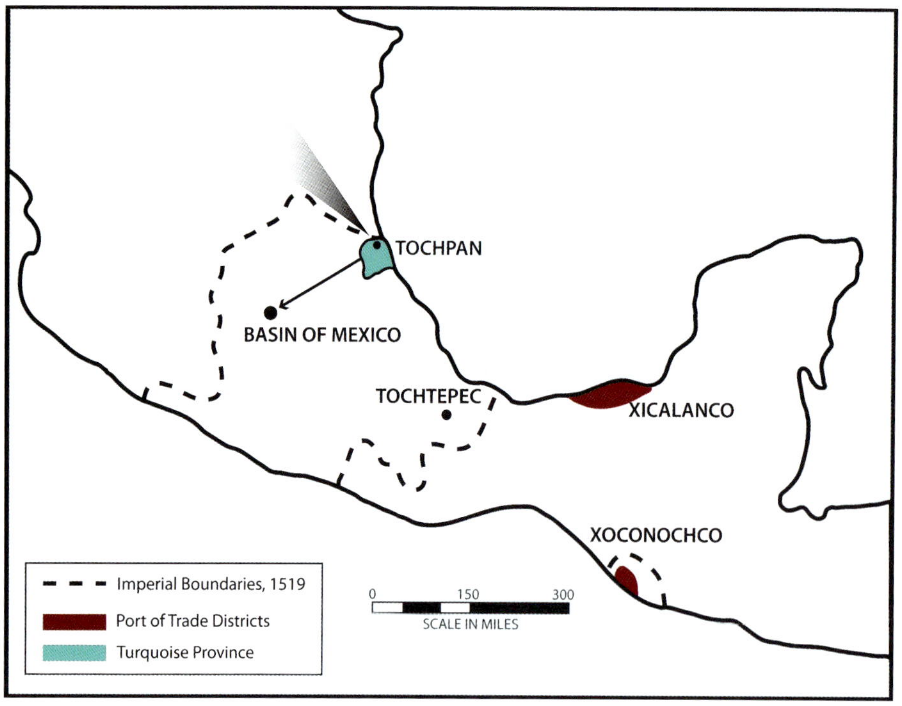

Figure 6. Possible trade route for turquoise into Tochpan from the north, and subsequent payment of tribute to Tenochtitlan (map by Jennifer Berdan).

(1975: 354), the royal emblem of Motecuhzoma Ilhuicamina (r. 1440–1468) was made from small *matlalxihuitl* (blue turquoise) stones, carried by merchants from the southern region of Huaxacac (present-day Oaxaca).

The *xiuhuitzolli* symbol was not taken lightly; in the Codex Boturini (Johansson 2007), the humble nomadic Mexica did not wear it, but the established rulers they encountered did. Similarly, the early rulers of Texcoco, until the famed Nezahualcoyotl, also lacked the diadem; and Mexica rulers before the imperial Itzcoatl were likewise denied this symbol (Sahagún 1993: f. 52r). And, interestingly, Mexica rulers after Cuauhtemoc – those holding office under Spanish rule – are shown divested of this royal symbol (Sahagún 1993: f. 51v). Sahagún's informants bestow it on all the rulers of Tenochtitlan, Texcoco and Tlacopan (the famed Triple Alliance, cf. Fig. 1), but only to the more recent rulers of the nearby city-state of Huexotla when it became attached to the Texcoco royal house (1950–1982, 8: ill. 28–54).

When new rulers were crowned, bestowal of this diadem, along with a fine turquoise nose plug, a blue cape adorned with small turquoise stones, and other royal insignia, was the defining moment in the coronation (Alvarado Tezozomoc 1975: 460, 573; Olko 2005: 113–136, 168; López Luján and Olivier 2009: 43–44). Spanish conquistadores encountered Motecuhzoma's diadem and nose plugs in his storehouse shortly after they arrived in Tenochtitlan (Sahagún 1950–1982, 12: 49).

Other nobles

Although the turquoise diadem is uniformly associated with Aztec rulers, apparently these crowns and other turquoise adornments (along with similarly prized tropical feathers, greenstones and gold) were also perquisites of other high-ranking nobles (*tecuhtli*). In the Codex Mendoza, the turquoise diadem is worn by rulers and judges, and also serves as an ideographic component of the placename glyph for Tecutepec, the *xiuhuitzolli* representing the essence of *tecuhtli*, or 'noble' (Berdan and Anawalt 1992, 3: f. 15v). In a legendary sense, the Aztecs considered turquoise mosaic shields as part of revered Toltec regalia, and valiant Toltec warriors were said to have worn turquoise ear plugs (Sahagún 1950–1982, 3: 22, 56). The socially ambiguous professional merchants (*pochteca*) presented turquoise mosaics in their extravagant banquets, and also adorned their soon-to-be-sacrificed bathed slaves with turquoise adornments (Sahagún 1950–1982, 9: 38, 59, 60). The ambitious merchants excepted, it was the hallmark symbol for Aztec royalty especially and nobility generally in the Basin of Mexico, but not, interestingly, among the powerful Tlaxcallans in the neighbouring valley to the east. Reinforcing this observation, Izeki (2008: 91) suggests that turquoise was becoming a particular symbolic attribute of the Mexica.

Given the rather high rate at which the aristocracy was growing among the Aztecs and their neighbours, how did they supply themselves with their essential status symbols, in this case, turquoise adornments? How were sufficient quantities of these supplies assured?

Turquoise tribute and trade

Despite the importance and intense use of turquoise in Aztec life, this precious material was not natively available in the Basin of Mexico and needed to be obtained through market exchange, foreign trade and imposed tribute. It has long been accepted that the turquoise used in Mesoamerica originated predominantly in the region of today's southwestern United States and less so from northern Mexico. Sahagún (1956, III: 334) mentions that fine turquoise or *teoxihuitl* came from a distance, although the direction is not specified (Saville 1922: 28). In the same passage, Sahagún also mentions that the

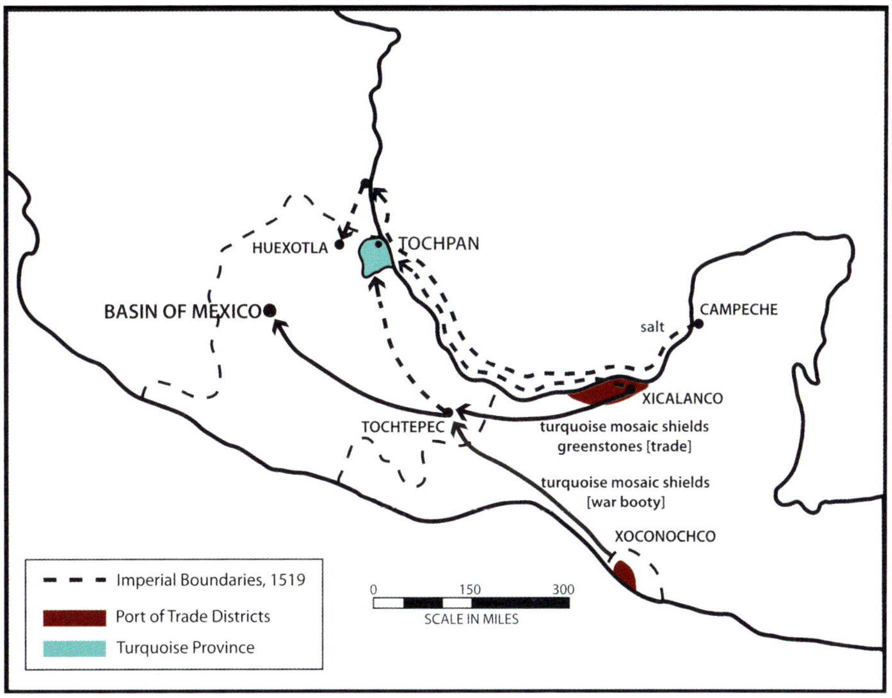

Figure 7. Possible movements of turquoise objects from southern Mesoamerica to Tochpan (map by Jennifer Berdan).

xiuhtomoltetl came from Guatemala and Xoconochco. There is no direct evidence that the Aztecs controlled the northern sources, either politically or economically, although they did maintain a political and commercial presence further south.

The Aztecs exacted tribute from conquered city-states grouped into 38 tributary provinces throughout central Mexico, and maintained asymmetrical relations with an additional 23 client states (Berdan et al. 1996). Among the 38 tributary domains were three provinces that paid part of their annual tribute in turquoise, although it has become quite obvious that turquoise stones and objects could be found virtually everywhere throughout and beyond the Aztec domain (and sold with other luxuries in the largest marketplace of the realm, at Tlatelolco). What is interesting about the Aztec tribute demand is its curious geographical distribution. The three provinces in question are Tochpan on the Gulf coast, Quiauhteopan in Guerrero, and Yoaltepec in Mixtec country (Fig. 4). How did the turquoise get to these areas, and in particular how did it appear in sufficient abundance that the Aztecs could demand it in tribute and expect to receive it on a regular, sustained basis?

When they conquered city-states (which eventually became clustered into provinces), the Aztecs required regular tribute payments from their new subjects. Their demands focused on raw materials and manufactured goods that were conspicuously available in the city-state. This does not necessarily mean that the materials had to be natively available in that area as there are several instances where tributary goods were obtained by trade across city-state and imperial boundaries (Berdan 1996). Was this the case with our oddly distributed turquoise? Or not?

Tochpan

Tochpan paid some of its tribute in one string of turquoise beads and two 'plates', disks (or round shields) of turquoise stones (depicted as mosaics), due annually. It also paid tribute in two strings of greenstone beads (Fig. 5). As currently understood, neither the turquoise nor the greenstone was locally available in the Tochpan region. But Tochpan did have a renowned market. And just inland, Huexotla served as a quasi-international trading centre; it was situated on major north–south and east–west trade routes and was an important entrepôt for the trade in salt (Berdan et al. 1996: 293). It is likely that turquoise entered this province through trade, and it must have done so in notable quantities. I see two plausible scenarios here.

One possible route was from the north, and a source on the border of the present-day states of Zacatecas and Coahuila has been suggested by Harbottle and Weigand (1992: 83–84). However, in a later article, Weigand and Weigand (2001: 190) mention that turquoise from Zacatecas, Coahuila and San Luis Potosí was 'low grade, very soft, and almost powdery'. If trade from the north was involved, perhaps Tochpan's turquoise tribute (necessarily of high quality), came from farther north. This would have provided a fairly direct route from the documented native sources of turquoise, and Tochpan could have been supplied through Huexotla or within its own market by long-distance merchants or down-the-line trade where the goods changed hands several times (Figs 6 and 7).

A second scenario applies to both the turquoise and greenstone tribute (the greenstone also poses some difficulties: that is, what was it doing in Tochpan?). Could it be that both the turquoise and greenstone objects arrived in this northern Gulf coast area via a different and rather unexpected route: by trade from the south? We do know that both of these prized stones were available in the trading district of Xicalanco to the south, where Aztec merchants were pleased to obtain them: among the goods acquired by those professional merchants (*pochteca*) for the Mexica ruler were greenstones and turquoise mosaic shields. Early Spanish explorers also mention the presence

a

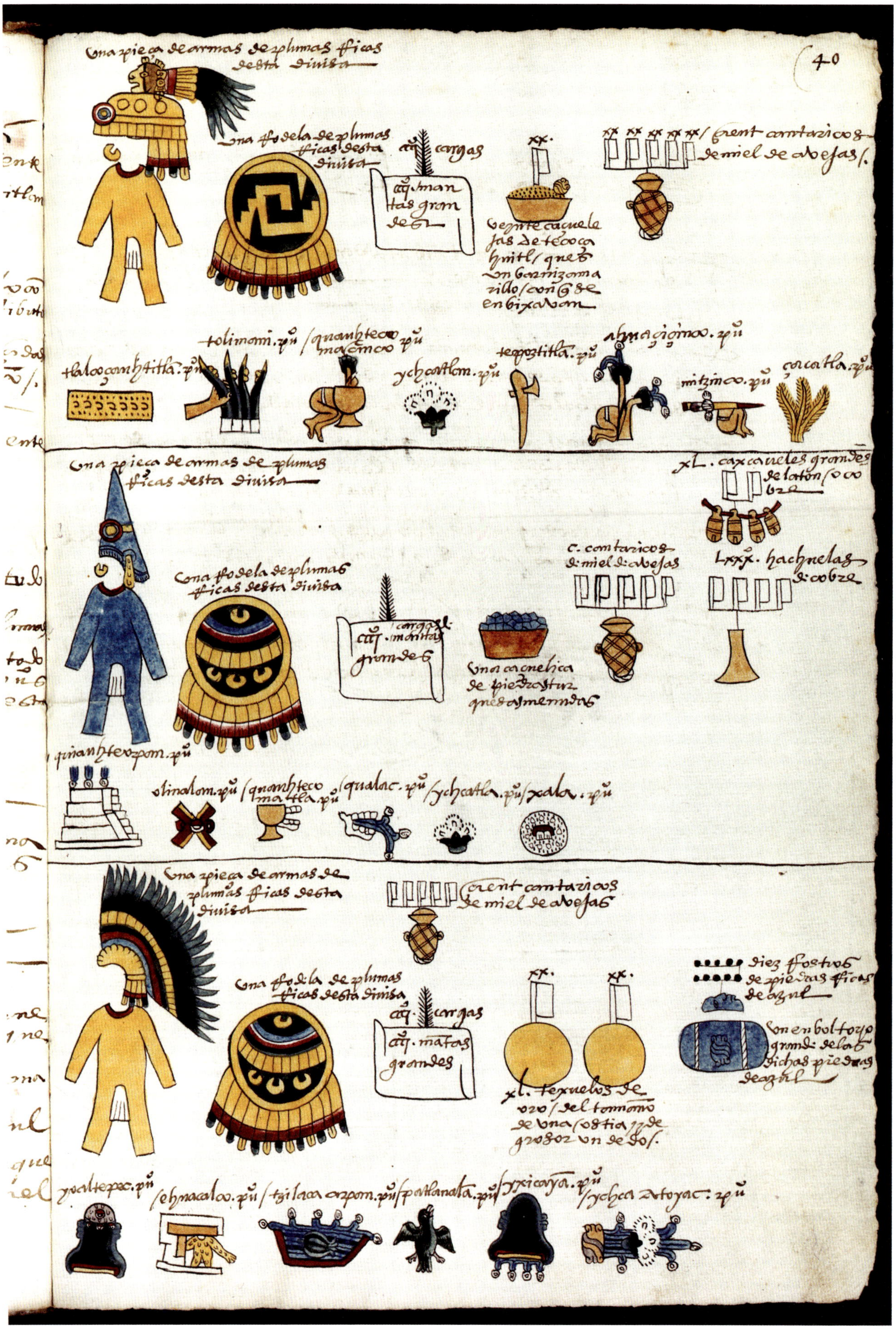

Figure 8. (a) The provinces of Quiauhteopan and Yoaltepec with their turquoise tribute (Codex Mendoza, f. 40 r, Bodleian Library, Oxford). (b) Black and white image with translated captions (from Berdan and Anawalt 1992, 4: f. 40r, bottom two provinces).

b

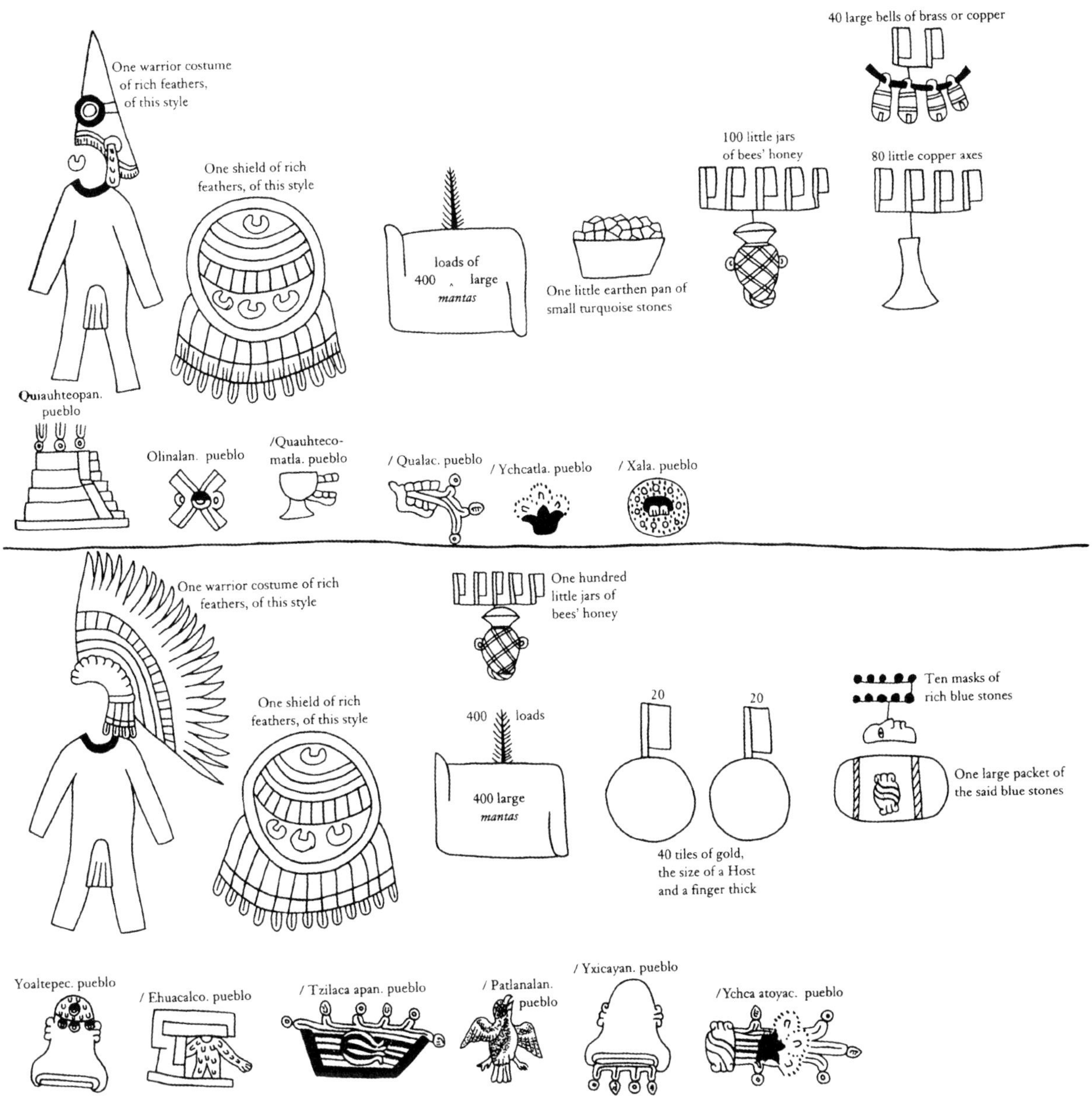

of stone mosaic work (and sometimes specifically turquoise stonework) in this lowland area of Mesoamerica, around present-day Tabasco and Veracruz (Saville 1922: 6–8; Martyr D'Anghera 1970: 19–20). And if traded goods move in clusters (as they most often appear to do), then it is also significant that at least some of the salt arriving in Huexotla was from Campeche, not far from Xicalanco.

Turquoise mosaic shields seem to have been quite abundant in the south, as they were included in the battlefield booty by Aztec warrior-merchants in the Xoconochco region on the Pacific coast. Greenstones, turquoise mosaics, salt (and multitudinous feathers) could have arrived in Tochpan from the south, either along the coast, or via inland routes through the trading centre of Tochtepec, as merchants from both Xicalanco and Xoconochco passed through there (Fig. 7).

In either scenario, or both, there must have been well-established and reliable trade networks for these luxuries to attract the attention of the Aztec tribute assessors, and those networks seem to have leap-frogged over several provinces. In any event, the Aztecs found Tochpan capable of providing these precious objects in regular, predictable tribute.

Quiauhteopan and Yoaltepec

The conquered provinces of Quiauhteopan and Yoaltepec in the south-central part of the empire may represent a somewhat different story (see Fig. 8). Quiauhteopan gave a bowl of small blue turquoise stones (*matlauac xiuitl*), while Yoaltepec paid a large packet of blue (turquoise) stones and ten blue-stone (turquoise) masks, annually (Berdan and Anawalt 1992, 3: f. 40r; Fig. 8). Farther to the east, and not recorded in the *Matrícula de Tributos* or Codex Mendoza, lay Quetzaltepec: when conquered, its tribute included riches such as cacao, precious feathers, greenstones and, of great interest to us, 'small and very rich fine turquoise (*teoxihuitl*)' (Alvarado

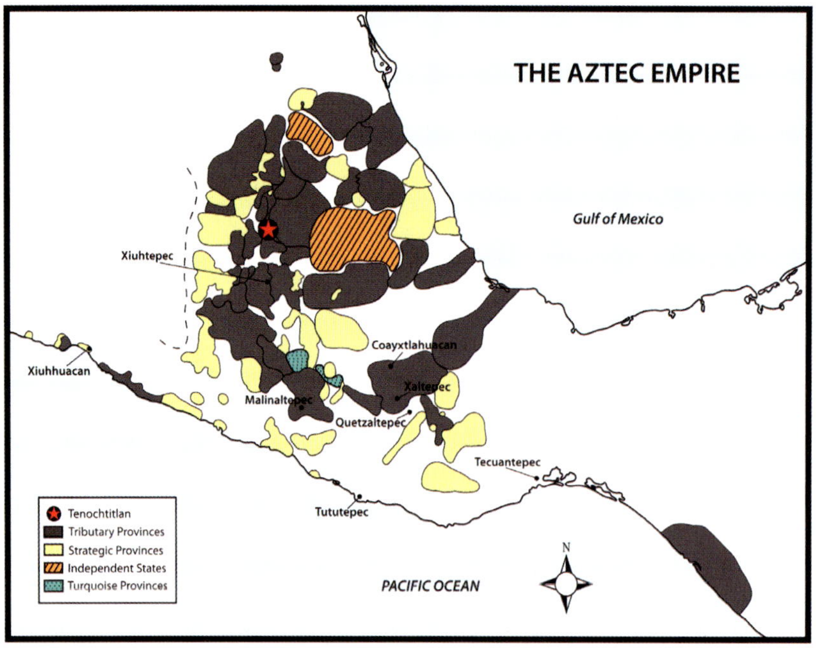

Figure 9. Turquoise from tributary provinces and other Aztec subjects in the southern empire (map by Jennifer Berdan).

Tezozomoc 1975: 609). Indeed, this whole southern part of the empire seems to have been a hotbed for turquoise and turquoise mosaics: conquered Coayxtlahuacan gave tribute in small blue (and green and 'dark' stones) for kingly crowns and adornments (Alvarado Tezozomoc 1975: 338), Xaltepec (near Tecuantepec) paid part of its tribute in very rich stone mosaic work (Alvarado Tezozomoc 1975: 600), several conquered towns around Tecuantepec paid some of their tribute in *teoxihuitl* stones for mosaic work and battlefield booty from those same towns included a military accoutrement called a *xiuhtezcatl*, a turquoise stone-decorated mirror (Alvarado Tezozomoc 1975: 543). A city-state near the Pacific coast and within the bounds of the Aztec Empire (in Cihuatlan province) was named *Xiuhhuacan* (Place That Has Turquoise) (Fig. 9). These Mixtec and Zapotec areas were the scene of highly competitive city-states, with aggressive alliance building through extravagant feasting and displays, all of which required the production of large quantities of elite wealth including, of course, turquoise (Pohl 2001, 2003).

Where did all this turquoise originate? Again, I offer some alternative scenarios. The first involves hypothesised trade from the north, through the land of the Tarascans or Purépecha. Sahagún tells us that *teoxihuitl* or fine turquoise came from a long distance, although he does not specify direction (see Fig. 9). Projected trade from the northern sources of turquoise could have moved along the Pacific coastal belt either inland or along the coast, through the Aztatlán cultural sphere and on into Tarascan territory. What happened next is unclear, as Pollard states that there is no direct evidence that Tarascan state merchants controlled trade in turquoise beyond their borders (2003a: 85). The turquoise trade may have bypassed the Tarascan world via water routes, disembarking at ports such as Huatulco (for Tututepec) farther south. The early colonial native ruler of Tututepec apparently was quite fond of turquoise, for among his possessions were several strings of various-sized turquoise beads and 'three rabbits carved in turquoise stone' (Pohl 2003: 322). Turquoise could have worked its way inland through long-distance merchants or via regional, down-the-line trade. If it did so, it is nonetheless a bit curious as to why it seemed to leap-frog over some conquered provinces, and emerge as Aztec tributary items at others (not unlike the case of Tochpan). Still, this continues to be the most popular explanation for the presence of turquoise in Guerrero, Chiapas and nearby areas (see Harbottle and Weigand 1992; Pollard 2003b: 57; Pohl 2003: 176, 322; Smith 2003: 124–125; Weigand 2008).

It is difficult to resist the temptation to suggest the possibility of some local sources of turquoise in areas other than northern Mexico and the American Southwest. A quote from Saville (1922: 28) makes the point that 'we cannot reject the opinion that ultimately ancient workings will be found at more than one site in Mexico'. We should also perhaps attend to a statement in Sahagún (1950–1982, 11: 223) that speaks rather immediately and intimately to the acquisition of turquoise, suggesting direct experience or knowledge by one or more of his informants: 'I search for turquoise. I excavate turquoise. I remove turquoise.' And it might be kept in mind that Sahagún does mention a turquoise mine near Tepotzotlan in the northwestern Basin of Mexico that provided Early Postclassic Tula with its supplies of common and fine turquoise (1950–1982, 10: 168). Another enigma, the genesis of the name of a town just south of the Basin of Mexico, *Xiuhtepec*, has yet to be resolved (Fig. 9).

Apparently, some of the earliest evidence for archaeological turquoise in Mesoamerica is located in the state of Guerrero (at Mezcala, 600 BC, cf. Harbottle and Weigand 1992: 80), and early evidence of turquoise has also been found at Chalcatzingo, Morelos (also dating to around 600 BC) (Weigand and Weigand 2001). This area saw its greatest interest in turquoise during the Postclassic period, especially the Late Postclassic (Harbottle and Weigand 1992: 81). The province of Quiauhteopan, rendering turquoise in tribute to the Aztecs, also gave those same overlords copper bells and

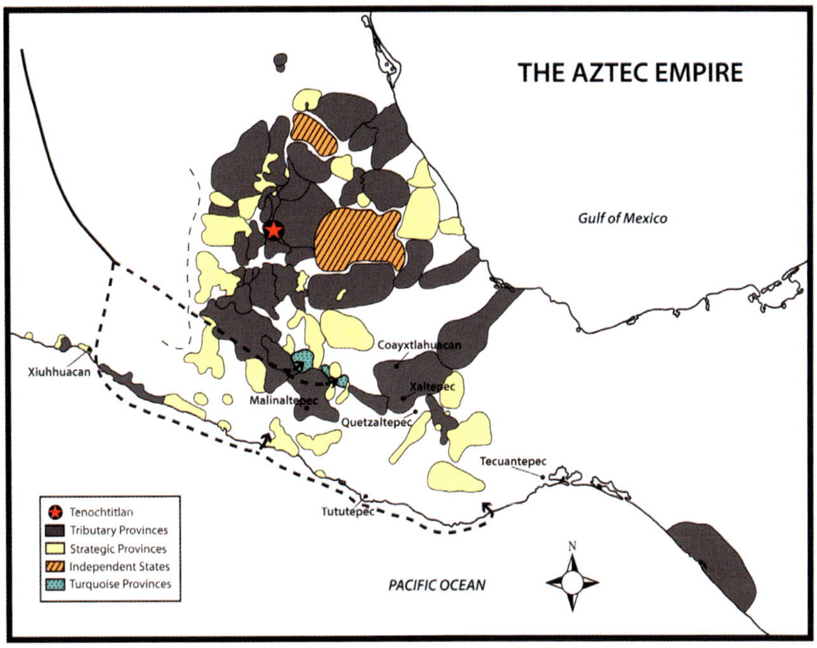

Figure 10. Posited turquoise trade from the north into the southern empire (map by Jennifer Berdan).

copper axes. While all these objects could have been clustered in trade, the co-occurrence of copper and turquoise is a notable coincidence given their geological affinities. If trade were involved, it would be coming from the Tarascan area (where, we have already seen, those merchants apparently did not traffic in turquoise beyond their borders). We might have to assume two different trading networks for these materials: copper from the Tarascans and turquoise from Tututepec on the coast. It may be that long-distance trade offers only a partial answer to the considerable presence of turquoise in the southern empire. I merely suggest another possibility – that of some as-yet undetected local sources.

Some final considerations

The Aztec aristocracy relied on turquoise as a primary symbol of its lofty status. This necessitated continual and predictable flows of this colourful stone into the heart of the Aztec realm: Tenochtitlan and nearby cities in the Basin of Mexico. These stones, and objects made from them, arrived in Tenochtitlan through tribute, market exchanges and the efforts of long-distance merchants. The specifics of these movements, indeed the sources of the stones themselves, continue to be hypothetical and perplexing.

An intriguing question revolves around the extent to which the Aztec pattern may be projected into the past. To what degree did any of these processes involving turquoise precede the development of the Aztec Empire? There is general agreement that, scattered and spotted remains excepted, turquoise in Mesoamerica is a Postclassic phenomenon. Beginning in the Early Postclassic (ca. AD 900–1200), turquoise became widespread throughout Mesoamerica, revealed in the large quantities of materials and artefacts as well as frequent iconographic representations at Tula, Toltec-period Chichen Itza and other centres (Harbottle and Weigand 1992; Pohl 2001; Taube 1994). The rapid spread of this prized material in the pre-Aztec Postclassic was undoubtedly due to the activities of merchants and traders. Some of these individuals may have been professional entrepreneurs transporting turquoise and other goods over long distances somewhat in the manner of Aztec *pochteca* (without, of course, actually being *pochteca*). Others may have been small-scale traders, moving turquoise from market to market and consumer to consumer. Merchants and markets preceded Aztec hegemony, and the Late Postclassic Aztec imperial tribute system relied to some extent on the dependability and predictability of these exchange mechanisms in conquered regions.

An additional consideration remains: the possibility that some blue or blue-green stones on extant artefacts may not be turquoise at all, but rather stones of similar appearance. This has been revealed with the famous (or infamous) Máscara de Malinaltepec (a Teotihuacan-style mask with non-Teotihuacan-style mosaic; Scott 2010), which is now understood to be composed primarily of amazonite, with turquoise tesserae on the nose and around the eyes only (Sánchez Hernández and Robles Camacho 2010). Further research along these lines may provide specific identification of stones on other extant artefacts. This, combined with efforts to identify geological sources of these precious materials, will provide an essential foundation on which to build a more thorough understanding of the availability, movements and meaning of turquoise in Aztec imperial life, and by extension, the role of high-level status symbols in complex ancient empires.

References

Alvarado Tezozomoc, F. (1975) *Crónica mexicana*. Commentary by M. Orozco y Berra. Mexico City: Editorial Leyenda.

Anders, F., Jansen, M. and Reyes García, L. (1991) *Códice Borbónico*. Graz: Akademische Druck-u. Verlagsanstalt.

Berdan, F. (1996) 'The tributary provinces', in F. Berdan, R. Blanton, E. Boone, M. Hodge, M. Smith and E. Umberger, *Aztec Imperial Strategies*. Washington, DC: Dumbarton Oaks Research Library and Collections, 115–135.

Berdan, F. and Anawalt, P. (1992) *The Codex Mendoza*, 4 vols. Berkeley: University of California Press.
Berdan, F., Blanton, R., Boone, E., Hodge, M., Smith, M. and Umberger, E. (1996) *Aztec Imperial Strategies*. Washington, DC: Dumbarton Oaks Research Library and Collections.
Bierhorst, J. (1992) *History and Mythology of the Aztecs: The Codex Chimalpopoca*. Tucson: University of Arizona Press.
Boone, E. (1983) *The Codex Magliabechiano*. Berkeley: University of California Press.
Harbottle, G. and Weigand, P. (1992) 'Turquoise in pre-Columbian America', *Scientific American* **266**(2): 78–85.
Hernández, F. (1959) *Historia Natural de Nueva España*, 2 vols (*Obras Completas*, vols 2 and 3). Mexico City: Universidad Nacional de Mexico (UNAM).
Izeki, M. (2008) *Conceptualization of 'Xihuitl': History, Environment and Cultural Dynamics in Postclassic Mexica Cognition*. BAR International Series 1863. Oxford: BAR.
Johansson, P. (2007) *Tira de la Peregrinación (Códice Boturini). Arqueología Mexicana*, edición especial serie codices, 26. Mexico City: Editorial Raíces/ Instituto Nacional de Antropología e Historia (INAH).
López Luján, L. (1994) *The Offerings of the Templo Mayor of Tenochtitlan*, B.R. Ortiz de Montellano and T. Ortiz de Montellano (tr.). Niwot: University Press of Colorado.
López Luján, L. and Olivier, G. (2009) 'La estera y el trono. Los símbolos de poder de Motecuhzoma II', *Arqueología Mexicana* **17**(98): 40–46.
Martyr D'Anghera, P. (1970) *De Orbe Novo*, F. MacNutt (tr.). New York: Burt Franklin.
Matrícula de Tributos (1982) Graz: Akademische Druck-u. Verlagsanstalt.
Olko, J. (2005) *Turquoise Diadems and Staffs of Office*. Warsaw: Polish Society for Latin American Studies and Centre for Studies on the Classical Tradition.
Pohl, J. (2001) 'Strategies for cultural and commercial exchange between Mexico and the American Southwest, 1100–1521', in V. Fields, V. Zamudio-Taylor and E. Gamboa Carrera (eds), *The Road to Aztlan*. Los Angeles: Los Angeles County Museum of Art, 86–101.
Pohl, J. (2003) 'Ritual ideology and commerce in the southern Mexican highlands', in M. Smith and F. Berdan (eds), *The Postclassic Mesoamerican World*. Salt Lake City: University of Utah Press, 172–177.
Pollard, H. (2003a) 'The Tarascan empire', in M. Smith and F. Berdan (eds), *The Postclassic Mesoamerican World*. Salt Lake City: University of Utah Press, 78–86.
Pollard, H. (2003b) 'West Mexico beyond the Tarascan frontier', in M. Smith and F. Berdan (eds), *The Postclassic Mesoamerican World*. Salt Lake City: University of Utah Press, 5–57.
Quiñones Keber, E. (1995) *Codex Telleriano-Remensis*. Austin: University of Texas Press.
Sahagún, B. de (1956) *Historia General de las Cosas de Nueva España*, K.A. Garibay (ed.), 4 vols. Mexico City: Editorial Porrúa.
Sahagún, B. de (1950–1982) *Florentine Codex: General History of the Things of New Spain*, A.J.O. Anderson and C.E. Dibble (ed. and tr.), 12 vols. Salt Lake City: University of Utah Press.
Sahagún, B. de (1993) *Primeros Memoriales by Fray Bernardino de Sahagún*, facsimile edition. Norman: University of Oklahoma Press.
Sánchez Hernández, R. and Robles Camacho, J. (2010) 'Mineralogía, petrografía y características lapidarias de la Máscara de Malinaltepec', in S. Martínez del Campo Lanz (ed.), *La Máscara de Malinaltepec*. Mexico City: Instituto Nacional de Antropología e Historia, 125–151.
Saville, M. (1920) *The Goldsmith's Art in Ancient Mexico*. New York: Museum of the American Indian Heye Foundation.
Saville, M. (1922) *Turquoise Mosaic Art in Ancient Mexico*. New York: Museum of the American Indian Heye Foundation.
Scott, S. (2010) 'La Máscara de Malinaltepec y sus compañeras', in S. Martínez del Campo Lanz (ed.), *La Máscara de Malinaltepec*. Mexico City: Instituto Nacional de Antropología e Historia (INAH), 65–109.
Smith, M. (2003) 'Key commodities', in M. Smith and F. Berdan (eds), *The Postclassic Mesoamerican World*. Salt Lake City: University of Utah Press, 117–125.
Taube, K. (1994) 'The iconography of Toltec period Chichen Itza', in H. Prem (ed.), *Hidden among the Hills: Maya Archaeology of the Northwest Yucatan Peninsula*. Möckmühl: Verlag Von Flemming, 212–246.
Weigand, P.C. (2008) 'Turquoise: formal economic interrelationships between Mesoamerica and the North American Southwest', in L. Webster and M. McBrinn (eds), *Archaeology without Borders*. Boulder: University Press of Colorado, 343–353.
Weigand, P.C and Weigand, A. (2001) 'A macroeconomic study of the relationships between the ancient cultures of the American Southwest and Mesoamerica', in V. Fields, V. Zamudio-Taylor and E. Gamboa Carrera (eds), *The Road to Aztlan*. Los Angeles: Los Angeles County Museum of Art, 184–195.

Author

Frances Berdan, Department of Anthropology, California State University, San Bernardino, USA

Mexican turquoise mosaics in Vienna

Christian Feest

In an article published in German in 1896, Oppel (1896) compiled information from the available literature on the then known 22 mosaic-encrusted objects from ancient Mexico, all of them from European collections and all believed to have been brought to Europe in the sixteenth century, although 16 of them had no documented history prior to the first half of the nineteenth century.[1] Over the past 114 years, substantial numbers of related objects have been excavated or looted from sites in Mexico (e.g. Saville 1922; Vargas 1989; González Licón and Márquez Morfín 1994; Martínez del Campo 2010). These have entered public and private collections on both sides of the Atlantic and offer an opportunity for stylistic comparisons with the historically collected pieces and the possible identification of regional styles. While additional information has come to light on a number of other historically collected Mexican mosaics until now considered lost (including two illustrated in seventeenth-century books), no further actual objects of this kind have been identified.[2] The skull and two animal heads in the Ethnologisches Museum Berlin were considered lost since the end of the Second World War. Although the small wooden figure of a Deity with Two Horns at the Museum für Völkerkunde in Vienna is usually not considered to be a mosaic, it is included here because of the amount and quality of its mosaic encrustation.[3]

This paper takes a closer look at three Mexican mosaics preserved in Vienna: the Deity with Two Horns, a turquoise mosaic shield, and an animal-head mosaic mirror frame, all of which have been held in Austrian Habsburg collections since the late sixteenth century. While all of them have been the subject of publications since the late nineteenth and early twentieth century, many historical and technical questions remain unanswered.

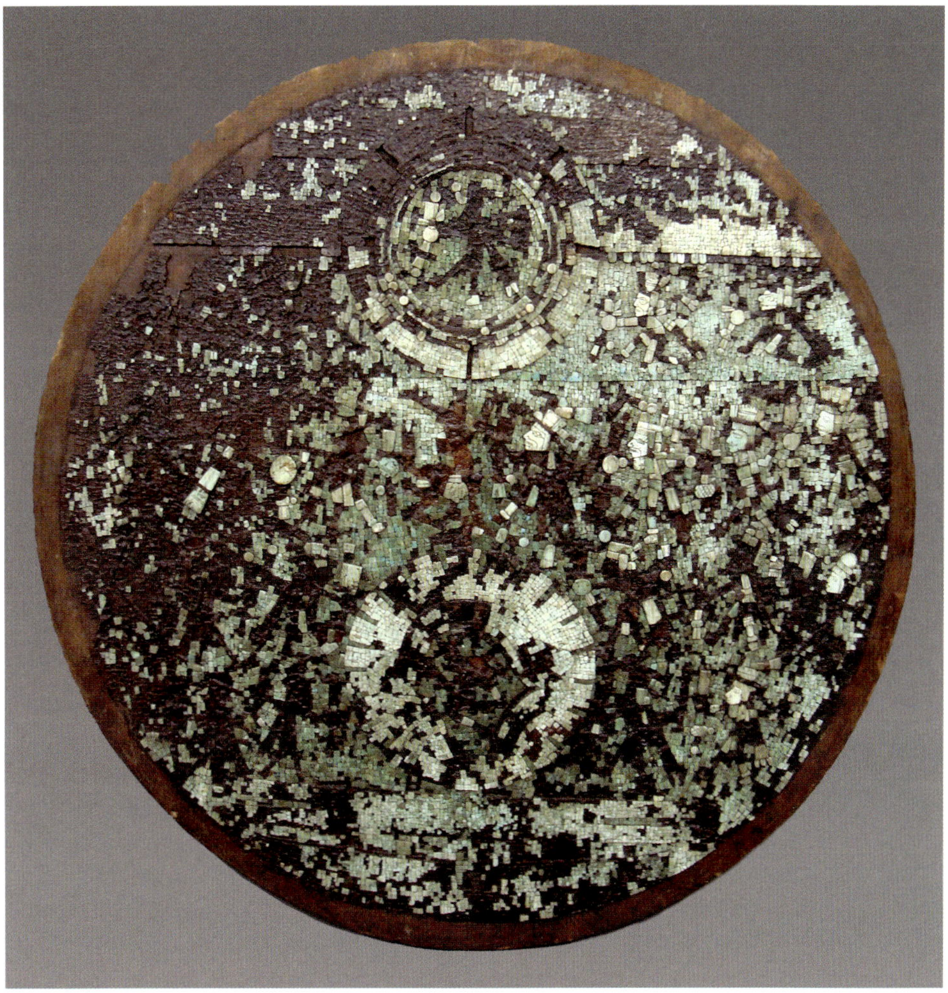

Figure 1. Turquoise mosaic shield. Museum für Völkerkunde Wien, cat. no. 43379. (Photo: Christian Feest, 2009.)

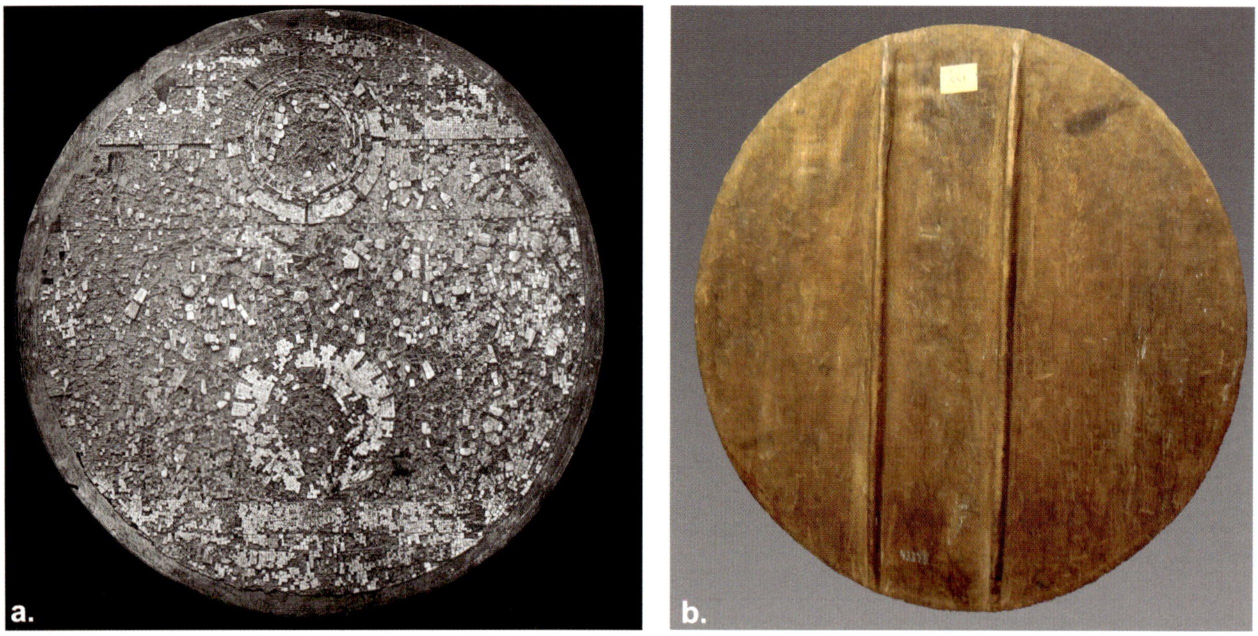

Figure 2. Turquoise mosaic shield. Museum für Völkerkunde Wien, cat. no. 43379. (a) The earliest illustration of the shield (after Heger 1892). (b) Rear view. (Photo: Christian Feest, 2009.)

Turquoise mosaic shield

The turquoise mosaic shield now in Vienna (Museum für Völkerkunde Wien, cat.no. 43372) was first recorded in the 1596 inventory of the Ambras castle collection of Archduke Ferdinand of Tyrol as 'A mosaic with little stones in it', to which the 1663 inventory adds the qualification 'perfectly round' (Fig. 1). In 1730 it was described as 'A round mosaic edged with leather, therein figures made of little stones.' It is also found in the inventories of 1613 and 1621, but is missing in the inventories of 1788 and 1818. No provenance is indicated until 1882 when, still at Ambras castle, the shield was for the first time identified as 'Mexican' (Ilg and Boeheim 1882), perhaps because some other objects taken from Ambras to Vienna in 1806 had been recognised as being of Mexican origin by Ferdinand von Hochstetter, whose work, however, was still unpublished in 1882, and who remained unaware of the existence of the items remaining in Tyrol. During a visit to Ambras in 1891, Zelia Nuttall (1891; 1892: 52) recognised the importance of this mosaic and identified it as a shield. The piece was transferred to Vienna later in the same year and subsequently first published by Franz Heger (1892: 384–390, pl. XVIII; Fig. 2a). Heger (1892: 380–384) provides transcriptions of all entries in the Ambras inventories relating to the shield as well as a short history of the Ambras collection, based on Sacken (1855; for a more extensive account see Scheicher 1979: 73–135). Later discussions of the shield include Nowotny (1960: 38–41, pls 4–7) and Feest (1990: 24–25, figs 1, 16–17).

The shield is formed of a completely circular wooden board, which measures 41.9 cm in diameter and from 0.7 to 1.2 cm in thickness (Fig. 2b). On its back there are two slightly asymmetrically placed vertical ridges, 1.2 cm wide rising to a height of 0.8 cm, and 8.4 cm apart. Since the ridges have no holes or other devices that could have been used for attachment, their function may have been primarily to prevent the board from warping (although this does not appear to be the case in the National Museum of the American Indian shield discussed below). In addition to rough tool marks, the lower part of the back has a fine engraved line running parallel to its edge at a distance of about 1.4 cm. Two such lines are also found on the front, one defining the edge of the mosaic, while the other one is placed halfway between this line and the edge (most clearly visible in the lower right section). It could be that these lines on the front and back defined the limits of the edging of leather referred to in 1730; there is no proof for Heger's suggestion that originally the whole back was covered with leather (Heger 1892: 388–389). A circular hole found recently near the lower left edge of the front side of the shield does not penetrate the wooden board; it does not appear on the illustration published by Heger, and must have been made between 1892 and 1960. At the lower right edge, a small portion of the wood must have been chipped off during the same period of time.

The mosaic is set on a base of resin and consists of flat pieces of turquoise of different sizes, shapes and hues. The resin, which according to Heger (1892: 384) 'burns over the fire with a pleasant odor and smoking flame', has been applied directly to the smoothened, slightly convex surface of the wood in layers of varying thickness of up to about 3 mm.[4] The resulting somewhat stepped surface already defines the basic outline of the design of the mosaic to be described below. Most of the resin is blackish brown in colour, but some of it is quite reddish.

The majority of the turquoise tesserae are rectangular, but there are also wedge-shaped, triangular and completely irregular forms.[5] They range in colour from white to shades of green and turquoise and in size from about 1 mm square to more than 5 mm; one of them has been identified as a broken turquoise bead (Nowotny 1960: 39). In addition, there are shaped and engraved tesserae (mostly pure white, some of which have the appearance of shell) that are used in the representation of 24 human figures, which are part of the composition. Some of these are much larger than the non-representational ones, and they are attached above the background formed by the others

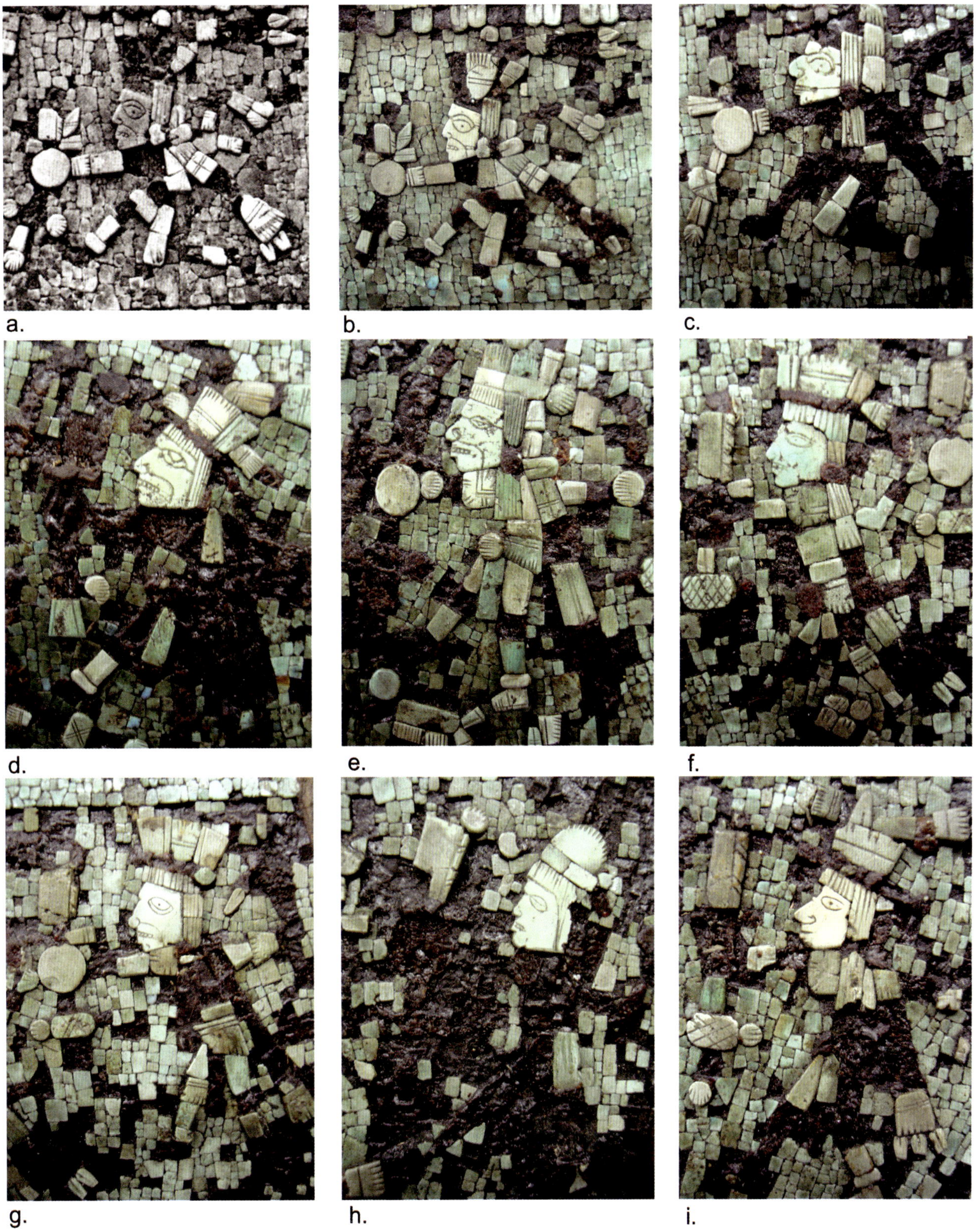

Figure 3. Turquoise mosaic shield. Museum für Völkerkunde Wien, cat. no. 43379. Details of representations of warriors: (a) top row, right of centre (after Heger 1892); (b) top row, right of centre, 2009; (c) top row, far right; (d) second row, first right of centre; (e) second row, second right of centre; (f) second row, third right of centre; (g) second row, fourth right of centre; (h) third row, far right; (i) fourth row, far right. (All photos except (a): Christian Feest, 2009.)

(Fig. 3a–i). Both Heger (1892: 385) and Nowotny (1960: 40) give the number of figures as 23, although Nowotny's redrawing clearly shows all 24 (see Pasztory 1983: 277).

Perhaps half of the tesserae have fallen off over the course of time, and some of the resin base (especially at the edges) has likewise been lost. A comparison of the present condition with the illustrations published by Heger (1892) and Nowotny (1960) indicates that the process is ongoing, especially in the most damaged part of the shield, the upper left quarter. Some of the more recent losses have been preserved separately from the object as it is difficult to locate their exact original position. But the comparison also shows that pieces were added, which

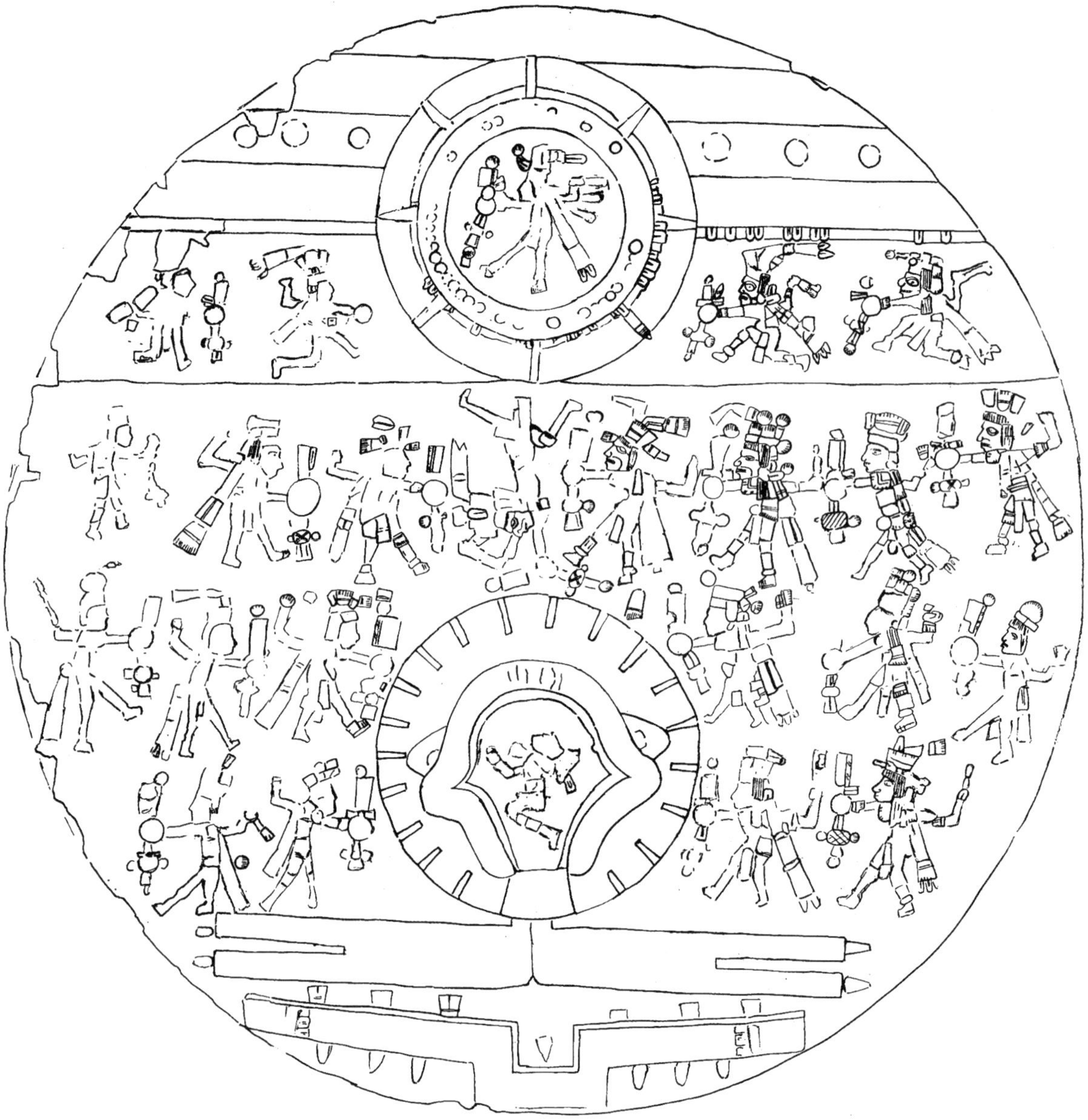

Figure 4. Turquoise mosaic shield. Museum für Völkerkunde Wien, cat. no. 43379. Details of representations of warriors (after Nowotny 1960).

do not show up in the 1892 and 1960 images, and some of them are clearly misplaced. Some of these losses and wrong replacements may be seen by comparing Figures 3a and 3b. As documented by photographs taken in June and October 1992, the shield apparently received some conservation treatment in the summer of 1992 from the Museum's then conservator, Norbert Kirchner, who unfortunately left no notes detailing his work.

Nowotny (1960: 39) speaks of 'many thousands of tesserae of which the mosaic was originally composed'. I made a count of the numbers in 14 well-preserved 1 × 1 cm squares in various parts of the shields, including areas with representational designs.[6] I encountered a range of from 13 to 39 tesserae per square centimetre with an average of a little over 25. Given that the mosaic originally covered an area of 1195 square centimetres, I estimate a total of approximately 30,000 tesserae.

The following description of the design is based, with some emendations, upon the account provided by Nowotny (1960: 39–40; see Fig. 4). The upper third of the shield (132 mm high) depicts five layers of the sky (1: 20 mm/green; 2: 20 mm/light green and slightly raised; 3: 17 mm/whitish; 4: 19 mm/whitish and slightly raised with a 4 mm lower edge of suspended leaf-shaped tesserae; 5: 52 mm/green) and an image of the sun, which is placed centrally across zones 2–5. The sun is flanked in zone 3 by three disks on each side, presumably representing stars, and in zone 5 by four human figures facing inwards; one figure is shown inside the sun, another is falling headlong down from the sun towards the earth. This part (256 mm high) is divided into an upper section with a green background and a lower section with a whitish background. Centrally placed at the lower edge of the upper section is an image of the moon crescent in the shape of an animal jaw from which water emerges and divides into two streams. There is one figure inside the moon and 17 others (7 in one row above the moon, 6 and 4 in two rows flanking it); all, except for the figure at the left of the bottom row, are facing and moving inwards. The whitish zone at the lower edge of

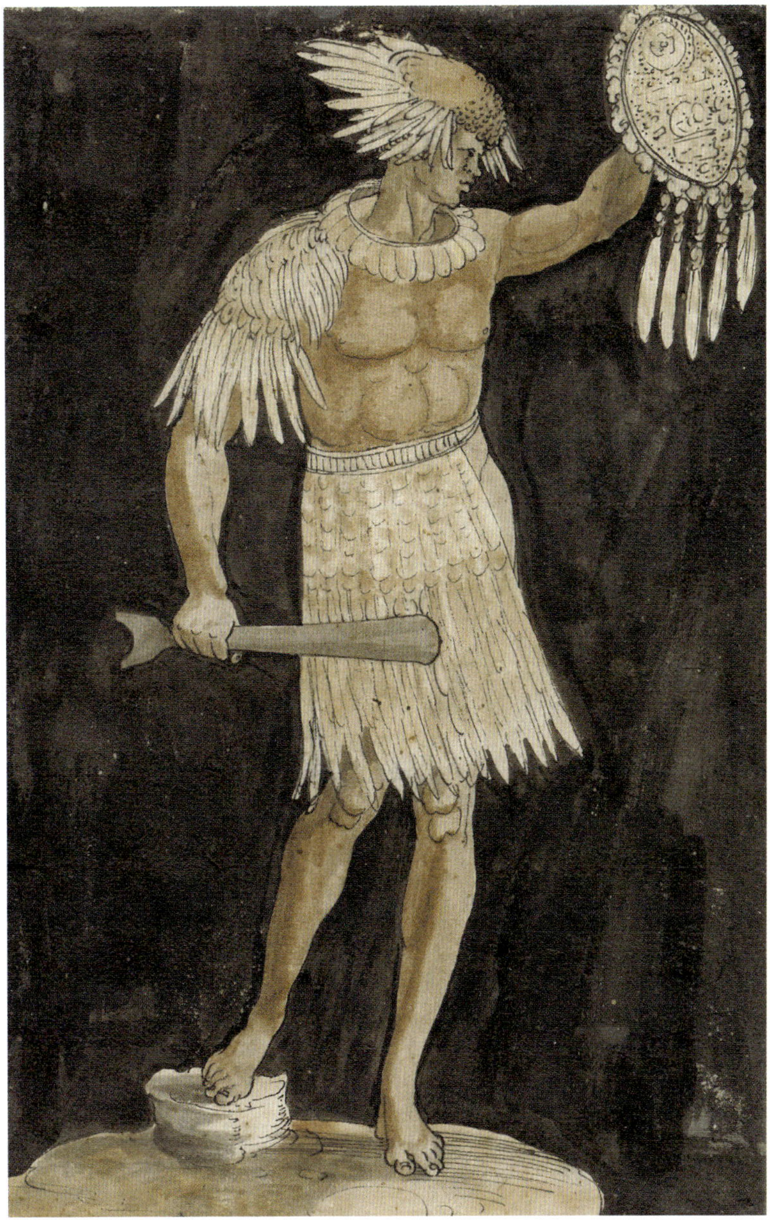

Figure 5. Black youth holding a shield similar to the Vienna turquoise mosaic shield. Pen and ink drawing with wash, attributed to Hans Burgkmair, ca. 1520, 23.5 × 16 cm. British Museum, Department of Prints and Drawings, SL,5218,128 (Sloane bequest). (© Trustees of the British Museum.)

the shield represents the jaws of the earth. The human(oid) figures carry an atlatl, a shield, two darts and an ensign; the figures differ from one another mainly in their headdresses, but only nine faces and five headdresses have been preserved. Nowotny concludes by referring to 'a famous gold ornament' from Tomb 7 in Monte Alban showing the same composition of sun, moon and earth jaw, 'a representation current in the ancient Indian art. A safe interpretation is impossible.'

It is not known when and from whom Archduke Ferdinand obtained the shield. Nowotny (1960: 38–39) suggested the possibility that it could have been the one given in 1524 by Charles V to his brother Ferdinand, father of the archduke: 'shield of flat wood, round, covered with various stones like turquoise and of other colours, in which are figured and worked in relief various figures of men monsters and birds and other figures, all in the said stone-work and garnished all around by a strip of thin gold held together with other pieces of gold like belts' (translated after the original given by Nowotny 1960: 27; Nowotny's own translation [1960: 38] is almost as erroneous as the abstract supplied by Boeheim 1888: XCII and cited by Heger 1906: 307). While the possibility cannot be excluded, the lack of prominent 'other colours' and of figures other than men makes it not very likely. The shipment of 1519, including 'Moctezuma's gifts' to Cortés, featured 'sixteen shields with stonework with their coloured feathers, which are suspended from their rims' (Nowotny 1960: 23). These may have included the one sent to Ferdinand in 1524 as well as six others given in 1523 to Archduchess Margaretha in Mecheln, after the objects sent in 1519 had been taken to the Spanish Netherlands in 1520: 'six shields with their feathers, decorated with blue stones ... (of these one was given [prior to 1530] to the said Seigneur de Ville [ambassador of the Duke of Lorraine, Anton II]' (Zimmermann 1885: CXX).

The likelihood that the Vienna shield was part of the shipment of 1519 is supported by a drawing attributed to Hans Burgkmair, executed possibly in Brussels around 1520, in

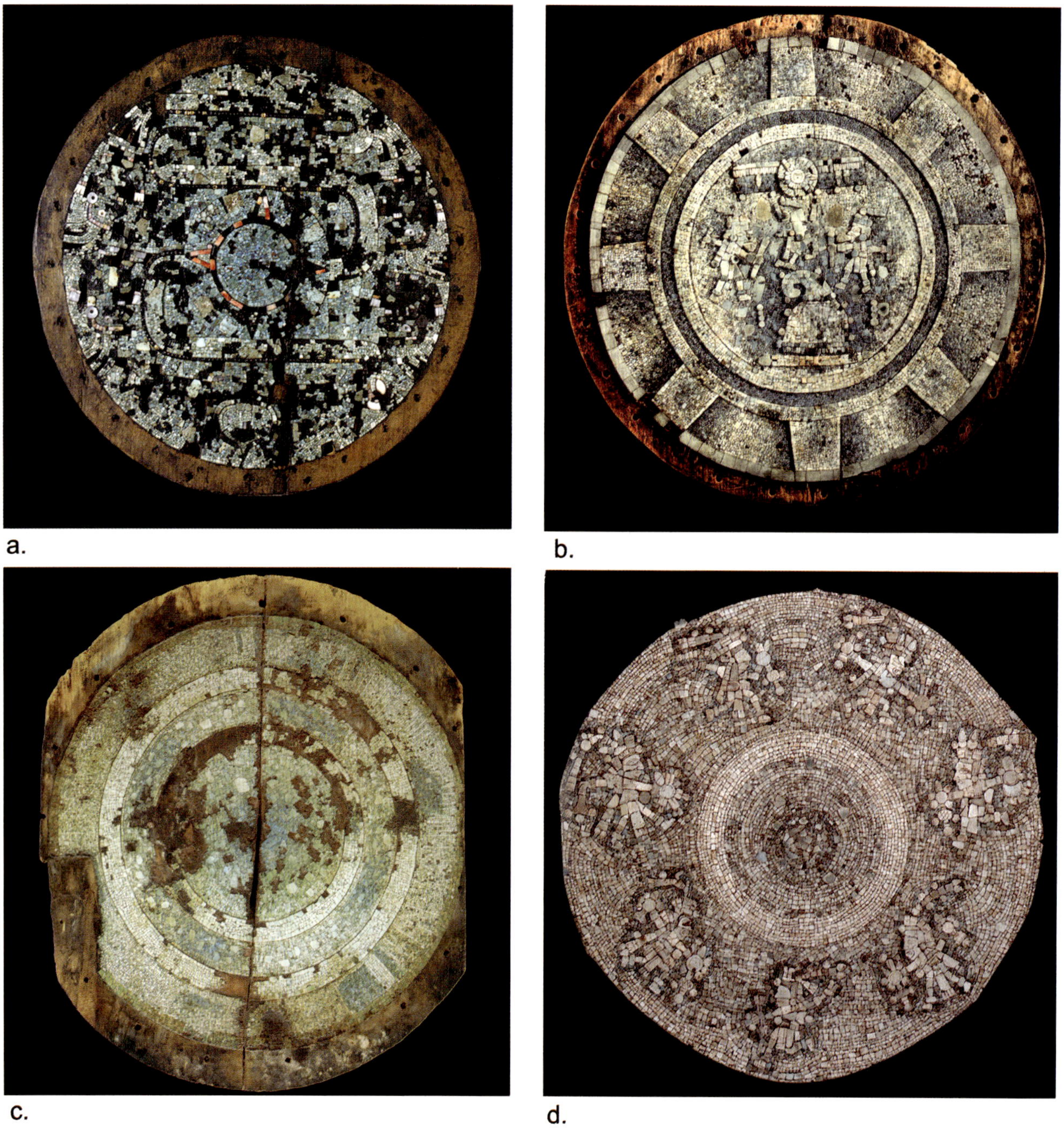

Figure 6. Comparative examples from other collections. (a) Mosaic shield (turquoise, shell, mother-of-pearl, gold on wood). British Museum, cat. no. ST.397a (© Trustees of the British Museum). (b) Mosaic shield (turquoise, found in a cave in Acatlán, Puebla. National Museum of the American Indian, Washington, DC, cat. no. 10.8708 (Carl A. Purpus coll.) (after McEwan et al. 2006). (c) Turquoise mosaic shield found in a cave near Tehuacan, Puebla. Musées Royaux d'Art et d'Histoire, Brussels, cat. no. AAM 68.11 (after Purin 2003). (d) Turquoise mosaic shield from Offering 99, Templo Mayor, México, D.F (after McEwan et al. 2006; see also Velázquez et al., in this vol.).

which an African-looking model is shown sporting various Brazilian and/or Mexican items of featherwork and holding a shield, which in size and composition very much resembles the object now in Vienna (Rowlands 1988: 187–188, pl. XXIII; Fig. 5). The shield shown in the drawing is edged with what appear to be plumes and has five feathers (possibly of gold) dangling from its lower rim.[7] Since the Burgkmair shield seems to be of about the same size as the Vienna shield and shares with it the basic design, it is tempting to assume that the drawing represents the actual shield now in Vienna.

As Oppel (1896: 11) pointed out, the shield in the British Museum (cat. no. ST.397a; see Carmichael 1970: 35; McEwan et al. 2006: 12, 59–66; Fig. 6a) is the only other historically collected Mexican turquoise mosaic shield to have survived of the great number that came to Europe in the course of the sixteenth century. It measures only 31 cm in diameter and also consists of a circular mosaic on a circular wooden board (made of pine), whose back is completely flat. It differs from the Vienna shield not only in its prominent use of shell, but also in the fact that its representations of human faces and bodies are made up of small tesserae rather than of shaped and incised elements, although a few such pieces are also found. Relief effects are produced by carving rather than by differences in the thickness of the layer of resin. A curious feature of the shield are numerous holes penetrating the wooden base, positioned

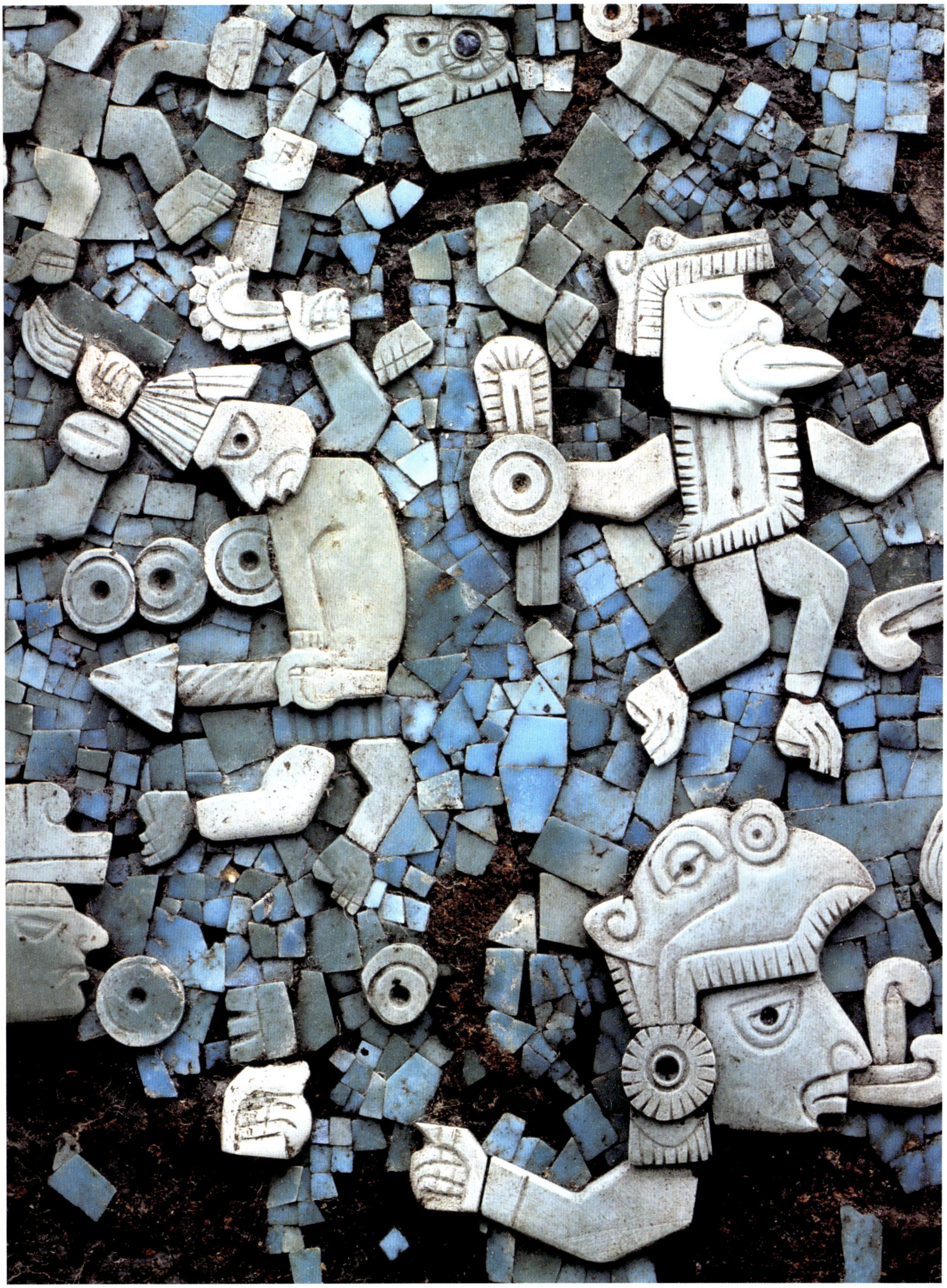

Figure 7. Detail from mosaic panel (turquoise, shell, obsidian on wood), Oaxaca. Museo de las Culturas de Oaxaca, cat. no. 10-362484 (after Solis 2004).

and shaped to coincide with design elements of the mosaic. The wooden rim features a series of holes encircling the mosaic at more or less regular distances, which probably served for the attachment of an edging.

Such holes are also found on two turquoise mosaic shields archaeologically recovered with their wooden bases, one in the National Museum of the American Indian (NMAI) in Washington, DC and one in the Musées Royaux d'Art et d'Histoire in Brussels. The NMAI shield (cat. no. 10.8708; Fig. 6b) was collected before 1921 by Carl A. Purpus and is supposed to have come from a cave in Acatlán, Puebla, together with seven fragments of other shields (cat. nos. 10.8701–8707; Saville 1922: 71–75; McEwan et al. 2006: 63; Scott 2010: 83–91).[8] Based on its provenance and style it has been identified as a Mixtec work. It is slightly larger than the British Museum piece (32.5 cm in diameter), considerably thicker (2 cm) than the shield in Vienna, and may thus be regarded to be of the same functional subtype as the British Museum shield, even though stylistically it is much closer to the Vienna shield. This is especially true of its use of shaped and incised elements, some of which are identical in form and iconographic function as those found on the shield in Vienna, while the central design elements protrude even more prominently from the background; it is only coincidental that, like the Vienna shield, the one at the NMAI features a scene showing a person falling from the sun. In common with the Vienna shield it features two vertical ridges on its back, but these have been drilled to form two pairs of holes, presumably for the attachment of strings for carrying the shield. This may suggest that the Vienna shield, which lacks both these holes and the ones alongside its rim, may be an unfinished specimen. If so, the identity of the Vienna shield with the Burgkmair drawing, which is both carried by a (however imaginary) person and has feathers attached to its rim, may be in doubt.

The NMAI fragments (Saville 1922: 76–79, pls XXIII–XXIX) are from shields ranging in diameter from 27.8 to 38.7 cm. All appear to have a very thick wooden base (1.3–2.5 cm), wider wooden rims with fewer or no holes, simple circular designs and fairly large tesserae. Shaped and engraved elements appear to be absent, and some relief is produced by the different thickness of the layers of resin. The presence of traces of amate paper on the parts not covered with mosaic has led Saville to suggest that these areas were once covered by feather mosaic (or, less likely, drawings on paper).

The shield in Brussels (cat. no. AAM 68.11; see Fig. 6c) was acquired in 1968 and is said to have come from a cave near Tehuacan, Puebla, where it was found together with fragments of other wooden shields without mosaics, an anthropomorphic mask and various mosaic-encrusted ornaments (Purin 2003; Gaida et al. 2003: XXV–XXVI; McEwan et al. 2006: 62). It is slightly bigger than the Vienna shield, measuring 45.5 cm in diameter, and has a proportionally much wider wooden rim. In many respects, it resembles the fragments at the NMAI (concentric circles, tesserae of widely differing sizes, including larger ones, no superimpositions of tesserae and no shaped and engraved elements, relief by means of different thickness of layers of resin), but there is a series of holes surrounding the mosaic, although they are more widely spaced than on the British Museum and NMAI shields.

A turquoise mosaic shield archaeologically recovered (and quite reliably reconstructed) from Offering 99 at the Templo Mayor of Tenochtitlan (Fig. 6d; Velázquez and Maríon 2006; see also Velázquez et al., in this vol., pp. 75–87) measures 28 cm in diameter and is reported to consist of about 15,000 tesserae. If the count is accurate this would give an average of only about six tesserae per square centimetre, just about a quarter of the number on the Vienna shield. It shows images of seven richly dressed males in motion made of shaped and incised elements, which once again are very similar in form and iconographic function to those on the Vienna shield. If the reconstruction is reliable they are not placed above the mosaic background of mostly white, brownish and some blue tesserae, but are enclosed by it; the generally circular arrangement of the background tesserae is adapted to follow the contours of the figures.

Similar work is found on a rectangular panel archaeologically recovered from a cave in the Cuicatec region of Oaxaca (Museo de las Culturas de Oaxaca, Oaxaca, cat. no. 10-362584; Steele and Snavely 1997; Solis 2004: 218–219; Martínez del Campo 2010: 35, fig. 9; Fig. 7). Here, the figures composed of shaped and incised elements are placed in front of the background, but the constituent shapes differ significantly from those on the Vienna, Templo Mayor and NMAI shields.

In summary, the Vienna shield differs from the others in the form of its wooden base (vertical ridges on the back and the lack of holes around the rim, which implies a different method of attaching the edging). It is similar in size to the Brussels shield, technically most similar to the one at the NMAI and the Oaxaca panel, and in the rendering of human figures very close to the Templo Mayor and NMAI shields. Its design is by far the most complex, with only the British Museum piece coming close to it with which it shares the deviation from the basic circular pattern on the other three shields. Even if the similarity with the design shown on the Burgkmair drawing is only suggestive, it is clearly the only shield with a documented sixteenth-century collection history.

Animal-head mirror frame

While the animal-head mirror frame (Museum für Völkerkunde Wien, cat. no. 43382; Fig. 8a) shares with the shield its sixteenth-century provenance from the Ambras collection, its major interest lies in the fact that it documents the early colonial survival of mosaic work in Mexico. It appears in the 1596 Ambras inventory as 'A head of an animal of varicoloured stones, which the heathen adore', to which the 1663 inventory adds 'the right ear of which has been broken off', thus clearly establishing the identity of the entry with our piece. In 1788 the inventory calls it a 'Dog's head', notes 'but the nose is of wax' and offers a first vague suggestion of its possible place of origin: 'That the head is an Egyptian Anubis, I dare not assert, but that it was used for the purpose of idolatry can hardly be doubted.' The 1818 inventory refers to it as 'A dragon's head glued together of glass and hard stones.' Franz Heger (1892: 380) found it in 1891 at Ambras castle among a group of Venetian works of glass and was the first to identify it as Mexican. It was transferred to Vienna in the same year and catalogued in the collections of the precursor of the present Museum für Völkerkunde Wien. Heger (1892: 380–384) has provided transcriptions of all inventory records as well as the most extensive physical description of the item (1892: 398–400, pl. XXII); other descriptions are by Nowotny (1960: 60–61, pl. 16) and Feest (1990: 25–26, figs 17, 18).

The small head of a predator (presently 9.7 cm in length, 8.5 cm in width and 5.2 cm in height) consists of a carved core

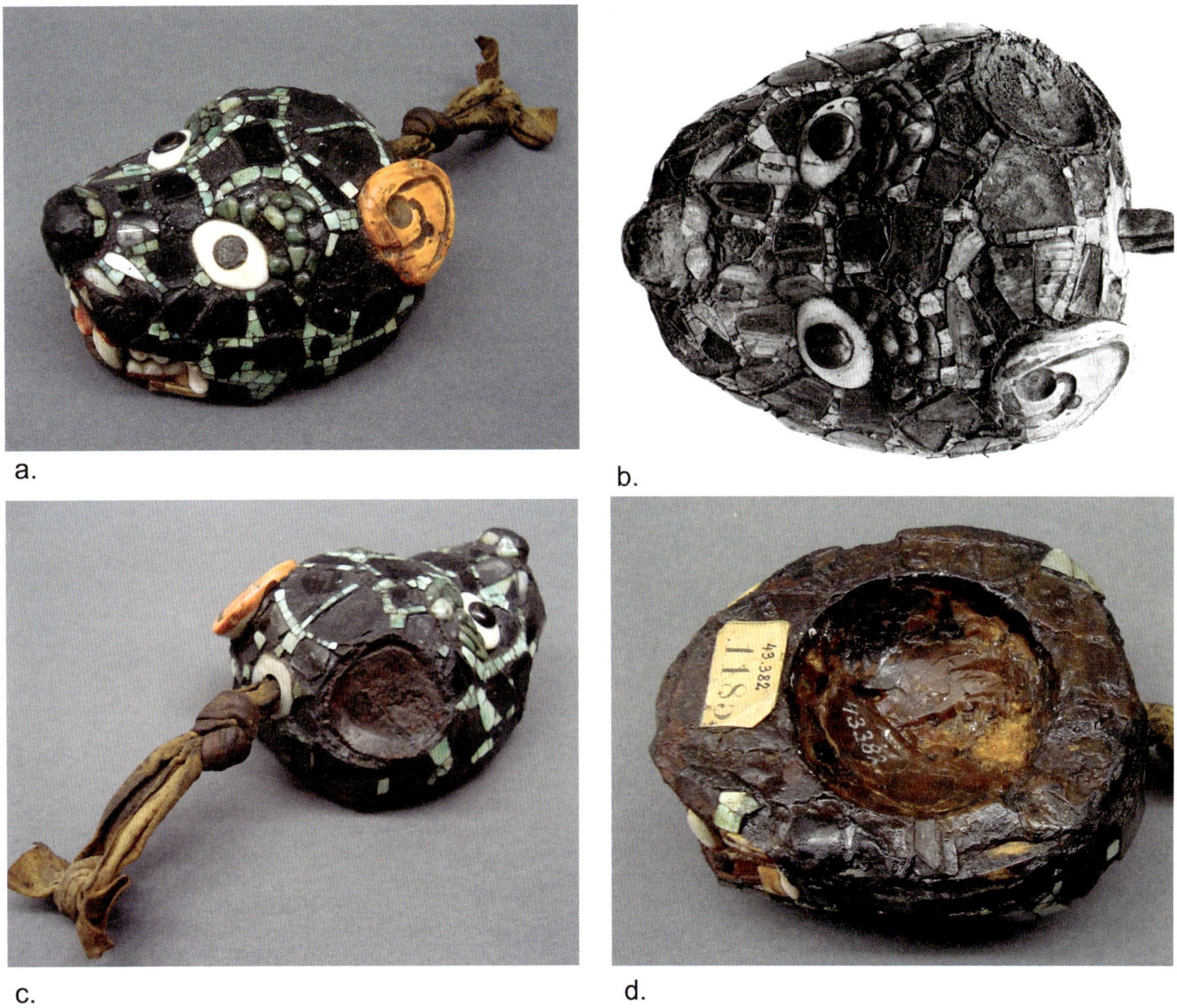

Figure 8. Animal-head mirror frame with mosaic (turquoise, jade, shell, glass on wood). Museum für Völkerkunde Wien, cat. no. 43382. (a) top view; (b) after Heger 1892; (c) rear view; (b) bottom view. (Photos: Christian Feest, 2009.)

of light coloured wood, whose outside is completely covered with a dark resinous paste into which irregular pieces of shell, green speckled jade, turquoise and thin shards of a greenish glass have been set to form a mosaic. The animal's left ear (the right having been lost prior to 1663) is made of a carved orange *Spondylus* shell; the eyes are of white shell to which an iris of glass or obsidian has been glued, which on its underside features a ring filled with white paste marking the pupil. Heger (1892: 399) describes the material of the iris as 'brownish-black glass', Nowotny (1960: 61) considers the 'plano-convex lens' to be made of 'greenish obsidian', and Feest (1990: 25) refers to it as 'greenish brown obsidian'. Except for the lateral incisors of red shell and the lower canines of glass, the teeth are also made of white shell, whereas the lips are formed by reddish and brownish pieces of shell. The tongue of red shell is attached to the mouth cavity by means of a piece of iron wire in such a way that it can move up and down freely. The gums and nostrils are indicated by red paint. The nose has a piece of jet set at its base and was originally flanked on each side by two pointed tesserae of mother-of-pearl and glass representing the whiskers. At the back of the head a double-knotted loop of leather with traces of red paint emerges from a circular piece of shell (Fig. 8c).

The underside of the core (Fig. 8d) has been hollowed out to form a semi-globular cavity of 5.5 cm in diameter, which is coated with a light yellow resin. The cavity was once surrounded by a circular frame covered with tesserae of turquoise and mother-of-pearl, of which only fragments have survived the centuries.

There is evidence for various past interventions by conservators. By 1788, the nose certainly had been repaired by means of wax, but it is not known what other changes may have occurred at the same time. A comparison of photographs shows that between the 1890s and 1960 the iris of the left eye as well as several tesserae were lost (see Fig. 8b). Heger (1892: 380, 400) describes it as a 'most beautiful mosaic' and says that only a few tesserae were missing (most of them at the frame of the cavity on the underside). He calls the earlier restoration of the nose 'somewhat awkward'. Conservation work undertaken in 1990 by Norbert Kirchner had become necessary when the mosaic began to lift off the wooden core, probably as a consequence of having been taken in 1970 to the *Before Cortés* exhibition at the Metropolitan Museum of Art in New York (Easby and Scott 1970: no. 294). At that time, the nose was significantly reshaped, making it less pointed, and some tesserae from an unknown source were added to the left eyebrow and the central part of the forehead.

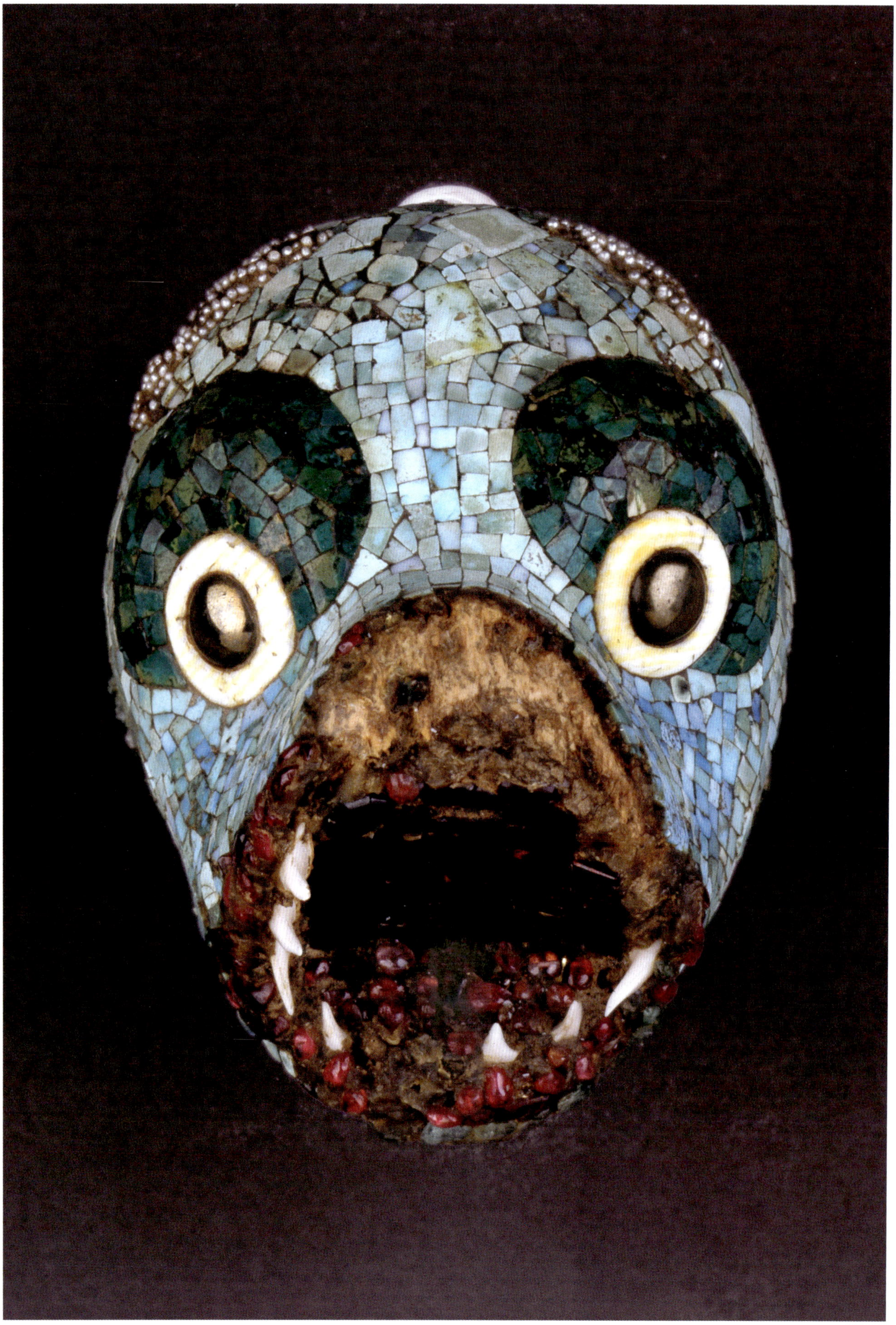

Figure 9. Animal head with mosaic (turquoise, malachite, shell, seed pearls, glass, shark's teeth, garnet on wood). (British Museum, cat. no. St.400a, © Trustees of the British Museum.)

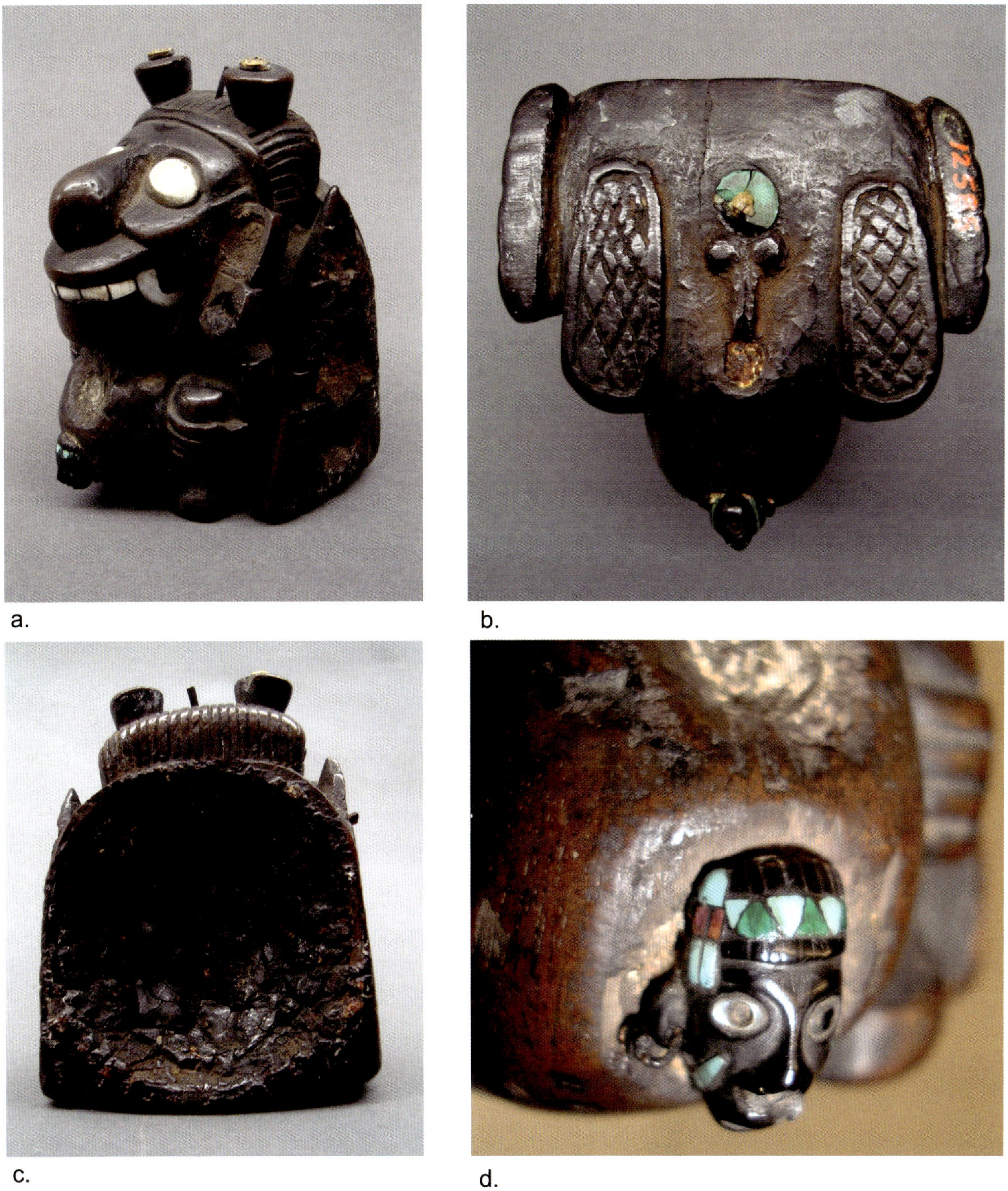

a. b. c. d.

Figure 10. Figure of the Deity with Two Horns. Museum für Völkerkunde Wien, cat. no. 12585. (a) Front view; (b) bottom view; (c) rear view; (d) detail of face with mosaic encrustation on the belly of the figure. (Photos: Christian Feest, 2009.)

The identification of the animal as a 'dog' or 'dragon' in the old inventories was placed in doubt by Heger (1892), who cautiously referred to it as the 'head of an animal' or 'predator'. Nowotny (1960: 60) returned to the identification as a dog 'with the pointed snout of the Mexican dog' – a questionable conclusion given the uncertainty about the original shape of the nose. Easby and Scott (1970) and Feest (1990) cautiously avoid a specific identification of the beast.

In 1788, Johann Primisser, then curator of the Ambras collection, first suggested that the cavity on the underside might once have held 'perhaps a metallic mirror or something of the like'. This was again proposed by Heger (1892: 400) and accepted by Nowotny (1960) and Feest (1990). Historical support for this suggestion comes from the 1598 inventory of the Munich Kunstkammer, which lists three very similar items: 'A dog's head covered with small plates of turquoise and other unknown stones, on the backside a mirror of cast metal'; 'a hostile-looking small dog's head, also covered with turquoise, on the back a cast mirror'; and 'yet another dog's head of a hound, as described before covered with little stones and with a cast mirror' (Fickler 2004, 1: 532–533). A broadly comparable item was already listed in 1541 as a gift from the future king Philip II of Spain to the Duque de Escalona ('a mirror from the Indies, made in the shape of the head of a small dog') (Martin Gonzalez 1990: 157).

The comparatively large number of such mirror frames in late sixteenth-century collections suggests the existence of a colonial cottage industry for the production of items, possibly already for a collector's market. In 1571, for example, on the occasion of the visit to Burgos by Anna of Austria on her way to marry Philip II in Segovia, a procession was held in her honour, featuring 'twenty-four Indians with shirts, trousers and mantles of coloured taffeta, playing balloon, ... and all those were wearing masks, very natural and very well painted, inlaid with many stones' (Anon. 1571: f. xxxvir). It seems highly unlikely that as many as 24 mosaic-encrusted masks had survived in Burgos from the time of the conquest of Mexico.

The Vienna animal-head mirror frame thus appears to be the only surviving example of a once-common type of artefact. In a wider perspective it may be compared with the Deity with Two Horns (see below) and especially the British Museum animal head, which is of about the same size, carved of light coloured wood (of the coral tree, see Cartwright et al., in this vol., pp. 7–14), encrusted with mosaic and fitted on the back with a cavity to hold a mirror (Carmichael 1970: 28–29, 35–36; McEwan et al. 2006: 82–85; Fig. 9). The presence of seed pearls, red garnets, emeralds, beryl, spinel and zircon in the mosaic, together with the use of wax in their attachment, at least suggests a major restoration, probably in the context of the European part of its history and possibly in Italy (the latter suggested by a small piece of yellow-green glass on the mosaic). The use of glass and iron wire (to attach a tongue in the animal's mouth) suggests a colonial origin of the Vienna mirror frame.

Deity with Two Horns

The wooden figure of a Deity with Two Horns (Museum für Völkerkunde Wien, cat. no. 12585; Fig. 10a) has a collection history entirely different from the two other pieces now in Vienna. When in 1881 it was transferred from the Imperial Schatzkammer (Treasury) in Vienna to the collections of the then Anthropological-Ethnographic Department of the Museum of Natural History in Vienna, it was regarded as a Maori idol from New Zealand. Heger (1906: 311) suggested a possible identity of the figure with an entry in the 1590 inventory of the Kunstkammer of Archduke Karl in Graz: 'A Moorish face with several turquoise and two large pearls, on it three precious stones a large pearl lost' (Zimmermann 1888: XXIX), which was first recognised and later rejected by Nowotny (1949; 1960: 68), who instead offered the rather unlikely alternative of two items on the list of Cortés's shipment of 1519. My research has shown that the 'Moorish face' survived in Graz until 1668, when it was described as one of 'seven bigger and smaller images of idols, of which one is of black wood and shaped like a lion's head. On its back inlaid with a piece of glass, and somewhat decorated with pearl, from its appearance it must have once been decorated with stones' (Wastler 1879–1880, 6: C). It was seen in Graz by a visitor between 1604 and 1617 and confused with an Indonesian kris, before being transferred to Vienna shortly after 1765 (Feest 1986: 184–185; 1990: 31).

Carved from a piece of dark wood, the small figure (9.1 cm high, 6.7 cm wide, 6.1 cm deep) represents a squatting humanoid male with an oversized animal head with human ears and two excrescences protruding from the head. Into each of these 'horns' have been driven a nail with a golden head cast in the lost-wax technique and a silver spike; another silver spike inserted between the horns may formerly have carried a further ornament. White shell inlays attached with resin are used to mark the oval eyes, the six incisors and the fangs. The sides of the figurine display round cavities, which may have held what the old inventories describe as 'pearls'. The protruding belly was once inlaid with a necklace and a butterfly-shaped pendant. The major reason for the claim that this figure is a turquoise mosaic stems from a small head of jet (1.4 cm high) with inlays of obsidian, mother-of-pearl, red shell, turquoise and malachite and formerly fitted with ear pendants (Fig. 10d). On the underside, the soles of the sandals are represented by oval fields with cross-hatched incisions, while a formerly inlaid cavity appears to have represented the individual's penis, and the anus is marked by a round and now broken piece of malachite attached with another spike (Fig. 10b). The back has a shallow spherical cavity (6.7 cm in diameter, 1.5 cm deep) and now only shows the resin, which had been used to attach a mirror (Nowotny 1949; 1960: 61–67; Feest 1990: 28–32; Fig. 10c).

Briefly discussed by Heger (1906) and by Lehmann (1922), who suggested that the figure represented Xolotl, it was first described in detail by Nowotny (1949 and again in 1960; see also the comment by Burland 1950). Identification of the figure has shifted from Aztec to Mixtec (or Gulf coast) and from the representation of Xolotl to that of a Mixtec Ñuhu or of an Aztec-Mixtec Deity with Two Horns (see Feest 1990: 29–30). Pasztory (1983: 269, 275), to whom we owe the name presently used, suggested a possible colonial origin of the figurine. This would be supported by the mirror being of glass (according to the 1668 inventory, unknown to Pasztory), but either the author of this list could have mistaken obsidian for glass, or a possible obsidian or glass mirror could have been replaced in Europe by one made of glass.

As in the case of the animal-head mirror frame, a description of a similar object is found in the 1598 inventory of the Munich Kunstkammer: 'another [idolatrous devil's face/ mascara] also carved from wood, on the backside with a mirror cast of metal, suspended from two straps' (Fickler 2004, 1: 530–531). The fact that Maria, Archduke Karl's wife, was the daughter of Albrecht V of Bavaria, the proud owner of the Munich Kunstkammer, could indicate that the two images came from the same source or may have even been a pair (of twins, of course, if the Xolotl theory has any merits).

Notes

1. The recorded history of three of these 16 items has since been extended back in time. Both the encrusted skull and the animal head, which had been transferred in 1885 from Braunschweig to Berlin, were part of the Wolfenbüttel-Salzdahlum Kunst-

kammer and are listed in the Salzdahlum catalogue of 1753; the skull was first noted in 1700, and both pieces were illustrated in watercolours perhaps between 1746 and 1768 and probably by Philipp Wilhelm Oeding (Walz 1994: 69–73). The bone rattle in Rome, transferred in 1878 from Bologna, was noted in Bologna in 1745 as a gift from Pope Benedict XIV (Laurencich Minelli 1992: 124; see also Laurencich Minelli, in this vol., pp. 165–172).

2. These include a mask catalogued in Florence in 1553–1555 (Lehmann 1906: 320), a knife handle illustrated by Ulisse Aldrovandi (1648: 156) and another one by Liceto (1634; cp. Lehmann 1906: 319 n. 19). What was described as a wooden sceptre by Luigi Pigorini, who had not seen it himself, owned in 1891 by the Duchess (Teresa Massimo) di Rignano (Heikamp 1976: 474–475), may be the British Museum double-headed serpent, which was obtained from the duchess in 1894 (Carmichael 1970: 36). For shields, animal heads and wooden figures see below, and also Caygill (in this vol., pp. 183–196).
3. In my first survey of Mexico and South America in the European *Wunderkammer*, I discussed the 22 pieces described by Oppel, but gave the total number as 23 (Feest 1985: 238–239), following Lehmann (1906) and Saville (1922), who had already included the Deity with Two Horns, which I had listed separately under 'Other Mexican objects'. The list of 'twenty-four mosaics ... known in 1915' given by McEwan et al. (2006: 24) enumerates only 19 items, although it overstates the number of masks (even if the inlaid skulls are counted as masks, there is no mask in Vienna and none in Washington that was historically collected), but misses the knife handles and shields.
4. Nowotny (1960: 39), apparently on the basis of his reading of the sources, identifies the resin as that of the 'bat tree' (*tzinacanquauitl*); his suggestion that a black drawing of the design was made on the wood before the application of the resin is likewise based on the literature and not upon inspection of the shield.
5. The identification of the material as turquoise was made for Heger by the curator of the mineralogical collections of the Museum of Natural History in Vienna, Fritz Berwerth. Ilg and Boeheim (1882) had guessed that it was 'nephrite (?)'.
6. I placed a 1 × 1 cm grid over the shield, with columns numbered from the centre outward (negative numbers to the left, positive numbers to the right) and rows designated with letters (upwards in upper case, downwards in lower case). The squares included in the count are M12, M11, I8, H6, H3, e5, l-1, n-2, n-3, a3, A-4, a-4, l-4, l-5.
7. See also Feest (1990: 24–25), who discusses other possibilities, including one of six shields among the 'Joyas de Yndias' of Charles V, described in 1561 as showing 'two moons of turquoise, decorated with gold, below which is gold'. However, the presence of gold on the face of the shield makes this an unlikely alternative.
8. See also data available via the National Museum of the American Indian website: http://www.nmai.si.edu/searchcollections/results.aspx?&catids=2&objtypeid=Hunting%2fFishing%2fWarfare%3a+Warfare|Shield&src=1-4 (accessed 28 April 2011).

References

Aldrovandi, U. (1648) *Museum Metallicum*. Bologna: Ferroni.

Anon. (1571) *Relación verdadera, del recibimiento, que la muy noble y muy mas leal ciudad de Burgos ... hizo a la Magestad Real de la Reyna nuestre señora, doña Anna d'Austria, primera de este nombre: passando a Segouia, para celebrar en alla su felicissimo casamiento con el Rey don Philipp nuestro señor, segundo de este nombre*. Burgos: Phellipe de Iunta.

Boeheim, W. (1888) 'Quellen zur Geschichte der kaiserlichen Haussammlung und der Kunstbestrebungen des durchlauchtigsten Erzhauses', Urkunden und Regesten aus der k.u.k. Hofbibliothek. *Jahrbuch der kunsthistorischen Sammlungen des allerhöchsten Kaiserhauses* **7**(2): XCI–CCCXIII.

Burland, C.A. (1950) 'Further comments on the wooden figure from Mexico', *American Antiquity* **15**(3): 251.

Carmichael, E. (1970). *Turquoise Mosaics from Mexico*. London: British Museum.

Easby, E.K. and Scott, J.F. (1970) *Before Cortés: Sculpture of Middle America*. New York: The Metropolitan Museum of Art.

Feest, C. (1985) 'Mexico and South America in the European *Wunderkammer*', in O. Impey and A. MacGregor (eds), *The Origin of Museums*. Oxford: Clarendon Press, 237–244.

Feest, C. (1986) 'Zemes Idolum Diabolicum: surprise and success in ethnographic *Kunstkammer* research', *Archiv für Völkerkunde* **40**: 181–198.

Feest, C. (1990) 'Vienna's Mexican treasures: Aztec, Mixtec, and Tarascan works from 16th century Austrian collections', *Archiv für Völkerkunde* **45**: 1–64.

Fickler, J.B. (2004). *Das Inventar der Münchner herzoglichen Kunstkammer von 1598*, P. Diemer with E. Bujok and D. Diemer (eds). Bayerische Akademie der Wissenschaften, Philosophisch-historische Klasse, Abhandlungen, N.F. 125. München: Verlag der Bayerischen Akademie der Wissenschaften.

Gaida, M., König, V. and Nicklisch, A. (2003) 'Katalog der zusätzlichen Leihgaben in Berlin und Bonn', in *Azteken*. Köln: Dumont, XIX–XXXII.

González Licón, E. and Márquez Morfín, L. (1994) 'Rito ceremonial prehispánico en las cuevas de la Cañada, Oaxaca', in H.B. Nicholson and E. Quiñones Keber (eds), *Mixteca-Puebla: Discoveries and Research in Mesoamerican Art and Archaeology*. Culver City, CA: Labyrinthos Press.

Heger, F. (1892) 'Altmexikanische Reliquien aus dem Schlosse Ambras in Tirol', *Annalen des k.k. Naturhistorischen Hofmuseums* **7**: 379–400.

Heger, F. (1906) 'Verschwundene altmexikanische Kostbarkeiten des XVI. Jahrhunderts, nach urkundlichen Nachrichten', in *Anthropological Papers written in Honor of Franz Boas*. New York, 306–315.

Heikamp, D. (1976) 'American objects in Italian collections of the Renaissance and Baroque: a survey', in F. Chiapelli (ed.), *First Images of America* (2 vols), vol. 1. Berkeley: University of California Press, 455–482.

Ilg, A. and Boeheim, W. (1882) *Das k.k. Schloss Ambras in Tirol*. Wien: Holzhausen.

Laurencich Minelli, L. (1992) *Bologna e il Mondo Nuovo*. Bologna: Grafis Edizioni.

Lehmann, W. (1906) 'Altmexikanische Mosaiken und die Geschenke König Motecuzomas an Cortés', *Globus* **90**: 318–320.

Lehmann, W. (1922) *Altmexikanische Kunstgeschichte*. Orbis Pictus 8. Berlin: E. Wasmuth.

Liceto, F. (1634) *Pyronarcha sive de fulminum naturae deque felssium origine*. Padua: Crivelari.

Martin Gonzalez, J. J. (1990) 'Obras artísticas de procedencia americana en las colecciones reales españolas siglo XVI', in *Relaciones Artisticas entre la Península Ibérica y América* (Actas del V Simposio Hispano-Portugués de Historia del Arte). Valladolid: Universidad de Valladolid, 157–162.

Martínez del Campo Lanz, S. (ed.) (2010) *La Máscara de Malinaltepec*. Mexico City: Instituto Nacional de Antropología e Historia (INAH).

McEwan, C., Middleton, A., Cartwright, C.R. and Stacey, R. (2006) *Turquoise Mosaics from Mexico*. London: British Museum Press.

Nowotny, K.A. (1949) 'A unique wooden figure from ancient Mexico', *American Antiquity* **15**(1): 57–61.

Nowotny, K.A. (1960) *Mexikanische Kostbarkeiten aus Kunstkammern der Renaissance*. Wien: Museum für Völkerkunde.

Nuttall, Z. (1891) 'Mitteilungen über einen aztekischen Federschild in Ambras', *Zeitschrift für Ethnologie* **23**: 485–487.
Nuttall, Z. (1892) 'On ancient Mexican shields', *Internationales Archiv für Ethnographie* **5**: 34–53, 89.
Oppel, A. (1896) 'Die altmexikanischen Mosaiken', *Globus* **70**: 4–13.
Pasztory, E. (1983) *Aztec Art*. New York: Harry N. Adams.
Purin, S. (2003) 'Shield with turquoise mosaic', in *Aztecs*. London: Royal Academy of Arts, 474–475.
Rowlands, J. (1988) *The Age of Dürer and Holbein: German Drawings 1400–1550*. London: British Museum Publications.
Sacken, E. Freiherr von (1855) *Beschreibung der Ambraser Sammlung*, 2 vols. Wien: W. Braumüller.
Saville, M. (1922) *Turquoise Mosaic Art in Ancient Mexico*. New York: Museum of the American Indian Heye Foundation.
Scheicher, E. (1979) *Die Kunst- und Wunderkammern der Habsburger*. Wien: Molden Edition.
Scott, S. (2010) 'La Máscara de Malinaltepec y sus compañeras', in S. Martínez del Campo Lanz (ed.), *La Máscara de Malinaltepec*. Mexico City: Instituto Nacional de Antropología e Historia (INAH), 65–109.
Solis Olguín, F. (2004) *I tesori degli Aztechi*. Milano: Mondadori Electa.
Steele, J.F. and Snavely, R. (1997) 'Cueva Cheve Tablet', *Journal of Cave and Karst Studies* **59**(1): 26–32.
Vargas Pacheco, E. (ed.) (1989) *Las máscaras de la cueva de Santa Ana Teloxtoc*. Serie Antropología 105. Mexico City: Instituto de Investigaciones Antropológicas, Universidad Nacional Autónoma de México (UNAM).
Velásquez, A. and Maríon, M.E. (2006) 'Turquoise mosaic disc from the Templo Mayor', in C. McEwan, A. Middleton, C.R. Cartwright and R. Stacey (eds), *Turquoise Mosaics from Mexico*. London: British Museum Press, 61.
Walz, A. (1994) *Seltenheiten der Natur als auch der Kunst: Die Kunst- und Naturalienkammer auf Schloss Salzdahlum*. Braunschweig: Herzog Anton-Ulrich Museum.
Wastler, J. (1879–1880) 'Zur Geschichte der Schatz-, Kunst- und Rüstkammer in der k.k. Burg zu Grätz', *Mittheilungen der k.k. Central-Kommission zur Erforschung und Erhaltung der Kunst- und Historischen Denkmale*, **N.F. 5**: CXXXVIII–CXLVI, **N.F. 6**: XXIX–XXXV, LV–LXII, XCVI–CV, CXLVIII–CLI.
Zimmermann, H. (ed.) (1885) 'Urkunden und Regesten aus dem k.u.k. Haus-, Hof-, und Staatsarchiv in Wien', *Jahrbuch der kunsthistorischen Sammlungen des allerhöchsten Kaiserhauses* **3**: LXXXII–CLVI.
Zimmermann, H. (1888) 'Urkunden, Acten und Regesten aus dem Archiv des k.u.k. Ministerium des Innern', *Jahrbuch der kunsthistorischen Sammlungen des allerhöchsten Kaiserhauses* **7**(2): XV–LXXXIV.

Author

Christian Feest, formerly of the Museum für Völkerkunde, Vienna, Austria

The symbolism of turquoise in ancient Mesoamerica

Karl A. Taube

Introduction

Among the most celebrated items of ancient Mesoamerican artistry and craftsmanship are turquoise mosaics, typically fashioned of minute tesserae of turquoise and other precious materials delicately glued onto a wooden backing (see Izeki 2008). However, along with metallurgy, turquoise did not become widespread in Mesoamerica until roughly AD 900, at the beginning of the Postclassic period. Deriving from the American Southwest, turquoise began to supplant jadeite as the most important gemstone in ancient Mesoamerica. Thus the Aztec Nahuatl term for jadeite, *chalchihuitl*, may derive from the words for sand, *xalli*, and turquoise, or *xihuitl* (Mena 1927: 7). According to Foshag (1957: 8), this possible etymology may be due to the natural properties of jadeite. In contrast to the rather smooth, glass-like surface of fractured turquoise, freshly broken jadeite tends to appear quite rough and grainy, due to the stone's cryptocrystalline structure. But while both turquoise and jade may have been linguistically linked in Nahuatl, the symbolic significance of turquoise was quite distinct from that of jade, a stone widely identified with water and maize in ancient Mesoamerica. In contrast, turquoise was very much a 'sky stone' and related closely to the sun, fire and meteors. Along with the Aztec god of fire, Xiuhtecuhtli, meaning 'turquoise lord', there was also the Xiuhcoatl, the turquoise fire serpent. The oddly segmented body of this creature probably relates to a widespread Mesoamerican belief that meteors convert into worms and caterpillars upon striking the earth (Taube 2000a). The Xiuhcoatl spear thrower wielded by Huitzilopochtli and other Aztec gods is a star-shooter of meteoric darts.

In this study, I discuss the symbolism of turquoise in relation to three major periods of ancient Mesoamerica: the Classic, Early Postclassic and Late Postclassic periods. Although during the Classic period (AD 250–900) there was widespread and extensive trade in Mesoamerica, turquoise is virtually unknown in the archaeological record. This is largely due to the fact that sources of this stone are absent in Mesoamerica proper, but rather occur far to the north in what is now Arizona and New Mexico of the American Southwest. Nonetheless, a number of symbolically important elements later ascribed to turquoise were already present in the Classic period. It is not until the Early Postclassic (AD 900–1250) that turquoise becomes widely present in Mesoamerica. During this period, the major political centre in central Mexico was Tula, Hidalgo, which had extensive trade contacts extending to the American Southwest and as far as Costa Rica. In Toltec iconography of Tula and Chichen Itza in Yucatan, there is elaborate imagery pertaining to turquoise ornaments and their attendant symbolism, much of which can be found with the later Aztec of Late Postclassic Mesoamerica (AD 1250–1521). It appears that many aspects of the 'cult of turquoise', including particular costume regalia, originated with the Early Postclassic Toltec. Rather than simply inheriting a more ancient symbolic complex, it is likely that the Aztec consciously evoked the more ancient Toltec with the stone, both being related to warriors, kingship and the revered ancestral past.

Classic period precursors to Postclassic turquoise symbolism

During the Early Classic period (AD 250–600) the great metropolis of Teotihuacan in the Valley of Mexico had extensive trade contacts in Mesoamerica, including the Pacific and Gulf coasts, Oaxaca and the Maya region. However despite the fact that Teotihuacan secured *Spondylus princeps* shell from the shores of west Mexico, there is no turquoise known at Teotihuacan, despite decades of extensive survey and excavation (see Turner 1992). Nonetheless, themes pertaining to the later symbolic significance of turquoise, including a supernatural creature pertaining to war, fire and meteors were already present at Teotihuacan. Perhaps the earliest and most grandiose portrayals of this being appear on the Temple of Quetzalcoatl, a structure dating to roughly AD 200. The façade features two types of tenoned heads, one clearly the plumed serpent and the other a quite schematic and angular head covered with a surface of platelets (Fig. 1a). Rather than a living creature, the head has no body and is a form of helmet mask fashioned of shell plaques, with probably some 150 of these massive heads on all sides of the structure, making it by far the most ambitious programme portraying mosaic work known in ancient Mesoamerica.

In a number of studies (Taube 1992b, 2000a) I have argued that this mosaic helmet mask is a basic Teotihuacan symbol of war – a theme entirely consistent with the dedicatory mass sacrifices in the pyramid foundations. In many cases, the victims were dressed and armed in Teotihuacan military garb (see Sugiyama 1989, 1992). A painted Teotihuacan-style wood panel, possibly part of a box, depicts a male wearing a

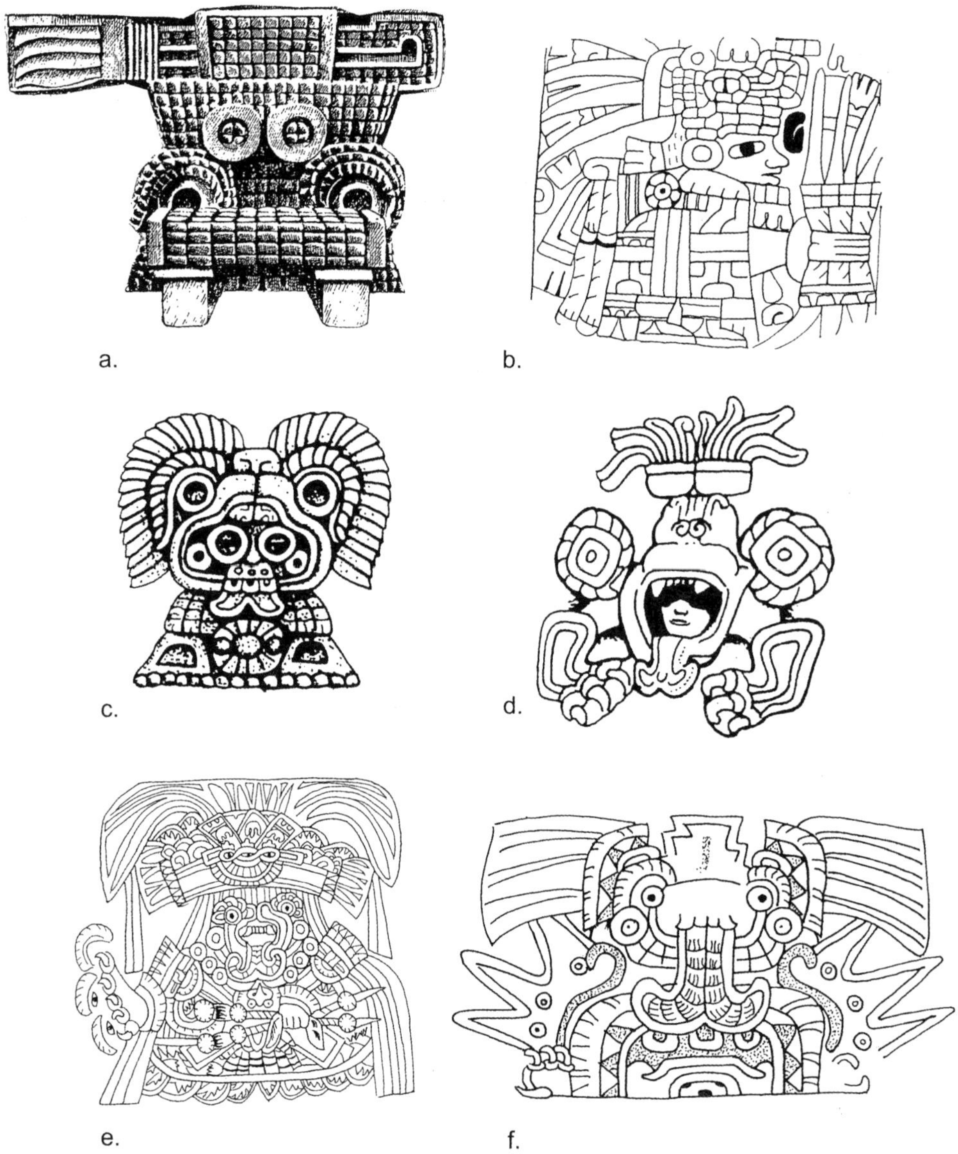

Figure 1. The Early Classic War Serpent of Teotihuacan: (a) War Serpent mosaic helmet headdress from the Temple of Quetzalcoatl, Teotihuacan (detail from Caso and Bernal 1952: fig. 184); (b) Teotihuacan figure wearing War Serpent headdress, detail of painted wooden slab (after Berrin and Pasztory 1993: no. 55); (c) ceramic figurine of Teotihuacan warrior bundle with helmet headdress of the War Serpent (from Taube 2000a: fig. 10.20a); (d) War Serpent with feline forelimbs, petalled eyes and bifurcated tongue, detail of ceramic censer lid from Xico (from Taube 2000a: fig. 10.11b); (e) mural portraying the War Serpent with petalled eyes and bifurcated tongue, Tepantitla, Teotihuacan (after Miller 1973: fig. 193); (f) Teotiuacan War Serpent with star signs on forelimbs and Xi sign atop the head, detail of stucco painted vessel (after Conides 2001: fig. 67b). (All drawings: K. Taube.)

notably similar headdress, although with plumed feline ears, a common characteristic of this being at Teotihuacan (Fig. 1b). In addition, ceramic figurines identified as warrior mortuary bundles commonly wear headdresses of this being (Fig. 1c; for Teotihuacan warrior bundles, see Taube 2000a: 306–309; Headrick 2007: 55–66).

Aside from the Temple of Quetzalcoatl and Teotihuacan figurines, similar examples of zoomorphic shell helmet headdresses appear with Classic Maya portrayals of figures dressed in Teotihuacan-style military costume (Fig. 2). Due to its clear serpent attributes in the Maya region, I termed it the War Serpent (Taube 1992b). In addition, Classic Maya texts explicitly labelled it as a serpent, with its epigraphic epithet being Uaxaklahuan Ub'ah Chan, or 18 Its Image/Head Snake (Freidel et al. 1993: 308–310). However, the faunal composition of this supernatural entity is more complex, as it also bears jaguar and even butterfly attributes. In many Late Classic Maya examples, it displays crenellated edgings also found with butterflies in both Classic and Postclassic central Mexico (Fig. 2e). Because of the combined serpent and butterfly traits I suggested that the War Serpent is in large part based on the concept of a supernatural caterpillar (Taube 2000a), a theme that accords well with Teotihuacan and Postclassic central Mexican concepts of the souls of warriors as butterflies, that is, the caterpillar and chrysalis symbolising the metamorphosis of the burning warrior bundle into the fiery butterfly soul (for the relationship of

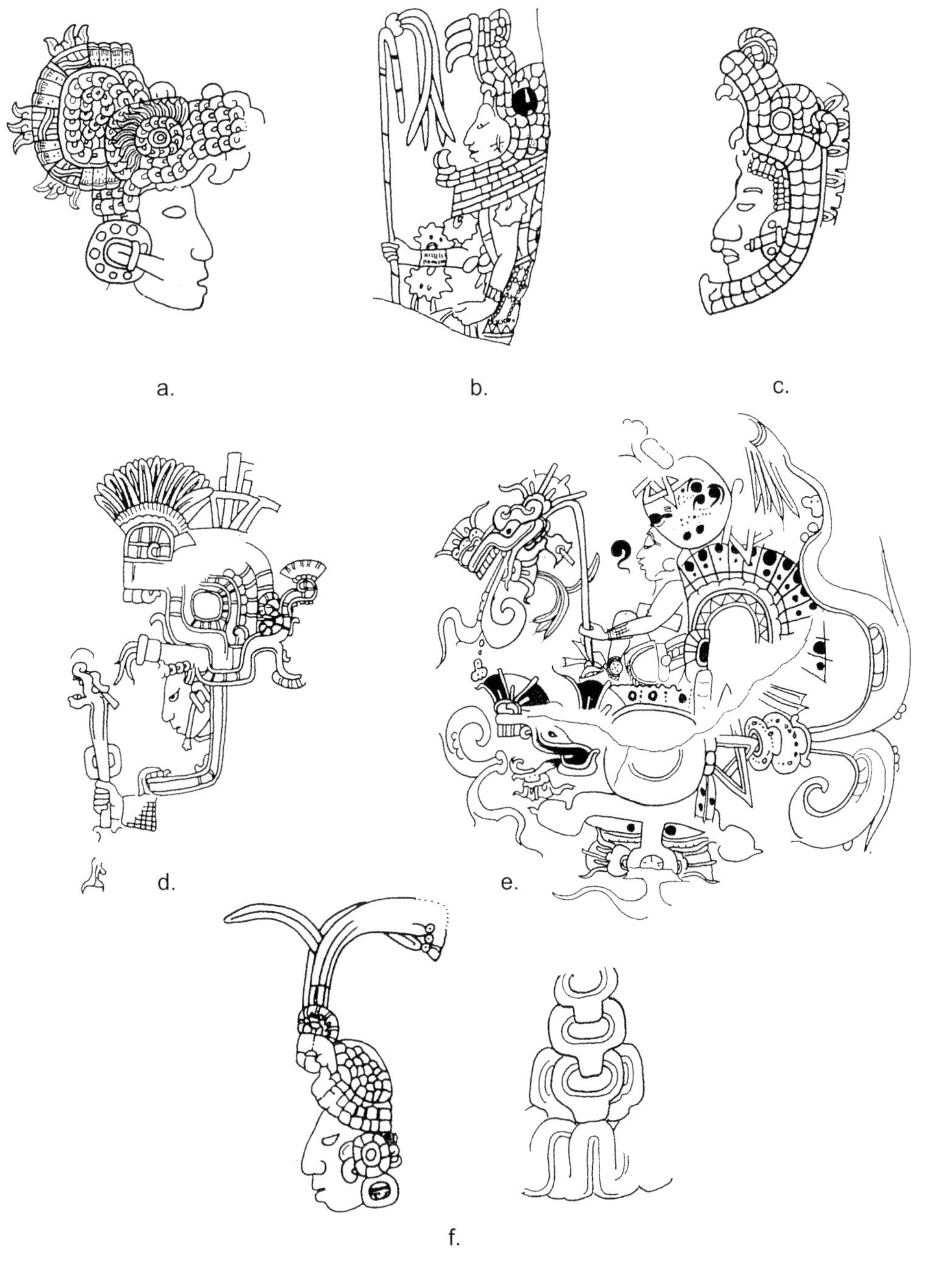

Figure 2. Classic Maya portrayals of the War Serpent: (a) Maya ruler Nun Yax Ayiin wearing War Serpent headdress with petalled eyes and burning feline ears (from Taube 2000a: fig. 10.2e); (b) Late Classic Maya figure wearing War Serpent mosaic helmet headdress (from Taube 1992b: fig. 6b); (c) Maya ruler with War Serpent helmet headdress, Lamanai Stela 9 (from Taube 1992b: fig. 6a); (d) Late Classic Maya king Chan Muwan wearing mosaic War Serpent headdress and wielding burning War Serpent spear thrower, Bonampak Stela 1 (from Taube 1992b: fig. 6d); (e) Late Classic Codex Style bowl of Maya figure with serpent spear thrower atop a burning War Serpent (from Taube 1992b: fig. 8d); (f) War Serpent headdress and burning rattlesnake tail worn by Maya ruler dressed in Teotihuacan-style military garb, detail of Stela 2, Naranjo (after Graham and von Euw 1975, II: 13). (All drawings: K. Taube.)

Teotihuacan butterflies to warrior souls, see Berlo 1983, 1984; Headrick, 2007; Taube 2000a, 2005, 2006).

Although the Classic Maya War Serpent is strongly serpentine, at Early Classic Teotihuacan it bears more attributes of the jaguar, including heavily clawed limbs, large ears and a toothy feline maw, although with a long, bifurcated serpent tongue as well as the feathered eyes commonly found with butterflies (Fig. 1c–f). In addition, for the Temple of Quetzalcoatl sculptures – the earliest known examples at the site – the creature has the large scroll behind the eye commonly found with Teotihuacan serpents along with a slightly upturned snout, not a trait found with Teotihuacan felines (see Caso and Bernal 1952: fig. 184). At present, however, there are no known Teotihuacan examples of the War Serpent with a serpent body, and for this reason, the creature could perhaps more accurately be considered a 'War Jaguar' for Teotihuacan. Nonetheless, for this discussion, the term War Serpent will be used for both Teotihuacan as well as Maya examples, with the caveat that the faunal characteristics probably vary somewhat between the two cultures, with the Teotihuacan form being

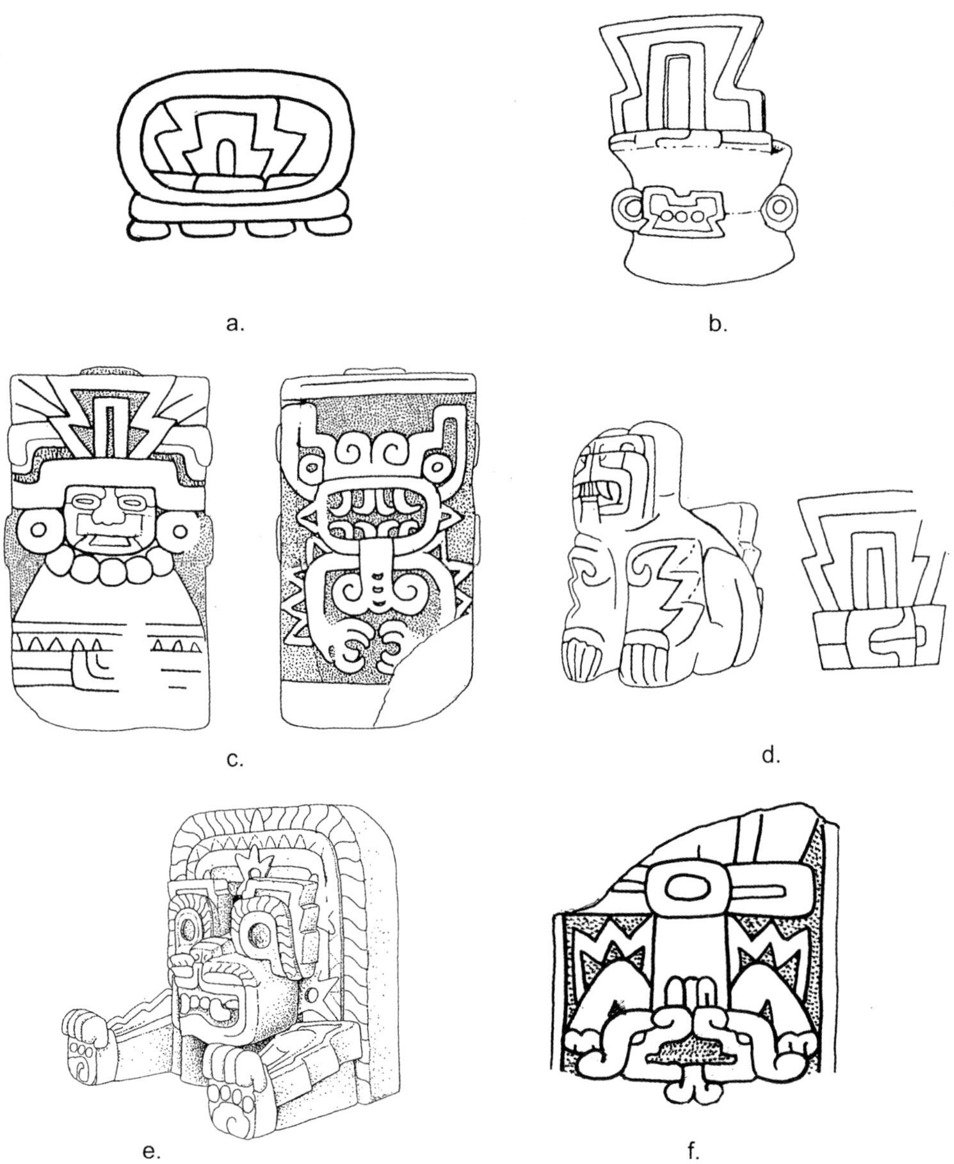

Figure 3. The Classic period Xi glyph and the Teotihuacan War Serpent: (a) Xi glyph with the coefficient of 9, detail of Teotihuacan Thin Orange vase from the Aztec Templo Mayor precinct (after López Luján et al. 2000: fig. 8.6a); (b) Xi sign headdress atop Late Classic censer from Xochitecatl (from Taube 2000b: fig. 32d); (c) Teotihuacan-style stela with figure wearing Xi sign headdress and War Serpent on the opposite side (after Harmer Rooke Galleries 1987: no. 37); (d) Teotihuacan alabaster sculpture of feline War Serpent (from Taube 2000b: fig. 32f,h); (e) Teotihuacan War Serpent emerging from starry feathered portal (after Matos Moctezuma and Solís Olguín 2002: no. 17); (f) monument portraying diving War Serpent with bifurcated tongue and serrated forelimbs (after Armillas 1944: 261). (All drawings: K. Taube.)

more feline and its Maya counterpart more serpentine. An excellent example of this creature at Teotihuacan appears in a mural from the Tepantitla compound (Fig. 1e). Along with a fanged jaguar muzzle, bifurcated serpent tongue and petalled butterfly eyes, it holds a bundle of darts in its human left hand and has a menacing, taloned jaguar paw as its right. A similar figure appears on a stucco-painted Teotihuacan vase, in this case with outstretched jaguar paws and a pair of large trefoil forms flanking the arms (Fig. 1f). These sharply pointed elements are clearly versions of Mesoamerican star signs, including examples known for the middle panels of the Late Classic South Ball Court at El Tajín (see Koontz 2009: 58, figs 3.13, 3.14). An Early Classic Teotihuacan censer lid attributed to the southern area of the Valley of Mexico depicts another example, in this case with prominent jaguar forelimbs and a pair of feathered elements on the snout probably denoting butterfly antennae (Fig. 1d). A human face projects from the maw of this being, indicating that the entire composition is an idealised version of the zoomorphic helmet headdress.

In a sharp departure from the frequent but erroneous identification of the helmet headdress on the Temple of Quetzalcoatl as Tlaloc, Alfonso Caso and Ignacio Bernal (1952: 113–114) suggested that it constitutes an ancestral form of the Aztec Xiuhcoatl, the turquoise fire serpent, an argument that I pursued in subsequent research, including the relation of Classic profiles of this being to Early Postclassic versions of the Xiuhcoatl (Taube 1992b; 2000a: fig. 10.10). Viewed *en face*, the War Serpent helmet headdress from the Temple of

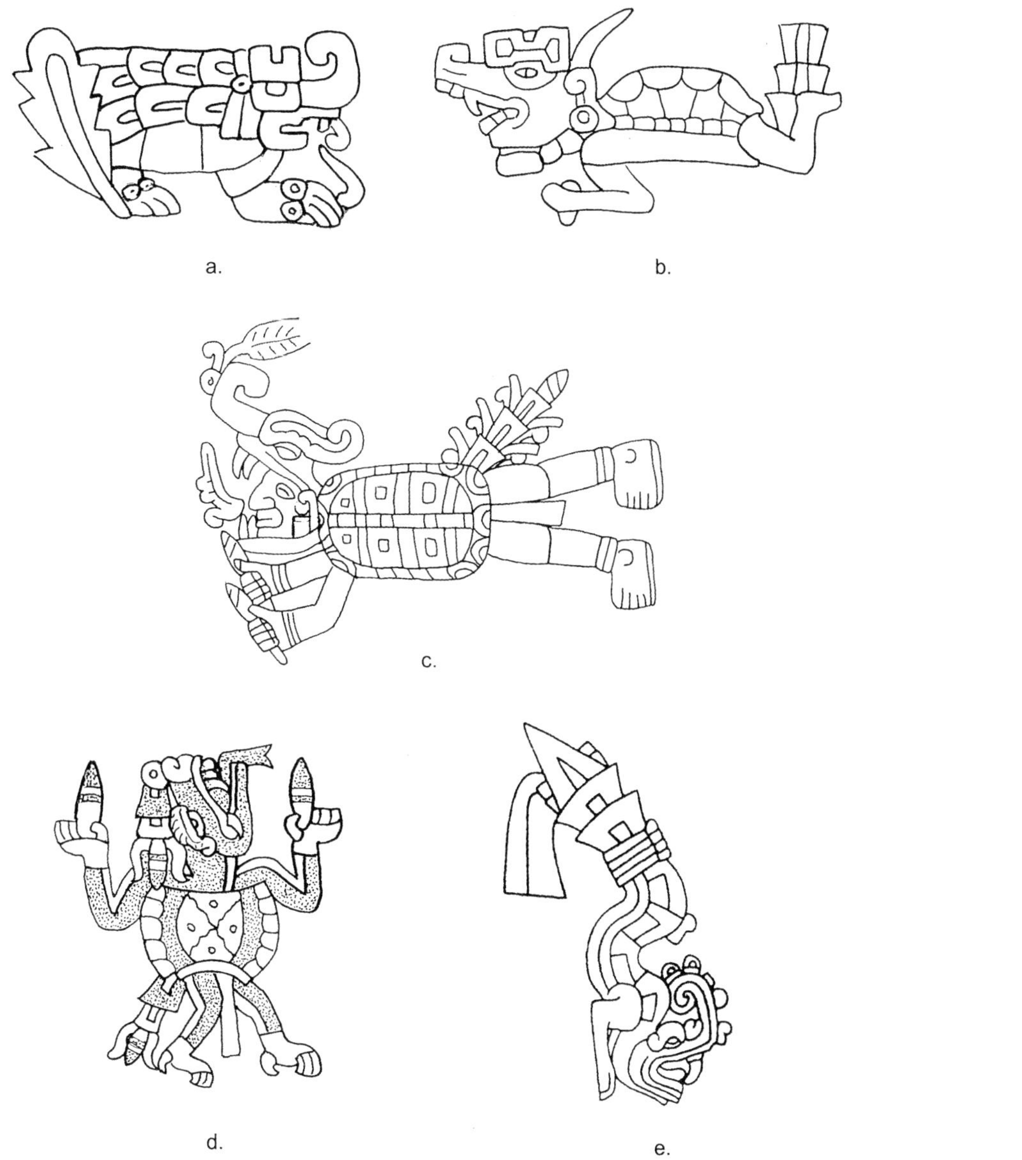

Figure 4. The Zapotec Xicani, Mixtec Yahui and the Aztec Xiuhcoatl: (a) Zapotec Classic period Xicani with turtle carapace and Xi sign tail (after Urcid Serrano 2001: fig. 4.107,2); (b) Late Classic Xicani with Xi sign tail, detail of Noriega Relief (after Urcid Serrano 2001: fig. 4.107,2); (c) flying Xicani or Yahui with burning Xi sign tail, Tomb 1, Zaachila (after Caso 1966: fig. 20); (d) Mixtec Yahui with burning Xi sign tail and another on its snout (Codex Nuttall, p. 12); (e) detail of Aztec monument of diving Xiuhcoatl with Xi sign tail, British Museum (from Taube 2000a: fig. 10.12g). (All drawings: K. Taube.)

Quetzalcoatl has a trapezoidal brow, creating in outline a stepped form resembling a pair of downward-pointing truncated triangles (Fig. 1a). This form probably denotes a still poorly understood Classic Mexican day sign first identified and named by Caso (1967: 175) concerning the writing of Late Classic Xochicalco. Terming it the Xi glyph, Caso noted the similarity of this stepped sign to the tail of the Late Postclassic Xiuhcoatl, a theme to be discussed in detail below (Fig. 4c–e and Fig. 13a).

Although Caso described Xi signs for Late Classic Mesoamerica, a Teotihuacan-style, Early Classic Thin Orange vessel excavated at the Aztec Templo Mayor indicates that this glyph was fully present at Teotihuacan (Fig. 3a). In this case and examples at Teotihuacan, the Xi glyph has a horizontal knotted element at the base, recalling knotted headdresses in Teotihuacan iconography (Fig. 3a and d). In addition, at Classic period Teotihuacan and Xochitecatl, it can be a headdress (Fig. 3b and c). For one Teotihuacan-style stela, a human figure wears this headdress on one side with a feline War Serpent on the other (Fig. 3c). Along with a long, bifurcated tongue, this War Serpent has sharply pointed elements lining its forelimbs, clearly a version of the trefoil star elements emanating out of the limbs of the aforementioned War Serpent appearing on an Early Classic Teotihuacan vessel (Fig. 1f). Resembling the serrated sides of the Xi sign, similarly marked limbs appear with other Teotihuacan examples of the War Serpent and may represent fiery brilliance (Fig. 3d–f).

At Teotihuacan, the Xi sign also appears with the War Serpent as a cranial element. Thus, for the aforementioned vessel scene with the War Serpent with probable star signs, it appears prominently on its brow (Fig. 1f). A remarkable sculpture from the Xala compound portrays this creature

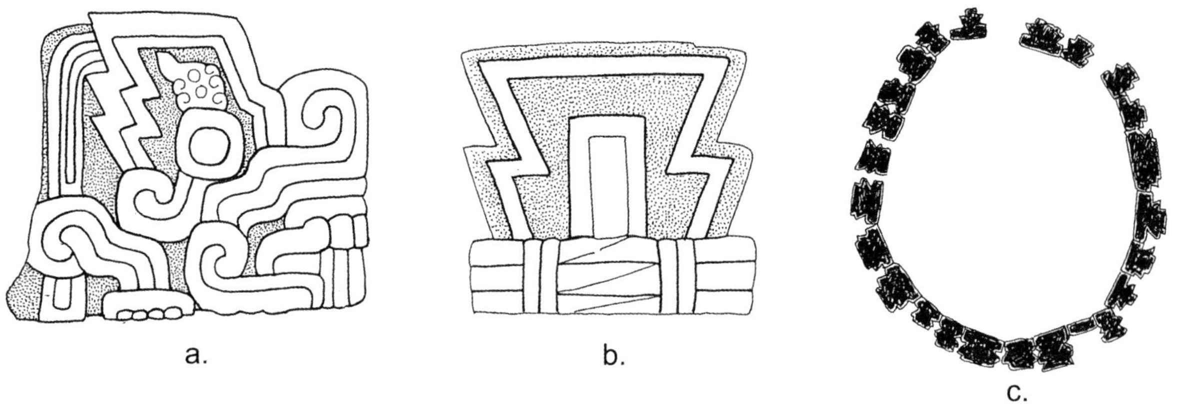

Figure 5. Monuments and turquoise from the Late Classic sites of Las Plazuelas, Guanajuato: (a) ball court sculpture of War Serpent with Xi sign in profile on brow (after Castañeda López 2008: 47); (b) Xi sign *almena* sculpture (after Castañeda López 2008: 45); (c) turquoise beads in form of Xi signs (after Arqueología Mexicana 2008: 25). (All drawings: K. Taube.)

with talons extended emerging from a feathered portal (Fig. 3e). The stepped ears of this creature form the Xi glyph, a convention recalling the mosaic helmet headdresses from the Temple of Quetzalcoatl. In addition, the serrated forelimbs of the Xala creature also depict the Xi sign, and this is also the case for a *tecalli* example, where it appears as the creature's tail with the aforementioned knot below (Fig. 3d).

A stone monument from Mexiquito – located on the Balsas River in Guerrero – portrays a diving, Teotihuacan-style War Serpent with the bifurcated tongue and serrated forelimbs, although here with elbows up in a diving avian position rather than in the extended limb and claws pose of the Xi sign from Teotihuacan (cf. Fig. 3f; Fig. 1d and f, Fig. 3e). Most probably, the Mexiquito monument depicts a diving, shooting star as the fiery War Serpent. Javier Urcid Serrano (2001: 202–204) notes that along with appearing as a day glyph in Zapotec writing, the Xi sign also serves as the tail of the Zapotec version of the Xiuhcoatl, the turquoise fire serpent of Late Postclassic central Mexico (Fig. 4a and b). Terming this creature Xicani, Urcid Serrano notes that it can be documented for the Zapotec as early as Monte Alban II (ca. 200 BC–AD 200). The Mixtec also had a form of the Zapotec Xicani, a being that commonly has a turtle carapace covering its body (Fig. 4c and d). Mary Elizabeth Smith (1973) noted that in the colonial Mixtec Codex Muro, it is glossed as *yaui* or *yavi*, and is clearly related to the term *yahui* appearing in the sixteenth-century Alvarado dictionary in reference to flying wizards (see Taube 2001). According to John Pohl (1994: 44), the Yahui is a supernatural form of fiery meteor or comet that flies through the sky, and in many cases it has a flaming tail. In addition, a tomb relief from Zaachila – a site with both Zapotec and Mixtec ancestry – portrays a flying Yahui or Xicani with flints in its hands as well as a burning tail tipped with a flint blade, a graphic depiction of a shooting meteor. Along with the Classic period War Serpent and the Aztec Xiuhcoatl turquoise serpent, the Zapotec Xicani and the Mixtec Yahui are also resplendent, meteoric beings (Taube 2001).

Massive stone examples of the Xi glyph stepped element and knot have recently been found at the Classic period site of Las Plazuelas, Guanajuato (Fig. 5b). Their style, including the rendering of the basal knot, strongly resembles the artistic canons of Teotihuacan. Aside from these sculptures, a monumental carving from the Plazuelas ball court features a supernatural creature with a crest in the form of the Xi sign in profile (Fig. 5a). A clear composite of serpent and jaguar, this creature constitutes a local version of the Classic period War Serpent. Along with a forelimb and feline canines, it has the upwardly turned snout commonly found with serpents in ancient Mesoamerica, including the mosaic headdresses from the Temple of Quetzalcoatl. Plazuelas also has one of the earliest documented examples of turquoise known for Mesoamerica, this being a series of beads carved in the form of the Xi sign (Fig. 5b). The site is located adjacent to the Lerma-Santiago river system that continues to the Pacific coast of Nayarit, a major trade corridor linking Mesoamerica to the American Southwest (Sauer 1932).

As has been mentioned, the Classic period War Serpent is also a fiery being that can appear with star elements and at times diving, suggesting that for the Classic period, the War Serpent is also a celestial embodiment of meteors. For Classic Maya examples, it can have a burning tail, including Naranjo Stela 2, which portrays a Maya king in Teotihuacan-style military costume with a War Serpent mosaic platelet headdress and a burning rattlesnake tail between his legs (Fig. 2f). A Late Classic, Maya 'Codex Style' bowl depicts a Teotihuacan warrior figure atop a War Serpent breathing fire with a burning rattlesnake tail and an inverted star sign below its belly (Fig. 2e). The warrior holds another burning War Serpent in his left hand, and as Andrea Stone (1989: 158) notes, this object probably denotes a spear thrower. At Late Classic Bonampak, Stela 1 portrays the ruler Chan Muwan dressed as the War Serpent while wielding another as his spear thrower (Fig. 2d). As in the case of the Naranjo stela and the Codex Style bowl, the serpent tail emits smoke or flames. Although serpent spear throwers remain to be documented at Teotihuacan, the mosaic War Serpent helmets from the Temple of Quetzalcoatl are topped by a horizontal element with one end feathered and the other curved (Fig. 1a and Fig. 6d). The horizontal form is notably similar to Teotihuacan portrayals of spear throwers, and it is likely that the entire motif depicts this weapon atop the War Serpent helmet (Fig. 6b–e). In terms of the later symbolism of turquoise in Mesoamerica, this is an important theme, as the Aztec turquoise fire serpent, or Xiuhcoatl, was

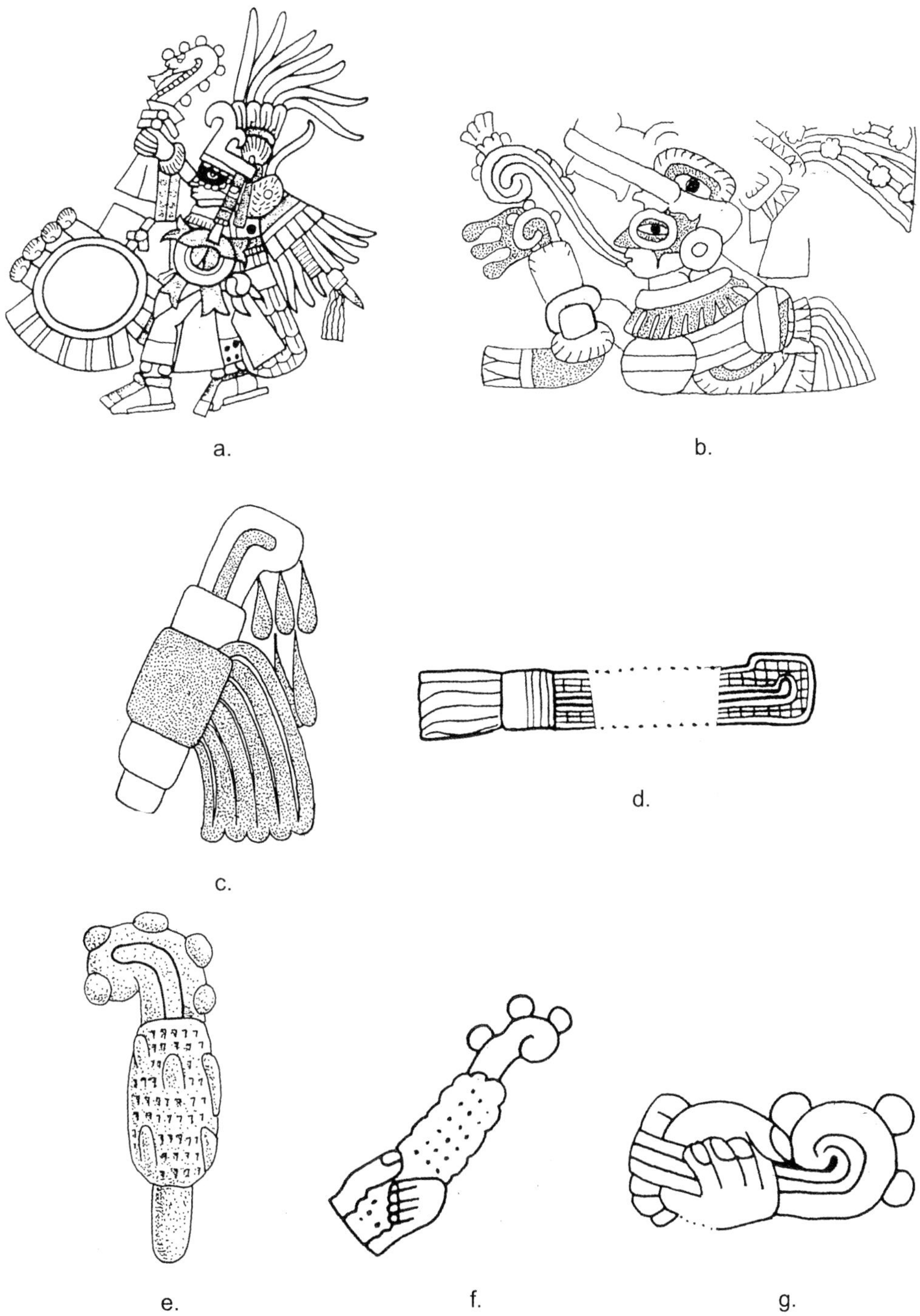

Figure 6. Aztec and Teotihuacan portrayals of spear throwers: (a) Aztec deity Huitzilopochtli wielding Xiuhcoatl spear thrower lined with stars (Codex Borbonicus, p. 34); (b) Early Classic warrior holding spear thrower tipped with blood and probable star, detail of Teotihuacan stucco-painted vase (after von Winning 1968: plate 310); (c) spear thrower with drops of blood falling from curving tip, detail of mural from Teotihuacan (after Gamio 1922, I: pl. 29); (d) possible spear thrower atop head of War Serpent helmet headdresses from the Temple of Quetzalcoatl (see Fig. 1a; after Caso and Bernal 1952: fig. 184); (e) Teotihuacan ceramic effigy of spear thrower with probable stars on tip (after Séjourné 1959: fig. 135); (f) Early Classic Teotihuacan-style spear thrower appearing on 'ball court marker' monument from El Mundo Perdiod, Teotihuacan (from Taube 2000a: fig. 10.17d); (g) Early Classic spear thrower appearing in text of Tikal Stela 31 (from Taube 2000a: fig. 10.17e). (All drawings: K. Taube.)

the pre-eminent spear thrower of Huitzilopochtli, a shooter of fiery meteoric darts (Fig. 6a).

The circular elements commonly appearing on Late Postclassic Aztec portrayals of the Xiuhcoatl serpent spear thrower as well as its snout denote stars, a motif that appears already in Early Classic Teotihuacan (Fig. 6a,b and e). One Early Classic Teotihuacan painted stucco vessel depicts a warrior wielding a spear thrower with a trefoil blood sign along with three circles denoting stars (Fig. 6b). The feathered butt of this weapon recalls the suggested spear thrower atop the War Serpent helmets from the Temple of Quetzalcoatl (Fig. 1a and Fig. 6d). A ceramic item from the same site depicts the star elements three-dimensionally at the curving tip of the spear thrower, and in texts from roughly contemporaneous Tikal

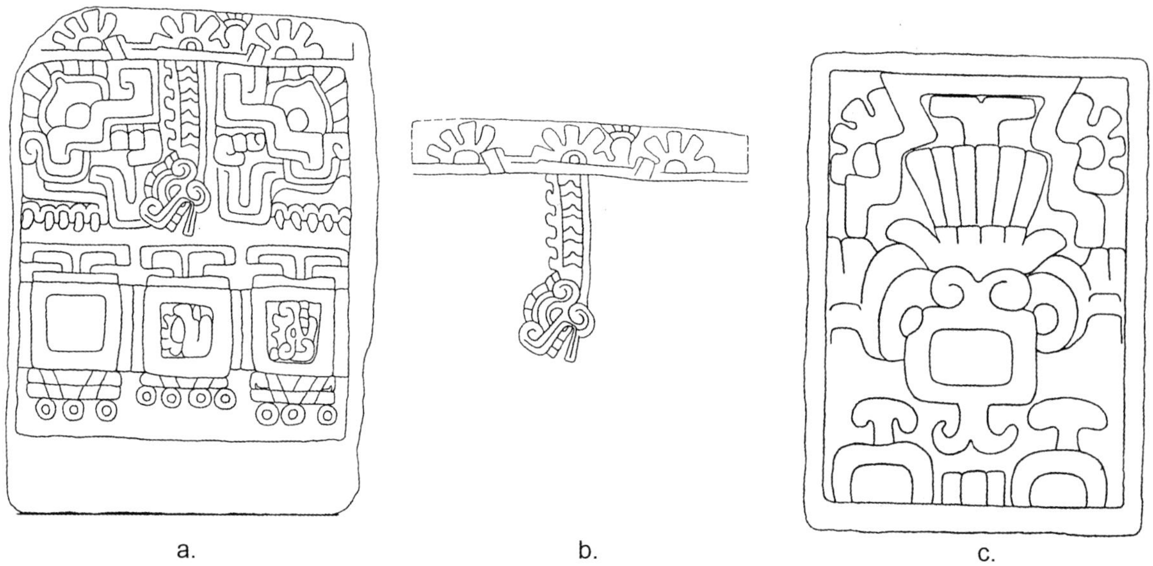

Figure 7. Late Classic monuments from Cola de Palma, coastal Oaxaca: (a) Monument 3 from Cola de Palma, note flanking War Serpents and descending serpent with similar head in upper portion of scene; (b) detail from Cola de Palma Monument 3 of serpent with segmented body descending from starry skyband; (c) Monument 6, Cola de Palma, note Xi sign atop head of front-facing War Serpent. (Drawings by Elbis Dominguez, courtesy of Javier Urcid Serrano.)

in Guatemala, Teotihuacan-style spear throwers have similar balls on their tips, again a probable reference to meteors (Fig. 6e–g). As in the English phrase 'shooting star', in ancient and contemporary Mesoamerican thought, meteors are fiery darts or arrows shot by celestial beings (Taube 2000a; Köhler 1989).

Monument 3 from Cola de Palma – a coastal Oaxacan site close to the border of Guerrero – portrays a pair of War Serpents closely resembling the aforementioned example from Las Plazuelas (Fig. 7a). Between the two creatures, there is a serpent with a segmented body as well as a similarly upturned snout and brow crest descending from a starry skyband (Fig. 7b). Urcid Serrano (2008) interprets this as a comet or meteor and describes the pair of flanking creatures on the Cola de Palma stela as the Xicani, that is, the Zapotec form of the Late Postclassic Mixtec Yahui and central Mexican Xiuhcoatl. Monument 6 from Cola de Palma portrays a frontal view of the same creature, complete with the long bifurcated tongue (Fig. 7c). Urcid Serrano (2008) notes that the element above this figure is the Xi sign flanked by two star elements.

Turquoise and the Early Postclassic Toltec

It is not until the end of the Late Classic period that turquoise becomes widespread in Mesoamerica (Weigand 1997). Thus, while there are rare glimmers of metal at Late Classic Maya sites (Bray 1977), there are no documented findings of turquoise aside from Chichen Itza, a Terminal Classic and Early Postclassic site which had strong and obvious connections with the distant site of Tula, Hidalgo. As will be noted, there was a virtual cult of turquoise at both sites, especially in the form of pectorals and back mirrors worn by warriors. For one mural painting of a Toltec warrior from Chichen Itza, no less than seven items of dress are coloured turquoise, with five of these probably being of the stone itself (Fig. 8a). In addition, a Toltec-style rock painting from Ixtapantongo, in the State of Mexico, portrays a warrior with a turquoise blue back mirror, nosepiece and pointed headdress, a clear precursor to the *xiuhuitzoli* turquoise crowns of Aztec kings (Fig. 8b).

The widespread appearance of both metal and turquoise in the Early Postclassic relates to a new quality of both materials, that of ready 'liquidation'. Unlike precious carvings of jade, bone or shell, objects of gold or copper could be readily re-melted into new pieces. A similar situation is also true for turquoise tesserae. In contrast to the vast amount of labour involved in carving an Olmec mask of solid jade, Postclassic artisans could easily sculpt wooden backings for turquoise mosaic surfaces. Although mosaics of jade, shell and mosaic are clearly present in Classic Mesoamerica (see McEwan et al. 2006: 15–17), turquoise tesserae are typically much smaller and finer, probably due to the fact that the stone occurs naturally in thin veins, in sharp contrast to the naturally occurring massive boulders and even multi-ton, solid pieces of jadeite in the Motagua Valley. In addition, once a turquoise mosaic object outlived its purpose, the stone tesserae could be readily dismantled from the wooden base to be used in another piece. Many turquoise mosaics, including a number of examples in the British Museum, show clear indications of carved pieces reused from previous items. Although precious and coveted, turquoise tesserae functioned much like 'reusable paint'.

Although turquoise was a new stone imported into the Maya area beginning ca. AD 900, the colour was already well known and celebrated in the region. For one, there is the famed 'Maya Blue' pigment of indigo dye and palygorskite clay traded widely in ancient Mesoamerica (see Houston et al. 2009: 65–66, 78–82). In Early Postclassic Toltec-style art, many male figures wear a small descending bird on their brow (Fig. 8a). When rendered in colour in bas-relief carvings and murals at Chichen Itza as well as the aforementioned rock painting at Ixtapantongo, the creature is turquoise blue (see Villagra Caleti 1954). Long ago, Eduard Seler (1902–1923, I: 690; V: 274) identified this particular bird as the lovely cotinga (*Cotinga amabilis*), a creature native to southeastern Mesoamerica, including the Maya area. In fact, Linda Schele

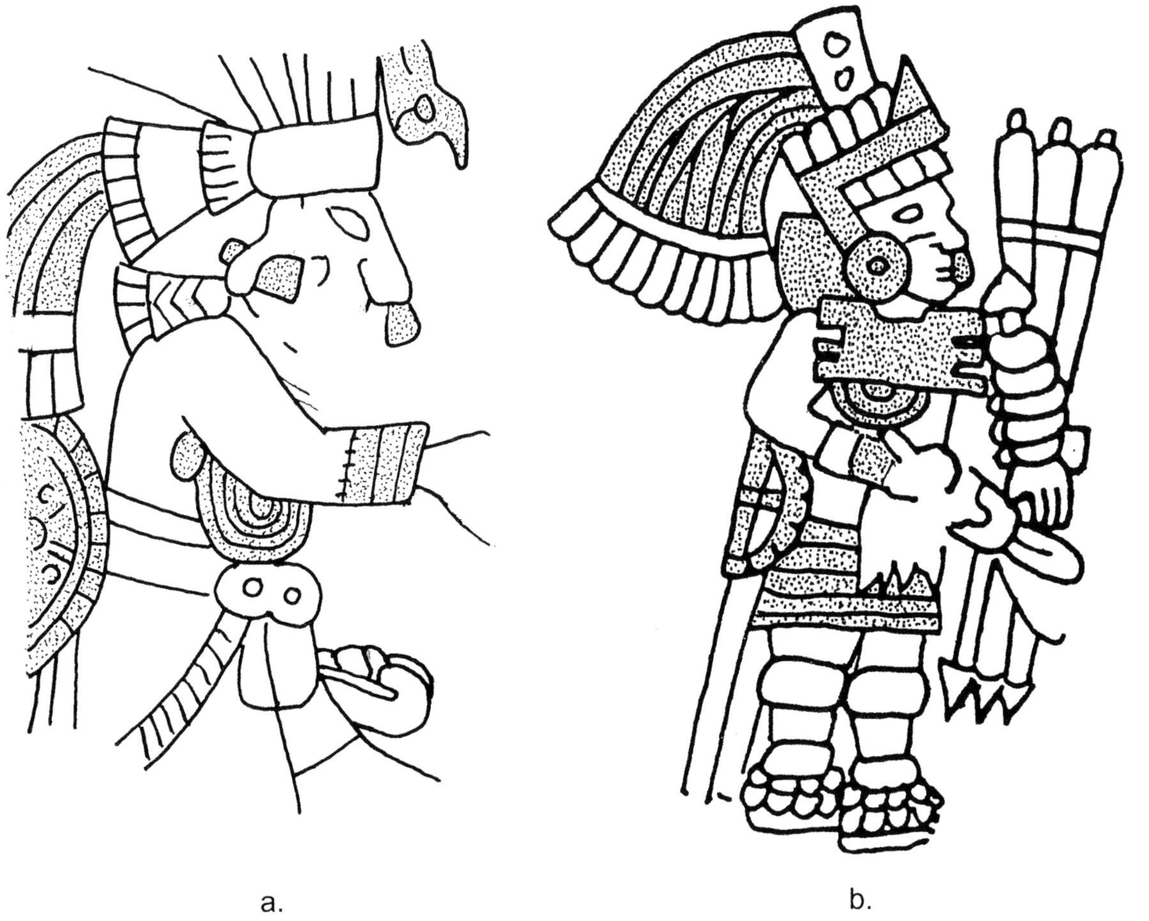

Figure 8. Early Classic paintings of Toltec warrior figures at Chichen Itza and in central Mexico (stippling denotes turquoise blue): (a) Toltec-style warrior wearing turquoise earspools, nosepiece and back mirror, note Xihuitzolli bird in headdress, mural fragment from the Temple of the Warriors, Chichen Itza (after Morris et al. 1931: plate 155a); (b) Toltec warrior wearing *xiuhuitzolli* crown, turquoise earspools, butterfly pectoral and back mirror, detail of rock painting at Ixtapantongo, State of Mexico (after Villagra Caleti 1954). (Both drawings: K. Taube.)

1997: 67–68, pl. 10, 25-6) noted that Late Classic Jaina-style figurines portray males wearing descending lovely cotingas on the brows of their headdresses, in this case painted in Maya Blue. The following is a brief description of the colouration of the lovely cotinga from a well-known bird guide: '"Electric" turquoise-blue overall with blackish-purple throat patch and deep purple patch on lower chest' (Howell and Webb 1995: 524). Known as the Xiuhtototl or 'turquoise bird' in Nahuatl, it commonly appears as a descending bird on the brow of the Late Postclassic, central Mexican god of fire, Xiuhtecuhtli, the 'turquoise lord'. In the Codex Borgia (Seler 1963), the bird can appear with the grid and dot motif denoting mosaic, such as turquoise, thereby blurring the distinction between its turquoise-coloured feathers and the stone. So important was the lovely cotinga at Chichen Itza that an ethnic enclave at the site, the Tutul Xiuh, was named after this bird, the Xiuhtototl (Mena 1927: 7; Tozzer 1941: 30, n. 159).

For Tula and Chichen Itza, one of the most prevalent ornaments of Toltec warriors is a broad, winged breast piece (Fig. 8b and Fig. 9a and b). In his early study of the iconography of Chichen Itza, Seler (1902–1923, V: 273) identified it as a plaque of jade mosaic, and a bas-relief from Chichen Itza portrays a warrior figure wearing this pectoral with fine cross-hatching, clearly denoting tesserae (Fig. 9b). However, although fashioned of mosaic, such Toltec-style chest pieces were of turquoise rather than green jadeite. Although not known at the time of Seler's pioneering research, remnants of polychrome paint on columns in the Temple of the Warriors area demonstrate that such pectorals are invariably turquoise blue, in contrast to colours of green marking jade or quetzal plumes. Thus in the carved columns from the Temple of Chak Mool, the Temple of the Warrior and the Northwest Colonnade, there are no less than 21 of these pectorals rendered in turquoise or Maya Blue (see Morris et al. 1931). In addition, versions of these items appear in subsequent Late Postclassic art where they are also turquoise blue, including examples worn by Xiuhtecuhtli or 'turquoise lord', the god of fire and time. Especially clear examples appear on a pair of Aztec censers in the Kislak collection in the United States Library of Congress, where the turquoise-coloured pectorals display the aforementioned motif denoting mosaic (Fig. 9c). In addition, there is an actual Late Postclassic example of such a pectoral fashioned of turquoise mosaic upon a wooden backing (see Berjonneau et al. 1985: pl. 179; Schildkraut 2000: no. 10). The centre of the piece portrays an armed male wearing an eagle headdress, again relating such objects to the souls of heroic warriors. In addition, the edges of the pectoral are inverted steps, recalling the Xi sign of Classic Mesoamerica.

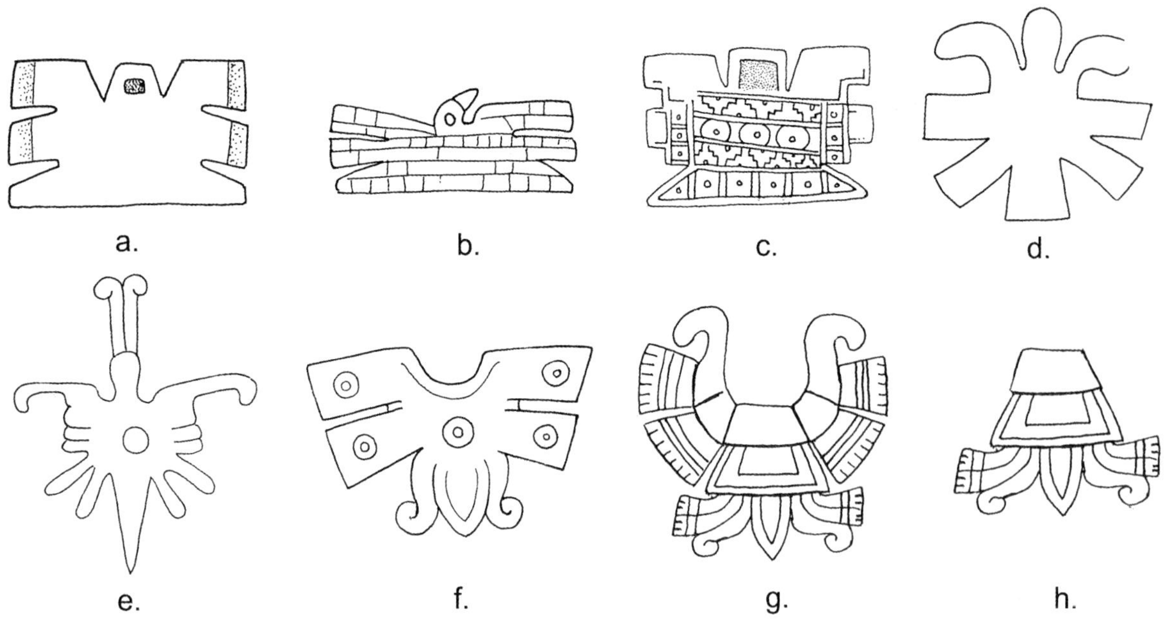

Figure 9. The Toltec butterfly pectoral and butterfly imagery in ancient Mesoamerica: (a) butterfly pectoral appearing on atlantean columns, Pyramid B, Tula; (b) butterfly pectoral with incisions denoting mosaic tesserae, detail of stone panel from Tula (after de la Fuente et al. 1988: no. 123); (c) Late Postclassic portrayal of turquoise blue butterfly pectoral with mosaic motif, detail of large ceramic censer (after Dunkelman 2007: no. 204); (d) Early Classic butterfly appearing in mural from Chacchoben, Quintana Roo (inverted for comparison, after Romero et al. 2001: 453); (e) Late Postclassic yellow butterfly probably denoting flame (Codex Borgia, p. 36); (f) Late Postclassic butterfly pectoral, detail of large Tlaloc censer (after Castañeda Delgado et al. 1997: no. 61); (g) gold butterfly nosepiece found in the Templo Mayor precinct (after McEwan and López Luján 2009: no. 34); (h) detail of *yacaxihuitl* form in centre of butterfly nosepiece. (All drawings: K. Taube.)

According to Eduard Seler (1902–1923, V: 298, 364), the Toltec mosaic breast piece denotes a butterfly. In Late Postclassic Mexico, butterfly wings often have angular tabs closely resembling the sides of the Toltec pectoral (Fig. 9e–g). In addition, a Maya mural portraying a textile design with butterflies from Chacchoben, Quintana Roo, indicates that this convention can be traced to the Early Classic period (Fig. 9d). A Late Postclasic censer from the Tehuantepec region depicts Tlaloc wearing an explicit butterfly pectoral with the abdomen flanked by flames of smoke (Fig. 9f). In addition, there are Late Postclassic gold nosepieces in the butterfly form, including an Aztec example excavated at the Templo Mayor (Fig. 9g). On close inspection, it can be seen that the central abdomen of this element forms the aforementioned Xi sign with a burning flint tail, as can be found with Late Postclassic examples of the Yahui and Xiuhcoatl (Fig. 9h; cf. Fig. 4d). This same central element also forms the *yacaxihuitl* turquoise nosepiece, an important emblem of Late Postclassic nobles (Fig. 14c–e). As Doris Heyden (1972: 4) notes, the stepped form of the *yacaxihuitl* derives from the tail of the Xiuhcoatl.

The most renowned Toltec examples of butterfly pectorals are on the famed atlantean columns placed atop Pyramid B (see Acosta 1961). Although in strong Toltec style, these massive sculptures also evoke military costume and weaponry from more ancient Teotihuacan, including darts and spear throwers, the shell platelet helmet and mirrors worn in the small of the back, a common costume element at Teotihuacan (see Taube 1992a). However, it is likely that one of the more subtle and profound links to more ancient Teotihuacan is the turquoise pectoral. As has been mentioned, at Teotihuacan as with the contact period Aztec, the souls of warriors were identified with both precious birds and butterflies. In both Teotihuacan and Aztec belief, they were fiery beings, with many Teotihuacan 'theatre censers' portraying butterflies rising from burning funerary bundles (Taube 2000a). Similarly, Aztec mortuary bundles of warriors, rulers and merchants were cremated, with the flames being closely connected to butterflies and the warrior soul (see Fig. 14c).

Although Toltec-style butterfly pectorals are generally absent in Late Postclassic Maya art, there are two likely versions in the Codex Dresden (Fig. 10a and b). Together with the horizontal elements appearing on the wings of Early Postclassic examples, they display versions of the Maya Yax sign signifying green or blue – hues covering the range of turquoise. The Codex Dresden contains other allusions to the symbolism and regalia of turquoise, including a detailed portrayal of the aforementioned Xiuhtecuhtli 'turquoise lord' of Late Postclassic central Mexico (Fig. 10d). Aside from displaying the horizontal facial marking of Xiuhtecuhtli, he has a descending Xiuhtototl bird in his headdress and a pectoral with the Aztec sign for turquoise, this being two half circles framing a central, vertical element (Taube and Bade 1991). More importantly, the accompanying text in this section of the Venus pages describes him phonetically as 'Chak' (red or great) *xi-wi-te-i*, a clear reference to Xiuhtecuhtli, the turquoise god of central Mexico (Fig. 10c). Moreover, a roughly contemporaneous mural from Tulum – on the turquoise shores of Quintana Roo – depicts the god with the horizontal yellow facial banding alternating with carefully rendered turquoise tesserae of varying size, indicating that such mosaics were present at this site (see Lombardo de Ruiz 1987: pl. 255). The Codex Dresden also depicts a Maya form of Quetzalcoatl wearing the jewellery and headdress of the Aztec wind deity Ehecatl-Quetzalcoatl (Fig. 10e). Known only for the Aztec and this example from the Codex Dresden, the headdress has two knots flanking a disk, this being again the Aztec turquoise sign (Fig. 10f).

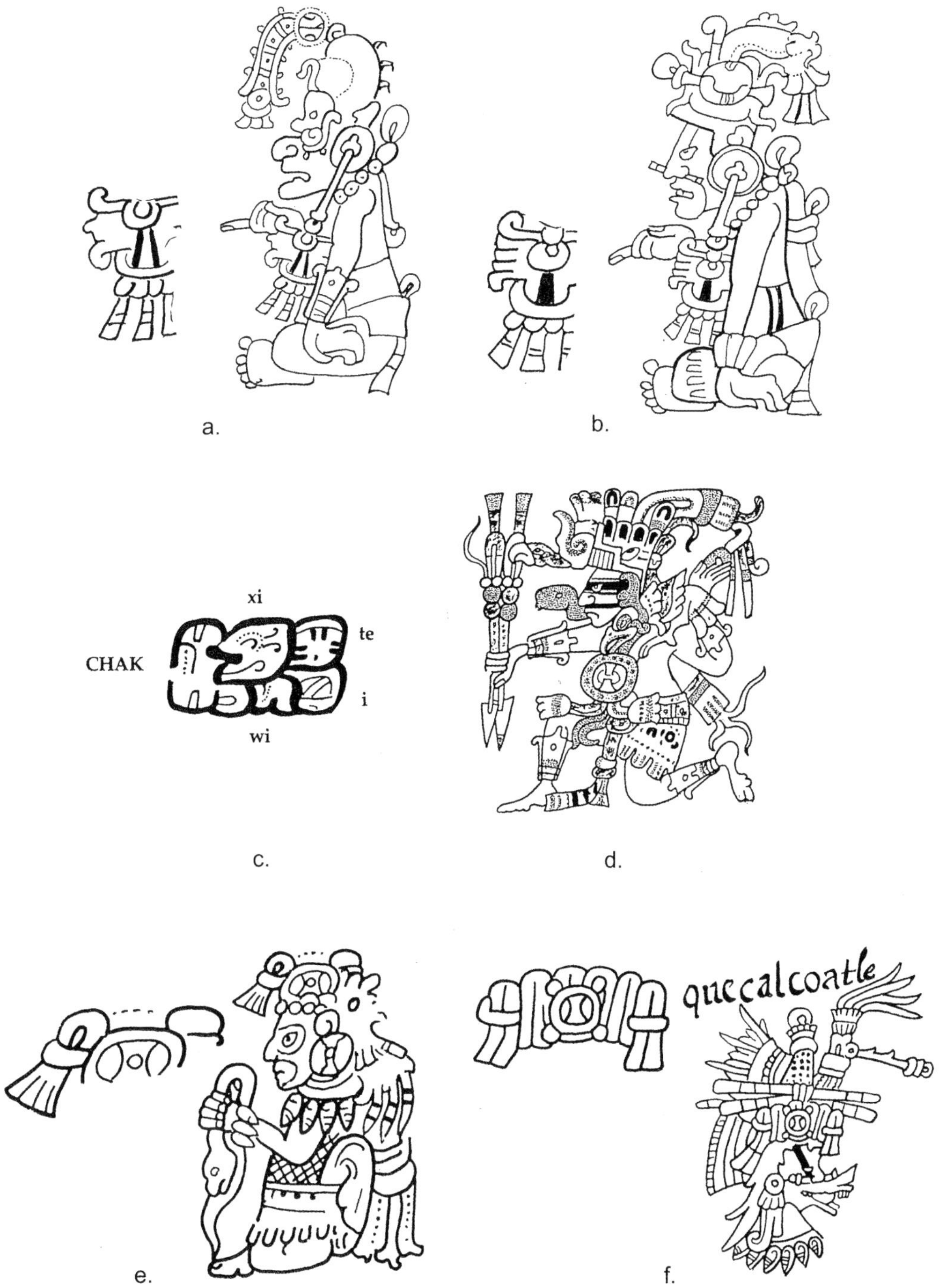

Figure 10. Turquoise symbolism and imagery appearing in the Late Postclassic Maya Codex Dresden: (a) Itzamnaaj with probable version of butterfly pectoral (Codex Dresden, p. 10c); (b) maize god with probable butterfly pectoral (Codex Dresden, p. 14a); (c) name glyph for Xiuhtecuhtli read CHAK xiwitei (Codex Dresden, p. 49) (from Taube and Bade 1991: fig. 1); (d) portrayal of Xiuhtecuhtli with Xiuhtototl browpiece and turquoise sign pectoral (Codex Dresden, p. 49) (from Taube and Bade 1991: fig. 1); (e) Maya version of Quetzalcoatl with serpent and quetzal on back wearing Aztec-style headdress of Quetzalcoatl (Codex Dresden, p. 4a) (from Taube and Bade 1991: fig. 5d); (f) Aztec portrayal of wind god Ehecatl-Quetzalcoatl with detail of headdress (Codex Telleriano-Remensis, f. 8v) (from Taube and Bade 1991: fig. 5f). (All drawings: K. Taube.)

For Toltec-style military regalia at Tula and Chichen, the most salient item of turquoise is a circular mirror worn on the back, known as *tezcacuitlapilli* among the later Aztec (Fig. 11c and d). In the case of the Toltec examples known for both Tula and Teotihuacan, such mirrors have a central pyrite surface surrounded by a crenellated frame of turquoise mosaic and other material containing four equidistantly placed serpents. As in the case of the butterfly pectorals, such devices allude to more ancient Teotihuacan with their back mirrors of finely fitted tesserae of iron pyrite worn by warriors (Taube 1992a). However, a noteworthy innovation still undocumented for Teotihuacan mirrors are the four Xiuhcoatl serpents appearing on the broad turquoise rim, with the most immediate comparison being earlier Classic Maya portrayals of four centipede

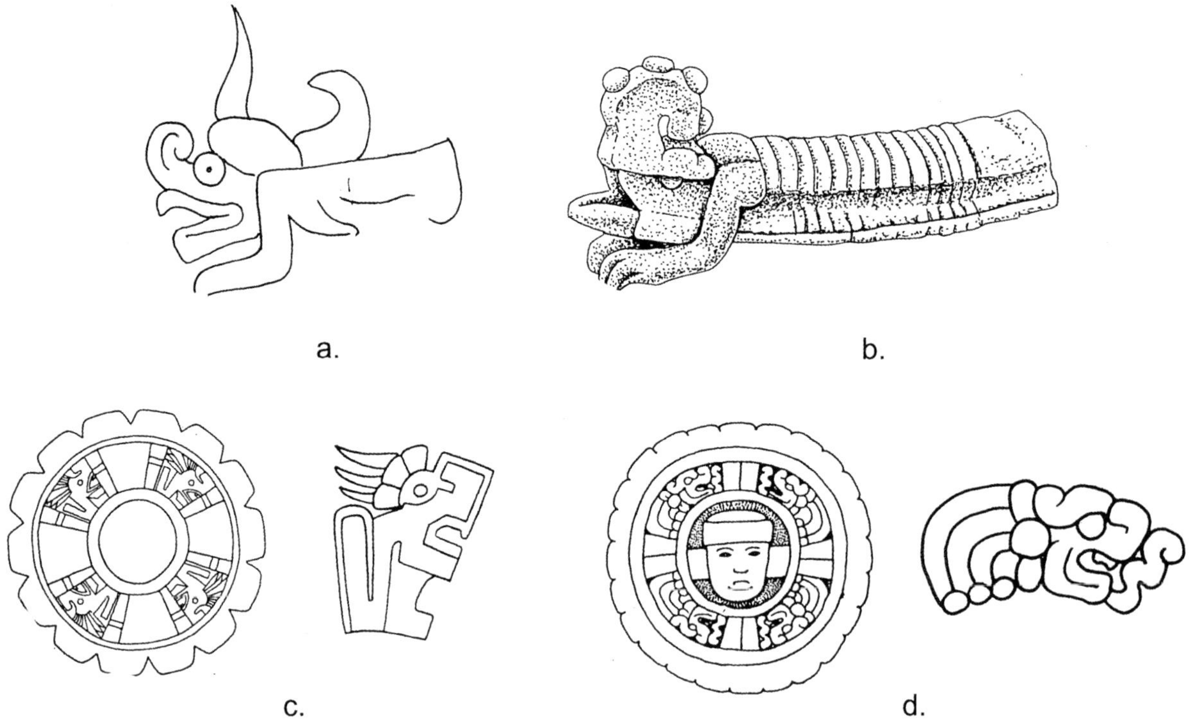

Figure 11. Early Postclassic portrayals of the Xiuhcoatl fire serpent at Chichen Itza and Tula: (a) glyphic sign of Xiuhcoatl with flames, detail of carved pier from Northeast Colonnade, Chichen Itza (after Morris et al. 1931: plate 96); (b) carved wooden image of Xiuhcoatl with stars on snout from Cenote of Sacrifice, Chichen Itza (after Fontaine and Purin 1993: 103); (c) turquoise and pyrite back mirror with detail of one of the four fire serpents (from Taube 1994: fig. 31a); (d) portrayal of back mirror with detail of one of four fire serpents, detail of atlantean column, Pyramid B, Chichen Itza (from Taube 1992a: fig. 12c). (All drawings: K. Taube.)

heads on crenellated solar disks (Taube 1994: 234, fig. 31d,e). In ancient Maya thought, centipedes were of the nocturnal underworld, but they were also creatures of the sun god, probably relating to the resplendent, dawn emergence of this being out of darkness (Taube 2003). With their forelimbs, upturned snouts and crested heads, the Xiuhcoatl serpents appearing on Toltec-style mirrors from Tula and Chichen Itza are virtually identical to the Late Classic creatures on Monument 3 from Cola de Palma, Oaxaca (Fig. 7a). But it is during the Early Postclassic that the War Serpent weds with turquoise, and therefore constitutes a critically important period for understanding the development of this being in Postclassic Mesoamerica. A glyph from the Northwest Colonnade at Chichen Itza portrays a Xiuhcoatl with an upturned snout and forelimbs and flames emanating from its body (Fig. 11a). In addition, a remarkable wooden sculpture from the Cenote of Sacrifice at Chichen Itza depicts this creature with stars on its upturned snout, along with a segmented body and forelimbs (Fig. 11b). The curving, star-studded snout not only recalls Aztec examples of the Xiuhcoatl, but also the aforementioned spear throwers of Early Classic Teotihuacan (Fig. 6a,b,e and g and Fig. 13d and g). Along with noting that this carving is an Early Postclassic form of the later Aztec Xiuhcoatl, Françoise Fontaine (in Fontaine and Purin 1993: 103) also calls attention to its oddly segmented body. As will be noted, this probably refers to the pupate, caterpillar nature of this fiery, meteoric being.

The turquoise mirrors of Tula and Chichen Itza are also known far to the north at Paquimé, or Casas Grandes, including one example with a copper backing that still retained turquoise tesserae (Di Peso 1974: fig. 255-2). For many years, Paquimé was considered to be an Early Postclassic site, that is, roughly contemporaneous with Tula and Chichen Itza. However, it is becoming increasing clear that it is actually Late Postclassic, although still abandoned before the coming of the Spanish in the early sixteenth century (Dean and Ravesloot 1993: 103). In addition, page 46 of the Codex Borgia, the final scene of the Middle Pages, portrays such a turquoise mirror emitting flames and surrounded by four burning Xiuhcoatl serpents, a clear allusion to the more ancient mirrors of the Toltec (Taube 1992a: 186; Fig. 14b).

The Aztec

In spite of their self-vaunted identity as fierce Chichimec nomads from the north, the Aztec, or Culhua-Mexica, were also keenly aware of their historic links to more ancient traditions of central Mexico, including of course the Toltecs of Tula, Hidalgo (see López Luján 2003). I would suggest that because of this, turquoise became the most esteemed form of royal regalia, including the *yacaxihuitl* nosepiece and the *xiuhuitzoli* crown, in striking contrast to the jade items of more ancient, Classic period kings. In two scenes of royal coronation appearing in the sixteenth-century Codex Duran, the pointed *xihuitzolli* crown is the item presented to the new emperor (for Codex Duran see Heyden 2000). This act of royal coronation recalls far earlier Late Preclassic and Classic Maya accession scenes of kings donning the jade 'jester god' (see Taube et al. 2010: fig. 43). In this regard, it is important to note that this turquoise item first occurs among the Early Postclassic Toltec. As noted before, the Ixtapantongo rock painting portrays a

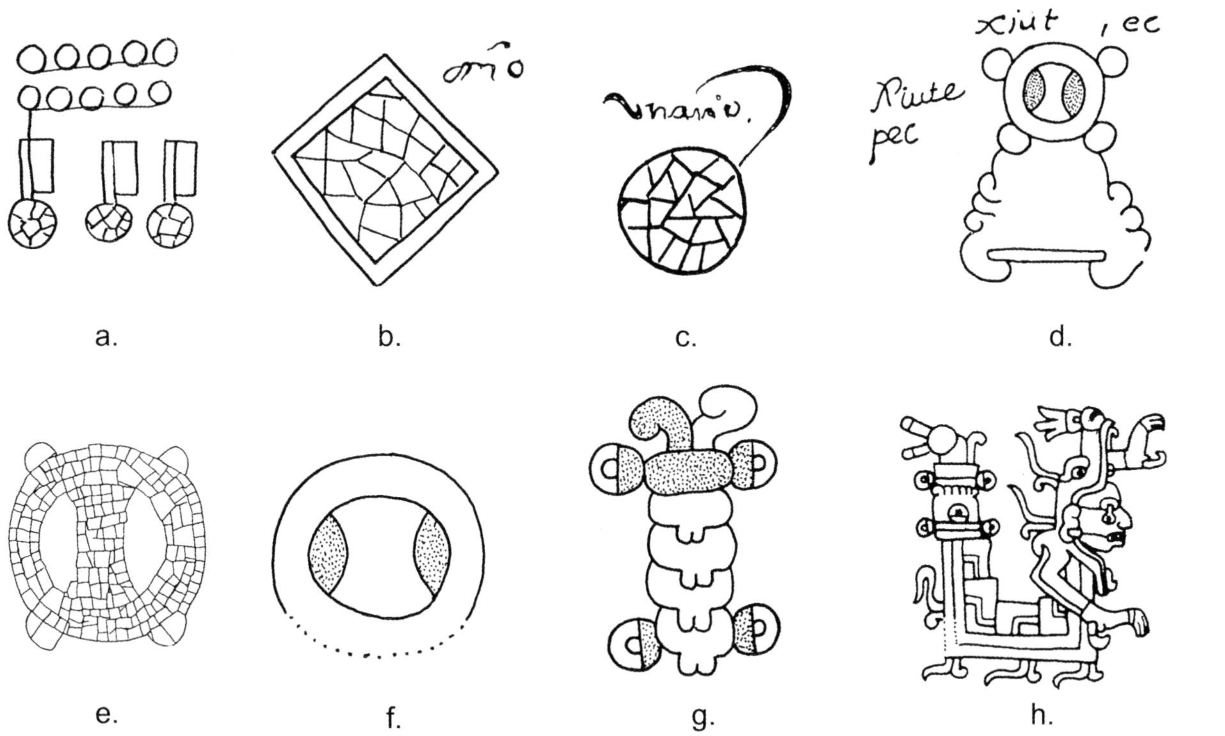

Figure 12. Turquoise signs and symbolism in Late Postclassic central Mexico: (a) Aztec notation for 70 years (Codex Mendoza, f. 71r); (b) turquoise mosaic sign for year accompanied by Spanish gloss '*año*' (Codex Kingsborough, f. 32v); (c) turquoise mosaic sign for year with Spanish gloss '*un año*' (Codex Osuna, f. 14v); (d) Aztec toponym for Xiuhtepec (*Matrícula de Tributos*, f. 3v); (e) Aztec turquoise mosaic in form of turquoise sign from the Templo Mayor cache (after Castillo Tejero and Solís Olguín 1975: pl. 41); (f) turquoise sign appearing on Aztec brazier (after Dunkelman 2007: no. 204); (g) burning rattlesnake tail with stars, detail of Aztec brazier (after Dunkelman 2007: no. 204); (h) burning fire serpent with truncated rattlesnake tail with stars (Codex Borgia, p. 44). (All drawings: K. Taube.)

Toltec warrior wearing such a device rendered in turquoise blue (Fig. 8b). Moreover, an atlantean sculpture with upraised arms from Tula portrays a figure wearing a blue-painted *xihuitzolli* (see de la Fuente et al. 1988: no. 25). In bas-reliefs from the Northwest Colonnade at Chichen Itza, there are no less than five examples of *xihuitzolli* painted turquoise blue (see Morris et al. 1931: pls 81, 103, 107, 127, 129). In the central scene from the Lower Temple of the Jaguars at Chichen Itza, both a plumed serpent and an adjacent armed woman wear *xihuitzolli* crowns clearly marked with mosaic patterns (Fig. 14a and b; see Maudslay 1889–1902, III: pl. 49). The great plumed serpent serves as an elaborate name for the principal figure in the scene, a figure wearing a mask of Quetzalcoatl – or in Yukatek Mayan K'uk'ulkan quite possibly a major historical ruler at the site (see Taube 1994: fig. 11).

Although we do not know the Toltec term for turquoise, for the Aztec it was *xihuitl*, a term that can have such varied meanings as 'year', 'comet (meteor)' and 'herb' as well as turquoise. It is clear that at least the meanings of year and shooting star do relate to turquoise. Thus among the Aztec, specific named years, such as 2 Reed or the following 3 Tecpatl, were denoted by placing them in rectangular turquoise frames. In addition, early colonial Nahuatl documents denote the year of 365 days by either a turquoise dot or a turquoise mosaic, either circular or rectangular in form (Fig. 12a–c). As has been mentioned, the Aztec also had a specific sign for the stone, this being two halved circles in a circular rim, often with four bosses at the edges (Fig. 10f and Fig. 12d–f). In Aztec toponyms appearing in the Codex Mendoza (Berdan and Anawalt 1992) and *Matrícula de Tributos* (Berdan and de Durand-Forest 1980), this sign is glossed in Latin orthography as *xiuhitl*, such as for the place name Xiuhtepec, or Turquoise Mountain (Fig. 12d). Remarkably, actual turquoise mosaic examples of such mosaics were excavated in the Templo Mayor precinct of Tenochtitlan (Fig. 12d). However, although this device clearly serves as the Aztec sign for turquoise, I know of no instance where it appears in a calendrical context denoting the 365-day year.

Although *xihuitl* is glossed as 'cometa' in early colonial dictionaries, Ulrich Köhler (1989: 289) notes that in colloquial Spanish, this term can refer to meteors as well as impressive but ponderously slow-moving comets. In support, Köhler (1989: 296) mentions that among contemporary Sierra Nahuat of Veracruz, *xihuitl* refers to meteors and meteorites. Given the significance of the Xiuhcoatl spear thrower as a star-shooter, it is likely that *xihuitl* did refer to meteors among the Aztec and other Nahuatl-speaking peoples of Late Postclassic highland Mexico. As has been mentioned, both the Classic period War Serpent and the Xiuhcoatl display butterfly and caterpillar traits, including the Early Postclassic wooden carving from the Cenote of Sacrifice at Chichen Itza, which has a segmented body recalling a worm or caterpillar along with its star-studded snout (Fig. 11b). In the *Matrícula de Tributos* and Codex Mendoza, there are explicit portrayals of caterpillars with butterfly faces and segmented bodies glossed with terms pertaining to the Nahuatl term *ocuila*, meaning 'caterpillar' or 'worm (Fig. 13a and b; see Taube 2000a: 288). Their bodies are essentially elongated Xi signs in profile, and clearly relate to those of Postclassic Xiuhcoatl, including even the massive example appearing on the Aztec Calendar Stone (Fig. 4e, Fig. 12h and Fig. 13c and g). Moreover, the prominent forelimbs of the War Serpent and the Xiuhcoatl also appear with Late

Postclassic portrayals of caterpillars, creatures with powerfully clinging thoracic or 'true' legs (e.g. Fig. 13a and b).

The visual correspondences between Mesoamerican portrayals of caterpillars and the turquoise fire serpent, or Xiuhcoatl, is striking, but there is also strong ethnohistoric and ethnographic evidence of meteors and meteorites being supernatural caterpillars or 'worms' in Mesoamerican belief. Although Köhler (1989: 295–296) does not mention the physical similarities of caterpillars to the Xiuhcoatl, he notes several important instances of meteorites being described as worms or caterpillars in colonial and contemporary Mesoamerica. Thus there is the following account in the Florentine Codex:

> It is said that the passing of a shooting star rose and fell neither ... without purpose nor in vain. It brought a worm [*ocuillo*] to something ... And of [the animal] wounded by a shooting star, they said: 'It hath ... been wounded by a shooting star; it hath received a worm (Sahagún 1950–1982, VII: 13).

Köhler (1989: 296) also relates the following belief among the modern Sierra Nahuat:

> where a *xihuitl*, the term applied both for meteor and meteorite, arrives on the earth, black caterpillars will appear, which are about 3 cm long and are piled together into a heap the size of a hand. These caterpillars are called *citlalocuile*, 'star-caterpillars,' as well as *citlalcuitlatl*, 'star shit,' ... and the heap looks indeed like excrement.

Following Köhler's ground-breaking research (Taube 2000a: 290–291), I found other examples of this striking belief, including this account supplied by Manuel Gamio (1922, 1: 316) for Nahuatl-speaking inhabitants in the Teotihuacan region: 'they say that in the place where a meteorite falls (*citlalcuitlatl* or *metzcuitlatl*) a blue worm forms' (translation by the author). Not only is *citlalcuitlatl* or 'star excrement' the same term termed used by Sierra Nahuat for meteorites, but the creature is blue, much like the turquoise fire serpent of Postclassic Mesoamerica.

The relation of shooting stars to caterpillars and worms continues among contemporary Maya peoples, including highland Chiapas, as can be seen in the following entry for *sansewal* in a dictionary of Tojolabal:

> Lightning, shooting star, light that appears at night in the mountains ...They are stars that fall as shooting stars. They are little worms of fire that move and are still not like stone, but rather like little black serpents, like little worms. In the sky they shine like a lamp. Upon falling to earth they divide (Lenkersdorf 1979: 312 [translation by the author]).

Recently, I came across the following description of shooting stars among Mopan Maya of northern Guatemala:

> God from time to time decides to show pity on the poor people and throws down to them some of these droplets of shining gold. These gold droplets are falling stars and called *ta'xülab* meaning 'dust of stars'. The devil, however, is in the clouds and catches the gold for himself. Instead of the gold, he throws down balls of worms, which are found time to time in the jungle (Ulrich and Ulrich 1982: 118).

Although this alone is a fascinating account, the authors did not mention the following entry in their previously published Mopan dictionary (Ulrich and Dixon de Ulrich 1979: 197) for the cited term *ta'xülab* – this being *pabellón de orugas* or 'caterpillar tent'. Moreover in Yucatecan languages, including Mopan, *ta'* signifies 'excrement', a phrase very similar to central Mexican terms for meteorites.

According to the *Historia de los mexicanos por sus pinturas* – one of the earliest and important colonial accounts pertaining to Aztec religion and belief – the Xiuhcoatl fire serpents reside in the fifth level of the sky, and from there come meteors (*cometas*) and other 'signs of the sky'. In a parallel and otherwise unique account in the Codex Vaticanus A, this fifth level of the sky is illustrated with stars shooting darts, a reference to the Nahuatl term for meteor, *citlalin tlamina* or 'star shoots dart' (Taube 2000a: 291). Mention has been made of the Aztec tutelary god Huitzilopochtli as a wielder of the star-shooting Xiuhcoatl spear thrower (see Fig. 6a). However, the preeminent being identified with the Xiuhcoatl is the fire deity Xiuhtecuhtli, or 'turquoise lord', who not only often wields this creature as his weapon, but also carries it on his back, much like a *nahual* or 'alter ego'. It is noteworthy that although the symbolically overlapping fire deity Huehueteotl, or 'Old God', can be traced to as early as Middle Formative Olmec art, Xiuhtecuhtli is an entirely Late Postclassic phenomenon, and remains undocumented for the known corpus of deities at Tula or Chichen Itza, probably because he is an embodiment of earlier 'ancestral' Toltec symbolism of turquoise pertaining to rulership and warfare. According to Nicholson (1971: 413), Xiuhtecuhtli 'served as the archetype of all rulers' and was explicitly evoked during Aztec rites of imperial coronation. However, in accession ceremonies, it was not only the god but the turquoise that he embodied in particular ornaments, such as the *xihuitzolli* crown and *yacaxihuitl* nosepiece. As noted by Olivier and López Luján (2009: 86), '[t]he *xihuitzolli* and other turquoise ornaments directly connected the sovereigns to the god of fire, Xiuhtecuhtli, the mother and father of all deities that resided in the centre of the universe'.

In contrast to the static and aged portrayals of Huehueteotl in Aztec monumental braziers, Xiuhtecuhtli is a virile warrior commonly wielding his Xiuhcoatl spear thrower. In the case of page 1 of the Fejérváry-Mayer Codex (Burland 1971), he is the victorious dispatcher of Tezcatlipoca, whose dismembered body is literally thrown to the four quarters of the world in a mythic episode of world creation. However, despite his strength and prowess, his turquoise accoutrements clearly allude to the more ancient Toltec, including his frequent *xihuitzolli* crown, *xiuhtototl* browpiece, and the stylised butterfly pectoral. It is a striking fact that in Aztec mortuary rites for dead warriors, effigy crematory bundles were fashioned with versions of the same garb of Xiuhtecuhtli and the earlier Toltec, including the *xihuitzolli* crown and *yacaxihuitl* nosepiece (Fig. 14c). For the cited example illustrated in the Codex Magliabechiano (Boone 1983), the effigy has an item not previously mentioned for the Toltec, this being

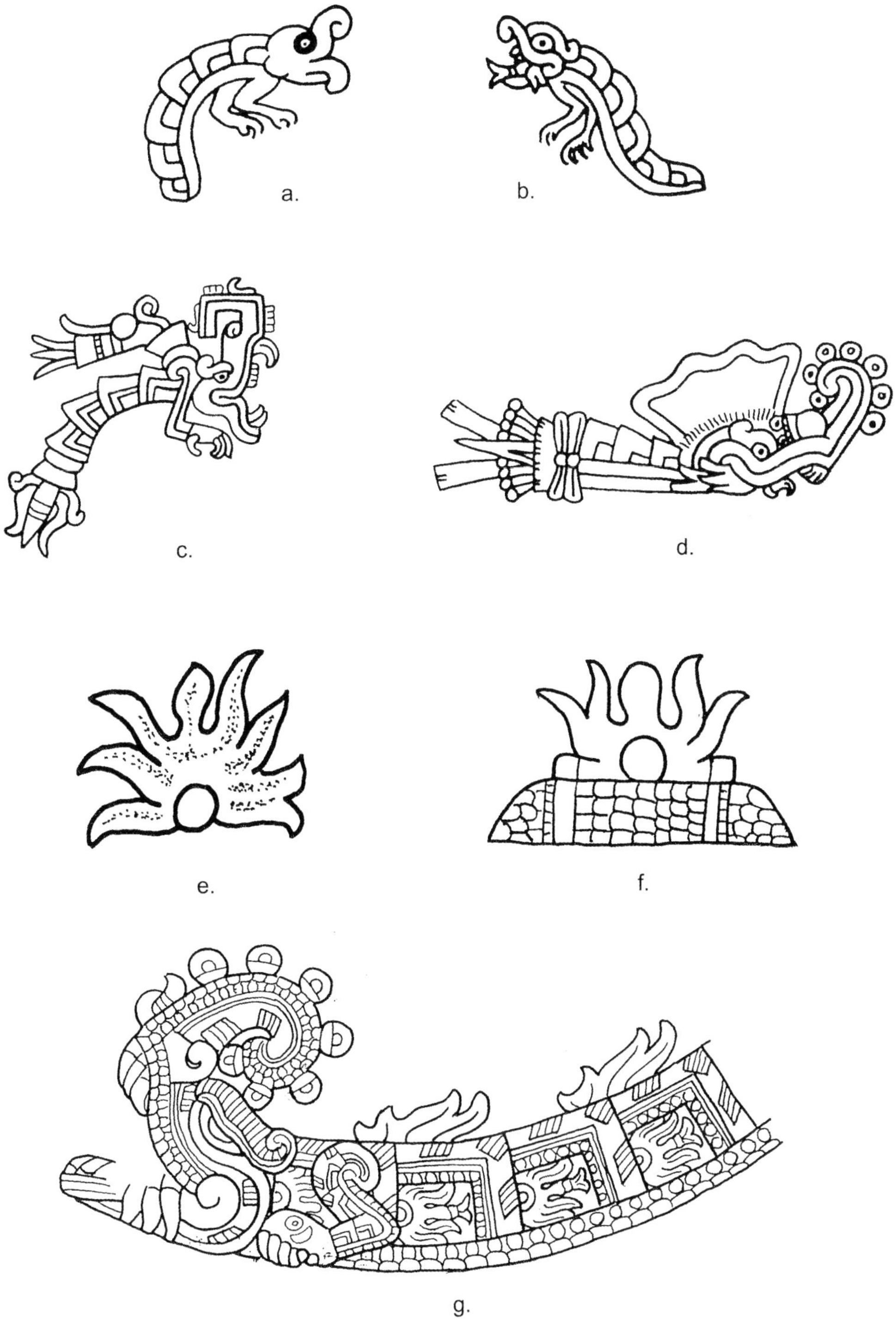

Figure 13. Late Postclassic and early colonial imagery pertaining to caterpillars, butterflies and the Xiuhcoatl: (a) caterpillar serving as sign for military title *tocuiltecatl* (Codex Mendoza, f. 65r); (b) caterpillar as toponymic sign for Ocuilan (Codex Mendoza, f. 10v); (c) Mixtec form of Xiuhcoatl with segmented body and prominent forelimbs (Codex Nuttall, p. 76); (d) Aztec Xiuhcoatl with segmented body and rudimentary butterfly wings (Codex Borbonicus, p. 20) (from Taube 2000a: fig. 10.13d); (e) butterfly flame with turquoise spot in lower centre (Codex Magliabechiano, f. 86r); (f) butterfly with turquoise spot rising out of bed of turquoise mosaic (Codex Borbonicus, p. 19); (g) Xiuhcoatl with butterfly flames on its segmented body, detail of Aztec Calendar Stone. (All drawings: K. Taube.)

the turquoise dog pectoral, or *xolocozcatl*. In fact, such items are relatively common in Toltec-style imagery from Chichen Itza, and for an example from the Northwest Colonnade with extant polychrome, it is light, turquoise blue (e.g. Morris et al. 1931: pl. 74). Although previously identified as a jaguar (Parsons 1980: no. 176), it is likely that a turquoise and shell pectoral in the collection of the Saint Louis Art Museum is an actual example of the *xolocozcatl*. The turquoise dog pendant appears twice on Codex Borbonicus page 10, here as a part of two sets of turquoise ornaments, these being the *xihuitzolli* crown, turquoise ear plugs and the *yacaxihuitl* nosepieces as well as the *xolocozcatl* pendants (Fig. 14d; see Nowotny and Durand-Forest 1974). These items flank the pole used for the Xocotlhuezti rites in honour of dead warriors, again linking these turquoise jewels to the souls of warriors (see Taube 2000a: 303–304).

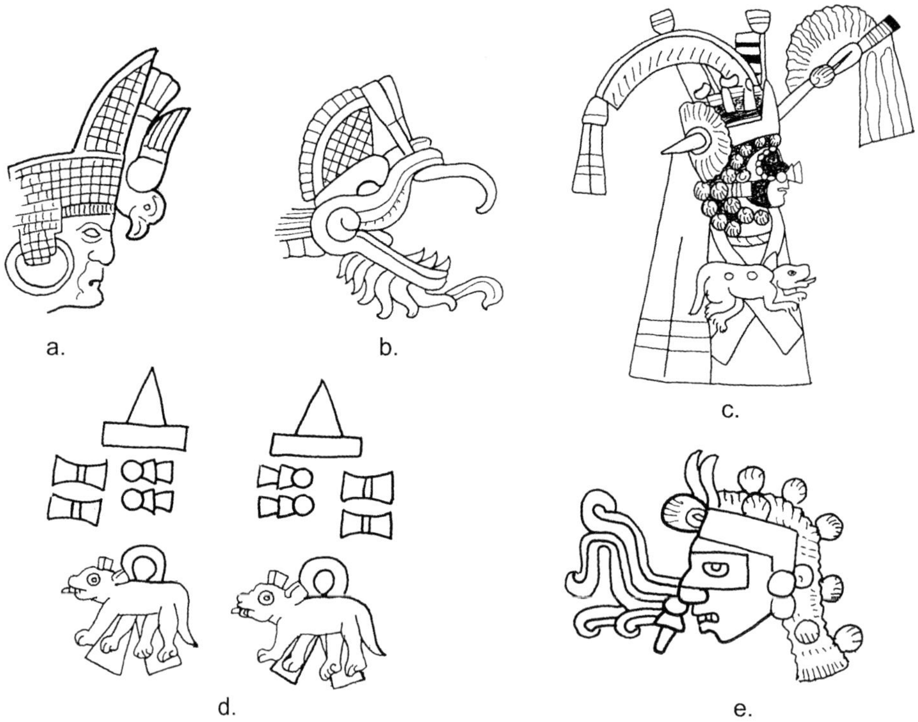

Figure 14. The *xihuitzolli* crown and related turquoise symbolism: (a) Early Postclassic female figure with mosaic *xihuitzolli*, Lower Temple of the Jaguars, Chichen Itza (after Maudslay 1889–1902, III: pl. 49); (b) plumed serpent wearing *xihuitzolli*, Lower Temple of the Jaguars, Chichen Itza (after Maudslay 1889–1902, III: pl. 49); (c) warrior effigy bundle wearing paper versions of the turquoise *xihuitzolli*, *yacaxihuitl* nosepiece and *xolocozcatl* dog pendant (Codex Magliabechiano, f. 72r); (d) two sets of turquoise ornaments placed by Xocotl pole (Codex Borbonicus, p. 10); (e) probable *cihuateotl* with *yacaxihuitl* exhaling smoke or flames before face (Codex Borgia, p. 46). (All drawings: K. Taube.)

A remarkable pair of large, Aztec-style censers in the Jay I. Kislak Collection in the Library of Congress, Washington, DC, bear directly on the relation of turquoise ornaments to the souls of deceased warriors (see Dunkelman 2007: n. 204). The figures are armed with shields and spear throwers, which they raise aggressively in their right hands. Not only do they wear clear forms of the turquoise butterfly pectoral (see Fig. 9c), but they also appear to have versions of the platelet helmets known for the great atlantean columns at Tula. Moreover, they also have *yacaxihuitl* nosepieces and large turquoise mosaic earspools, with the skirt and even the limbs of one of the figures painted turquoise blue. References to turquoise also appear on the censer sides, which bear repetitive examples of the circular Aztec turquoise sign (Fig. 12f). Especially striking motifs are the burning rattlesnake tails ornamenting their clothing (Fig. 12g). Lined with four stars, these fiery elements may refer to meteors, recalling burning rattlesnake tails in Classic Maya portrayals of the War Serpent (Fig. 2e and f). A burning fire serpent on page 44 of the Codex Borgia has a similar but more truncated rattlesnake tail, again flanked by four stars (Fig. 12h). Displaying shut eyes and slack, open mouths, the armed censer figures appear to be dead, and with their prominent skirts, it is likely that they are Cihuateteo, the fierce warrior women who died heroically in the act of childbirth or in war and whose charge is to carry the sun from its zenith to its setting in the west (see Sahagún 1950–1982, 6: 162–163). For the aforementioned Toltec-style turquoise mirror framed by four fire serpents appearing in the Codex Borgia, four skirted women also flank the mirror and serpents. Wearing the *yacaxihuitl* nosepiece of nobles and heroic warriors, these are probably also Cihuateteo, who were also referred to as *cihuapipiltin*, or 'princesses'. The two women standing to the right side of the burning mirror and Xiuhcoatl serpents have flames or smoke emanating from their faces and nosepieces, probably denoting the fiery nature of their fierce souls as well as the turquoise (Fig. 14e).

Conclusions

Although a stone of wholly non-Mesoamerican origin, turquoise became a pivotally important material in Postclassic Mexican religious thought. In sharp contrast to the Formative and Classic symbolic complex of jade, pertaining to maize, water and agricultural abundance, turquoise related more to warfare, fire, meteors, and the souls of heroic warriors and kings, much of which can be traced to still earlier traditions of Teotihuacan. The Toltec apparently were the first to articulate this precious stone to more ancient Teotihuacan war symbolism as well as developing a new, complex body of turquoise mosaic regalia, including the *xihuitzolli* crown, winged breast plaques, nosepieces and petalled back mirrors rimmed by four Xiuhcocoa, or 'turquoise fire serpents'. In both Toltec and later Aztec belief, the Xiuhcoatl embodied meteoric fire, and as such, appeared in the form of supernatural caterpillars, beings that represented the pupate form of the butterfly souls of warriors. In Late Postclassic central Mexican thought, the Xiuhcoatl was a creature closely identified with the fire god, Xiuhtecuhtli, a being also closely related to kingship and warrior souls as well as to turquoise symbolism and ornaments of the earlier Toltec. For the Late Postclassic Xiuhtecuhtli complex, there is the obvious question of the relation of this blue and inert stone to fire. As Patrick Johansson K. (in this vol., pp. 135–144) notes, this clearly derives from the colour of the strongest and most intense portion of a fire, that is, the 'heart of the flame'. Thus in the Codex Magliabechiano, turquoise-coloured disks form the lower centre of yellow, rising butterflies (Fig. 13e). On page 19 of the Codex Borbonicus, such a flaming creature rises out of a bed of turquoise mosaic (Fig. 13f). However, by far the most impressive scene of fiery butterflies is the Aztec Calendar Stone, where they are framed in the segmented body elements as the living embodiment of turquoise (Fig. 13g). Far from being a passive and inert stone, turquoise was intensely alive, and symbolised the souls of warriors and the fiercely blue heart of the pulsing flame.

References

Acosta, J.R. (1961) 'La indumentaria de las cariátedes de Tula', in M. León-Portilla (ed.), *Homenaje a Pablo Martínez del Río en el Vigésimoquinto Aniversario de la Primera Edición de Los Orígenes Americanos*. Mexico City : Instituto Nacional de Antropología e Historia (INAH), 221–228.

Anders, F. (1987) *Codex Zouche-Nuttall*. Graz: Akademische Druck-u. Verlagsanstalt.

Armillas, P. (1944) 'Mexiquito, gran ciudad arqueológica en la cuenca del Río de las Balsas', *El Mexico Antiguo* **6**(7–8): 261–262.

Arqueología Mexicana (2008) 'Guanajuato: Historia y arqueología', *Arqueología Mexicana* **16**(92): 24–25.

Berdan, F. and Anawalt, P. (1992) *The Codex Mendoza*, 4 vols. Berkeley: University of California Press.

Berdan, F. and de Durand-Forest J. (1980) *Matrícula de Tributos*. Graz: Akademische Druck-u. Verlaganstalt.

Berjonneau, G., Deletaille, E. and Sonnery, J.L. (1985) *Rediscovered Masterpieces of Mesoamerica: Mexico – Guatemala – Honduras*. Boulogne: Editions Arts.

Berlo, J. (1983) 'The warrior and the butterfly: central Mexican ideologies of sacred warfare and Teotihuacan iconography', in J. Berlo (ed.), *Text and Image in Pre-Columbian Art: Essays on the Interrelationship of the Visual and Verbal Arts*. BAR International Series 180. Oxford: BAR, 79–117.

Berlo, J. (1984) *Teotihuacan Art Abroad: A Study of Metropolitan Style and Provincial Transformation in Incensario Workshops*, 2 vols. BAR International Series 199. Oxford: BAR.

Berrin, K. and Pasztory, E. (1993) *Teotihuacan: Art from the City of the Gods*. London and New York: Thames and Hudson.

Boone, E. (1983) *The Codex Maglabechiano*: Berkeley: University of California Press.

Bray, W. (1977) 'Maya metalwork and its external connections', in N. Hammond (ed.), *Social Process in Maya Prehistory: Studies in Honor of Sir Eric Thompson*. New York: Academic Press, 365–403.

Burland, C.A. (1971) *Codex Fejérváry-Mayer*. Graz: Akademische Druck- u. Verlagsanstalt.

Caso, A. (1966) 'The lords of Yanhuitlan', in J. Paddock (ed.), *Ancient Oaxaca: Discoveries in Mexican Archaeology and History*. Stanford, CA: Stanford University Press, 313–335.

Caso, A. (1967) *Los Calendarios Prehispánicos*. Mexico City: Universidad Nacional Autónoma de México (UNAM).

Caso, A. and Bernal, I. (1952) *Urnas de Oaxaca*. Mexico City: Memorias del Instituto Nacional de Antropología e Historia (INAH), 2.

Castañeda Delgado, P., Bernal, A.M., Solís, F., Carmona, M. and Montero Alarcón, M. (1997) *Tesoros de México: Oro Precolumbino y Plata Virreinal*. Seville: Fundación El Monte.

Castañeda Lopez, C. (2008) 'Plazuelas, Guanajuato', *Arqueologia Mexicana* **16**(92): 44–47.

Castillo Tejero, N. and Solís Olguín, F.R. (1975) *Ofrendas Mexicas en el Museo Nacional de Antropología: Corpus Antiquitatem Americanensium*. Mexico City: Instituto Nacional de Antropología e Historia (INAH).

Conides, C. (2001) 'The Stuccoed and Painted Ceramics from Teotihuacan, Mexico: A Study of Authorship and Function of Works of Art from an Ancient Mesoamerican City', PhD dissertation, Colombia University (unpublished).

de la Fuente, B., Trejo, S. and Gutierrez Solana, N. (1988) *Escultura en Piedra de Tula*. Mexico City: Universidad Nacional Autónoma de México (UNAM).

Dean, J.S. and Ravesloot J.C. (1993) 'The chronology of cultural interaction in the Gran Chichimeca', in A.J. Woosely and J.C. Ravesloot (eds), *Culture and Contact: Charles C. Di Peso's Gran Chichimeca*. Albuquerque: Amerind Foundation and the University of New Mexico Press, 83–103.

Di Peso, C.C. (1974) *Casas Grandes: A Fallen Trading Center of the Gran Chichimeca, Vol. 2: The Medio Period*. The Amerind Foundation Publication No. 9. Flagstaff, AZ: Northland Press.

Dunkelman, A. (2007) *The J. Kislak Collection at the Library of Congress*. Washington, DC: Library of Congress.

Fontaine, F. and Purin, S. (1993) *Métropoles Mayas*. Brussels: Foundation Europalia.

Foshag, W.F. (1957) *Mineralogical Studies on Guatemalan Jade*. Smithsonian Collection **135**(5). Washington, DC: Smithsonian Institution.

Freidel, D., Schele, L. and Parker, J. (1993) *Maya Cosmos: Three Thousand Years on the Shaman's Path*. New York: William Morrow and Company.

Gamio, M. (1922) *La Población del Valle de Teotihuacan*. Mexico City: Secretaría de Agricultura y Fomento.

Graham, I. and von Euw, E. (1975) *Naranjo: Corpus of Maya Hieroglyphic Inscriptions* **2**(1). II. Cambridge, MA: Peabody Museum of Archaeology and Ethnology, Harvard University.

Harmer Rooke Galleries (1987) *Auction Catalog* XXVII. New York: Harmer Rooke Galleries.

Headrick, A. (2007) *The Teotihuacan Trinity: The Sociopolitical Structure of an Ancient Mesoamerican City*. Austin: University of Texas Press.

Heyden, D. (1972) 'Xiuhtecuhtli: Investidor de soberanos', *Boletín del INAH*, época II, **3**: 3–10.

Heyden, D. (2000) *El Templo Mayor de Tenochtitlan en la Obra de Fray Diego Durán*. Mexico City: Instituto Nacional de Antropología e Historia (INAH).

Houston, S., Brittenham, C., Mesick, C., Tokovinine, A. and Warinner, C. (2009) *Veiled Brightness: A History of Ancient Maya Color*. Austin: University of Texas Press.

Howell, S.N.G. and Webb, S. (1995) *A Guide to the Birds of Mexico and Northern Central America*. Oxford: Oxford University Press.

Izeki, M. (2008) *Conceptualization of 'Xihuitl': History, Environment and Cultural Dynamics in Postclassic Mexica Cognition*. Oxford: Archaeopress.

Köhler, U. (1989) 'Comets and falling stars in the perceptions of Mesoamerican Indians', in F. Aveni (ed.), *World Archaeoastronomy*. Cambridge: Cambridge University Press, 289–299.

Koontz, R. (2009) *Lightning Gods and Feathered Serpents: The Public Sculpture of El Tajin*. Austin: University of Texas Press.

Lenkersdorf, C. (1979) *Diccionario Tojolabal – Español*, 2 vols. Mexico City: Editorial Nuestro Tiempo.

Lombardo de Ruiz, S. (ed.) (1987) *La Pintura Mural Maya en Quintana Roo*. Mexico City: Instituto Nacional de Antropología e Historia (INAH).

López Luján, L. (2003) 'The Aztecs' search for the past', in E. Matos Moctezuma and F. Solís Olguín (eds), *Aztecs*. London: Thames and Hudson, 22–29.

López Luján, L., Neff, H. and Sugiyama, S. (2000) 'The 9-Xi vase: a classic Thin Orange vessel found at Tenochtitlan', in D. Carrasco, L. Jones and S. Sessions (eds), *Mesoamerica's Classic Heritage:*

From Teotihuacan to the Great Aztec Temple. Niwot: University Press of Colorado, 219–249.
Matos Moctezuma, E. and Solís Olguín, F. (2002) *Aztecs*. New York: Harry N. Abrams.
Maudslay, A.P. (1889–1902). *Biologia Central-Americana: Archaeology*, 5 vols. London: R.H. Porter and Dalau and Company.
McEwan, C. and López Luján, A. (2009) *Moctezuma: Aztec Ruler*. London: British Museum Press.
McEwan, C., Middleton, A., Cartwright, C. and Stacey, R. (2006) *Turquoise Mosaics from Mexico*. Durham: Duke University Press.
Mena, R. (1927) *Catálogo de la Coleccíon de Objetos de Jade*. Mexico City: Talleres Gráficos del Museo Nacional de Arqueología, Historia y Etnografía.
Miller, A.G. (1973) *The Mural Painting of Teotihuacán*. Washington, DC: Dumbarton Oaks Research Library and Collections.
Morris, E.H., Charlot, J. and Axtel Morris, A. (1931) *The Temple of the Warriors at Chichen Itzá, Yucatan*. Washington, DC: Carnegie Institution of Washington Publications 406.
Nicholson, H.B. (1971) 'Religion in pre-Hispanic central Mexico', in R. Wauchope (ed.), *Handbook of Middle American Indians*, Vol. 10, part 1. Austin: University of Texas Press, 395–446.
Olivier, G. and López Luján, L. (2009) 'Images of Moctezuma and his symbols of power', in C. McEwan and L. López Luján (eds), *Moctezuma: Aztec Ruler*. London: British Museum Press, 78–91.
Parsons, L.A. (1980) *Pre-Columbian Art: The Morton D. May and The Saint Louis Art Museum Collections*. New York: Harper and Row.
Pohl, J. (1994) *The Politics of Symbolism in the Mixtec Codices*. Vanderbilt Publications in Archaeology no. 46. Nashville, TN: Vanderbilt University.
Romero R., M.E., Flores Gutiérrez, D. and Mora-Echeverría, J. (2001) 'De cuentas de avatares: un calendario de Venus en Chacchoben, Quintana Roo', in L. Staines Cicero (ed.), *La Pintura Mural Prehispánica en México: Área Maya* **2**(4). Mexico City: Universidad Nacional Autónoma de México (UNAM), 447–459.
Sahagún, B. (1950–1982). *Florentine Codex: General History of the Things of New Spain*, A.J.O. Anderson and C.E. Dibble (tr.). Santa Fe, NM: School of American Research.
Sauer, C. (1932) *The Road to Cíbola*. Ibero-Americana 3. Berkeley: University of California Press.
Schele, L. (1997) *Hidden Faces of the Maya*. Mexico City: Impetus Comunicacíon.
Schildkraut, L. (2000) *The Merrin Gallery: Precolumbian Art*. New York: The Merrin Gallery [see www.mayavase.com K8293].
Séjourné, L. (1959) *Un Palacio en la Ciudad de los Dioses*. Mexico City: Instituto Nacional de Antropología e Historia (INAH).
Seler, E.E. (1902–1923) *Gesammelte Abhandlungen zur Amerikanischen Sprach und Alterthumskunde*, 5 vols. Berlin: Ascher & Co.
Seler, E.E. (1963) *Comentarios al Códice Borgia*. Mexico City: Fondo de Cultura Económica.
Smith, M.E. (1973) 'The relationship between Mixtec manuscript painting and the Mixtec language: a study of some personal names in Codices Muro and Sánchez Solís', in E.P. Benson (ed.), *Mesoamerican Writing Systems*. Washington, DC: Dumbarton Oaks Research Library and Collections, 47–98.
Stone, A. (1989) 'Disconnection, foreign insignia, and political expansion: the warrior stelae of Piedras Negras', in R.A. Diehl and J.C. Berlo (eds), *Mesoamerica after the Decline of Teotihuacan*. Washington, DC: Dumbarton Oaks Research Library and Collections, 153–172.
Sugiyama, S. (1989) 'Burials dedicated to the old temple of Quetzalcoatl at Teotihuacan, Mexico', *American Antiquity* **54**(1): 85–106.
Sugiyama, S. (1992) 'Rulership, warfare, and human sacrifice at the ciudadela: an iconographic study of feathered serpent representations', in J.C. Berlo (ed.), *Art, Ideology, and the City of Teotihuacan*. Washington, DC: Dumbarton Oaks Research Library and Collections, 205–230.
Taube, K. (1992a) 'The iconography of mirrors at Teotihuacan', in J.C. Berlo (ed.), *Art, Polity, and the City of Teotihuacan*. Washington, DC: Dumbarton Oaks Research Library and Collections, 169–204.
Taube, K. (1992b) 'The temple of Quetzalcoatl and the cult of sacred warfare at Teotihuacan', *Res: Anthropology and Aesthetics* **21**: 53–87.
Taube, K. (1994) 'The iconography of Toltec period Chichen Itza', in J.H. Prem (ed.), *Hidden in the Hills: Maya Archaeology of the Northwestern Yucatan Peninsula. Acta Mesoamericana* **7**. Möckmühl: Verlag von Flemming, 212–246.
Taube, K. (2000a) 'The turquoise hearth: fire, self sacrifice, and the central Mexican cult of war', in D. Carrasco, L. Jones and S. Sessions (eds), *Mesoamerica's Classic Heritage: From Teotihuacan to the Great Aztec Temple*. Niwot: University Press of Colorado, 269–340.
Taube, K. (2000b) *The Writing System of Ancient Teotihuacan*. Ancient America 1. Barnardsville and Washington, DC: Center for Ancient American Studies.
Taube, K. (2001) 'Yahui', in D. Carrasco (ed.), *The Oxford Encyclopedia of Mesoamerican Cultures*. New York: Oxford University Press, 359–60.
Taube, K. (2003) 'Maws of heaven and hell: the symbolism of the centipede and serpent in classic Maya religion', in A. Ciudad Ruíz, M.H. Ruz Sosa and M.J. Iglesias Ponce de León (eds), *Antropología de la Eternidad: la Muerte en la Cultura Maya*. Madrid and Mexico City: Sociedad Española de Estudios Mayas and El Centro de la Cultura Maya, 405–442.
Taube, K. (2005) 'Representaciones del paraíso en el arte cerámico del clásico temprano de Escuintla, Guatemala', in O. Chinchilla and B. Arroyo (eds), *Iconografía y Escritura Teotihuacana en la Costa Sur de Guatemala y Chiapas. U tz'ib*, Serie Reportes **1**(5): 33–54.
Taube, K. (2006) 'Climbing Flower Mountain: concepts of resurrection and the afterlife in ancient Teotihuacan', in L. López Luján, D. Carrasco and L. Cué (eds), *Arqueología de Historia del Centro de México: Homenaje a Eduardo Matos Moctezuma*. Mexico City: Instituto Nacional de Antropología e Historia (INAH), 153–170.
Taube, K. and Bade, B.L. (1991) 'An appearance of Xiuhtecuhtli in the Dresden Venus pages', *Research Reports on Ancient Maya Writing* 35. Washington, DC: Center for Maya Research.
Taube, K.A., Saturno, W., Stuart, D. and Hurst, H. (2010) 'The murals of San Bartolo, El Petén, Guatemala, Part 2: the West wall', *Ancient America* **10**. Barnardsville, NC: Center for Ancient American Studies.
Tozzer, A. M. (1941) *Landa's Relación de las Cosas de Yucatan*. Papers of the Peabody Museum of American Archaeology and Ethnology, Vol. l8. Cambridge, MA: Harvard University.
Turner, M.H. (1992) 'Style in lapidary technology: identifying the Teotihuacan lapidary industry', in J.C. Berlo (ed.), *Art, Ideology, and the City of Teotihuacan*. Washington, DC: Dumbarton Oaks Research Library and Collections, 89–112.
Ulrich, E.M. and Dixon de Ulrich, R. (1976) *Diccionario Maya Mopan/ Español, Español/ Maya Mopan*. Guatemala City: Instituto Lingüistico de Verano.
Ulrich, E.M. and Ulrich, R. (1982). *Stories of the Sacred, Serious, Sensational and Silly from Mopan Maya Texts*. Guatemala City: Instituto Lingüistico de Verano.
Urcid Serrano, J. (2001) *Zapotec Hieroglyphic Writing*. Washington, DC: Dumbarton Oaks Research Library and Collections.
Urcid Serrano, J. (2008) Personal communication.
Villagra Caleti, A. (1954) 'Pinturas rupestres "Mateo A. Saldaña" Ixtapantango, Edo. de México', *Caminos de México* **9**.
von Winning, H. (1968) *Pre-Columbian Art of Mexico and Central America*. New York: Harry N. Abrams.
Weigand, P.C. (1997) 'La turquesa', *Arqueología Mexicana* **5**(27): 26–33.

Author

Karl A. Taube, Department of Anthropology, University of California, Riverside, USA

Teoxihuitl: turquoise in Aztec thought and poetry

Patrick Johansson K.

Introduction

A precious stone, which magnified garments, statues, masks and jewels in pre-Columbian times, turquoise was also a conceptual gem of Aztec thought. The Nahuatl word for turquoise, *xihuitl*, according to the indigenous informants of the Spanish friar Bernardino de Sahagún, means 'the herb which is sprouting', but its meaning varies according to the context in which it is used.

It is related to a colour between blue and green, to fire, to Huitzilopochtli as Xiuhpilli, 'the turquoise prince', the sun god, to different birds, to beauty, to the year and more generally to time. In a literary context, it glitters in a myriad of metaphors and poetic expressions. In pictorial books, its function ranges from a simple phonetic sign to a complex visual significance.

Expressive oral gems

As the varied physical aspects of the gem depended on the type of soil in which it matured, the different symbolic values of the stone, the image, the word and the concept, were produced by the generic specificity of the text that contained them:

- Verbal relation (*tenonotzalli*): The most direct verbal reference and verbal expression concerning turquoise was the one that described the gemstone and how it was extracted. In this literary genre, the stone would become a word, a speech act, and eventually an idea but still closely related to the material context of the stone.
- Proverbs (*machiotlahtolli*) and moral speeches (*huehuetlahtolli*): In proverbs and moral speeches turquoise had different metaphorical meanings according to the context but it was generally related to beauty and moral value.
- Riddles (*zazanilli*): In riddles, the symbolism as well as the real physical properties of the gemstone was metaphorically referred to in order to induce the answer.
- Myths (*tlamachiliztlahtolzazanilli*): Interiorised qualities of the gemstone as well as different aspects of its symbolism were determinant in myths, not only as metaphors but as an essential part in the narrative construction of meaning.
- Shamanistic texts (*nahuallahtolli*): In shamanistic texts, the word for turquoise and the stone itself were used in a ritual and performance context.
- Songs (*cuicatl*): In songs, the colourful sounds of the word, the reference to the stone as well as all the potential conceptual allusions were placed on a metaphorical loom to weave poetry.

Pictorial expression

Carved in stone, painted on walls, vessels or on the pages of pictographic books, the image of turquoise could simply refer to the gem or be conceptually significant. An example of the first is found in the list of tributes of the Codex Mendoza (pl. XLII; see Berdan and Anawalt 1992 and Berdan, in this vol., pp. 91–102) where tesserae and blue masks expressed the fact that a city conquered by the Aztecs had to tribute turquoise raw material and artefacts made of turquoise.

Used as a sign, the image of turquoise could have different values and its role in visual syntax was considerably more complex. In this case, its iconographic function revealed an intricate network of concepts.

According to the available verbal and pictorial documents and after a brief analysis of their contents, this article has been divided into seven parts that seem to match the essential conceptual features of turquoise. The texts are quoted in both their original version and their translation to give an idea of the colourful Nahuatl phrases used.

1. The precious stone in its mother's womb

One of the primary characteristics of precious stones is that they mature in the earth's womb and that their extraction was culturally perceived as a birth. As far as turquoise is concerned, it matured in a cave, *oztotl* (a mine).

> *Auh quen ca xihuitl, ca oztoio, itlan huitz, itzalan huetztinemi in teuxihuitl, in huel icucic, in huel popoca: niman ie iehoatl, in mitoa: xihuitl anozo tlapalteuxihuitl.*
>
> And how is it with the turquoise? It comes out of a cave. From within is removed the fine turquoise, the mature, the one that smokes; and that is called turquoise or ruby.
>
> *Códice Florentino, Libro XI, capítulo 8*

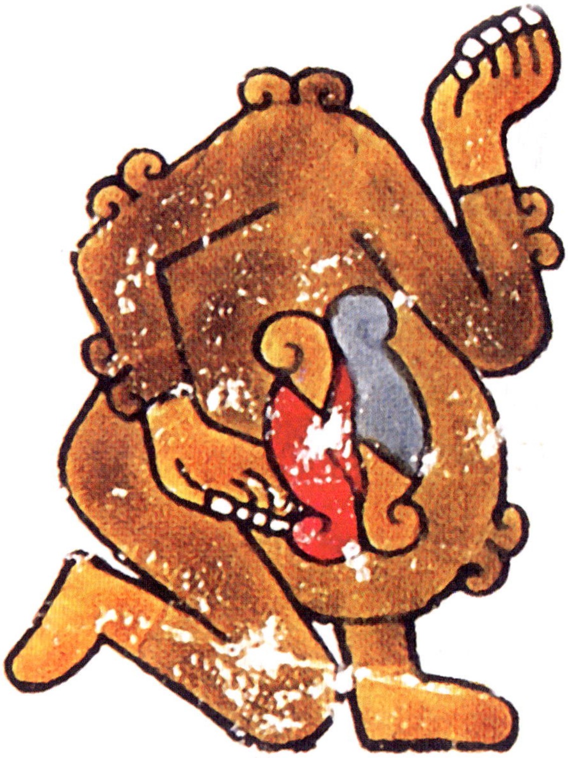

Figure 1. The precious turquoise in its mother's womb. Codex Vindobonensis, plate 41 (detail). Used with permission.

The Nahuatl word, *oztotl* (cave) also refers to the human mother's womb where the foetus is growing. The term for 'pregnant', *otztic*, actually comes from *oztotl* (Fig. 1).

Mother earth *tlalli* is also Tlaltecuhtli, 'the earth lord or lady', and more especially Mictlan the place of the dead, the realm of Mictlantecuhtli, a divine impersonation of death. In the very centre of the earth lies Xiuhtecuhtli, the 'turquoise lord', the god of fire and time. His dwelling place is referred to as Tlalxicco, generally translated as '(the place of) the navel of the earth' (see also part 4 for a possible translation of this expression).

The father and mother of the gods is the fire, the 'old god' *huehueteotl*, the 'old one of the four directions', *nauhyohuehue*, that is to say the one who is in the middle of a square, within the earth (Fig. 2.).

> *Copuchtia, quitzcactia in teteu innan in teteu inta in tlalxicco (tlal-xiuh-co) onoc, in xiuhtetzaqualco maquitoc, in xiuhtotoatica mixtzatzacuilitica in huehue teutl in aiamictlan in xiuhtecuhtli.*
>
> And he placeth him to the left, he provideth him the obsidian sandals, he the mother of the gods, the father of the gods, who resideth in the turquoise place (or navel) of the earth, who is set in the turquoise enclosure, [enclosed] with the waters of the lovely continga, enclosed with clouds – Ueueteotl, he of the misty netherworld, Xiuhtecutli.
>
> *Códice Florentino, Libro VI, capítulo 17*

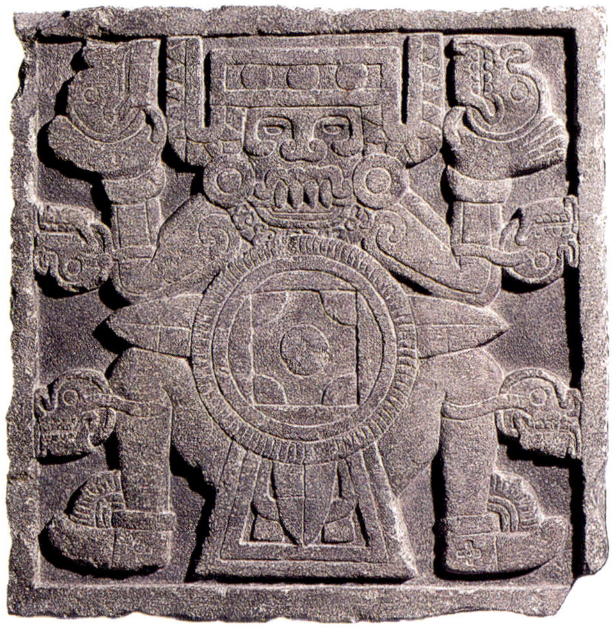

Figure 2. Tlaltecuhtli and the central quincunce. Museum of Templo Mayor, Mexico.

As we will see in part 4, this quincunx might be assimilated to fire and turquoise. The relation of turquoise to fire within the earth appears in the way indigenous experts detected the presence of the precious stone underground:

> *in canin iuhqui poctontli moquetza, huel quitta, in catleoatl maiauhiotitica, iehoatl o, in tlazotli: Auh intlacatle ipan aci: in canin poctontli moquetza, intla zan tlalnemiuhian, ic quimati ca uncan ca in tlazotli in tetl.*
>
> Wherever they can see that something like a little smoke [column] stands, that one of them is giving off vapor, this one is the precious stone. Something appears if it is only barren where the little [column of] smoke stands, thus they know that the precious stone is there in the earth.
>
> *Códice Florentino, Libro XI, capítulo 8*

The practical function of the operation is clear but we should not omit its religious aspects: watching the ground eastward at sunrise, to observe the exhalation of the earth, a tiny column of smoke standing where the stone is supposed to be, would probably generate a symbolic network of associated ideas. The squared centre of the world where Nauhyohuehue stands and the four-sided igneous realm of the old god of fire resemble, somehow, the squared turquoise tesserae.

2. Tesserae, mosaic and man

An important physical aspect of turquoise is that the stone is found in small pieces or tesserae. These tesserae were described by the native informants of Sahagún as follows:

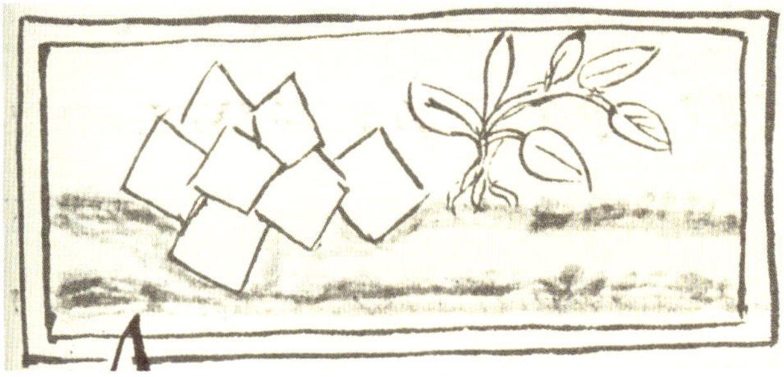

Figure 3. *Xihuitl*. The herb and the turquoise tesserae. Florentine Codex, book XI, chapter 8 (detail). Used with permission.

Auh inin xihuitl, amo cenquizqui, zan cacaiacaticac, iuhquin xalli, patlachtontli, patlachpipil, amo tlaquaoac, zan poxaoac.

And this turquoise, it is not in one piece. It is just fragmented, like sand, small and flat, small and flat, not hard, just soft.

Códice Florentino, Libro XI, capítulo 8

They are small, flat, soft and relatively squared, properties that are going to be symbolically processed in different texts and 'con-texts'. In order to give shape to and represent something, the pieces of turquoise were stuck in place on a support generally made of wood, and sometimes on skulls or bones.

Inic monequi, inic tlaqualnextilo, zan momana, mozaloa, zan ic tlaixtzaqualo.

It is required for use in adorning; they are placed, glued on, they are assembled onto a surface.

The fact that it was glued onto a support, and the material on which it was glued, was highly significant. Providing some sentences about turquoise as examples, the native informant said:

nixiuhtemoa, nixiuhtataca, nixiuhquixtia Nixiuhzaloa, nixiuhtzaqua: nicxiuhtzaqua in vapalli, in teixiptla.

I search for turquoise. I excavate turquoise. I remove turquoise. I glue turquoise. I affix turquoise. I affix turquoise to the beam, to the image.

Códice Florentino, Libro XI, capítulo 8

These sentences, though artificially pronounced on demand of the Spanish friar, contain the blue, soft verbal gem corresponding to the precious stone. As for the image obtained by assembling and sticking the little pieces of turquoise (tesserae), its manufacture was considered as a birth. As a matter of fact, 'to give a shape' in Nahuatl is *tlacatia*, which is phonetically very near *tlacati*, 'to be born', a similarity that is meaningful.

In a mortuary context, recording the dead princes and meditating about his own destiny after death, a poet said:

Cuix nixiuhchimalli? Aya Oc ceppa nozaloloz In niquizaz? in ayatica niquimilolo Aya Tlalticpac huehuetitlan niquimilnamiqui! Ohuaya Ohuaya.

Am I perchance a shield of turquoise Shall I be affixed and glued on so that I go out again? I am shrouded in a henequen cloak here on earth, near the drums Oh I remember them!

Romances de los Señores de la Nueva España (Garibay 1993), fol. 6

Is man like a handful of tesserae being glued together and shaped for a new life?

3. The etymology of the word(s) and the 'etymography' of the glyph

When asked about the origin of the words *chalchihuitl* and *xihuitl*, Sahagún's indigenous informant said that the first 'comes from nowhere' (*acampa quiza in itoca*). As for *xihuitl*, he said the following:

Inin xihuitl itech quiza in itoca xihuitl, in ixoatoc: ipampa in itlachializ amo cenca quiltic, zan achi micqui: iuhquin amo cenca mavizio, zan nel achi ixtlileoac.

The name of this turquoise comes from the herb which lies sprouting; because its appearance is [not very coloured] not very herb-green, just a little dull; As if it were not very estimable. It is actually a little dark-surfaced.

Códice Florentino, Libro XI, capítulo 8

The illustration corresponding to this definition in the Florentine Codex (Anderson and Dibble 1970) shows herbs and tesserae thus linking two words and two concepts (Fig. 3).

In the mythological and iconographic contexts of the Codex Boturini (Johansson 2007), one of the *mimixcoas* called Xiuhnel, 'dispersed turquoise' (tesserae), is referred to through an anthroponymic glyph composed of a turquoise mosaic and four squared tesserae (Codex Boturini, plate IV).

The mosaic, a circular disk and the four squared tesserae, might have composed a meaning beyond the immediate name reference. Sources also refer to a special kind of turquoise called *teoxihuitl*.

Inin itoca itech quiza in teutl, ioan in xiuitl, zan quitoznequi iiaxca, itonal in teutl, ioan, q.n. cenca mahuizio; ipampa acan cenca neci, canin zan quenman in neci: inin teuxiuitl cenca mahuizio, in itlan in itech motta amo cenca mahuizio: auh in achi hueca nezi, huelticeoac, iuhquin xiuhtototl, nelli iuhquin popoca.

The name of this come from *teotl* [god] and *xiuitl* [turquoise], which merely means that it is the property, the lot, of the god; and it means that it is much esteemed, because it does not appear anywhere very often. It seldom appears anywhere. This fine turquoise is much esteemed. And when it appears some distance away , it is quite pale, like the lovely cotinga, verily as if smoking.

Códice Florentino, Libro XI, capítulo 8

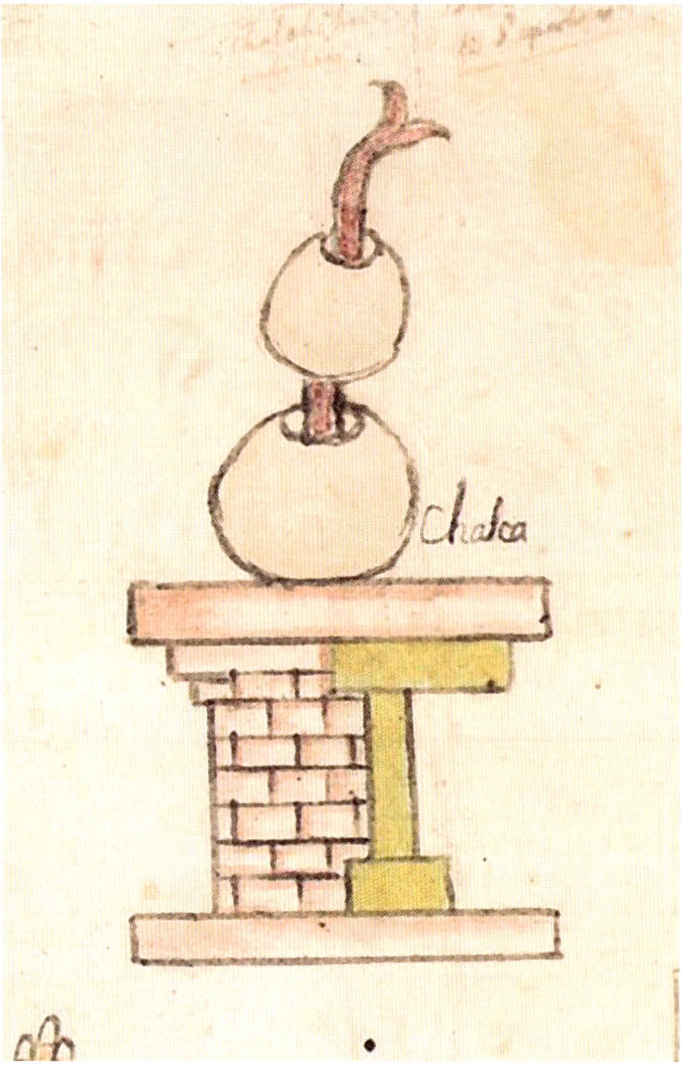

Figure 4. Two perforated gourd vessels compose the glyph 'Chalco'. Codex Azcatitlan, plate 3 (detail). Used with permission.

Although *teoxihuitl* can be translated in general terms as 'god's turquoise', the god *par excellence* is the sun and, as we will see in the next section, turquoise is closely associated with the sun.

- *Xihuitl*: If the term is entirely Nahuatl, according to the native informant of Sahagún and to the illustration, the precious stone's name comes from the herb *xihuitl* and thus is conceptually linked to what grows in the earth. Before we go any further it will be interesting to consider the etymology of the Nahuatl name for 'jade', *chalchihuitl*, a precious stone similar to turquoise in different aspects.

Chalchihuitl, 'jade': If a smoky aspect seems to be an essential property of turquoise, 'breath' is characteristic of jade:

> *Auh no yoan inic quiximati inin tlazotetl, uncan ca:*
> *muchipa tlacecelia, quilmach inin chalchihuitl ihiio;*
> *auh in ihiio cenca cecec, tlacamaoani: ihuin in motta,*
> *in mana chalchihuitl.*
>
> And thus do they know that this precious stone is
> there: [the herbs] always grow fresh; they grow green.
> They say this is the breath of the green stone, and its
> breath is very fresh; it is an announcer of its qualities.
> In this manner is seen, is taken the green stone.
>
> *Códice Florentino, Libro XI, capitulo 8*

It is therefore probable that both jade and turquoise shared the term *xihuitl* referring to the herb even though it became *chihuitl* in *chalchihuitl* due to the phonetic evolution of the word.

The common belief is that *chalchihuitl* comes from *xalxihuitl*, that is to say *xal(li)*, 'sand', and *xihuitl*, 'herb' or 'precious stone'. A phonetic evolution of [sh] to [tch] is possible in Nahuatl but [sh] is a relatively stable consonant and it is doubtful that both [sh] would become [ch] thus evolving from *xalxihuitl* to *chalchihuitl*. But if the first word began with a [tch], it is probable that the second would soon change from [sh] to [tch] following the well-known process of assimilation. *Chalxihuitl* would have become *chalchihuitl*.

Now if the meaning of *xihuitl* is clear, where would *chal-* come from? There are two possibilities:

- It could come from the adjective *chaltic* or *chachaltic*, 'hard'. *Chalxihuitl* would originally have been a 'hard precious stone' (differing from the turquoise which is soft, according to indigenous texts).

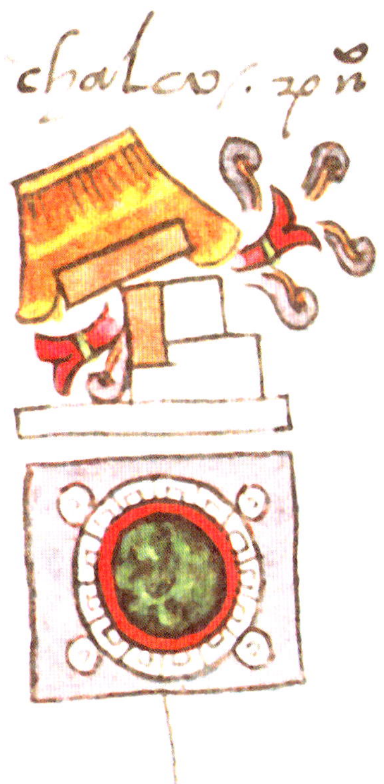

Figure 5. The image of jade in the glyph of Chalco. Codex Mendoza, plate IV (detail). Used with permission.

- It could also have come from the verb *chalia*, 'to inaugurate', that is to say 'to perforate'. Jade stones were often perforated to make necklaces or ear plugs. In the Codex Azcatitlan, the glyph for Chalco (Fig. 4) is composed of two gourd vessels being 'perforated' (*chalia*). This would tend to confirm this second hypothesis. There could be a simple phonetic relation between the perforated gourd vessel (*chalia*) and *chal* (*chihuitl*). But a conceptual tie could also exist between the two.

The traditional glyphs for jade (Fig. 5) and turquoise (Fig. 6) are somehow similar. They are composed of a main circle with four small circles disposed around and forming a square around the main circle. The glyph representing a *chalchihuitl* is white, red and green and set within a square. The glyph for turquoise is blue and red.

The four small circles around the main one probably evoke the fire god as 'the old one of the four (directions)', Nauhyohuehue:

Tla huiyan, tla tictotoca nahui acatl milinticâ.
Tla xihuiqui yn notâ nahui acatl milintica,
xiuhtli cozauhqui milintica, teteo ynan, teteo ynta
nauhcampa tlemuchitl yca tlatlalpitztica.

Come take it away, 4-Reed lit up
Come here you my father 4-Reed who glitters in the dark
Turquoise, yellow burning candle
mother of the gods, father of the gods.
who are popping fire all over.
Hernando Ruiz de Alarcón (1982), p. 162

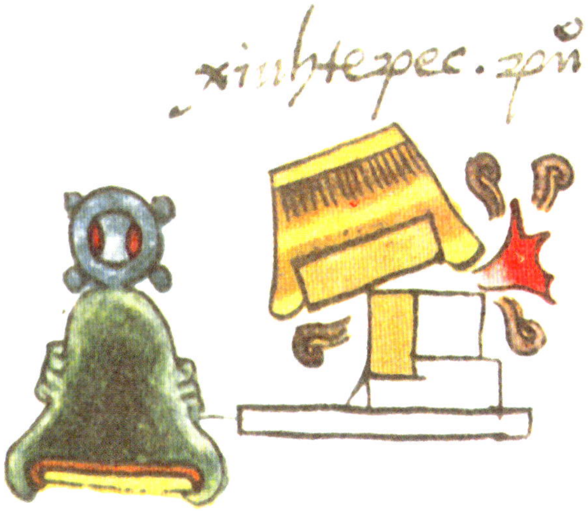

Figure 6. The image of turquoise in the glyph of Xiuhtepec. Codex Mendoza, plate VI (detail). Used with permission.

We should remember that 4-Reed is the calendrical name for fire. The four small circles around the central main one remind us also of the Sun stone where the four eras are located around a central circle which features the wrinkled face of the fire god (and/or the sun) (the Sun stone, Mexico's Museum of Anthropology).

The integration of a square formed by the four eras and the central circle could have been reproduced in the glyph for *xihuitl* thus expressing the idea of 'year' (*xihuitl*) and more generally 'time'. Ultimately the difference between jade and turquoise could have been expressed through the colours.

4. Turquoise, fire and smoke

The smoky aspect of turquoise, its colours, the etymology of the word and the 'etymography' of its corresponding glyph tend to associate turquoise with fire and smoke.

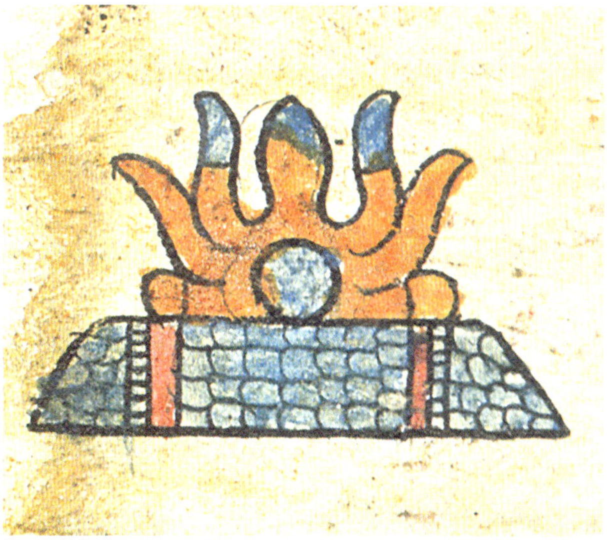

Figure 7. Tlalxiuhco 'the earth's place of turquoise'. Codex Borbonicus, plate 19 (detail). Used with permission.

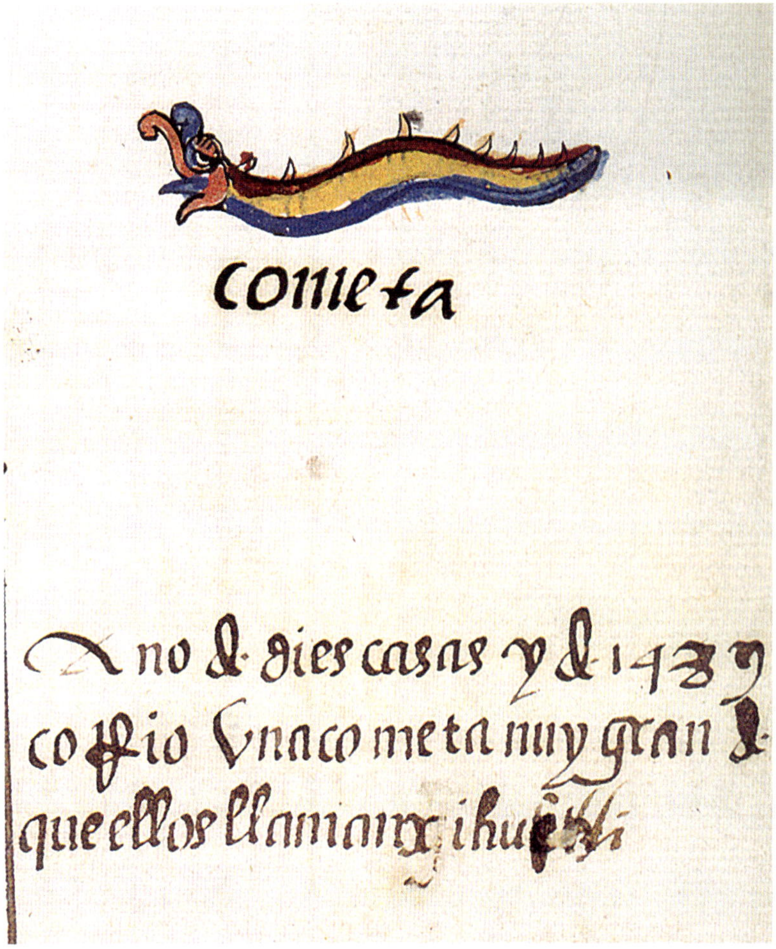

Figure 8. *Xiuhtli.* The comet. Codex Telleriano-Remensis, fol. 39v (detail). Used with permission.

Evoking the enthronement of an Aztec king, a text of the Florentine Codex establishes a link between fire, turquoise and rulership:

> *Copuchtia, quitzcactia in teteu innan in teteu inta in tlalxicco (tlal-xiuh-co) onoc, in xiuhtetzaqualco maquitoc, in xiuhtotoatica mixtzatzacuilitica in huehue teutl in aiamictlan in xiuhtecuhtli.*
>
> And he placeth him to the left, he provideth him the obsidian sandals he the mother of the gods, the father of the gods, who resideth in the turquoise place (or the navel) of the earth, who is set in the turquoise enclosure, [enclosed] with the waters of the lovely cotinga, enclosed with clouds –Ueueteotl, he of the misty netherworld, Xiuhtecutli.
>
> *Códice Florentino, Libro VI, capítulo 17*

The 'turquoise enclosure' (*xiuhtetzaqualco*) is the place of fire which is surrounded by a turquoise blue wall (Fig. 7).

Now coming back to etymological matters, the expression *tlalxicco*, 'in the navel of the earth', could very well have been, once, *tlalxiuhco*, 'in the place of turquoise'. The Nahuatl phonological system allows this possibility – we find *xippilli* instead of *xiuhpilli*, *toximmolpilli* for *toxiuhmolpilli*, *xippoalli* for *xiuhpoalli*, etc., where the *-uh* is assimilated to the next consonant which is therefore duplicated. In the eventual case of *tlalxiuhco*, *-uh* would have been assimilated to *-c* and would have become *tlalxicco*, subsequently being reinterpreted as 'in the navel of the earth', which is also highly significant. However, the 'turquoise enclosure' and the 'turquoise bird water' suggest that *tlalxicco* should be translated as if it were *tlalxiuhco*, 'the earth's place of turquoise' where stands Xiuhtecuhtli, the fire god (see Fig. 7).

In Figure 7 the three colours of fire – blue, yellow and red – are present but the turquoise platform, the turquoise extremities of the flames, and the turquoise disk in the very centre of the fire establish an obvious relationship between fire and turquoise.

In the mortuary context of the incineration of a corpse, an *atequilizcuicatl* 'song of water pouring' expresses this relationship as well:

> *Yn tlachinolxochitl, chimalxochitl,*
> *On cuecuepontoc.*
> *Tlatlatzcatimanía,*
> *Iyacaxochitl ontzetzelihui,*
> *Anquizo yehuatl*
> *Ye ic contzaquaco teocuitlatla ya*
> *Ye noconana ya xiuhtlacuilolía(n)*
> *¡Ye napiloltzin icnoconmemeya ha nohueyohuan!*
>
> The flower of fire, the shield flower
> open their corollas.
> in bursting flames
> The ill smelling flowers are scattered

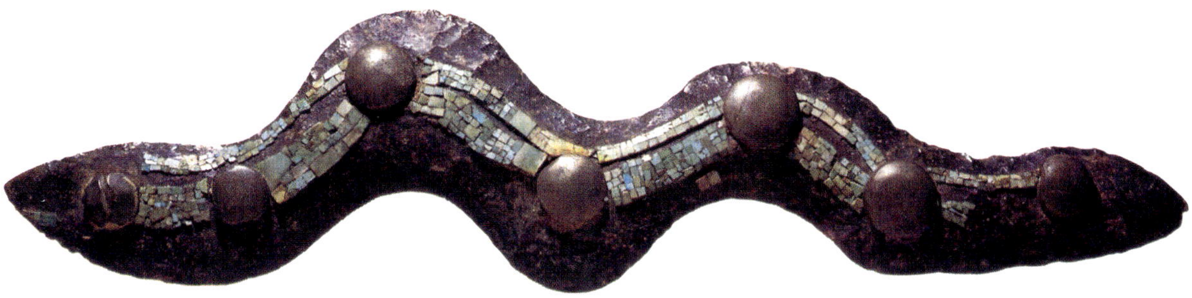

Figure 9. A *xiuhcoatl*-shaped comet of silex, turquoise and pyrite. Museum of Templo Mayor, Mexico.

You enclose him in the golden burning place *ya*
ye I take him from the turquoise painting place
ye my small jar of sadness pours sorrowful waters
ha my friends.
Cantares Mexicanos, fol. 58r

The golden burning place and the turquoise painting place are present in a representation of the god Xiuhtecuhtli (Codex Magliabechiano, plate 46).

In a shamanistic context in which fire is supposed to kill ants, the relationship of fire to turquoise is clear:

Tlacuel, xoxouhqui tlamacazqui, xiuhpapatlantzin,
Tle (t)axtica? Tla xocontotocati in puputecatl.

Hey you, blue priest, flyer of turquoise!
What are you doing? Take away the ants, chase them off.
Hernando Ruiz de Alarcón (1982), p. 169

Smoke. As we saw it, smoke is a constituent part of fire as well as an attribute of fine turquoise *teoxihuitl*.

In a shamanistic ritual that propitiates the capture of a deer with a net, the earth, the moon, the smoke, the Milky Way, the sun, the fire and the turquoise prince Nanahuatzin are summoned:

Tla xihuiqui, nonan tlalteuctli, notâ ce tochtli [...]
tezcatl ca hualpopocatimani,
nonan citlalcueye,
notâ totonametli, tlamacazqui nanahuatzin,
xiuhpiltzintli.

Come, you my mother earth, you, my father One Rabbit [...]
Mirror that is smoking
My mother with a dress of stars
My father: the sunbeam, the high priest Nanahuatzin,
The prince of turquoise.
Hernando Ruiz de Alarcón (1982), p. 165

Turquoise-fire in the sky. The comet *xiuhtli* or *xihuitl* is clearly related to turquoise through the word that refers to it as well as through its representation in Nahuatl iconography (Figs 8 and 9). As we saw it in this fourth section turquoise is intimately related to fire.

5. Time is blue

From within the earth (*tlalxicco* or *tlalxiuhco*), fire was mythologically brought to an existential dimension *Tlalticpac* ('on the earth') so that there could be movement, time and life.

5.1 The expansion of the sky

After the fourth era, a deluge, Tota and Nene made a promethean fire to roast fish. The column of smoke literally took the still-missing cardinal point 1-*Tochtli* (1-Rabbit) to its place, the south, and by doing so permitted the expansion of the sky (*ilhuicatl*), a subsequent creation of the domestic fire by Tezcatlipoca in 2-Reed and the beginning of time (Johansson 2008).

Every 52 years, in the last few hours of 1-Rabbit, time died, was incinerated (Codex Borbonicus, plate 34) to be born again in 2-Reed after a new fire was drilled on a sacrificed victim's breast. If a child was born that day, the name he was given contained *xiuh-*, 'year' or 'turquoise'. Year and turquoise are symbolically fused in the Nahuatl term *xihuitl*. As for the sun, another myth establishes how he was born by throwing himself into a fire.

5.2 Xihuitl: turquoise, year and time

The turquoise lord; the fire, the turquoise prince: the sun, and the mythically generated time symbolised by the sky, are blue. A riddle expressed it this way:

Zazan tleino, xoxouhqui xicaltzintli, mumuchitl ontemi?
Aca quittaz tozazaniltzin, tlacanenca: ilhuicatl.

What is it? a turquoise-blue bowl filled with popcorn?
For the one who will guess our little riddle it can't be anything but the sky.
Códice Florentino, Libro VI, capítulo 42

Time will be round and blue like the turquoise blue-bowl sky. Once the sky and the earth had expanded, Time could begin:

Maxiqualyancaqui o yn nonohualcatl on tepanecatl
yehuaya yancuic xihuitl cuepponiya ypan nicmati ye
noyol nicuicanitl yiehuaya niquehuaz oo huaye ynic
tzintic in ilhuicatl yn ica manaloc in tlalticpac yio
ohuiya.

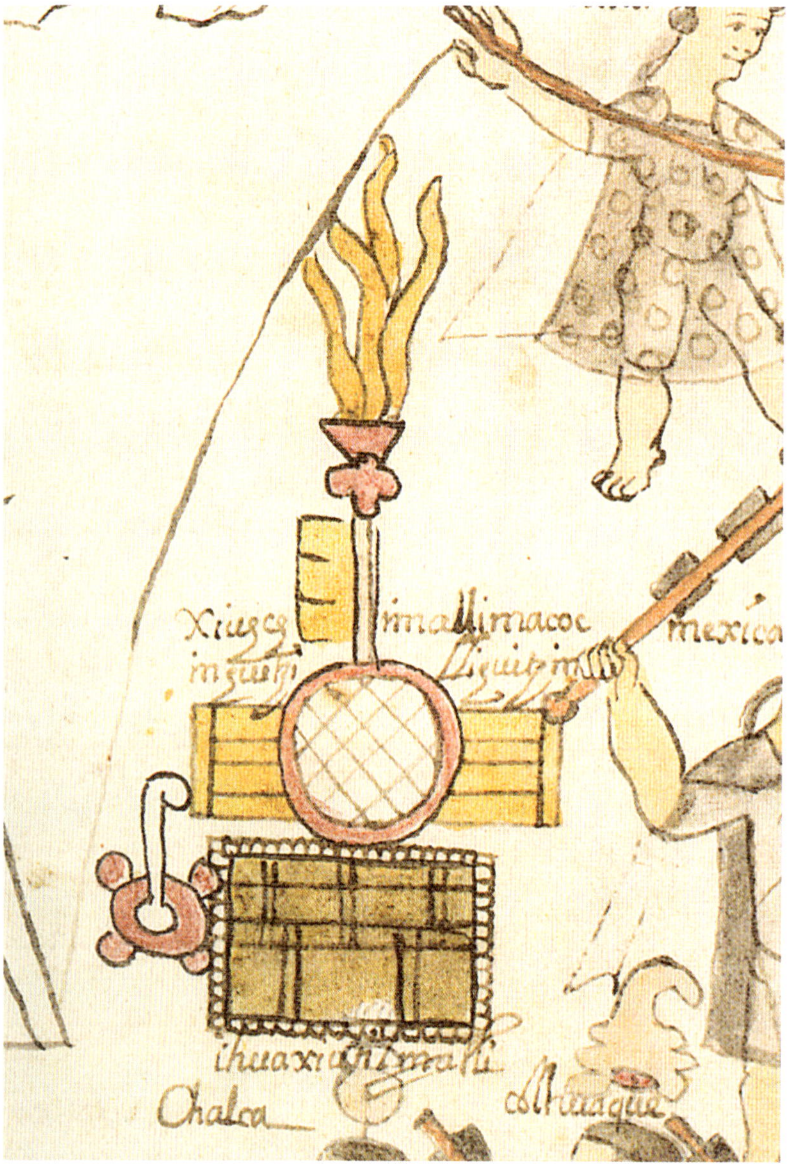

Figure 10. 'The turquoise shield and the turquoise cape.' Codex Azcatitlan, plate 10 (detail). Used with permission.

> Come here and listen you nonohualcatl, you tepanecatl *yehuaya* A new year opens its corolla *ya* My heart knows, I am a poet *yehuaya* I raise my song *huaya* Thus the sky is created and the earth is extended *iya ohuiya.*
>
> *Cantares Mexicanos, fol. 41r*

Turquoise and turquoise blue colour will be the iconographic way to express a date and a period of time. In a glyph of Codex Xólotl (plate 3) the turquoise mosaic disk represents the year and the little blue circle the number '1. The sign 'flint' is uncoloured. In another glyph in the same plate of the same codex, however, each turquoise mosaic disk represents 20 years, while the 12 small circles are 12 years, summing 52 years. A turquoise blue square often frames the year sign in Nahuatl iconography. In a period of 52 years, it is time as such that frames the world with turquoise (Codex Mendoza, plate I). In a 20-day period, that is to say a month, *meztli*, the precious turquoise displays a variety of colours (Codex Mendoza, plate XIX). In Codex Mendoza, days are sometimes referred to as if they were months (Codex Mendoza, plate LVIII), with colourful turquoises.

The integration of a circle and the numeric exponent '4' seems to be omnipresent in the iconography of turquoise and time. A representation of the course of the sun in the Maya book *Chilam Balam of Chumayel* seems to confirm it (*Chilam Balam of Chumayel*, p. 66; see Mediz Bolio 1998).

The blue sky, the course of the sun and time were closely related to turquoise. Its most magnificent expression is probably the famous double-headed turquoise serpent in the British Museum collection.

6. Turquoise, sun and rulership

As we expressed it before, the election and enthronement of a new *tlahtoani* was closely related to fire. The sun, *xiuhpilli* 'the prince of turquoise', is somehow the fire moving in the sky, the son of Xiuhtecuhtli the god of fire and among the Aztecs, Huitzilopochtli the ruler.

Actually the verb that expresses the course of the sun *tlatoca* is phonetically very close to the word 'to govern': *tlahtoca*. The *tlahtoani* is in fact the image of the sun on earth. When Huitzilihuitl, the second king of Mexico-Tenochtitlan, was installed as a *tlahtoani* it was said: 'we are under the protection of our god Huitzilopochtli of whom you are the image' (Durán 1967, II: 52). A Nahuatl proverb, evoking the possible death of a *tlahtoani*, states:

¿Cuix oc ceppa xiuhchimaltonatimomanaz? anozo
¿Cuix oc ceppa xiuhchimaltonatimomanaquiuh?
Quitoznequi: yn tlein cualli yn tlein yectli opoliuh
ocenmayan poliuh ocenmayan ihtlacauh. In iuhqui
ahca huei tlatoani omic, opoliuh, anozo ohualquiz.

Will perchance the turquoise shield appear shining?
Or: will perchance the turquoise shield come to
appear shining?
That is to say: that which is good, that which is
righteous has vanished; it has forever vanished; it is
forever corrupted. It is as [when] some great ruler has
died, has vanished, or has been disgraced.
Códice Florentino, Libro I, Addendum II

When the Mexicas reached Temazcaltitlan, Huitzilihuitl (the first) was given the insigniae of kingship: the turquoise shield and the turquoise cape (Fig. 10).

Xiuhchimalli maco mexica In Huitzilihuitzin, ihuan
xiuhtilmatli.

The Mexicas gave Huitzilihuitzin a shield of
turquoise, a cape of turquoise.
Códice Azcatitlan, plate 10

It is interesting to observe that the term 'turquoise' is not expressed through the colour (blue) but with the image of a smoking turquoise.

The turquoise diadem and the turquoise lip plug will become the emblem of kingship (*tlahtoani* or *tecuhtli*) for the Aztecs (Codex Mendoza, fol. 15v).

7. Beauty and virtues of turquoise

The adjective *tlazo-*, 'precious', which is a component of *tlazotetl*, 'precious stone' referred not only to beauty but also to value and love. Expressing beauty, turquoise was generally accompanied by jade. It was used for example, to qualify a beautiful speech:

Huel chalchiuhtic, huel teuxiuhtic, huel acatic, huel
ololiuhqui
Inin tlatolli, itechpa mitoaya: in aquin cenca huel
tecutlatoa, tenonotza:mitoaya. Cenca mahuíztic inic
otlato:
Iuhquin chalchíhuitl, iuhquin teuxíhuitl.

Like a jade, like a turquoise, like a reed or round are
you words. Thus it was said when somebody spoke
as a lord, when he addressed the people, it was said:
what he says is very important. It is like jade. It is like
turquoise.
Códice Florentino, Libro VI, capítulo 43

The inexorability of destiny was compared to a precious stone and accepted as such:

Zan nicchalchiuhmati, nicteuxiuhmati in quexquich
nopan yauh, in nopan quiza.

Only as jade, as turquoise do I consider what is falling
on me, what is happening to me.
Olmos (1993), capítulo 8

The description of some beautiful blue birds show different hues of the colour:

Xihuitzilli:
Macitica, motquitica texotic: xiuhtototic, ticeoac
teuhxiuhtic, xippopoca, teuxippopoca.

Costa Hummingbird
It is entirely, completely light blue like a cotinga,
pale like fine turquoise. It smokes like turquoise, it
smokes
like fine turquoise.
Códice Florentino, Libro XI, capítulo 12

The destiny of a dying woman (or a woman about to be sacrificed) is expressed in terms of turquoise:

Beautiful women
Ti mahuiztlachihualla
monequetzca moxiuhtoz quetzalpetlapan
ton icaca Aya

You, marvellous creature
In the place where you stand, in a quetzal feathered
mat.
You will become a turquoise (or an herb)
There you stand *Aya.*
Romances de los Señores de la Nueva España
(Garibay 1993), fol. 7r

This poem of Tlaltecatzin of Cuauhchinanco could very well have been addressed to a woman who had died in childbirth (*Cihuateotl* or *Cihuatetehuitl*) and would then take the sun every afternoon from its zenith back to its mother's womb at sunset. As her male counterpart she would have had a sun-like destiny.

As the Aztec god Huitzilopochtli, turquoise matured in its earthly mother's womb, taking shape to symbolise the sun, the blue sky and time, the latter being more specifically impersonated by the Xiuhcoatl, 'the serpent of turquoise'. It is also closely related to fire and rulership, and ultimately to beauty and virtue.

References

Anderson, A.J.O. and C.E. Dibble (eds) (1970) *Florentine Codex*, 12 vols. Monographs of the School of American Research. Santa Fe, NM: The School of American Research and the University of Utah.

Berdan, F. and Anawalt, P. (1992) *The Codex Mendoza*, 4 vols. Berkeley: University of California Press.

Cantares Mexicanos (1994) Manuscript in Biblioteca Nacional de México, Mexico (in Nahuatl).

Códice Azcatitlan (1995) Paris, Bibliothèque Nationale de France/ Société des Américanistes (facsimile).

Códíce Borbónico (1981) México: Editorial Siglo XXI.

Códice Florentino (1979) (Testimonios de los informantes de Sahagún) (facsimile). Gobierno de la República Mexicana, Mexico City: Giunte Barbera.

Codice Magliabechiano (1970) Graz: Akademishe Druck und Verlagsanstalt.

Códice Vindobonensis (1992) Mexico City: Akademishe Druck und Verlagsanstalt, Fondo de Cultura Económica.

Códice Xólotl (1980) Mexico City: Instituto de Investigaciones Antropológicas, Universidad Nacional Autónoma de México (UNAM).

Durán, D. (1967) *Historía de las Indias de Nueva España e Islas de Tierra Firme*, 2 vols. Mexico City: Editorial Porrúa.

Garibay, K.Á.M. (ed.) (1993) *Romances de los Señores de la Nueva España*, '*Poesía Náhuatl*', Vol. 1: *Paleografía*, 2nd edn. Mexico City: Universidad Nacional Autónoma de México (UNAM).

Izeki, M. (2008) *Conceptualization of 'Xihuitl': History, Environment and Cultural Dynamics in Postclassic Mexica Cognition*. BAR International Series 1863. Oxford: BAR.

Johansson, P. (2004) *Machiotlahtolli: la Palabra-Modelo. Dichos y Refranes de los Antiguos Nahuas*. Mexico City: McGraw-Hill.

Johansson, P. (ed.) (2007) *Códice Boturini o Tira de la Peregrinación. Arqueología Mexicana* [special issue: Códices].

Johansson, P. (2008) 'And the flint stone became a rabbit ... The creation of the South and the origin of time in the Aztec *Legend of the Suns*', *Indiana* **2**: 77–99.

Mediz Bolio, A. (tr.) (1998) *Libro de Chilam Balam de Chumayel*. Mexico City: CONACULTA.

Olmos, fray Andrés de (1993) *Arte de la Lengua Mexicana*, M. León-Portilla and A. León-Portilla (ed. and tr.). Mexico City: Instituto de Cooperación Iberoamericana (facsimile).

Quiñones Keber, E. (1995) *Codex Telleriano-Remensis*. Austin: University of Texas Press.

Ruiz de Alarcón, H. (1982) 'Tratado de supersticiones', in P. Ponce, Sánchez de Aguilar et al., *El Alma Encantada*. Mexico City: Fondo de Cultura Económica, 127–223.

Author

Patrick Johansson K., Institute of Historical Research, University of Mexico, Mexico City, Mexico

Turquoise and squash blossom: a Pueblo dialogue of the long run

Peter M. Whiteley

Introduction

As both substance and symbol, turquoise (Fig. 1) figures prominently in the Native North American Southwest, especially among the Pueblo Indians and their ancestors (e.g. Pogue 1915).[1] There are clear continuities between Pueblo thought systems and Mesoamerican ideas, and strong hints of direct sharing, albeit varying with relative complexities of social scale (cf. Parsons 1933). Lévi-Strauss (e.g. 1963) has made a compelling case for dualism in indigenous American thought– in social organisation, architecture, ritual, aesthetics and mythology. The principal dyadic complement to turquoise in the Pueblo imagination is squash, and its associated symbolic variations: notably, scarlet macaws and *Spondylus* or other red shell or stone. This pairing appears to go deep into the past, and reflects long-term patterns of trade.

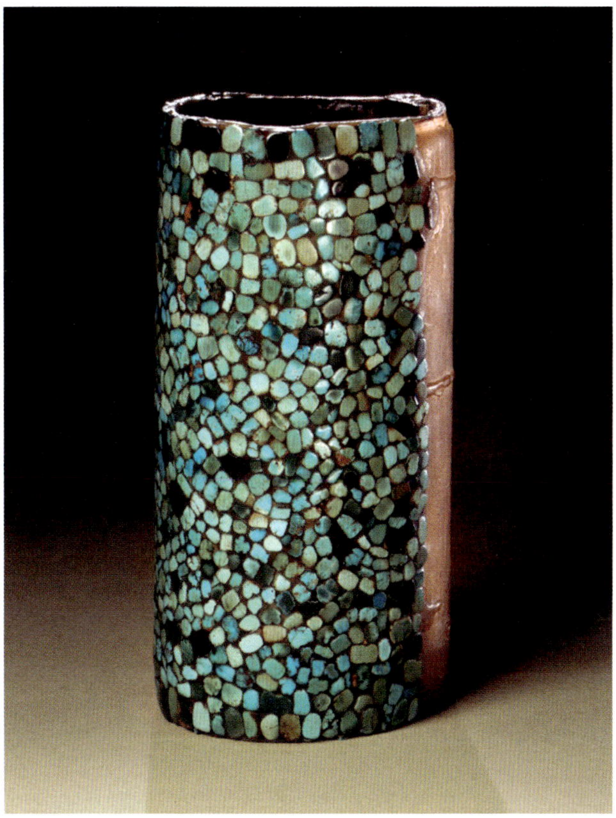

Figure 1. Cylindrical basket overlaid with mosaic turquoise. Pueblo Bonito, Chaco Canyon, ca. 11th century AD. American Museum of Natural History (image No. 1452).

The modern Pueblos comprise six language groups: Tiwa, Tewa and Towa of the Tanoan family, Keres and Zuni, both linguistic isolates, and Hopi, a Uto-Aztecan language (Fig. 2). None of the individual Pueblo languages is mutually intelligible, yet the Pueblos share much material and ideational

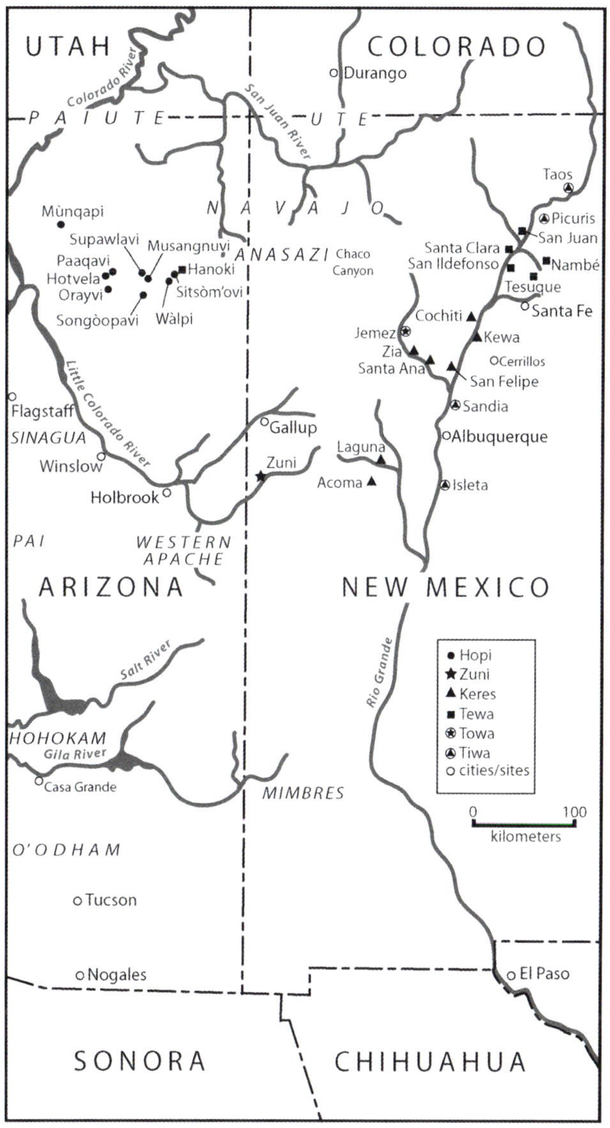

Figure 2. Present Pueblos, with general locations of non-Pueblo peoples (Navajo, Western Apache, Paiute, Ute, Pai, O'odham) and archaeological cultures (Anasazi, Hohokam, Sinagua, Mimbres). (Figure by Peter Whiteley. Drawing: Jennifer Steffey, used with permission.)

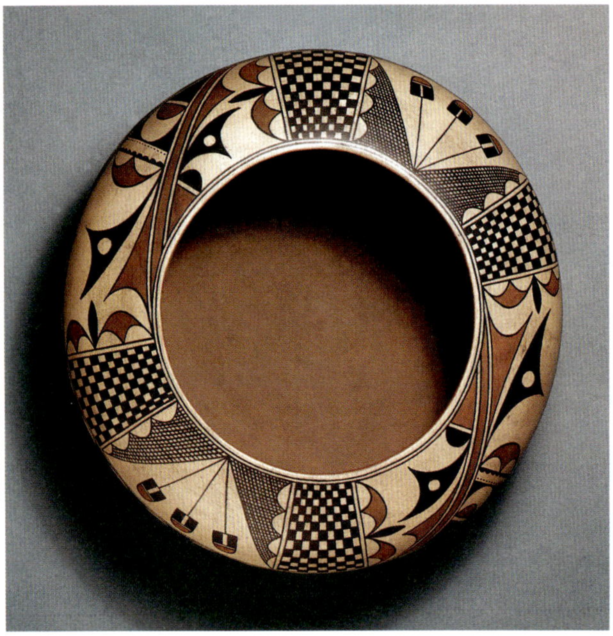

Figure 3. Ceramic bowl, Maria Martinez , designed by Julian Martinez (San Ildefonso Pueblo). American Museum of Natural History (50.2/6583).

culture in common such as maize-beans-squash agriculture and the Katsina religion, and identify themselves distinctly from other local indigenous peoples – Navajos, Apaches, Utes, Paiutes, O'odham (Pimans), Comanches and Pai. Of these, the Navajo come closest culturally, with many ideas – including an emphasis on turquoise – borrowed from the Pueblos, after their arrival (with other Apacheans) in the American Southwest about 600 years ago.

Archaeologically, the apogee of the prehistoric Pueblo 'great house' system lies at Chaco Canyon, with its celestially aligned, dramatic public architecture and incipient stratification. Specialised production at Chaco featured turquoise, traded to Mesoamerica in exchange for scarlet macaws and copper bells. By the fourteenth century AD, great houses had developed into larger, more widely dispersed sites that became the historic Pueblos – first recorded by Spanish explorers in the 1530s.

Colours, directions, gemstones and cosmology

In Pueblo symbolism, dualities replicate and multiply: two becomes four and six, four becomes sixteen: 'the tetramerous aspect of Tewa symbolic classification' is 'relentless and pervasive' (Ortiz 1969: 22) (Fig. 3). The Tewa world is bounded concentrically by four sacred mountains, four flat-topped hills and four shrines at the outskirts of the village. Village space is organised into four quarters, four plazas; ritual performances follow sequences and arrangements of four. For the Western Pueblo Hopi, the pattern is similar. For both Eastern and Western Pueblos, the cosmos is organised into four quarters: for both Tewa and Zuni, directions are cardinal, while for Hopi, they are inter-cardinal, marked by sunrise and sunset at the Winter and Summer solstices, and associated with sacred mountains. The Keres have both the four-corner sunrise-and-sunset points and the intermediate cardinal directions (White 1942: 80–83).

Within the tetradic cosmos, Pueblo orientation is inward towards the centre (Ortiz 1972: 142–143). Centres or 'earth navels' are multiple, interlinking important spatial and cosmic divisions – like 'airports' according to one Tewa elder (Ortiz 1969: 24). The Tewa centre of centres – the 'earth navel earth middle place' lies at the heart of a village in the plaza shrine. The quadripartite cosmic plane articulates on this centre point. Space also extends in a vertical axis mundi linking zenith and nadir. The Pueblo cosmos is thus like a three-dimensional star, pivoting on its centre of gravity. To the four directions (or eight for the Keres) of this level, one, two or three more may be emphasised – up, down and centre. A general three-fold division into cosmic layers is subdivided, again with variations: Zuni has four layered underworlds and four upper worlds, sandwiching present space-time (e.g. Tedlock 1979). For the Hopi, we are living in the Fourth World, an epoch/space that succeeds three prior, lower worlds, through which people ascended, in one variation up a ladder whose beams and rungs are fastened with strings of turquoise (Voth 1901: 99). 'Earth navels', quadrants and layered worlds are all found in Mesoamerican cosmology.

A Hopi example

Each Hopi direction has an associated colour, domesticated plant, flower, bird, tree or shrub, game animal, prey animal, coloured cloud and gemstone or other hard substance (Fig. 4). Each is also associated with a deity or other spirits, including chiefs of the six directions, cloud chiefs, Warrior Katsinas and such (Parsons 1939: 365–368). Oppositions run in parallel across world quadrants: southwest is to southeast as bluebirds are to scarlet macaws, as turquoise (the colour) is to red, as bean plants are to squash and as mountain sheep are to antelope. To some degree, there is symbolic interchange within a series: a bird of the appropriate direction may stand in for a plant, game animal, or cloud of the same quadrant, especially vis-à-vis an opposite quadrant. Turquoise and bluebirds are partial symbolic synonyms (and homonyms: see below); similarly, scarlet macaws are the bird variant of squash and *Spondylus* shell, so to speak. Oppositions between elements among axes may shift according to context: blue-green, for example, may be opposed to yellow *or* red *or* white. Blue-green may combine with white in opposition to yellow and red, as a southwest–northeast axis to northwest–southeast; yellow and red may also substitute for each other, in opposition to blue-green or white. Other Pueblo models show variations on this theme (White 1942 [Keres]; Tedlock 1979 [Zuni]; Ortiz 1969, 1972 [Tewa]). But even among regional or contextual variations, there remain major congruences. Moreover, the general articulation of directions and colours is found from the American Southwest deep into Mesoamerica (Mathews and Garber 2004) and in the prehistoric and protohistoric 'Mississippian' agricultural tradition of the American Southeast too (Knight 2004).

Turquoise as gemstone represents Hopi southwest. Other Hopi hard substances are more obscure, but references in a key ritual song are Keresan, not Hopi words, illustrating how cosmological ideas are shared across Pueblo linguistic boundaries (Voth 1901: 133–334; White 1942: 82, n 9).[2] As for Nahuatl, its Uto-Aztecan relative, the Hopi word

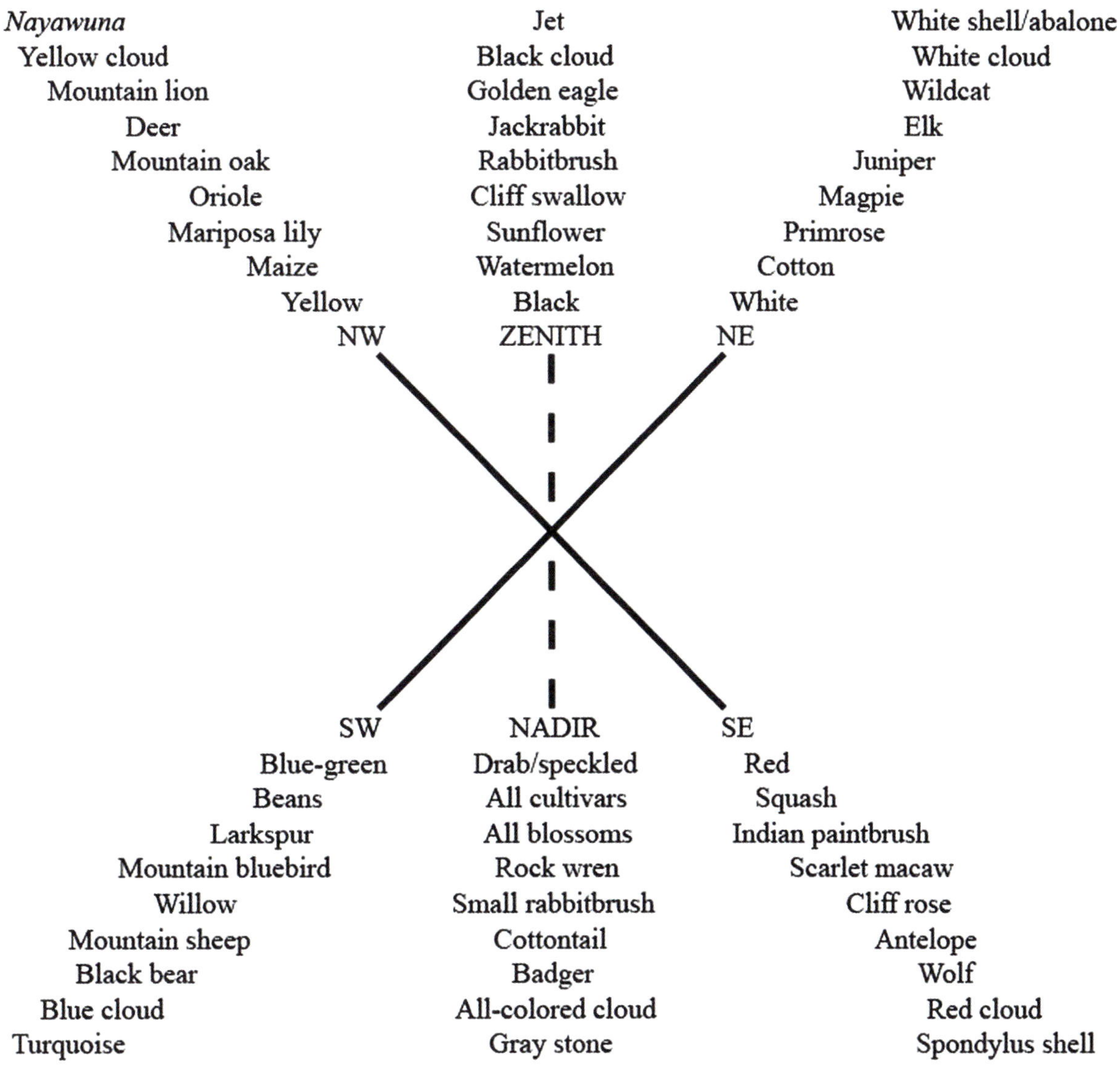

Figure 4. Hopi directional associations. (Image: Peter Whiteley.)

for turquoise the gem differs from the colour term. Hopi *tsorposi*, turquoise, literally refers to *tsooro*, the mountain bluebird (*Sialia currucoides*), and *poosi*, 'seeds' – quite distinct from *sakwa-*, the salient colour term (Fig. 5). This parallels the Mexica correlation of turquoise with the blue cotinga, and 'turquoise seeds' also appear in other Uto-Aztecan folklore (below). Several Hopi directional stone/shell names correlate with names for deities in Keresan such as 'Abalone Man' (Boas 1928: 280). The same may be true also for turquoise, although 'Turquoise Man' is recorded as a personified deity, a partner to 'Salt Woman', only at Zuni (Parsons 1939).

The cosmological arrangements apply to time as well as space. Each Hopi epoch is marked with its own colour, shown on painted bows that Katsina spirits still bring for little boys at Powamuy ('Bean Dance') in February and Niman ('Home Dance') in July. Similarly for the Keres: at Acoma, the earliest underworld layer is white, succeeded in space-time by red, blue-green and yellow (White 1962: 110).

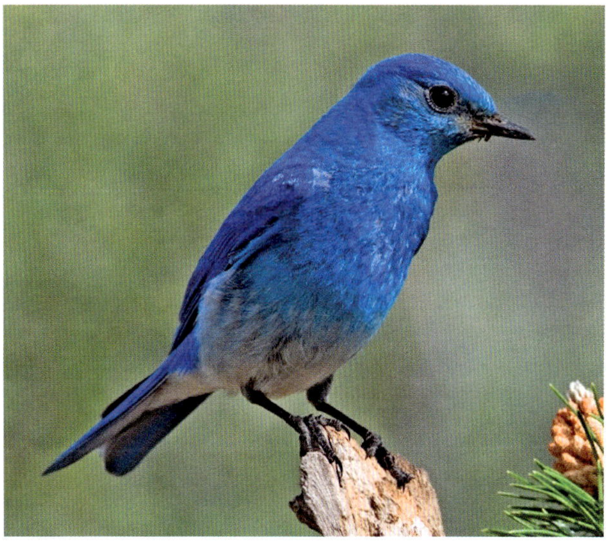

Figure 5. *Tsooro* (Hopi), the mountain bluebird (*Sialia currucoides*). (Photo: Elaine R. Wilson, used with permission.)

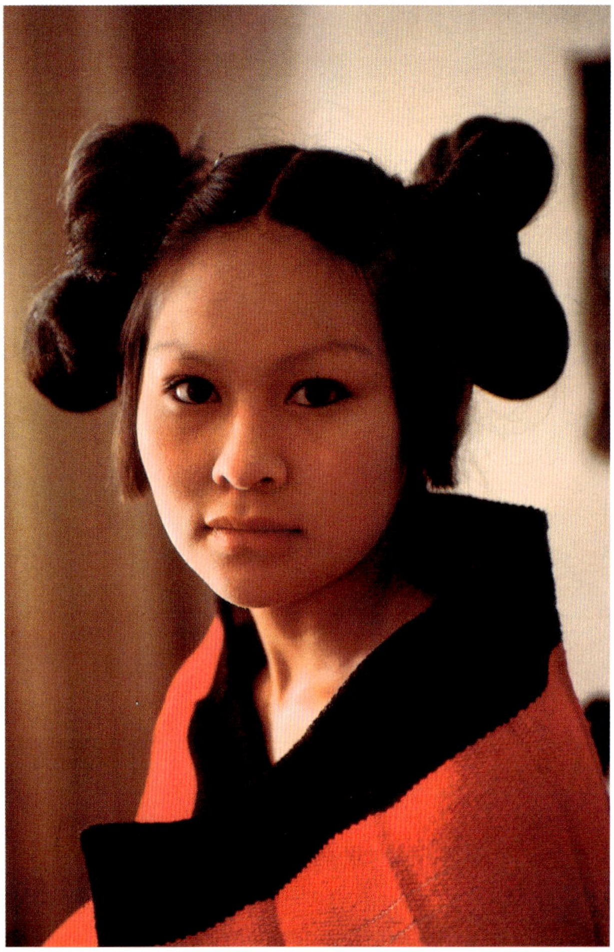

Figure 6. Hopi girl with squash-blossom coif. (Photo: Susanne Page, used with permission.)

Turquoise in myth and ritual

In Pueblo and Navajo cultures, turquoise is ubiquitous in myth and ritual, as well as economics, aesthetics and adornment. Turquoise continues to be used as an offering, both in flake form – just as recorded by the Coronado expedition in 1540 – and as ground together in powder with white shell and cornmeal. The animate Zuni and Hopi Sun loves to receive the gifts of turquoise flakes, cornmeal, prayer feathers and sometimes prayersticks adorned with turquoise beads. He reciprocates by sending rainclouds over the springs and fields to renew life on earth for plants, insects, animals and human beings.

In Zuni belief, the Sun disk is carried across the sky by the 'Holder of the Bearer of Light' seated on a 'huge turquoise' and wearing many turquoise necklaces. Within the upper Zuni cosmic layer, 'stands the deep blue Mountain of Turquoises, the reflected light from whose shining surface gives the sky its rich colour' (Hill 1947: 43). In Navajo thought, turquoise features prominently in creation mythology: when people first emerged onto the surface of this earth, all was water: in order to drain the earth, they had to dig channels – with turquoise shovels tipped with red coral (Hill 1938: 11–12). In the absence of light, First Man fashioned the sun, smoothing out a large turquoise disk. As with Zuni, the sun is carried across the sky by a young man, although for Navajos (equestrians since the seventeenth century) the carrier rides a beautiful turquoise stallion. Navajo Turquoise Boy and Turquoise Girl are deities sent at creation to live in Mount Taylor, Turquoise Bead Mountain, whose sacred colour is turquoise blue. Sky-blue turquoise, as in some Pueblo contexts and as for turquoise and jade among the Mexica, is the colour of living water and rain.

In Pueblo and Navajo values, turquoise represents high status and material wealth, and may operate as special-purpose money. In a Hopi ceremony, for example, ritual clowns must bargain for their lives with Warrior Katsinas, who threaten to execute them for their licentious behaviour. At the climax, the Clown chief offers the Warrior chief a bag of turquoise as a mortgage on the clowns' lives. The Warrior chief accepts and the clowns are permitted to live.

Time and society

Thus as colour, as gemstone, and as deity, turquoise takes its place in a scheme of Pueblo cosmological, ritual, aesthetic and economic values. As with world epochs, colours also correspond with the Pueblo individual's life-cycle, and are especially represented by Hopis with coloured corn ears. At birth a perfect ear of white corn (*tsotsmingwu*, the 'corn mother') is given to an infant when s/he receives names from paternal 'aunts'. White for Hopi is the colour of the northeast, sunrise and 'white dawn'. At death, a perfect ear of *blue* corn is presented, and final names bestowed by members of the clan that first named the child – completing a cycle of social and symbolic oppositions: west/southwest (turquoise blue) is the direction followed by souls of the departed to Maski, the house of the dead.

Dualism reverberates through social structures too. Tewa and Keres Pueblos are organised into two socio-ceremonial divisions: all Tewas belong to either the Winter or Summer 'moiety' (half), each centred in a moiety house. Two caciques – Winter chief and Summer chief – rule the village for half the year each, switching at the equinoxes (Ortiz 1969). For Keres, the moieties are Turquoise and Squash, centred in dual kivas of the same names. The association shows that, for the Pueblos in general, turquoise is the colour of winter, orange/red (squash) that of summer: bluebirds are to macaws as ice is to fire. And although the directions differ and there are no parallel ceremonial moieties, turquoise is the colour of cold for the Hopi also. Should a man bring turquoise-coloured moccasins to wear for a summer dance he would be roundly chastised: *sakwatootsim*, 'blue-green shoes', attract cold forms of moisture – winter clouds, snow and fog.

Male and female

If turquoise is opposed to squash as winter is to summer, as icicles are to flower blossoms, and as bluebirds are to macaws, the metaphor extends also to gender differences. The association of squash blossoms with maidenhood is well known for Hopi: 'squash-blossom' hairstyles express emergent fecundity (Fig. 6). This idea is widely represented in material culture, including designs on basketry and ceramics, on ritual altars, and on Katsina masks. With the development of silver jewellery in the late nineteenth century, it was extended to 'squash-blossom' necklaces.

As squash blossoms are female, turquoise is their male counterpoint. The interrelation of male, as turquoise, and female, as both red and white shell, is directly seen in the Navajo story of the origin of sexuality. When the Navajo first

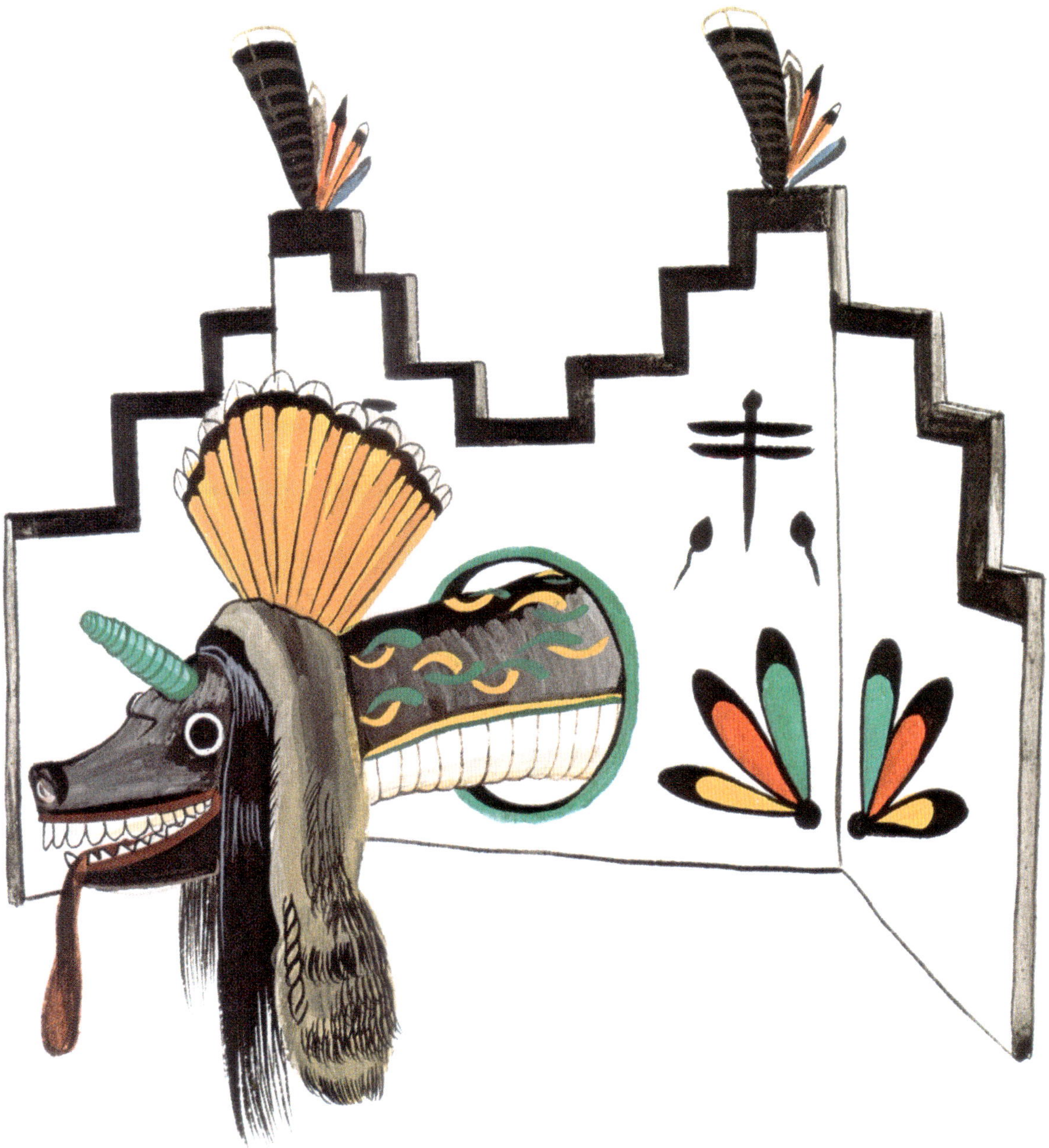

Figure 7. Kolowisi, Zuni plumed water serpent, emerging through an altar screen in ritual puppetry. Painting by Duane Dishta (Zuni Pueblo) ca. 1968. (Courtesy Barton A. Wright, used with permission.)

emerged, they had no sexual organs. First Woman decided that men and women needed something to attract them to each other:

> ... she fashioned a penis of turquoise. Then she rubbed loose cuticle from a woman's breast and mixed it with yucca fruit, which she put inside the turquoise penis ... Next she made a vagina of white shell. Into the vagina she placed a clitoris of red shell. Then she rubbed loose cuticle from a man's breast and mixed it with yucca fruit, which she placed in the clitoris. And she combined herbs with various kinds of water and placed that mixture deep inside the vagina. That way pregnancy would occur ... She placed the vagina on the ground. Next to it she placed the penis. Then she blew medicine upon both of them from her mouth. And she spoke these words to the penis:
> 'Now think!' she said to it.
> 'Think about the one to your left.'
> The penis did as it was told, and its mind extended a great distance. [1st Woman does the same with the vagina]
> And to both of them, ... First Woman said these words:
> 'Now shout!' she said to the two of them together (Zolbrod 1984: 55).

The turquoise phallus reappears as the horn of Zuni Kolowisi and Hopi Paalölöqangw, the giant feathered-serpent, represented in ritual puppetry and other ceremonial contexts (Fig. 7).

Turquoise as a symbol of masculinity occurs also as a theme of war and hunting. Zuni Sun gave his turquoise rabbit-stick (a boomerang) to the War Twins to kill earthly monsters. For Western Apaches, whose mythology has also been influenced

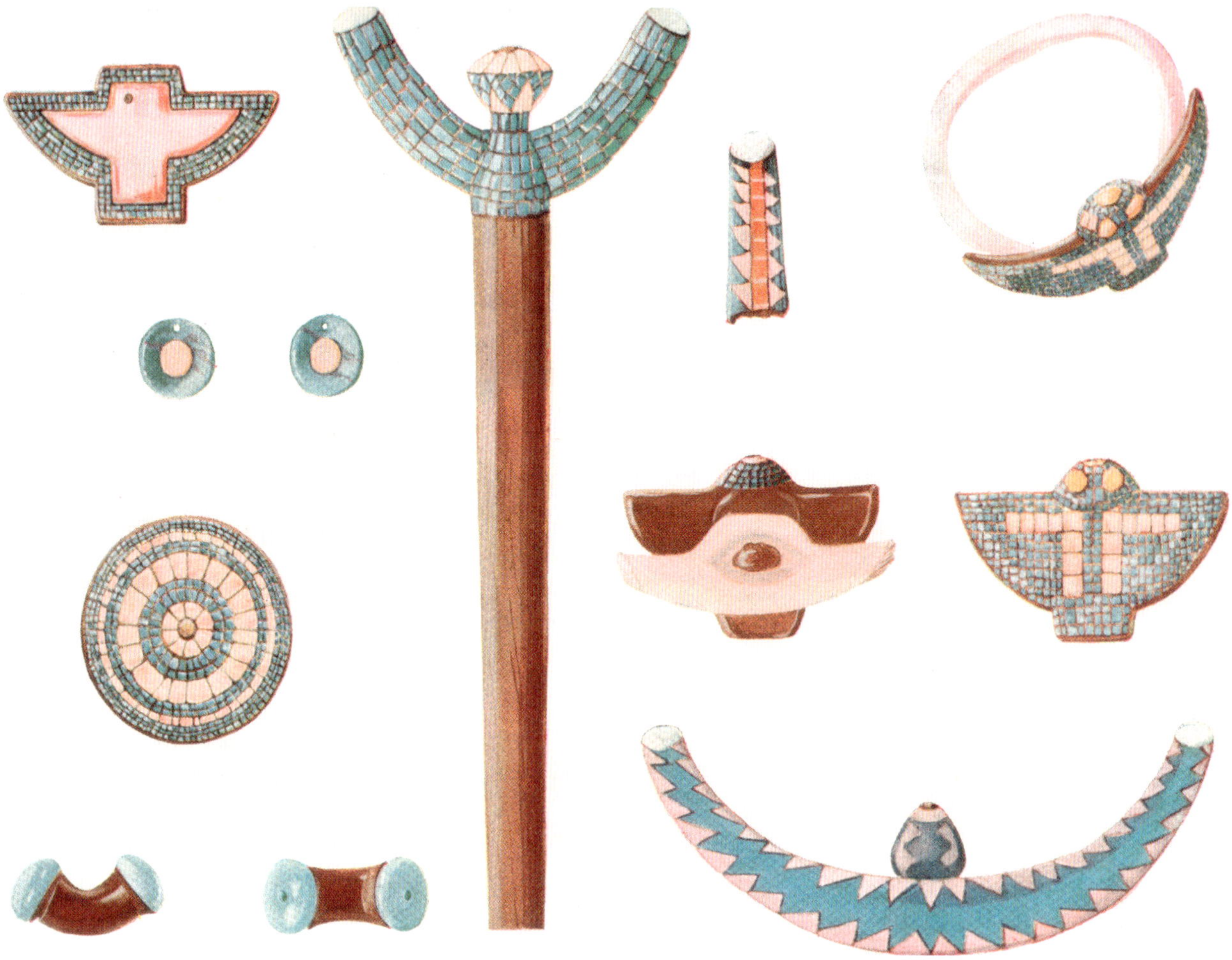

Figure 8. Artefacts from Ridge Ruin, Arizona, twelfth century AD. (From 'Burial of an early American magician', J.C. McGregor, 1943, *Proceedings of the American Philosophical Society* **86**: 2, pl. II, used with permission.)

by the Pueblos, Sun gave the War Twins a turquoise sword for the same purpose. At Jemez Pueblo, bluebird feathers are offered to war deities with lightning war-god sticks. These ideas seem directly akin to Mexica mythology where several war deities, including Huitzilopochtli, wield spears of turquoise. One of the clearest Pueblo archaeological examples of this martial theme occurs in the 'Magician's Burial' near Flagstaff, Arizona, a twelfth-century Sinagua site ancestral to the Hopi. Hopis identify the turquoise-rich regalia – 'equaled only by some of the fine mosaic work from Middle America' (McGregor 1943: 288) – as denoting the Motsmongwi, head of the Warriors' society (Fig. 8).

The conjunction of turquoise and squash as emblems of gender and sexuality appears to be an old idea in Pueblo thought (Fig. 9). Squash-coloured shells overlaid with mosaic turquoise are found at Anasazi, Sinagua and Hohokam sites (Liu 2005). For overlays with tesserae – turquoise, coral and jet – *Spondylus princeps* ('squash shell') remains the shell of choice at Kewa (Santo Domingo Pueblo). Suggesting widespread association with fertility throughout Mesoamerica and the northern Andes, *Spondylus* was offered to springs and, powdered, spread over fields to attract rainwater; the Moche of Peru also preferred it for overlay – sometimes with turquoise (Donkin 1998: 292–299) (Fig. 10).

Similarly, Pueblo squared flat earrings often feature a centrepiece of *Spondylus* or abalone shell against a turquoise mosaic ground. In another variation on this interplay, a Zuni Medicine society ceremony features young girls decorated with many strands of shell and turquoise beads. They are presented with substances representing the directional colours to grind on a metate (Hill 1947: 45): turquoise, red coral, carbonised corncobs (black) and abalone shell – the elements combining to create fertility. The same interplay of blue-green and red or yellow stones, with or without the addition of white and black, remains pervasive in Pueblo representations, including contemporary jewellery.

Turquoise in Pueblo history

Zuni and Kewa figure most centrally in the historic Pueblo mining of turquoise, and the manufacture of turquoise into beads, tesserae and other forms. Both these Pueblos provide the strongest links back to the use of turquoise and shell by the Chaco Anasazi, and remain to this day the most active Pueblos in turquoise jewellery production (and Kewa also of shell beads, or *heishi*).

Prior to consolidation (into one pueblo) in the late 1600s, Zuni comprised six or seven towns and was a major source of the turquoise trade for Indians of southern Arizona and Sonora (Fig. 11). Zuni – or the 'seven cities of Cibola' – was an early allure to Spanish conquistadors. In 1536, Cabeza de Vaca and Esteban, originally an African slave, delivered news from Sonora of turquoise, 'brought from the distant north, where people lived who dwelt in very large houses, and with whom they occasionally traded for the green stones in exchange for parrot plumes' (Bandelier 1890: 42). In 1539, guiding Friar Marcos de Nizza, Esteban briefly penetrated to the Zuni town of Kiakima, losing his life in the process. En route, the party learned from Opatas and Sobaipuri Pimas of a wealth of turquoise and buffalo robes traded far to the south by the Zunis. Zuni houses, it was reported, had, 'many decorations made with turquoises in the principal doorways' and Zunis wore cotton garments and 'girdles of turquoises' (Bandelier 1890: 134). Native Sonorans also described travelling regularly to Zuni to acquire turquoise and buffalo robes – trade patterns that were apparently long-established.

The following year, guided by Nizza's report, Francisco Vázquez de Coronado led a large expedition throughout Pueblo country. They were disappointed by the lack of hoped-for gold and silver, but reported turquoise in quantity, especially among the Zuni (e.g. Winship 1896: 549). Echoing Moctezuma's gifts of turquoise artefacts to Charles V via Cortés (Kunz 1890: 63), turquoise was presented to Coronado at several Zuni towns. In turn, Coronado sent to Viceroy Antonio de Mendoza two painted cloths, a buffalo robe, 'some turquoises, two earrings of the same ... [and] some plates decorated with these turquoises' (Winship 1896: 562).

Zuni thus appears as a major regional centre for extant indigenous trade in turquoise. This reflects an extensive trade structure that persisted throughout the colonial period:

> ... until 1859, the New Mexico Pueblos made annual trading expeditions to Guaymas and into the heart of Sonora, bartering buffalo robes, piñon, meat, and other products for iridescent conch-shells, and the bright plumes of the parrot (Bandelier 1890:177).

Kewa (Santo Domingo) traditional territory includes the same Cerrillos mines that probably supplied Chaco Canyon (see Magnus, in this vol., pp. 221–224). Historically, Cerrillos was associated with the old Tano Pueblos of the Galisteo Basin (abandoned by the late eighteenth century AD), and then with Kewa, which took in the remnant Tano refugees in the 1780s–90s (White 1935: 17). The Tano pueblo of San Marcos was nearest to the mines, and its name in Tewa (proximate to Tano), Kunyak'ongwikeyi, means 'Turquoise Pueblo ruin' (Harrington 1916: 551).

Shortly after their arrival in 1598, Spanish colonists named Cerrillos 'Las Minas de Chalchihuite' ('the chalchihuitl mines') or 'Cerro [Mount] Chalchihuitl' – names used into the early twentieth century (Harrington 1916: 494; Blake 1899). As Nahuatl for both jade and turquoise – and unlike any terms for turquoise in Pueblo languages – *chalchihuitl* undoubtedly derives from Nahuatl-speakers who accompanied the colony (cf. Bandelier 1890: 21). These same Mexicas may also have seeded a 'Montezuma cult' that provided one impetus for the Pueblo Revolt of 1680 (which drove out the colonists for 12 years) and continues to receive expression – for example, in the Rio Grande Pueblos' Matachines festival (cf. Parmentier 1979). For the Pueblos, 'chalchihuitl' became mystically associated with the apotheosised Mexica king, who would one day return to restore sovereignty to all Native peoples. In 1880, for example, Kewa men, protesting nascent American mining at Cerrillos, asserted that turquoise was sacred and 'must not go

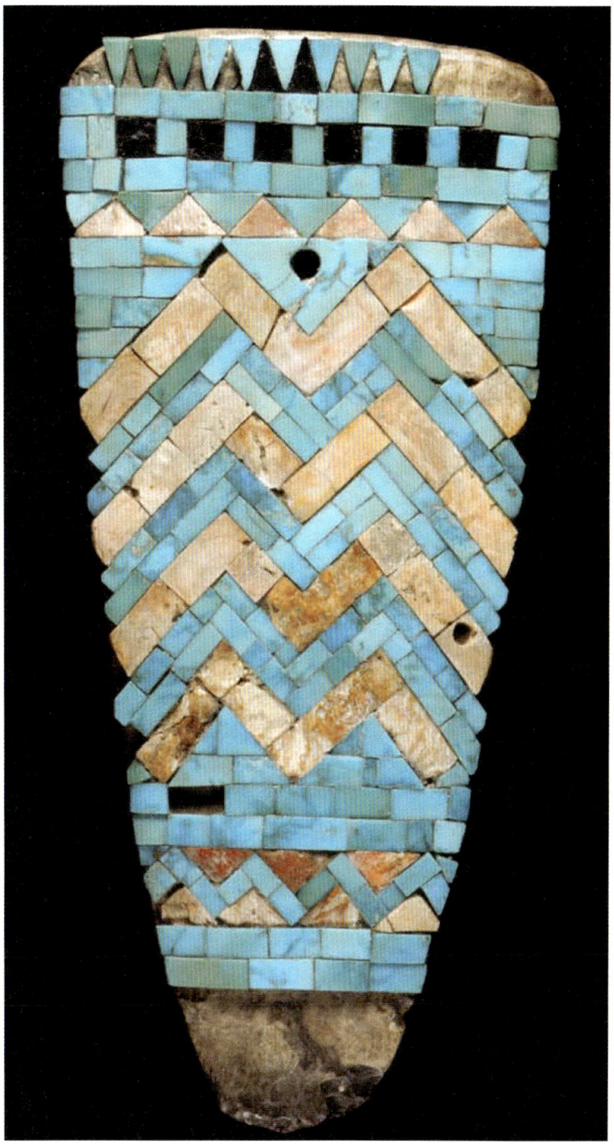

Figure 9. Flint blade with mosaic turquoise, shell, and jet overlay, Poncho House, SE Utah, ca. AD 1000–1200. Courtesy of the Peabody Museum of Archaeology and Ethnology, Harvard University (No. 23-7-10/A5633).

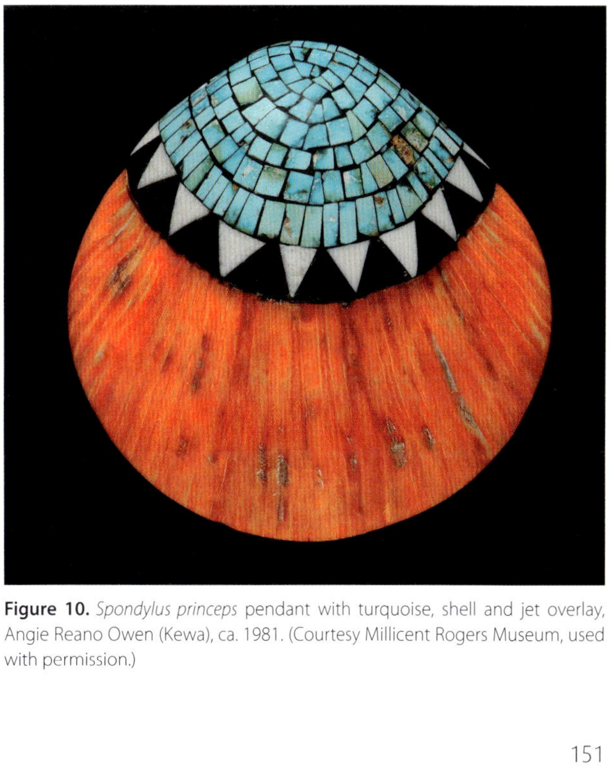

Figure 10. *Spondylus princeps* pendant with turquoise, shell and jet overlay, Angie Reano Owen (Kewa), ca. 1981. (Courtesy Millicent Rogers Museum, used with permission.)

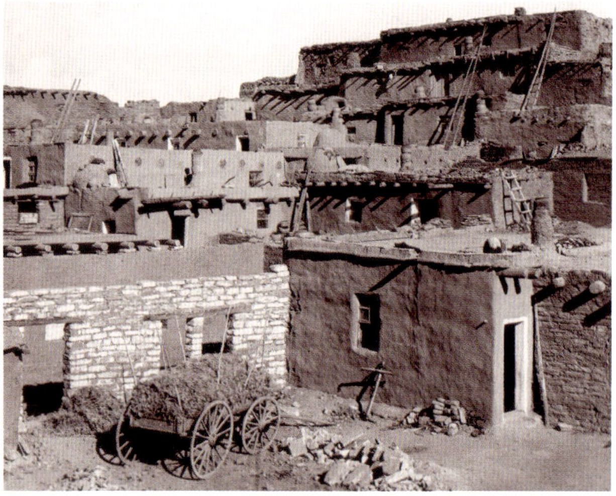

Figure 11. Zuni Pueblo, A.C. Vroman, ca. 1899. Seaver Center for Western History Research, Los Angeles County Museum of Natural History (V-2361 (1070)).

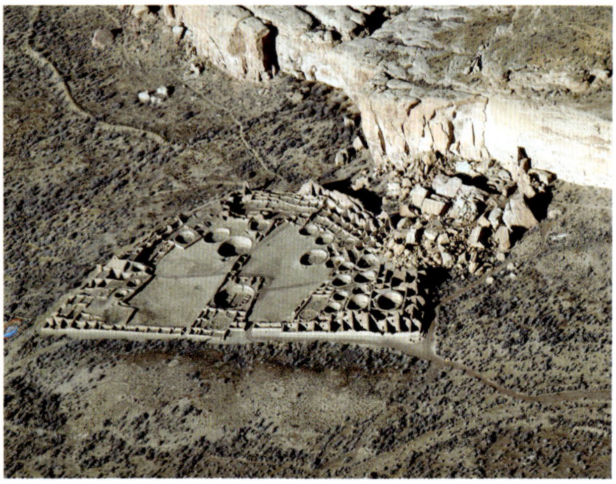

Figure 12. Pueblo Bonito, Chaco Canyon, 2009. (Photo: Robert E. Adams. Courtesy Robert E. Adams, Albuquerque, New Mexico, used with permission.)

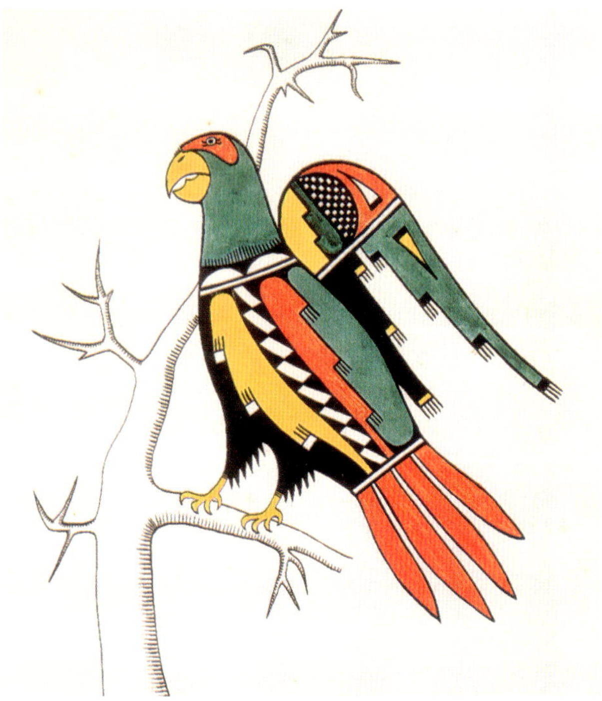

Figure 13. 'Fantastic Bird – Parrot' by Awa Tsireh (San Ildefonso Pueblo), ca. 1922–25. School of Advanced Research, Indian Arts Research Center (No. P188). Used with permission.

into the hands of those whose saviour was not Montezuma' (Kunz 1890: 61). 'Chalchihuitl' remained a common term in nineteenth-century Pueblo and Navajo Spanish for turquoise (and only turquoise, not similar minerals: Blake 1899).

Mining at Cerrillos began ca. AD 800 and reached a peak from AD 1150 to 1650 (Pueblo of Santo Domingo 1978: 323). Both timing and scale would confirm extensive trade with Mesoamerica. William P. Blake, the first geologist to record the mines in 1857, was amazed:

> On reaching the locality I was struck with astonishment at the extent of the excavation. It is an immense pit with precipitous sides of angular rock, projecting in crags ... From the top of the cliff, the excavation appears to be 200 feet in depth and 300 or more in width ... This great excavation is made in the solid rock, and tens of thousands of tons of rock have been broken out ... It seems hardly possible that such an amount of rock could have been removed by men without the aid of powder and machinery (Blake 1858: 227–229).

Blake (1858: 230) also reported clear evidence of ongoing Pueblo mining, confirming Kewa oral history that turquoise mining was continuous throughout the colonial period.[3] After American commercial mines opened in the 1880s, Kewa men staged a series of raids. In 1910, a large party raided the Tiffany mine in a well-coordinated effort: 'The Indians', it was reported, 'still claim the turquoise mines from which their forefathers took turquoise centuries ago' (Tyler 1964: 185).

Structures of the long run

The dialogue of turquoise and red shell goes far back into the past. Kewa jewellery of today, both in style and technique, descends from very similar manufactures found among the Anasazi. At Hopi and Zuni, construction of houses and kivas (ceremonial chambers) involves dedication offerings of turquoise and shell, especially at the building corners, just as under kiva pilasters at Chaco Canyon (Mathien 2001: 112–113). Like earth navels and tetradic cosmologies, this idea too was widespread: similar building-offerings of *Spondylus* shell were made at Monte Alban, Tikal and Teotihuacan (Donkin 1998: 299).

Chaco Canyon, especially the Pueblo Bonito site, is the pre-eminent centre of turquoise artefacts in the prehispanic American Southwest: the amounts there (minimally, 56,000 pieces at Pueblo Bonito) far outweigh other sites, and it was a centre of bead and artefact production, shipping finished pieces south to the Valley of Mexico in exchange for scarlet macaws, copper bells and chocolate (e.g. Mathien, 2001; Crown and Hurst 2009) (Fig. 12). Some archaeologists even contend that Chacoan great-house society arose in response to the Mesoamerican need for turquoise (Weigand and Harbottle 1993). Weigand proposed an enduring trade structure linking south to north, which outlived particular state regimes. Expansion of complex sociocultural forms into the American Southwest, its 'Mesoamericanisation', thus reflects a structure of the long run, in Braudel's sense. The exchange chain linked the Valley of Mexico, via western Mexico, to the Hohokam, Mimbres and the Chaco Anasazi. Colonial history recited above suggests trade persisted long after the great-house period itself.

The use of hard, bright gemstones as indices of sacredness and high social status is global and cross-cultural (Clark 1986). In Mesoamerica, a transition from jade to turquoise in this regard seems to have occurred around 1200 years ago (Taube, in this vol., pp. 117–134). Structured valuing of hard bright objects is associated with strongly hierarchical and/or hierarchising social formations; elites demonstrate their power with the symbolism of glittering, durable emblems that broadcast *permanence*. This status value in contexts both ritual and secular was a major engine driving long-term interchange between central Mexico and the American Southwest. It appears to be no coincidence that AD 800 marks the beginning of prehistoric mining at Cerrillos.

O'odham (Pima) stories of Casa Grande, a Hohokam great house on the Gila River, hint at memories of broad regional interaction (Fig. 13). Turquoise, it is said, was originally discovered nearby, where a blue-tailed lizard was descending into the earth, when Morning Green was chief.[4] Morning Green distributed so much turquoise that news of it reached the Sun at his eastern house over the mountains. Sun coveted the beautiful gemstone and sent Parrot to obtain it from the Casa Grande chief. At first, Morning Green's daughters tried to tempt Parrot with squash seeds, watermelon seeds, charcoal and boiled maize kernels:

> Parrot was obdurate and would not taste it, but, noticing a turquoise bead of blue-green colour, he swallowed it; when the two daughters of the chief saw this they brought to him a number of blue stones, which the bird greedily devoured. The girls brought valuable turquoise beads, which Parrot ate; then he flew away. The girls tried to capture him, but without success. He made his way through the air to the home of the Sun in the East, where he drank an emetic and vomited the turquoises, which the Sun god distributed among that people which reside near his house of rising, beyond the eastern mountains. This is the reason, it is said, why these people have many stone ornaments made of this material (Fewkes 1912: 46–47).

The interplay of macaws and turquoise in the story is reminiscent of the old trade structure (Fig. 14). Scarlet macaws were traded into Chaco for turquoise, and remain ritually important in the Pueblos – despite the fact that macaws do not occur naturally anywhere nearby. Moreover, the O'odham emphasis on turquoise 'seeds' echoes the etymology of Hopi *tsorposi*, 'bluebird seeds', suggesting a widespread mythological theme.

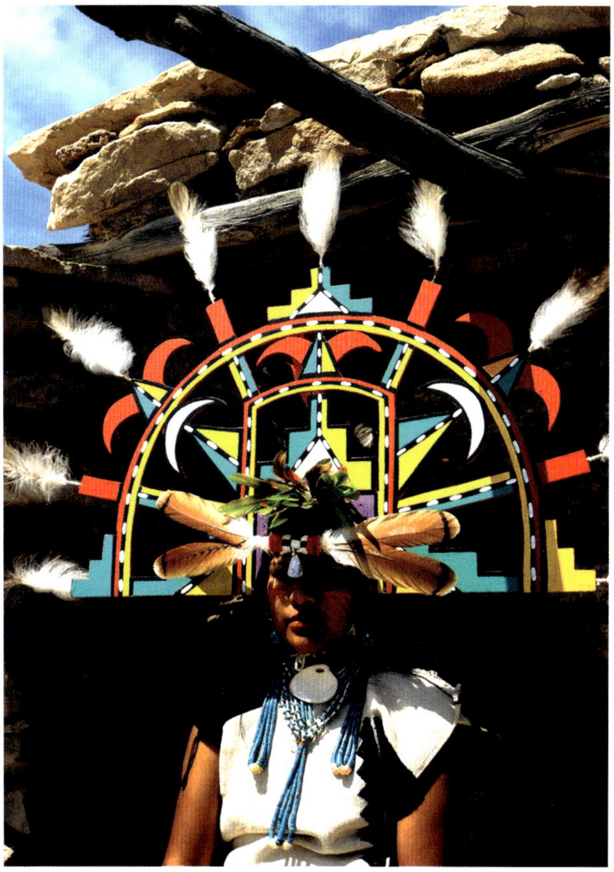

Figure 14. Hopi butterfly dancer (Carol Dawahoya) in turquoise and shell jewellery, with tablita painted in the six directional colours (with parrot, hawk, and eagle feathers – a macaw tail feather stands at the rear). (Photo: Jerry Jacka, 1980, used with permission.)

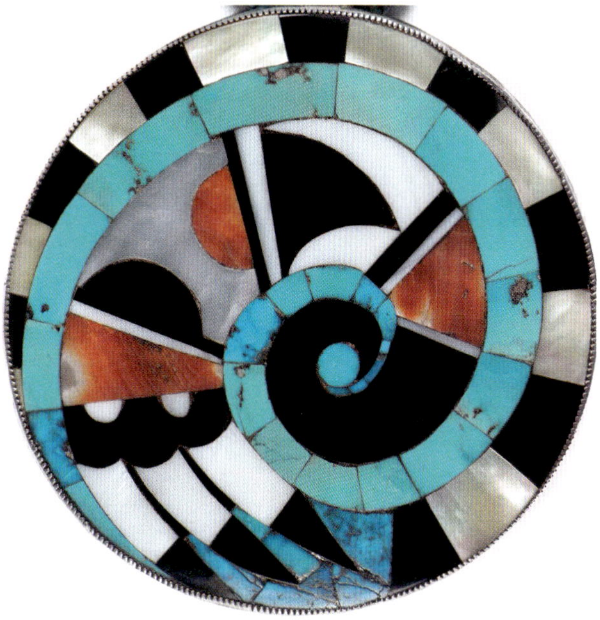

Figure 15. Parrot motif pendant, Lambert Homer, Sr. (Zuni Pueblo), turquoise, *Spondylus*, jet, white shell, ca. 1950. Challis Thiessen Collection. (Photo: Arch Thiessen, used with permission.)

Conclusions

The ubiquity of turquoise in the Pueblo Southwest and its high value suggests this long-run structure figured prominently – along with maize-beans-squash agriculture, and associated Mesoamerican symbolism – in the unification of disparate Pueblo peoples (Fig. 15). Turquoise is powerful: it buys wealth, marks status, is a compensable ritual offering to the gods and spirits, and orients the universe in space and time. Paired with squash or *Spondylus* shell ('squash shell') – or their equivalents, notably scarlet macaws – it is a major axis of symbolic polarities in Pueblo thought, representing male and female, day and night, ice and fire, hard and soft, a supreme symbol of the transformation of nature by culture. The hard-substance tetrad of turquoise, white shell, red shell and jet was evidently an articulating force in the sociocultural imaginary of states, statelets and the Puebloan chiefdoms of the larger Mesoamerican interaction sphere deep into the past.

Notes

1. I am most grateful to Stephen Plog for his thoughtful comments on this chapter.
2. *Spondylus princeps* as the primary hard, red substance of the Hopi southeast is my inference. Red stone or other types of red/orange shell serve the same function.
3. In 1983, with Florence Ellis, I conducted field research at Kewa on traditional land use and learned directly from elders of historic and ongoing interests in the Cerrillos mines (cf. Pueblo of Santo Domingo 1978: 334–336).
4. As in other Uto-Aztecan languages, the same word is used for blue and green. The name may thus contain an implicit reference to turquoise the colour.

References

Bandelier, A.F. (1890) *Contributions to the History of the Southwestern Portion of the United States*. Cambridge: Cambridge University Press.

Blake, W.P. (1858) 'The Chalchihuitl of the ancient Mexicans', *American Journal of Science and Arts* **II**(25): 227–232.

Blake, W.P. (1899) 'Aboriginal turquoise mining in Arizona and New Mexico', *The American Antiquarian* **21**(5): 278–284.

Boas, F. (1928) 'Keresan texts', *Publications of the American Ethnological Society* **8**(1–2).

Clark, G. (1986) *Symbols of Excellence*. Cambridge: Cambridge University Press.

Crown, P. and J. Hurst (2009) 'Evidence of cacao use in the prehispanic American Southwest', *Proceedings of the National Academy of Sciences* **106**(7): 2110–2113.

Donkin, R.A. (1998) *Beyond Price: Pearls and Pearl-Fishing—Origins to the Age of Discoveries*. Memoirs of the American Philosophical Society 224. Philadelphia, PA: American Philosophical Society.

Fewkes, J.W. (1912) 'Casa Grande, Arizona', *Annual Report of the Bureau of American Ethnology* **28**: 25–179.

Harrington, J.P. (1916) 'The ethnogeography of the Tewa Indians', *Annual Report of the Bureau of American Ethnology* **29**: 29–636.

Hill, G. (1938) 'The use of turquoise among the Navajo', *The Kiva* **4**(3): 11–14.

Hill, G. (1947) 'Turquoise and the Zuni Indian', *The Kiva* **12**(4): 42–52.

Knight, V.J. (2004) 'Ceremonialism until A.D.1500', in R. Fogelson (ed.), *Handbook of North American Indians 14, Southeast*. Washington, DC: Smithsonian Institution, 734–741.

Kunz, G.F. (1890) *Gems and Precious Stones of North America*. New York: Scientific Publishing.

Lévi-Strauss, C. (1963) *Structural Anthropology*. New York: Basic Books.

Liu, R.K. (2005) '*Spondylus* in Precolumbian, historic and contemporary Southwest jewelry', *Ornament* **28**(3): 60–66.

Mathews, J.P. and J.F. Garber (2004) 'Models of cosmic order: physical expression of sacred space among the ancient Maya', *Ancient Mesoamerica* **15**: 49–59.

Mathien, F.J. (2001) 'The organization of turquoise production and consumption by the prehistoric Chacoans', *American Antiquity* **66**(1): 103–118.

McGregor, J.C. (1943) 'Burial of an early American magician', *Proceedings of the American Philosophical Society* **86**(2): 270–298.

Ortiz, A. (1969) *The Tewa World: Space, Time, Being and Becoming in a Pueblo Society*. Chicago: University of Chicago Press.

Ortiz, A. (1972) 'Ritual drama and the Pueblo world-view', in A. Ortiz (ed.), *New Perspectives on the Pueblos*. Albuquerque: University of New Mexico Press, 135–161.

Parmentier, R. (1979) 'The mythological triangle: Poseyemu, Montezuma, and Jesus in the Pueblos', in A. Ortiz (ed.), *Handbook of North American Indians 9, Southwest*. Washington, DC: Smithsonian Institution, 609–622.

Parsons, E.C. (1933) 'Some Aztec and Pueblo parallels', *American Anthropologist* **35**(4): 611–631.

Parsons, E.C. (1939) *Pueblo Indian Religion*. Chicago: University of Chicago Press.

Pogue, J.E. (1915) 'The turquois: a study of its history, mineralogy, geology, ethnology, archaeology, mythology, folklore, and technology', in *Memoirs of the National Academy of Sciences*, Vol. XII, Part II. Washington, DC: National Academy of Sciences.

Pueblo of Santo Domingo (1978) 'Additional findings of fact', *Indian Claims Commission* **42**: 319–344.

Tedlock, D. (1979) 'Zuni religion and world view', in A. Ortiz (ed.), *Handbook of North American Indians 9, Southwest*. Washington, DC: Smithsonian Institution, 499–508.

Tyler, H.A. (1964) *Pueblo Gods and Myths*. Norman: University of Oklahoma Press.

Voth, H.R. (1901) *The Oraibi Powamu Ceremony*. Field Columbian Museum Publication 61. Chicago: Field Columbian Museum.

Weigand, P.C. and Harbottle, G. (1993) 'The role of turquoises in the ancient Mesoamerican trade structure', in J.E. Ericson and T.G. Baugh (eds), *The American Southwest and Mesoamerica: Systems of Prehistoric Exchange*. New York: Plenum Press, 159–177.

White, L.A. (1935) *The Pueblo of Santo Domingo, New Mexico*. Memoirs of the American Anthropological Association 43. Arlington, VA: American Anthropological Association.

White, L.A. (1942) *The Pueblo of Santa Ana, New Mexico*. Memoirs of the American Anthropological Association 60. Arlington, VA: American Anthropological Association.

White, L.A. (1962) 'The Pueblo of Sia, New Mexico', *Bureau of American Ethnology Bulletin* **184**.

Winship, G.P. (1896) 'The Coronado expedition', *Annual Report of the Bureau of American Ethnology Bulletin* **14**: 329–613.

Zolbrod, P. (1984) *Diné Bahane': The Navajo Creation Story*. Albuquerque: University of New Mexico Press.

Author

Peter M. Whiteley, American Museum of Natural History, Division of Anthropology, New York, USA

The exotica of the Settala Museum and other northern Italian collections

Antonio Aimi

Introduction

The research conducted since the 1970s into the exotica in the Italian eclectic collections that in the sixteenth and seventeenth centuries contained specimens from 'exotic countries', has the merit of having defined for the first time a field of study that has yet to be fully understood. Until then, research was limited to a rather meagre list of the extra-European objects in various collections that ignored issues of correct cultural attribution, contextualisation in the broader picture of European collecting, and reflections about the notion of the 'other' that started with the voyages of discovery. Initially, this research examined the Medici collections in Florence (Heikamp 1972, 1976; Laurenich and Ciruzzi 1981; Ciruzzi 1983),[1] the Giganti, Aldrovandi and Cospi collections in Bologna (Laurencich 1983, 1984, 1985), and the Settala Museum in Milan (Aimi 1983; Aimi et al. 1984, 1985; Aimi and Bassani 1986). It helped to shed light on a legacy that effectively had been erased given that in those cities the threads linking the old collections to the modern museums had over the course of time been broken.[2] By contrast, studies of the exotica that had reached Rome in the sixteenth and seventeenth centuries were not about the rediscovery of a forgotten heritage. Many of the city's cultural institutions, such as the Society of Jesus and the Pigorini Museum (housed in the Palazzo del Collegio Romano until 1975 and from which it had inherited the ethnographical collections) had either a direct or an indirect relationship with the collections of the past. It is perhaps for this reason that, apart from Rivosecchi's book on Atanasio Kircher, research about the Italian capital initially focused on the objects rather than the general context of collecting in Renaissance and Baroque Rome (Rivosecchi 1982).[3] During the same period, alongside works that tended to examine both collections and exotica and, in particular, on exotica of American origin, Ronald Lightbown (1969) and Ezio Bassani (1977)[4] were conducting research exclusively into collections of oriental and African objects in collections and museums throughout Italy.[5] Bassani's research became not only a complete inventory of African art in Italy, but also an overview that enabled a reconstruction of collecting in the sixteenth and seventeenth centuries that complemented other authors' research. More general research has been conducted in parallel with these monothematic studies through to the present day. Collectively, this corpus has examined both the phenomenon of eclectic collecting and poorly researched individual collections that revealed unknown or little-known nuclei of exotica.[6] This has made it possible to draw up a map of eclectic Renaissance and Baroque collections that has not only contributed to the definition of a detailed overview of places such as Rome (Lo Sardo 2001; Nocca 2001; Germano and Nocca 2001), Bologna (Laurencich 1992; Simili et al. 2001), and Milan (Violani 1988; Albonico 1990; Aimi 1994; Morandotti 2005, 2008), but has also drawn attention to apparently minor centres such as Como (Pavoni et al. 1983; Zimmermann et al. 1985), Mantua (Franchini et al. 1979; Brown 1985), Pisa (Garbari et al. 1991), and Naples (Stendardo 1991; Aimi 1993). Interestingly, contrasting conclusions emerge from research that provides a survey of the collecting of exotica unthinkable just 30 years ago. On the one hand, it would seem to be possible to make a number of general observations regarding this phenomenon and its implications. On the other hand, the lack of data regarding cities such as Venice and Genoa, and the opportunity offered by the Settala Museum in Milan to see the collection through the 'working tools' of the collector should invite us to proceed with great caution. In this paper I shall present new data regarding the collecting of exotica to show how correct understanding of the very nature of a collection is all too frequently conditioned by mere chance.

The fact that, to date, no research into collecting has brought to light significant collections of exotica in Venice and Genoa is certainly surprising given that the two cities were just the kind of places where it would be logical to find greater numbers of pieces from 'exotic' countries. Indeed, as is well known, Venice and Genoa were mercantile, economic and cultural centres of the first order that were particularly attentive to geographical discoveries. In particular, it should be remembered that for almost a century, the Ligurian city entertained a privileged relationship with Spain, which thanks to its dominions outside Europe, was the obligatory crossroads for the traffic of extra-European objects (Braudel 1982: 140–155). Venice instead saw the publication of the greatest number of Italian editions of chronicles of the Conquest and travel journals such as Ramusio's three volumes (1550–1559) that represented the most important collection of their kind in Europe prior to the publications edited by Hakluyt and De Bry.[7] Taking Venice as an example that is much better known than that of Genoa, it is very probable that important nuclei of exotica existed here too, as suggested by the information emerging from the research conducted on its inventories. Objects identified to date are however of little significance. For

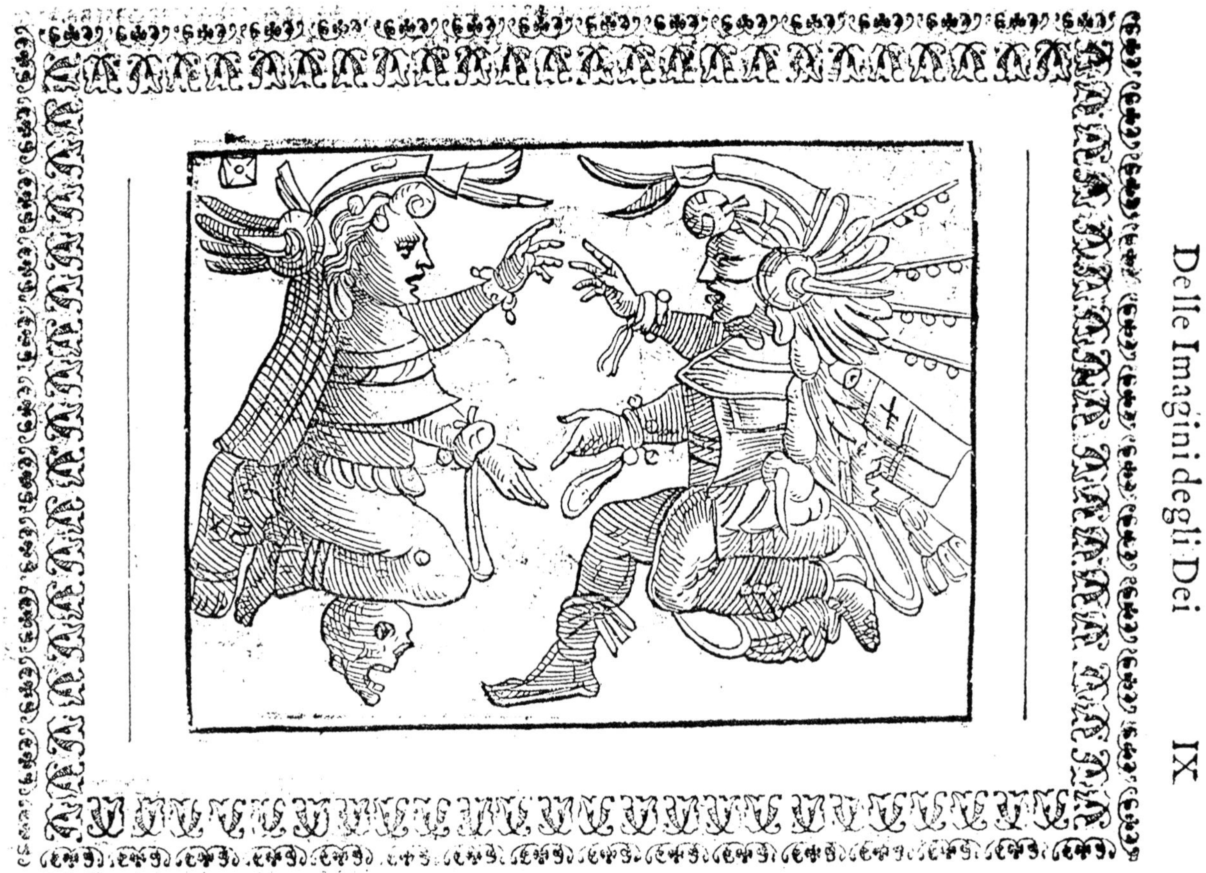

Figure 1. Lorenzo Pignoria, *Seconda parte delle imagini degli dei indiani* (*Second part of the images of the Indians*). The engraving reproduces reasonably faithfully the drawing at the bottom of f. 2v of the Codex Vaticanus 3738, inverting right and left (Pignoria 1615: 9). (Photo: Antonio Aimi, used with permission.)

example 'the mosaic of birds' feathers' in the Petrobelli collection (Hochmann 2008: 15), the 'flask of Indian walnut' in the Grimani Calergi collections (inventory by Vincenzo Grimani Calergi 1647 in Benussi 2007: 338), the 'Turkish and Indian things' in the Helman collection (Mason 2008: 287–288), the 'strange objects coming from the Indies' in the Vendramin collection (Pomian 1989: 88),[8] an Owo bracelet (Bassani 2000: 170–171), the 'vases of Indian walnut' (Cipollato 1961: 226–231), the porcelain and a copy of the Codex Vaticanus 3738 in the Contarini collection (Ambrosini 1982: 18). The same can be said about the collections on the 'Venetian Terraferma' (i.e. the mainland), which, although better documented, have no ethnographical sections of particular note.[9] The only exception is that of Pignoria, who stands out among the other collectors not so much for the quality or quantity of his pieces (yet to be studied in depth), but for the publication of a preliminary comparison between the Egyptian divinities and those of the 'Eastern and Western Indies' (Pignoria 1615). For this work, that anticipated a methodology far ahead of its time and an attention to specifically ethnographical themes, Pignoria managed to obtain a copy of the images of the Codex Vaticanus 3738, and a number of 'Mexican' pieces from the collection of the Duke of Bavaria (Pignoria 1615: 23) (Fig. 1).[10] He failed, however, in his attempt to obtain similar copies from other, probably 'Venetian', collections.[11] Despite all this information we are still not able to explain whether the lack of important exotica sections in the Venetian collections is due to an absence of objects, the scattering suffered by the collections over time, or the interest of individual scholars who, *pour cause*, may have privileged traditional works of art, or indeed a combination of all three factors.

What is certain, however, is that there was at least one collection in Venice with a number of remarkable pieces. It was where it would have been most logical to find it: the home of Giovanni Battista Ramusio. He writes about it in a passage that seems to have so far escaped the attention of all those who have studied multifaceted Renaissance intellectual. It reads:

> One also read in the said history by Signor Gonzalo [Fernández de Oviedo], the form and manner in which with a number of hieroglyphic images they describe their stories and mark the memory of their kings of Mexico, that are certain figures of animals, flowers and men shown in diverse acts and manners: as seen in those books that the said Signor Gonzalo sent as a gift to V.E. [Giacomo Fracastoro] and myself, in past years, full of many and various figures and curiosities (Ramusio [1556] 1985: 11).[12]

From these sentences it would appear that among the objects sent by Oviedo there were a number of Mexica or early colonial period codices of, at least in part, annalistic content. All traces of these codices have been lost and any theory as to their fate would be pure conjecture.

The situation in Turin, a city lacking relationships with the world beyond Europe and that, before becoming the capital

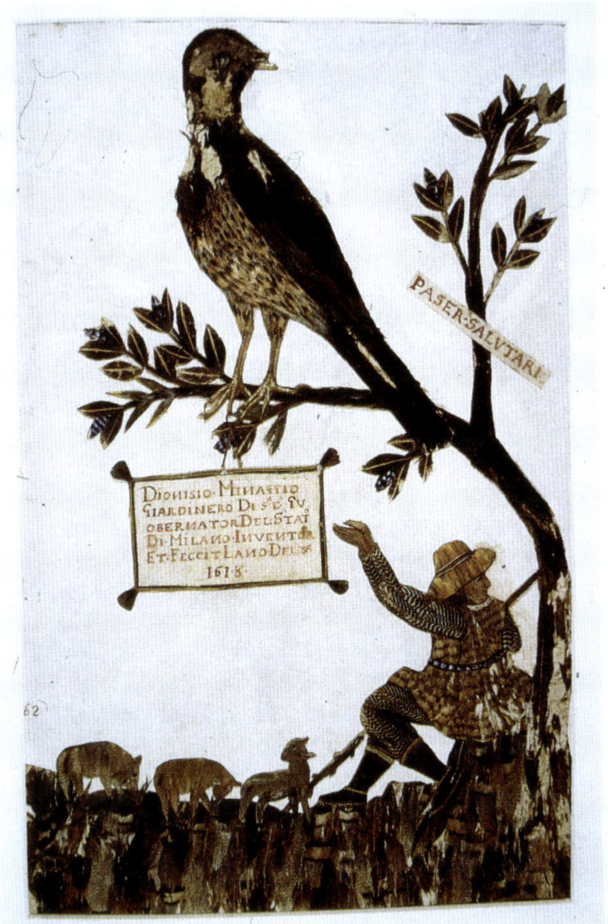

Figure 2. Dionisio Minaggio, *Libro di piume*. The first page, in the original order, of the book. In the centre, the author's 'signature'. (Photo: Antonio Aimi, used with permission.)

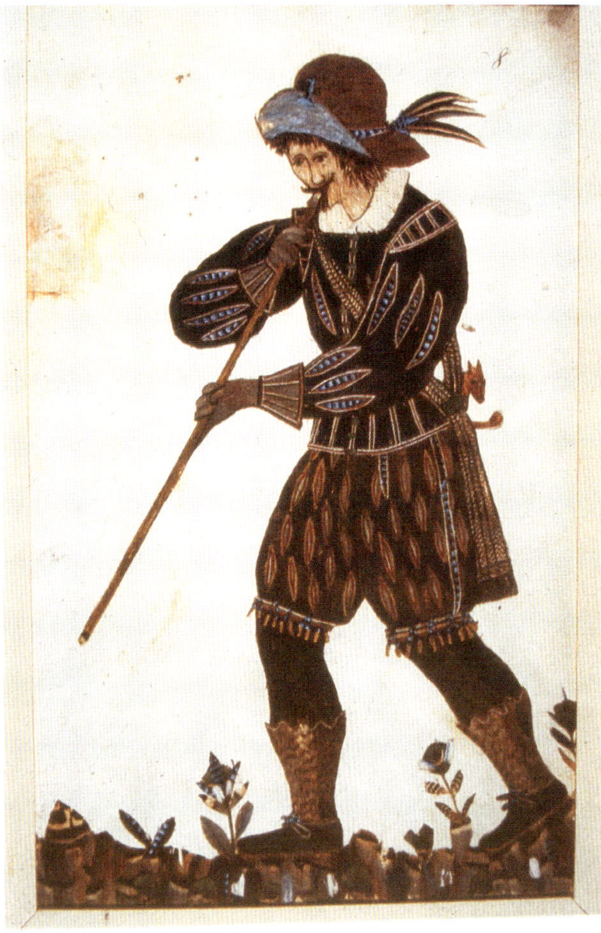

Figure 3. Dionisio Minaggio, *Libro di piume*. Man with a blowpipe, the original order of the page is unknown. (Photo: Antonio Aimi, used with permission.)

of the Duchy of Savoy (1561), had only played a relatively marginal role in the history of Italy, was to a certain extent the opposite of that in Venice. Turin was a place where objects from the 'exotic' countries should have logically been found in small quantities. Indeed, it appears that the ducal collections of the sixteenth and seventeenth centuries had nothing spectacular except certain important exotica: the Sapi-Portuguese spoon and oliphant (Bassani 2000: 166–167), a pendant from the Iranian tradition (Bollati and Messina 2009: 129), a Moghul cup in silver and nephrite, and an Indian knife handle from Sri Lanka (Inventory of Palazzo Madama, Museo Civico d'Arte Antica, inv. 383/PM e 192/AV).[13] To these objects we may add the nucleus of eight Indian and Indonesian pieces in the Armeria Reale, the arrival of which in the Sabaudian collections is uncertain. If we consider pieces that have been lost – 'Turkish-esque' vases and arms, 'Chinese-esque' boxes, and other chinoiseries (Bava 1995: 265–332; Pinto n.d.) – it is clear that the court of Savoy was not completely disinterested in objects from countries outside Europe. In any case, the contrast between Venice, the first, great and wealthy commercial capital of a global economy (Braudel 1982: 101–123) and the small, warrior duchy, spanning the Alps is glaring. However, the lacunae in the Venetian collections and archives prevent further reflection.

Roughly midway between Venice and Turin in geographical, economic and commercial terms lays Milan, the capital of a wealthy state that became a Spanish dominion around 30 years after the discovery of America. This new political situation had significant consequences for collecting, even though the loss of independence and autonomous relationships with the extra-European world was in part compensated by the presence of an aristocracy related to the Iberian nobility, artists and craftsmen who had been in the service of the Catholic kings. A certain number of examples of exotica in the Milanese collections and the 'importation' from Mexico of the technique of 'painting' with birds' feathers are testimony to this particular situation (Fig. 2). Indeed, in the years prior to 1618, a certain Dionisio Minaggio, a Lombard 'gardener of Sa. Ea. Gubernador of the State of Milan', 'painted' a *Libro di piume* with 156 plates 'of which 112 with a prevalently ornithological subject, 16 with hunting scenes, 14 with masks from the Comedy of Art, eight with players and their instruments, five with scenes of various genres and of trades' (Violani 1988: 15). Nothing is known of the life of Minaggio, not even where he learnt the art of 'painting' by cutting and gluing feathers onto a paper support, but the presence among his subjects of two parrots from Mexico and figures in exotic costume (including a hunter with a blowpipe) suggest that he was a man with experience of America who returned home with certain 'tools of the trade' (Fig. 3). It is in any case difficult to imagine him in an Amenteca art workshop anywhere other than in colonial Mexico. The art of Dionisio Minaggio would probably not have generated any particular wonderment in Milan as there were already a number of important objects

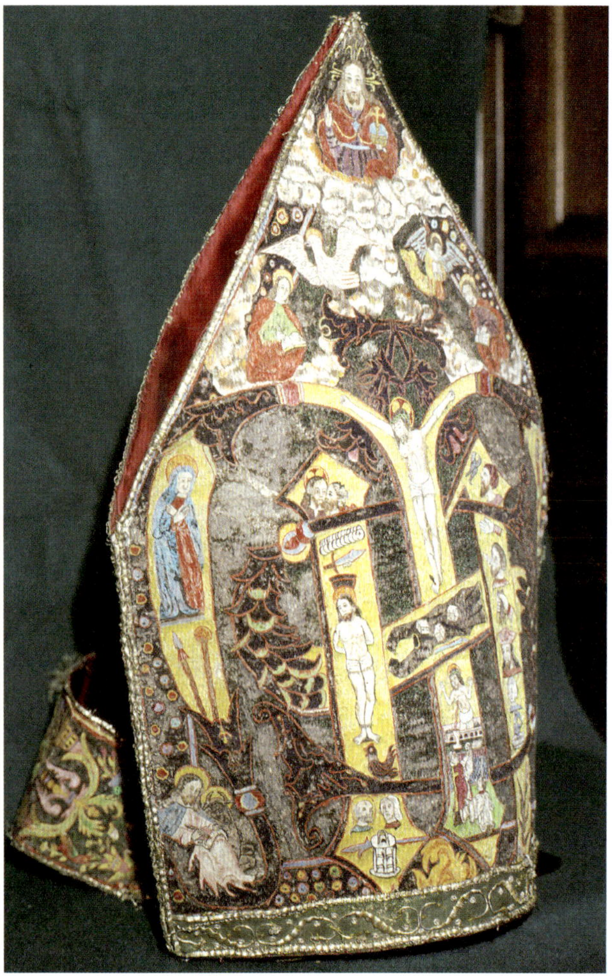

Figure 4. The penitential mitre of San Carlo painted with birds' feathers. Museo del Duomo, Milan. (Photo: Antonio Aimi, used with permission.)

created with feathers in the city, for example: the penitential mitre of San Carlo and a 'painting of feathers of Birds of India ... that opens in three pieces' belonging to Anne of Austria, fourth wife of Philippe II (MSBA, *Descrizione*... 1627: 3r) (Fig. 4). However, the vicissitudes of the collecting of exotica in Milan are particularly interesting not only due to the presence of these objects and the story of the 'gardener of Sa. Ea. Gubernador', but also because the documentation regarding the Settala Museum provides us with an unexpected and extremely interesting view of the collector's attitude with regard to the exotica in what is the city's most important eclectic collection. It is worth remembering that the notes by Manfredo Settala on the great illustrated codices allow us to revise our image of the collection. The Settala Museum was in Schlosser's opinion 'the closest to the characteristics of the German collections' in which 'the devilries and curiosities of the North' dominated (Schlosser [1908] 1974: 105)(Fig. 5).[14] Schlosser's judgement was based solely on the Latin edition of the catalogue published in 1664 and perhaps on the celebrated engraving by the painter Cesare Fiori showing a view of the Museum, the fruit of a compromise between reality and the demands of composition (Terzago 1664)(Fig. 5). The Austrian scholar was unaware of the two Italian editions of the catalogue that appeared in 1666 and 1677 and which did not include Terzago's numerous erudite references and gave a more popular tone to the book inserting (in the 1677 edition) those works that had enriched the collection in the meantime (Scarabelli 1666, 1677). We should also note that these are mere details, as the three catalogues substantially provide an image of a collection heavily influenced by an overwhelming Baroque taste. Schlosser's opinion of the collection may appear excessive and it is certainly incorrect to see proof of Italian 'rationality' and 'good sense' in the collections from Florence and Bologna.[15] However, no one would have questioned his verdict on the Settala Museum had the analysis of the five illustrated codices not revealed that the collector observed and studied the objects with a totally different eye to those of the printed catalogues discussed by Schlosser and taken up by many others.[16] In many cases Settala's captions show criteria of comparison and study of materials that were fairly modern for the time. They belie errors and commonplaces present in the scientific literature of the period and in the printed catalogues that the collector, abstaining from declaring his own point of view, had entrusted to figures better aligned with the editorial customs and the taste of the time. This can

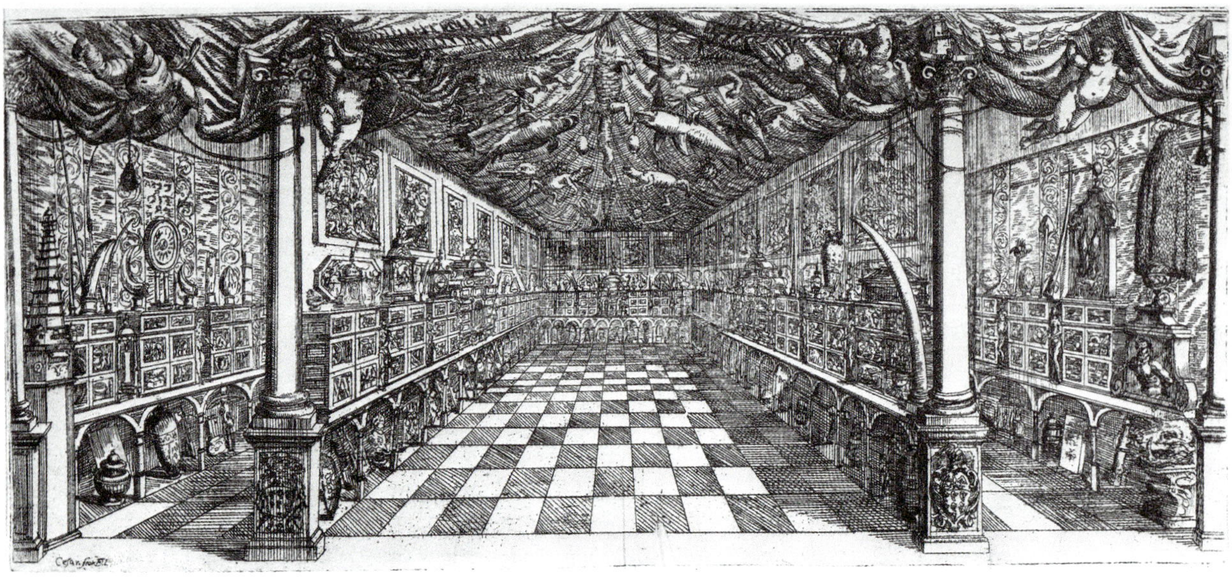

Figure 5. Cesare Fiori, 1666, *View of the Settala Museum in Scarabelli*. Top right: the Tupinambà cloak and two clubs from eastern Brazil, probably 'Tapuya' and/or Tupinambà. (Photo: Antonio Aimi, used with permission.)

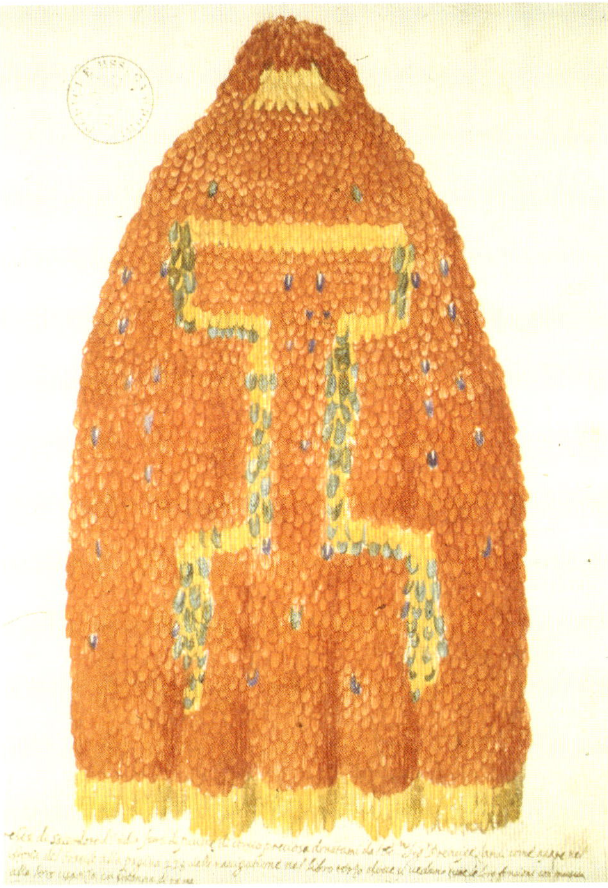

Figure 6. The Settala Museum's Tupinambà cloak. Bottom: the handwritten caption by Manfredo Settala. (MSBE [Settala Museum] 21: 5r; used with permission.)

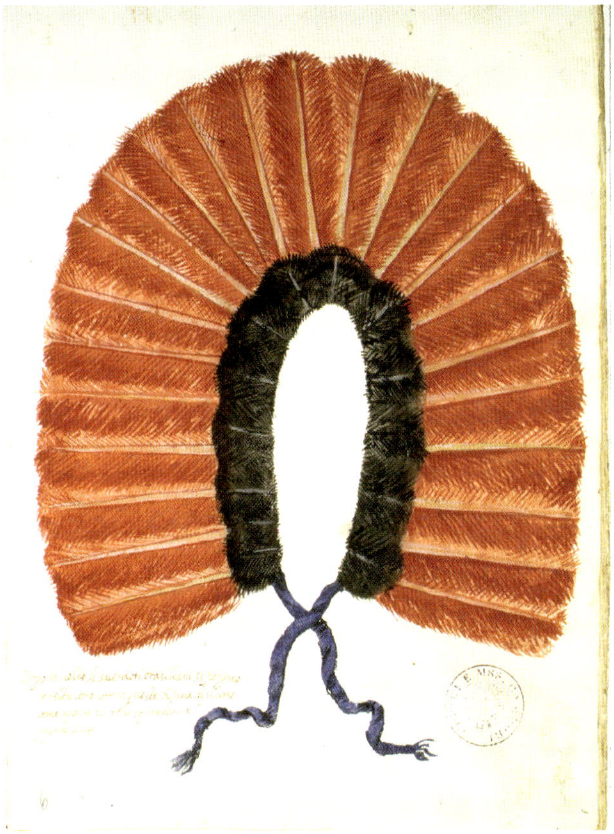

Figure 7. The Settala Museum's Tupinambà diadem. Bottom left: the handwritten caption by Manfredo Settala. (MSBE [Settala Museum] 21: 6r; used with permission.)

be seen in the research conducted by Vincenzo De Michele, Alessandro Morandotti and the present author for the exhibition *Musaeum Septalianum* (Aimi et al. 1984). Furthermore, they show that the Museum was the result of collecting and technical-scientific experimentation that were by then professional rather than being subordinate to other ends or the 'amateurism' of collecting conceived and conducted as a hobby. This approach, evident throughout the Museum, was particularly remarkable in the section dealing with technical and scientific instruments, as well as that of the exotica. This last section, which is of particular interest to us here, included objects from America, the Near East, Sub-Saharan Africa, India and China. The American subsection comprised the nucleus of items of Tupinambá ritual costume received as a gift from Prince Landi, numerous weapons, hammocks and other products of the Amerindian peoples and the societies of the early colonial period. There were also a few pieces from Mexico, most of them being craft objects from the early colonial period. These collections did not seem to include turquoise mosaics from the Americas – the only two turquoise pieces present in the collection came from present-day Iran (De Michele 2009).

In contrast with what we think happened in almost all the European collections of the time (if we continue to base our assumptions on sources such as printed catalogues), ethnographic objects in the Settalian collection rather than being generically placed among the curiosities were appropriately identified according to their specific characteristics and subdivided into thematic areas using a comparative approach: vases, weapons, and so on. For the 'classification' of the objects, Manfredo Settala relied on the existing literature, the information provided by the donors of the objects, and the help and expertise of missionaries passing through Milan. These were largely Jesuits, among whom there were figures of a certain intellectual weight such as Cristoforo Borri, Gian Piero Maffei, Giacinto de Magistris, Gian Filippo de Marini and Alfonso de Ovalle. Furthermore, in contrast with his contemporaries Kircher or Pignoria, who attempted to explain the characteristics of geographically distant cultures (e.g. India and China) using the interpretative categories employed in the study of ancient societies (e.g. Egypt), Settala did not venture into precipitous theorising, preferring to limit himself to the study of individual objects, what today would be called material culture, avoiding the symbolic interpretation of the object typical of the sixteenth and seventeenth centuries, and even present in the work of scientists such as Aldrovandi.

Exemplary documentation of this modus operandi is provided by the captions written by Settala to accompany the drawings of a group of Tupinambá ritual costumes comprising a cloak of feathers (the only surviving item), a crown and a belt in feathers and a rattle with the seeds of *Thevetia neriifolia* Juss (or *Thevetia ahouai* D.C.) (Figs 6–9):

> Cloak of a priest of India made of crow's feathers given to me by His Excellency Prince Landi as in the Historia del Brasile on page 228 of the Navigations in the third book where one see all their functions with music in their usages in copper print (MSBE [Settala Museum] 21: 5r).

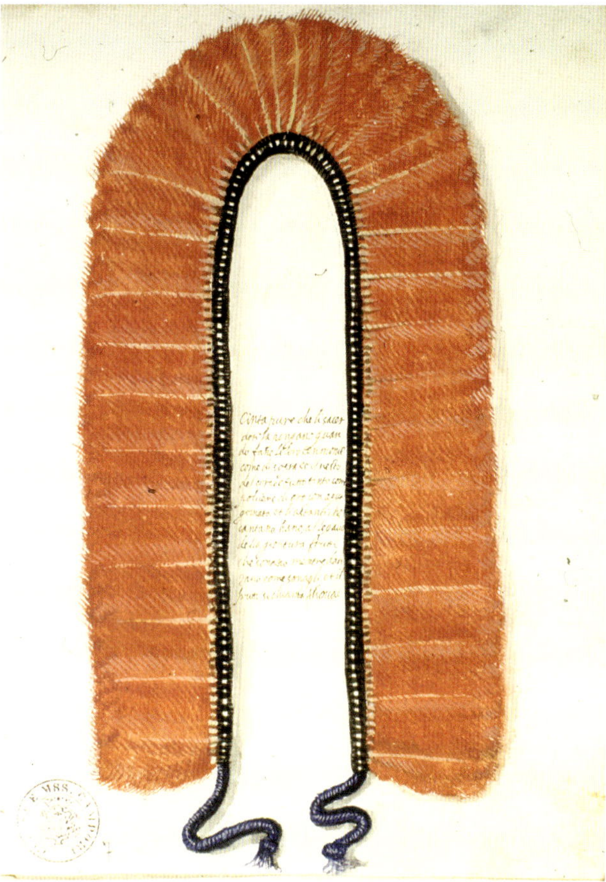

Figure 8. The Settala Museum's Tupinambà belt. Centre: the caption by Manfredo Settala. (MSBE [Settala Museum] 21: 7r; used with permission.)

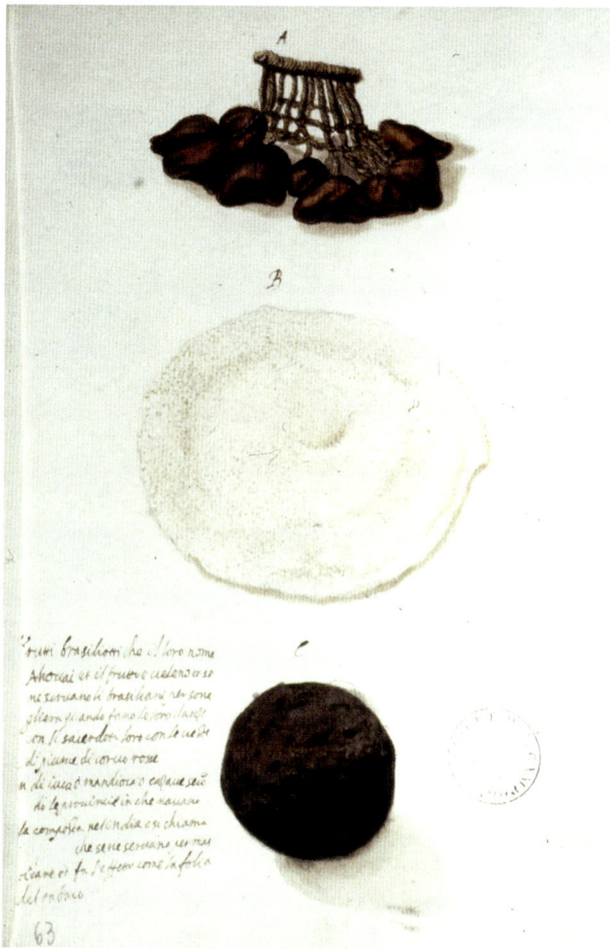

Figure 9. The Settala Museum's Tupinambà rattle with seeds. (MSBE [Settala Museum] 21: 63r; used with permission.)

> After the cloak the Brazilian priests place on their heads as a crown this *bizaria* of feathers as it appears in the above mentioned folio at page 228 (MSBE [Settala Museum] 21: 6r).

> Belt that the Brazilian priests wear when they conduct their ceremonies as above with the rest of the body all coloured with water paste and gold dust and those present who sing have tied to their joints fruit that sound while they dance like rattles and the fruit is called Ahouai (MSBE [Settala Museum] 21: 7r).

> (Rattle) Brazilian fruits by the name of Ahouai and the fruit is poison and is used by the Brazilians as a rattle when they dance with red crow cloaks (MSBE [Settala Museum] 21: 63r).

The engraving to which Settala refers is found on page 228 of the third volume of the *Historiae Antipodum* by Theodore De Bry (1590–1630) and illustrates a passage in the *Navigatio in Brasiliam* by Léry in which he speaks of the Tupinambá dances (Fig. 10). The print depicts three *pagé* in the act of directing the ritual dance of a group of men arranged in a circle around them (Léry 1590–1630: 228). As in the same volume by De Bry, the passages by Staden in which he describes the sacrifices of prisoners are illustrated by a series of plates in which a shaman appears dressed in almost identical clothing. It is clear that Manfredo Settala, on the basis of his own valuations or information received from Landi, felt confident enough to identify his ethnographic objects with those represented by Léry and not those by Staden. Leaving aside the validity of this attribution and his observations, which were actually correct apart from the unfortunate reference to gold dust,[17] it seems to me to be important to note that it cannot be other than the fruit of a comparative type of approach.[18]

These characteristics clearly show how Manfredo Settala used his collection and its function as he saw it. They would have remained completely unknown had he not recorded his observations in the captions to the illustrated catalogues, had they not been rediscovered, and if they had not been analysed with the attention they deserve. Perhaps, in conclusion, all that remains for us is to accept that our view of collecting in the sixteenth and seventeenth centuries is based on extremely fragile foundations and all too frequently conditioned by the whims of chance.

Acknowledgements

Many individuals have helped me complete this work and while I cannot list them all, I would like to thank them sincerely. Particular thanks, however, go to Giuliana Clacani who allowed me to obtain in record time an all but untraceable text, to Stefania Mason for her precious notes regarding the situation in Venice and to Clelia Arnaldi, Ariela Bollati, Simonetta Castronovo and Enrica Pagella for the data regarding the Sabaudian collections.

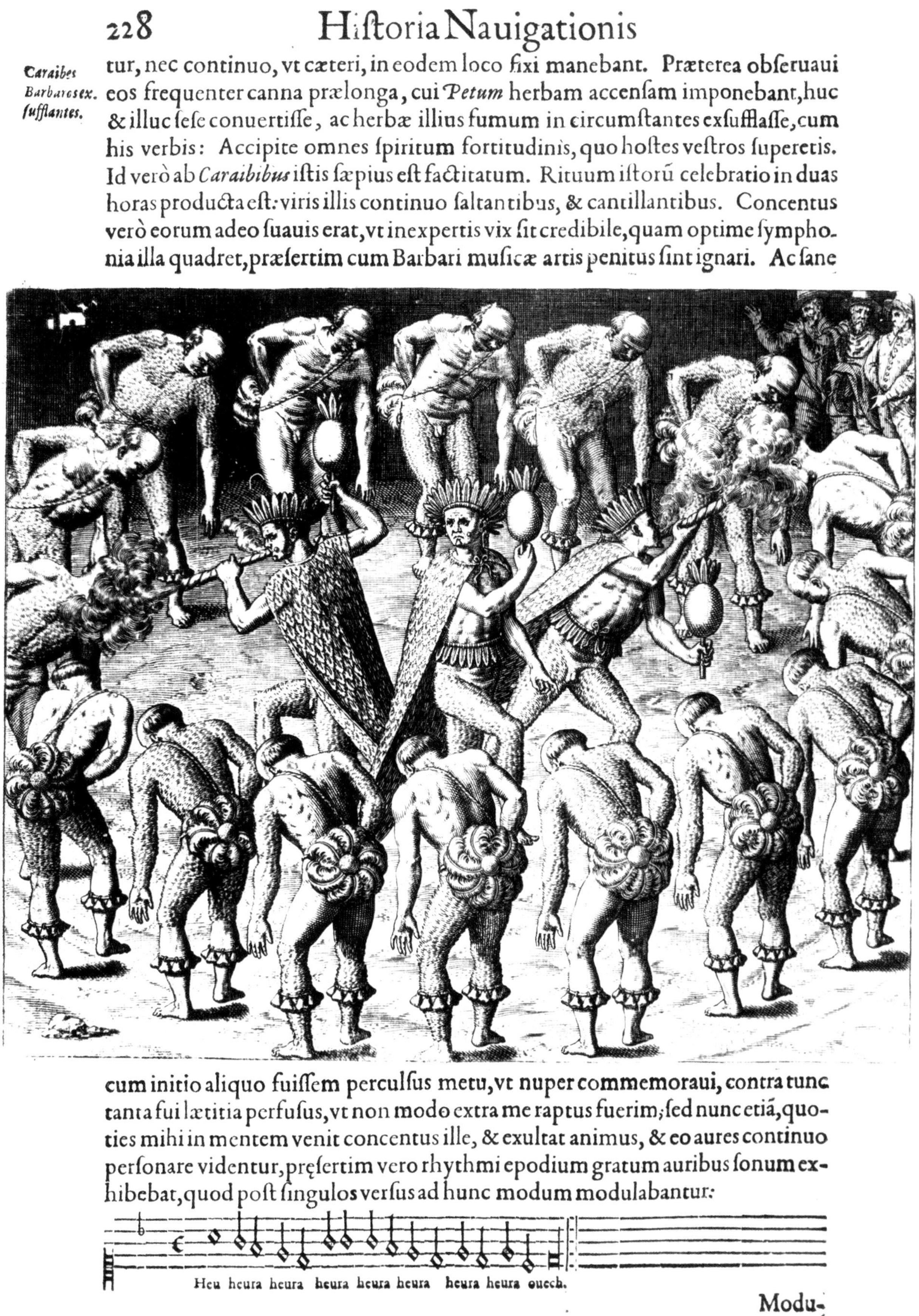
228 Hiſtoria Nauigationis

Caraibes Barbaros ex. ſufflantes.

tur, nec continuo, vt cæteri, in eodem loco fixi manebant. Præterea obſeruaui eos frequenter canna prælonga, cui *Petum* herbam accenſam imponebant, huc & illuc ſeſe conuertiſſe, ac herbæ illius fumum in circumſtantes exſufflaſſe, cum his verbis: Accipite omnes ſpiritum fortitudinis, quo hoſtes veſtros ſuperetis. Id verò ab *Caraibibus* iſtis ſæpius eſt factitatum. Rituum iſtorũ celebratio in duas horas producta eſt: viris illis continuo ſaltantibus, & cantillantibus. Concentus verò eorum adeo ſuauis erat, vt inexpertis vix ſit credibile, quam optime ſymphonia illa quadret, præſertim cum Barbari muſicæ artis penitus ſint ignari. Ac ſane

cum initio aliquo fuiſſem perculſus metu, vt nuper commemoraui, contra tunc tanta fui lætitia perfuſus, vt non modo extra me raptus fuerim, ſed nunc etiã, quoties mihi in mentem venit concentus ille, & exultat animus, & eo aures continuo perſonare videntur, pręſertim vero rhythmi epodium gratum auribus ſonum exhibebat, quod poſt ſingulos verſus ad hunc modum modulabantur:

Heu heura heura heura heura heura heura heura ouech.

Modu-

Figure 10. Theodore De Bry, Tupinambà ritual dance led by three shamans wearing items similar to those in the Settala Museum (Léry in De Bry 1630: 228). Manfredo Settala refers to this engraving in the captions to the preceding drawings. (Photo: Antonio Aimi, used with permission.)

Notes

1. Following this initial research, numerous other studies of the collecting of exotica have been published but only the most important are cited in this article.
2. Naturally, the reality of the erasure of the collecting tradition in these three cities was not changed by the research that occasionally examined random groups of objects. In the case of Milan, the most emblematic example of this interruption, which in Bologna and Florence was to an extent attenuated by the presence of the universities (through to the 1920s, the university of the Lombard capital was the University of Pavia), the clearest evidence of initiatives incapable of modifying the situation is proved by the 'resurrection' of the Settala Museum (Fogolari 1900; Ratti 1906; Galbiati 1926, 1928) and by the relatively more recent research by Carla Tavernari (1976). This kind of *damnatio memoriae* was actually ratified by a text from the Ambrosiana Library, the institution that had inherited the Settala Museum. Angelo Paredi, in his role as 'prefect' (director) of the Ambrosiana, wrote that all the objects of the Settala Museum had been destroyed in the bombing of 1943 (Paredi 1969: 21). Of course, this was a lie of which the director of the Milanese library was well aware. The reasoning behind this behaviour is difficult to understand, although it may be observed that the thesis clearly favoured the alienation of a number of Settalian objects in clear violation of the existing legislation (Aimi 1992: 220–221).
3. The research closest to the themes of collecting is that of Rocchi concerning the objects in the Coleti collection (Rocchi 1983).
4. In this case too, mention should be made of the most significant of the numerous publications by the scholar from Varese, noting that his research into Italian collecting culminated in the volume examining the situation on a European level (Bassani 2000).
5. An invaluable exception to the Africanist approach to these studies is represented by the volume on the Polynesian objects collected by Cook (Bassani 1982).
6. A number of these research projects highlighted new aspects of eclectic collecting within the Italian context of the sixteenth and seventeenth centuries (Basile 1987; Battisti 1989). Other studies did not venture beyond a superficial description of the phenomenon that failed to consider the merits of the object or, more importantly, its value and significance in the context of that period (Lugli 1983; Olmi 1985, 1992; Nobili 1998).
7. In addition to Columbus, Cabot and the role of the Genoan bankers at the Spanish court, the presence of Genoa and Venice in the Mediterranean East, the intensity of the relationship between these cities and the 'other' worlds is demonstrated by the participation of about 20 Ligurians in the conquest of Mexico (Thomas 2001), and in the work of the 'Venetian' Vincenzo Paletino, who, returning from America, wrote an essay on the legitimacy of the Conquest.
8. An entire catalogue of the Vendramin collection, the twelfth, was devoted to the exotica. Unfortunately it is among those lost (Borenius 1923; Favaretto 1994: 268; Pomian 2004: 76).
9. Mention should be made of the research into the following collections: Gualdo in Vicenza (Pomian 1989: 89–93), Moscardo and Calceolari in Verona (Favaretto 1994: 269–270), Mantova Benavides and Pignoria in Padua (Favaretto 1972; 1994: 26; Pomian 1989: 100–101). Bassani, furthermore, notes a Sapi-Portuguese salt cellar in the Marsiletti collection in Vicenza (Bassani 2000: 239). In this case too, only the most important and recent publications are cited.
10. Although the anonymous author of the presentation of *Il Manoscritto messicano 3773* declares that the work of Pignoria contains both the drawings of the Codex Vaticanus 3773 and those of the Codex Vaticanus 3738, Pignoria's images and his own declarations seem to indicate that he drew only on the second.
11. Even though Pignoria himself (1615: 23) declares that he obtained his images, originally belonging to Cardinal Marco Antonio Amulio, from Ottaviano Malipero, he might have seen the copy of the same codex in the Contarini collection.
12. With regard to the scientific contact, friendship, exchanges of gifts and commercial relationships between Ramusio and Fernández de Oviedo see the declarations of the reporter himself (Fernández [1557] 1959, I: 35, 305; III: 334 etc.) and Gerbi's observations that examine the entire Venetian network and its correspondents (Gerbi 1975: 220–230 etc.).
13. The presence of these last objects in the ducal collections of the sixteenth and seventeenth centuries is confirmed by Clelia Arnaldi and Simonetta Castronovo of the Museo Civico d'Arte Antica, Turin.
14. The valuation of the Viennese scholar has recently been reprised by Giuseppe Olmi (1985, 1992) for whom the Settala Museum and the Cospi and Moscardo collections 'represent in some senses a retrogression when compared with the collections of the sixteenth-century naturalists' (Olmi 1985: 12–15).
15. The Austrian scholar writes: 'This rational, practical, sensible people ... to who we may attribute mathematics as a natural science always experienced an intimate abhorrence of the romanticisms of the witches' cauldron' (Schlosser [1908] 1974: 102).
16. Three of the seven 'volumi mezzani' that constituted the Museum's illustrated catalogue are conserved in the Ambrosiana Library in Milan, two are in the Estense Library in Modena while two have not been traced. All seven were alienated from the Ambrosiana at an unknown date. The three that have returned to the Milanese library were traced and repurchased from a German dealer early in the twentieth century.
17. The reference to gold is probably a true *lapsus calami* due perhaps to an implicit comparison with the Tupinambá body decorations and those of the ritual of the Muisca sovereigns to whom we owe the 'original' El Dorado myth. In any case, nothing in the Settalian documentation suggests a transposition among the Tupinambá of the El Dorado myth or any reference to 'gold'.
18. Following the drafting of this article, a new book (Rovetta et al. 2010) was published that presents a reconstruction of the story of the Museo Settala that not only ignores many of the earlier studies (nonetheless used, in certain cases, without acknowledgement of the source – see, for instance: Buono 2010: 48–51) and the state of abandonment and neglect in which the collection's pieces were kept (including the fabrics from the ancient Congo kingdoms and the Tupinambá cloak mentioned in this article), but actually justifies the illegal sale of many items in the 1960s and 70s in open violation of Italian law including the so-called '*Schiavo Incatenato* or 'Enchained Slave' which, after remarkable adventures was purchased by Milan's Civic Collection of Applied Art. For a brief review of this book see the article published in *Il Giornale dell'Arte* by those (Alessandro Morandotti, Vincenzo de Michele and the present author) who in the early 1980s had rescued the surviving pieces of the Museo Settala from the embarrassing situation in which they were to be found (Aimi et al. 2011).

References

Manuscript sources

Inventory of Palazzo Madama, Museo Civico d'Arte Antica, Turin
MSBA *Il Museo Settala* , Biblioteca Ambrosiana, Milan (Z 387-389 sup.)
MSBA *Descrizione.di vari ritratti* .... 1627, Biblioteca Ambrosiana, Milan S.P. II 262 fasc. 16

MSBE *MSS Campori [Il Museo Settala]*, Biblioteca Estense, Modena γ H 1.21-22

Printed sources

Aimi, A. (1983) 'Il Museo Settala: i reperti americani di interesse etnografico', *Archivio per l'Antropologia e la Etnografia* **113**: 167–186.
Aimi, A. (1992) 'L'occasione perduta del Museo Settala', *Bulletin des Musées Royaux d'Art et d'Histoire* **63**: 213–222.
Aimi, A. (1993) 'Napoli e l'America', in L. Laurencich (ed.), *Tesori dalle Ande, Centro Europeo Mostre*, La Spezia, 23–29.
Aimi, A. (1994) 'Belli da lontano: collezionismo e reperti etnografici americani: Il caso di Milano', in *Andando más más se sabe, Atti del Convegno: 'La scoperta dell'America e la cultura italiana', CNR*. Rome: Bulzoni, 217–247.
Aimi, A. and E. Bassani (1986) 'Mirabilia Orbis', *Kos* **24**: 39–53.
Aimi, A., De Michele, V. and Morandotti, A. (1984) *Musaeum Septalianum*. Firenze: Giunti Marzocco.
Aimi, A., De Michele, V. and Morandotti, A. (1985) 'Towards a history of collecting in Milan in the late Renaissance and Baroque periods', in O. Impey and A. MacGregor (eds), *The Origins of Museums*. Oxford: Clarendon Press, 24–28.
Aimi, A., De Michele, V. and Morandotti, A. (2011) 'Che cosa nasconde quel catalogo?', *Il Giornale dell'Arte* **312**: 28.
Albonico, A. (1990) *Il Cardinale Federico 'Americanista'*. Rome: Bulzoni.
Ambrosini, F. (1982) *Paesi e Mari Ignoti*. Venice: Deputazione.
Basile, B. (1987) *L'Invenzione del Vero*. Rome: Salerno.
Bassani, E. (1977) *Scultura Africana nei Musei Italiani*. Bologna: Calderini.
Bassani, E. (1982) *Cook. Polinesia a Napoli nel Settecento*. Bologna: Calderini.
Bassani, E. (2000) *African Art and Artefacts in European Collections 1400–1800*. London: British Museum Press.
Battisti, E. (1989) *L'Antirinascimento*. Milan: Garzanti.
Bava, A.M. (1995) 'La collezione di oggetti preziosi', in G. Romano (ed.), *Le Collezioni di Carlo Emanuele I di Savoia*. Turin: CRT, 265–332.
Benussi, P. (ed.) (2007) 'Appendice documentaria', in L. Borean and S. Mason (eds), *Il Collezionismo d'Arte a Venezia: Il Seicento*. Venice: Marsilio, 335–382.
Bollati, A. and Messina, V. (2009) *Collezioni del Museo Civico d'Arte Antica di Torino: Cammei, Intagli Paste Vitree*. Turin: Palazzo Madama.
Borenius, T. (1923) *The Picture Gallery of Andrea Vendramin*. London: Medici Society.
Braudel, F. (1982) *I Tempi del Mondo*, Turin: Einaudi.
Brown, C.M. (1985) *La Grotta di Isabella d'Este*. Mantua: Arcari.
Buono, A. (2010) 'Mantello Tupinambá', in A. Rovetta (ed.), *Collezioni Settala e Litta Modignani, Pinacoteca Ambrosiana VI*. Milan: Intesa San Paolo (Mondadori Electa), 48–51.
Cipollato, M.T. (1961) 'L'eredità di Federico Contarini: gli inventari della collezione e degli oggetti domestici', *Bollettino dell'Istituto di Storia della Società e dello Stato Veneziano* **3**: 221–253.
Ciruzzi, S. (1983) 'Gli antichi oggetti americani nelle collezioni del Museo Nazionale di Antropologia e Etnologia di Firenze', *Archivio per la Antropologia e la Etnologia* **113**: 151–165.
De Michele, V. (2009) Personal communication.
Favaretto, I. (1972) 'Andrea Mantova Benavides: inventario delle antichità di casa Mantova Benavides – 1695', *Bollettino del Museo Civico di Padova* **61**: 35–169.
Favaretto, I. (1994) '"Una lucertola sive altro animale dell'America ...": il fascino di terre lontane nelle collezioni venete tra XVI e XVII secolo', in A.A. Caracciolo (ed.), *Il Letterato tra Miti e Realtà del Nuovo Mondo*. Rome: Bulzoni.
Fernández de Oviedo, G. [1557] (1959) *Historia general y natural de las Indias I–V*. Madrid: Atlas, BAE.
Fogolari, G. (1900) 'Il Museo Settala', *Archivio Storico Lombardo* **14**: 58–126.
Franchini, D., Margonari, R., Olmi, G., Signorini, R., Zanca, A. and Tellini Perina, C. (1979) *La Scienza a Corte*. Rome: Bulzoni.
Galbiati, G. (1926) 'Antichità sudamericane all'Ambrosiana', *Le Vie d'Italia e dell'America Latina* **9**: 1027–1030.
Galbiati, G. (1928) 'Notizie sulle antichità sudamericane del Museo Settala all'Ambrosiana', in *Atti XXII Congresso Internazionale degli Americanisti*, 509–520.
Garbari, F., Tongiorgi Tommsi, I. and Tosi, A. (1991) *Il Giardino dei Semplici*. Pisa: Cassa di Risparmio.
Gerbi, A. (1975) *La Natura delle Indie Nove*. Milan and Naples: Ricciardi.
Germano, A. and Nocca, M. (eds) (2001) *La Collezione Borgia*. Naples: Electa.
Heikamp, D. (1972) *Mexico and the Medici*. Florence: Edam.
Heikamp, D. (1976) 'American objects in Italian collections of the Renaissance and Baroque: a survey', in F. Chiappelli (ed.), *First Images of America I*. Berkeley: University of California Press, 455–482.
Hochmann, M. (2008) 'Le collezioni veneziane nel rinascimento: storia e storiografia', in M. Hochmann, R. Lauber and S. Mason (eds), *Il Collezionismo d'Arte a Venezia: dalle Origini al Cinquecento*. Venice: Marsilio, 3–39.
Laurencich, L. (1983) 'Oggetti americani studiati da Ulisse Aldrovandi', *Archivio per l'Antropologia e la Etnografia* **113**: 187–206.
Laurencich, L. (1984) 'L'indice del Museo di Giganti. museologia scientifica. Interessi etnografici e ordinamento di un museo cinquecentesco', *Museologia Scientifica* **1**: 191–242.
Laurencich, L. (1985) 'Museography and ethnographical collections in Bologna during the sixteenth and seventeenth centuries', in O. Impey and A. MacGregor (eds), *The Origins of Museums*. Oxford: Clarendon Press, 17–23.
Laurencich, L. (1992) (ed.) *Bologna e il Nuovo Mondo*. Bologna: Grafis.
Laurencich, L. and Ciruzzi, S. (1981) 'Antichi oggetti oggetti americani nelle collezioni del Museo Nazionale di Antropologia e Etnologia di Firenze: due mantelli di penne dei Tupinambà', *Archivio per la Antropologia e la Etnologia* **111**: 121–142.
Léry, J. de (1590–1630) 'Navigatio in Brasiliam ...', in T. De Bry (ed.), *Historiae Antipodum sive Novi Orbis*. Frankfurt.
Lightbown, R. (1969) 'Oriental art and the Orient in late Renaissance and Baroque Italy', *Journal of the Warburg and Courtauld Institutes* **32**: 228–279.
Lo Sardo, E. (ed.) (2001) *Il Museo del Mondo*. Rome: De Luca.
Lugli, A. (1983) *Naturalia et Mirabilia*. Milan: Mazzotta.
Mason, S. (2008) 'Carlo Helman (or Elman)', in M. Hochmann, R. Lauber and S. Mason (eds), *Il Collezionismo d'Arte a Venezia: dalle Origini al Cinquecento*. Venice: Marsilio, 287–288.
Morandotti, A. (2005) *Milano Profana nell'Età dei Borromeo*. Milan: Electa.
Morandotti, A. (2008) *Il Collezionismo in Lombardia*. Milan: Officina libraria.
Nobili, C. (1998) 'Oggetti precolombiani dell'antico Messico e camere delle meraviglie italiane: il caso del Museo Nazionale Preistorico Etnografico Luigi Pigorini di Roma', in F. Busatta and S. Busatta (eds), *Da Montezuma a Massimiliano*. Padova: Comune di Padova, 6–67.
Nocca, M. (ed.) (2001) *Le Quattro Voci del Mondo*. Naples: Electa Napoli.
Olmi, G. (1985) 'Science – honor – metaphor: Italian cabinets of the sixteenth and seventeenth centuries', in O. Impey and A. MacGregor (eds), *The Origins of Museums*. Oxford: Clarendon Press, 17–23.
Olmi, G. (1992) *L'Inventario del Mondo*. Bologna: Il Mulino.
Paredi, A. (1969) 'Storia dell'Ambrosiana', in A. Falchetti (ed.), *La Pinacoteca Ambrosiana*. Milan: Neri Pozza, 11–23.
Pavoni, R. (ed.) (1983) *Collezioni Giovio: le Immagini e la Storia*. Como: Musei Civici.

Pignoria, L. (1615) 'Seconda parte delle imagini degli dei indiani', in V. Cartari, *Le Vere e Nove Imagini delli Antichi*. Padua: Tozzi, 1–63.

Pinto, S. (ed.) (n.d.) *Musei d'Arte a Torino: Cataloghi e Inventari delle Collezioni Sabaude*. Torino: Soprintendenza per i Beni Artistici e Storici del Piemonte.

Pomian, K. (1989) *Collezionisti, Amatore e Curiosi*. Milan: Il Saggiatore.

Pomian, K. (2004) *Dalle Sacre Reliquie all'Età Moderna*. Milan: Il Saggiatore.

Ramusio, G.B. (1550–1559) *Delle Navigationi et Viaggi I–III*. Venice: Giunti.

Ramusio, G.B. [1556] (1985) 'Discorso di messer Gio. Battista Ramusio sopra il terzo volume delle Navigazioni e Viaggi', in G.B. Ramusio (ed.), *Navigazioni e Viaggi V*. Turin: Einaudi, 5–17.

Ratti, A. (1906) 'La risurrezione di un museo milanese (Il Museo Settala)', *Rendiconti R. Istituto Lombardo di Scienze e Lettere* **39**: 1011–1020.

Rivosecchi, V. (1982) *Esotismo in Roma Barocca*. Rome: Bulzoni.

Rocchi, C. (1983) 'A collection in the Pigorini Museum in Rome: G.D. Coleti's Ecuadorian objects', *Archivio per la Antropologia e la Etnologia* **113**: 227–242.

Rovetta, A. (ed.) (2010) *Collezioni Settala e Litta Modignani, Pinacoteca Ambrosiana VI*. Milan: Intesa San Paolo (Mondadori Electa).

Scarabelli, P.F. (1666) *Museo ò Galeria adunata dal Sapere, e dalla Studio Del Sig. Canonico Manfredo Settala*. Tortona: Viola.

Scarabelli, P.F. (1677) *Museo ò Galeria adunata dal Sapere, e dalla Studio Del Sig. Canonico Manfredo Settala*. Tortona: Viola.

Schlosser, J. von [1908] (1974) *Raccolte d'Arte e di Meraviglie del Tardo Rinascimento*. Florence: Sansoni.

Simili, R. (ed.) (2001) *Il Teatro della Natura di Ulisse Aldrovandi*. Bologna: Compositori.

Stendardo, E. (1991) 'Ferrante Imperato: il collezionismo naturalistico a Napoli tra '500 e '600 ed alcuni documenti inediti', *Atti e memorie dell'Accademia Clementina* **28–29**: 43–79.

Tavernari, C. (1976) 'Manfredo Settala, collezionista e scienzato milanese del '600', *Annali dell' Istituto e Museo di Storia della Scienza* **1**: 43–61.

Terzago, P.M. (1664) *Musaeum Septalianum*. Tortona: Viola.

Thomas, H. (2001) *Quién es Quién de los Conquistadores*. (n.p.): Salvat.

Violani, C. (1988) *Un Bestiario Barocco*. Milan: Museo Civico di Storia Naturale.

Zimmermann, T.C. (ed.) (1985) *Atti del Convegno: 'Palo Giovio: il Rinascimento e la Memoria'*. Como: Società Storica Comense.

Author

Antonio Aimi, Università degli Studi di Milano, Milan, Italy

A new glance at Bologna's sixteenth- and seventeenth-century museums and their Mexican items

Laura Laurencich-Minelli

A Mexican present to Clemente VII in Bologna

Contrary to its apparently marginal role in the politics of the time, Bologna played a significant part in sixteenth-century European international relations on at least three occasions. On 24 February 1530, Charles V of Spain was crowned emperor by Pope Clemente VII (Giulio de' Medici) in the basilica of San Petronio in Bologna in order to ratify the new European order established by the Peace of Barcelona (20 June 1529). Two years later, between December 1532 and February 1533, the pope and the emperor had a second meeting in Bologna to affirm the results of this political agreement. Twelve years later Bologna once again took centre stage in international European relations by hosting during three years (19 March 1545–1 April 1548) the eighth session of the Trento Council. These three events brought to Bologna exceptional participants and international personalities who contributed to the enrichment of the town's antiquities collections with gifts from the New World. Dominican friar Leandro Alberti reports that as soon as the emperor left Bologna on 3 March 1533, a Spanish Dominican by the name of 'Domenico who was coming from the New Indies', offered to Pope Clemente VII a magnificent Mexican gift composed of objects that he described in a list that is new to Americanist scholars (Alberti [1541] 2006: 629–630). The items mentioned are quite extraordinary. There are 'two coverlets made and woven with feathers ...; some books very nicely painted which looked like hieroglyphs by which they understood each other as we do by letters ...; some very thick masks decorated by turquoises through which [Friar Domenico] said that the demons were speaking to them ...; a yellow stone knife two fingers wide and one foot long with the handle all covered with turquoise'. Then he mentions some stone knives that 'cut like razors' as well as 'stoles' and other Christian priests' vestments 'nicely worked with those feathers'. Friar Leandro ends the list by mentioning that he received some items that were part of this present. These were 'some codices, some knives, together with the big knife used to kill and to sacrifice men to their idols' although he specifies that he gave 'the big knife, a codex and one of those knives similar to a razor' to Mr Giovanni Achillini to decorate his museum.

Leandro Alberti's list not only opens up new perspectives on the history of the relationships between Mexico and Bologna, Rome and Florence (being Pope Clemente VII, Giulio de' Medici), but also raises interesting questions about how long and where Alberti and Achillini's objects were kept in Bologna, and the identity of the Spanish Dominican friar by the name of 'Domenico who was coming from the New Indies'.

In a 'handbook' for erudite travellers, Friar Leandro tells us that the scholar Giovanni Achillini had one of the best antiquities collections in Bologna, and that it was open to friends and visitors (Alberti 1568: ff. 335v–363r). Consequently, the Mexican ceremonial knife, a codex, and one of the knives, similar to a razor that Mr Giovanni had received as a gift were kept in an outstanding collection alongside antiquities from Rome, ancient medals and coins, musical instruments and portraits of Roman emperors. However, Friar Leandro does not mention the whereabouts of his Mexican items. We do not know of any antiquities collections that belonged to San Domenico church (Bologna) in whose monastery he lived. As a Dominican friar, Leandro Alberti could not keep any material goods, nor could he hoard any riches according to the rules established by the saint that founded the order. This could be proof that he may have given both codices and knives to some of his friends. Current research about these objects suggests that the Spanish Dominican by the name of 'Domenico who was coming from the New Indies' was Friar Domingo de Betanzos, who left New Spain to go to Rome in order to ask for the separation of Mexican Dominicans from the Provincia de Santa Cruz de Hispaniola in 1532. After this visit he went to Bologna in order to meet the pope on 3 March 1533. On this occasion he bestowed upon Clemente VII his American gift with the aim of securing his agreement on the reduction of the higher Dominican clergy's sojourn in the New Indies (Laurencich-Minelli 2012a; Laurencich-Minelli and Domenici forthcoming; Remesal [1619] 1988: 83–84).

Antonio Giganti and Ulisse Aldrovandi's Aztec objects

The interest in collecting by Bologna's educated classes resulted in the creation of two museums towards the latter part of the sixteenth century: the museums of Antonio Giganti and Ulisse Aldrovandi. Antonio Giganti (1535–1598) was a learned man who lived in close contact with some of the Roman Catholic

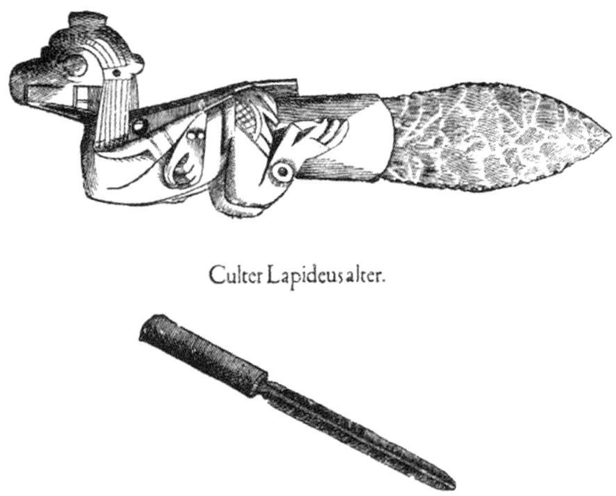

Figure 1. Above: ceremonial knife; below: everyday knife (Aldrovandi 1648: 156, 157) (The British Library).

Church's most powerful clergy and their cultural circles. He had been employed as a secretary to two individuals who took part in the Trento Council – Bishop Ludovico Beccadelli (from 1550 to his death in 1572) and Cardinal Gabriele Paleotti, Archbishop of Bologna (1580–1589). Monsignor Ludovico Beccadelli was a scholar who Friar Leandro considered as one of Bologna's best-known humanists as well as being the owner of an outstanding collection of antiquities (Alberti 1568: 364–365).

Cardinal Gabriele Paleotti was among those who reformed Bologna's clergy according to the new ideas that emerged from the Trento Council. He too was the owner of a collection of antiquities, as well as being a member of one of the city's most prized academies that had emerged among the circles connected to several of Bologna's high prelates. Antonio Giganti played an important role in these academies because he was well acquainted with the ecclesiastic elite that flourished in Bologna after hosting the eighth session of the Trento Council. It was following this event that Antonio Giganti began to assemble his museum around 1549 to which he later added the Beccadelli collection.

Giganti's museum epitomises the interest of Bologna clergy's in collecting, which prospered as a consequence of the events that had taken place in the town during the first half of the sixteenth century. In his effort to organise his museum, Antonio Giganti compiled an 'Indice' (index) in 1586. This was an inventory of the objects that combined item descriptions alongside sketches of their position on the walls of his *studium*. From this we learn that his museum contained about a thousand objects, and that Giganti was attempting to apply the *ars memoriae* ('art of memory') to its displays. This quintessentially Renaissance notion aimed at capturing simultaneously, both in the *studium* and in the mind of the viewer, the whole world's knowledge comprising both *naturalia* (items from the natural world) and *artificialia* (manmade objects) (Giganti [1586] 1984; Laurencich-Minelli 1984; 1985: 17–19; 2003: 27–40; 2005:121–122).

Ulisse Aldrovandi (1522–1605) was a *protomedico* (the head of Bologna's physicians and pharmacists) and a well-known *professor de fossilibus, plantis et animalibus* (professor of fossils, plants and animals) at the University of Bologna. He was in close contact with Italian and European scholars alike. He knew the Spanish physician Nicolò Monardes from Seville, who was well known for his documentation of Mexico's natural history, as well as collectors such as Tommaso Cavalieri in Rome and, of course, Francesco I and Ferdinando I de' Medici in Florence.

Dr Ulisse Adrovandi began collecting his first items around 1549 as a means of documenting his research. His museum was born out of this collection that he reports as having held about '7000 geological and paleontological items ... plus some very nice objects from the Indies'.

Adrovandi epitomises academic collecting within the context of the University of Bologna during the second half of sixteenth century. Furthermore, he was also Antonio Giganti's neighbour and friend. Their respective museums, housed on the same block, complemented each other as we learn from the *cataloga* (Aldrovandi's manuscripts stored in the University Library of Bologna (BUB)). Aldrovandi often availed himself of Giganti's collection to write dissertations on items that he himself did not own. A comparison between Giganti's inventory and Aldrovandi's *cataloga* reveals that in 1597, when Antonio Giganti retired to Fossombrone, Aldrovandi acquired his collection (Aldrovandi 1599, 1648 and Aldrovandi, BUB Mss. 6, 21–24, 26, 38, 46, 70, 76, 91, 92, 136). Aldrovandi's *cataloga* also tells us about Dr Ulisse's friendship with Ferdinando I de' Medici, and about the reciprocal exchange of items and new museographical data (Laurencich-Minelli 1980, 1982, 1983).

Giganti and Aldrovandi's interest in Aztec culture and the natural history of the Americas ultimately shows a degree of consistency despite the two scholars' different scientific standards and collection sizes. Antonio Giganti and in particular Ulisse Aldrovandi regarded their Mexican cultural and natural items as curiosities but, at the same time, they saw them as inspiration and reference for further research that eventually led to later experimental investigations. Around 1567 Aldrovandi was planning a scientific expedition to Mexico and the Americas in order to directly document both natural specimens and cultures that fell outside classical canons and therefore could not be studied through the classics (Tugnoli-Pattaro 1981).

Leandro Alberti's new list enables us to trace back to 1533 the history of some Aztec objects from the Giganti-Aldrovandi collections. For example, if we take Aldrovandi's two knives, the ceremonial one and his 'everyday knife' (Fig. 1), they not only match Giganti's two knives but also the two knives that Leandro Alberti is said to have given to Mr Giovanni Achillini together with a codex 'to decorate his *studium*' (Aldrovandi 1648: 156–157; Giganti [1586] 1984: 211, 233). Consequently, Aldrovandi's knives can be dated back to Clemente VII's present. Before reaching Aldrovandi's *studium*, these items had enriched both Achillini and Giganti's collections. Although we can reconstruct these objects' interesting history, the actual artefacts are unfortunately lost. Aldrovandi's 'nine Indian idols' match perfectly in both number and description to Giganti's 'Stone idols from the New Word' (Aldrovandi 1648: 540). From this we can surmise that they must have been kept by Antonio Giganti before becoming part of Aldrovandi's *studium* although, at present, we cannot offer any better suggestion about their

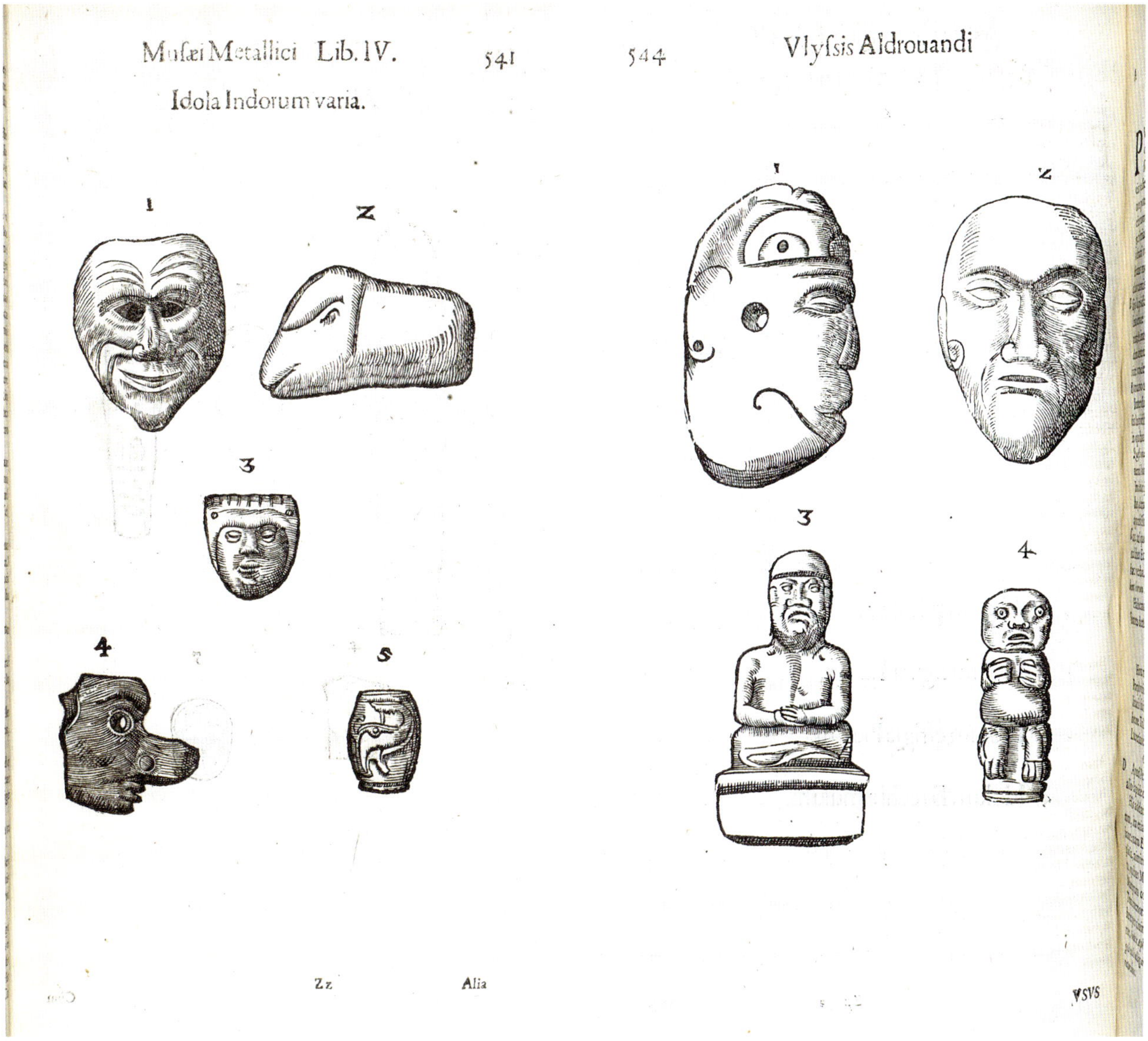

Figure 2. Idols from the New World (Aldrovandi 1648: 541, 544) (The British Library).

provenance. These nine idols are also unfortunately lost (Fig. 2).

The wooden Aztec mosaic mask described and illustrated by Aldrovandi does not match any of Giganti's items mentioned in his inventory (Aldrovandi 1648: 550). Instead it seems to coincide with one of the 'thick masks decorated in turquoises' from Leandro Alberti's list (Fig. 3). Presumably this object reached Bologna in 1533, but it was not kept in the town. New data emerging from my recent study of Medici cabinet's inventories suggest that Clemente VII gave all the turquoise masks he had received in Bologna to a Florentine museum that Cosimo I (1519–1574) had established by then in Palazzo Vecchio. Additional clues about these transactions come from the British Museum and Pigorini Museum's Aztec masks whose sixteenth-century history leads back to Florence. Quite possibly, Ferdinando I de' Medici (1549–1609) may have given the mask to his friend Aldrovandi knowing his interest in Mexican cultures. The location of the remaining Mexican codices given to Friar Leandro (and subsequently to Achillini from the present to Clemente VII in Bologna) still remains an open-ended question.

Antonio Giganti showed the same interest in ancient writing systems as Monsignor Beccadelli. In the inventory of his collection he mentioned 'a stone fragment with some characters brought by a Portuguese traveller from the Indies to Portugal' (perhaps some Maya writing), 'a copy of two pages of a book from the New World that look like hieroglyphs' and 'a map from the New World made out of rough but polished bark' (a pre-Columbian map) (Giganti [1586] 1984: 236–237). These items are not mentioned in the extensive collection of Aldrovandi's manuscripts, nor do they match Alberti's list. This suggests that either Aldrovandi did not get the hieroglyphic codices from Giganti or, if they had ended up in his collections, they were not interesting enough for him to include them in his *cataloga*. Indeed, Aldrovandi was first and foremost a naturalist and not interested in the interpretation of Preclassic 'alphabets'. He was rather keen on documenting the different types of techniques that were used by different cultures in a variety of writing systems (see his unfinished *Bibliologia* 1579–1582, Aldrovandi, BUB, Ms. 83). However, there is a possibility that these hieroglyphs may have ended up in the hands of one of the University of Bologna's

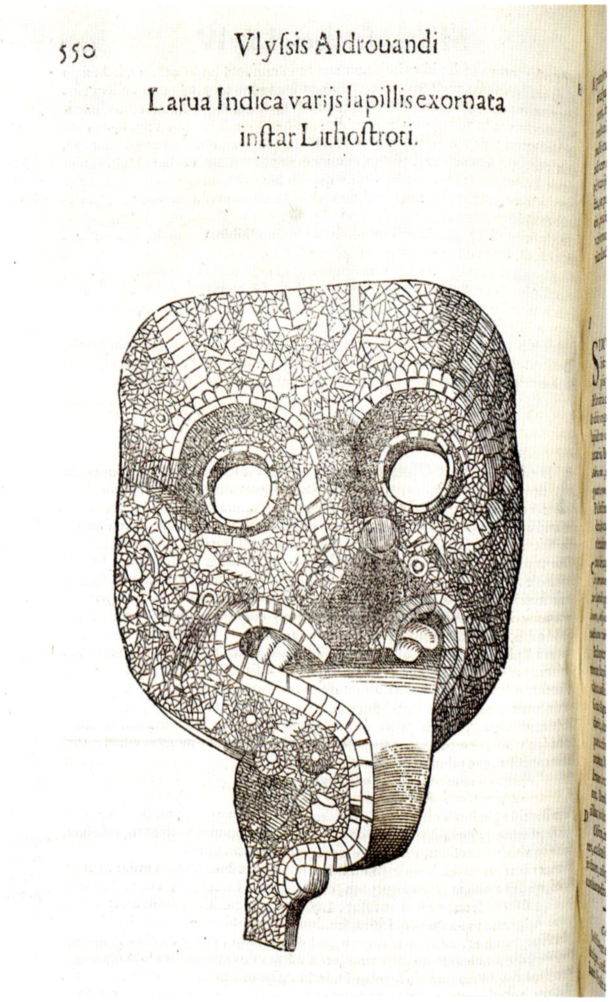

Figure 3. Mexican mask (Aldrovandi 1648: 550) (The British Library).

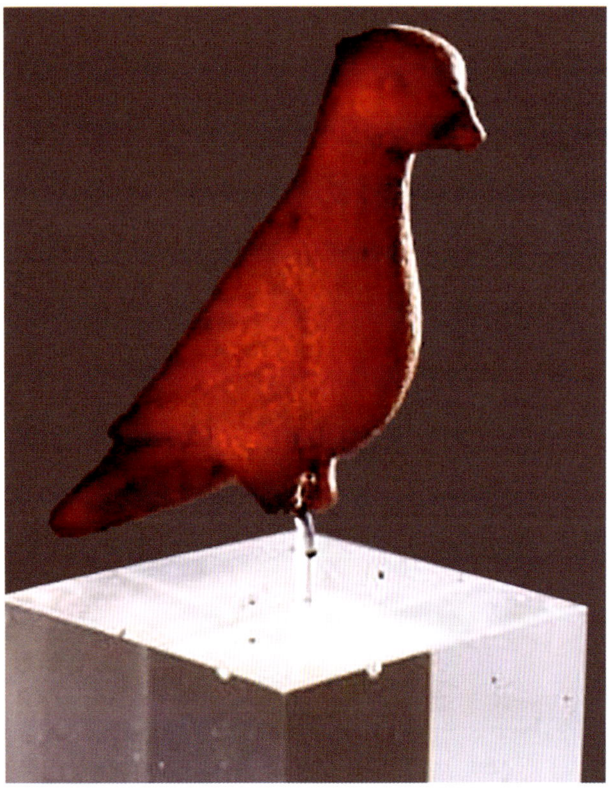

Figure 4. The humming bird, now in the Museo Medievale, Bologna. (Photo: Riccardo Vlahov, used with permission.)

public lecturers who had been trying to decipher pre-Greek writing since 1520 following an established tradition in the study of hieroglyphs in this town. More detailed research on Aldrovandi's manuscripts may reveal some clues regarding this suggestion.

Some Aztec objects in Bologna's museums of the seventeenth century

The history of Bologna's museums and their Aztec items took a new turn in the seventeenth century. Two years before Ulisse Aldrovandi's death in 1605, he left his museum and his extensive library to the Senate of Bologna. The whole collection, then separated from his library, was subsequently transferred to the Palazzo Pubblico in 1617 where it was arranged in six rooms managed by a keeper who opened it to visiting scholars. This early example of a public museum was admired, and by that time was still receiving bequests.

Bologna's interest in exotic objects inspired Marchese Ferdinando Cospi (1609–1686) to collect some outstanding Mexican objects for his museum. Marchese Ferdinando belonged to the senatorial Cospi family, that is, Bologna's high aristocracy. His mother was Costanza de' Medici, the grand-niece of the cardinal Alessandro de' Medici. Consequently Ferdinando grew up at the Medici's court in Florence, and his museum received many items from the Medici. Later on, when Ferdinando Cospi decided to give his museum to Bologna he placed it in the 'Palazzo Pubblico' alongside that of Aldrovandi (1657). He commissioned the engraver Giuseppe Mitelli to portray it in 1658 and asked Dr Lorenzo Legati to write two catalogues (1667, 1677). Ferdinando clearly wanted to give a scientific character to his museum following Aldrovandi's model, while Legati's 1677 catalogue, dedicated to Ferdinando de' Medici, suggests that Cospi's museum was more a combination of Aldrovandi's scientific museum and the Medici's collections. In 1680 an anonymous inventory was added to the Cospi Museum: the *Inventario semplice* (Anon. 1680).

Cospi's Aztec objects

From Legati's two catalogues and the *Inventario semplice* we learn that the Cospi Museum contained really exceptional Aztec items, and we can surmise that the marchese and his friends perceived them as exotic antiquities. For example, in both catalogues Dr Legati refers to the two outstanding ceremonial knives with wooden handles decorated in mosaic as two idols in the guise of a sphinx, but only in the 1677 catalogue does he mention these two 'sphinxes' as coming from the Indies, without explaining whether from the East or the West Indies (1677: 477–478). Oddly he failed to recognise the use of the two knife handles despite the availability of a very good example of a sacrificial Mexican knife in the adjacent Aldrovandi Museum (cf. Fig.1). His misinterpretation of these objects was probably due to his interest in the objects' appearances rather than in their function. In other words, his concern with the figure crouched like a sphinx prevented him from considering that it could hold a blade in its hands. These two handles may be the surviving part of the 'sacrificial tools' that

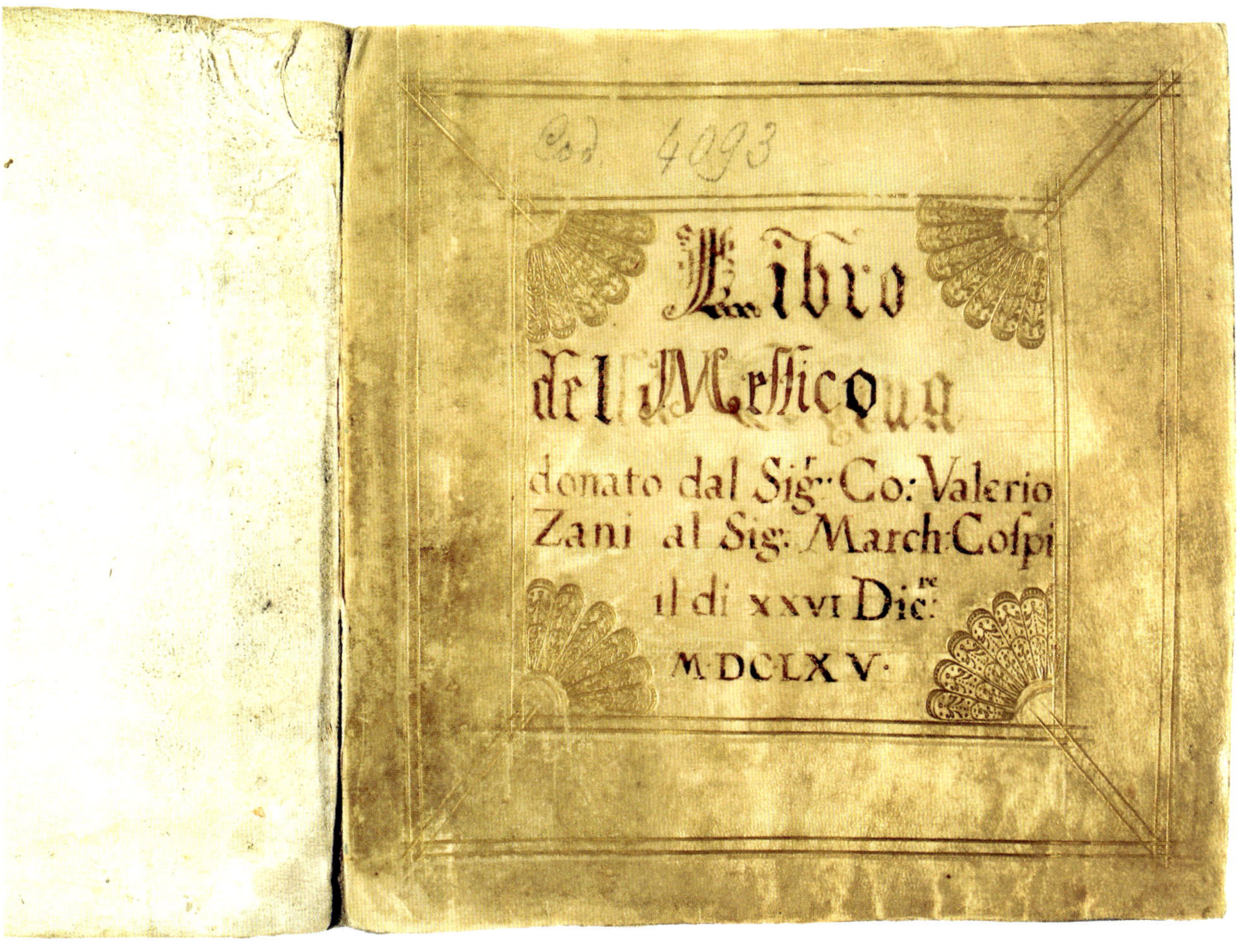

Figure 5. The cover of Codex Cospi (Biblioteca dell'Università di Bologna).

Betanzos gave to Clemente VII in 1532 while in Rome (Dávila Padilla 1596: 73–74; Laurencich-Minelli and Domenici forthcoming). At present, no conjecture can be made about the provenance of the small and delicate Aztec amber humming bird. It might have been acquired after the compilation of the Legati catalogue of 1677, but certainly before the compilation of the inventory of 1680, where it is registered among the *Cose peregrine, naturali e artificiali*, that is among the curiosities made by nature and by man (Fig. 4).

The Codex Cospi

The Cospi Museum had been housed in the 'Palazzo Pubblico' for eight years when it received a Mexican codex as a bequest by Count Valerio Zani. On the seventeenth-century Bolognese parchment cover of the codex that replaced the Precolumbian one, we read: *Libro della China donato dal Sig. Co. Valerio Zani al Sig. March. Cospi il dì XXVI Dic.re MDCLXV* (Chinese book presented by Count Valerio Zani to the Marchese Cospi on December 26th, 1665), later amended to the *Libro del Messico* (the Mexican book) (Fig. 5). The codex was described as an 'Indian book' in Legati's 1667 catalogue but this title was amended to the *Libro del Messico* as a result of his 'scientific' research during the preparation of the 1677 catalogue, for which he compared the Codex Cospi's hieroglyphs with the ones in the Codex Vindobonensis kept in the Würmian Museum and reproduced by Olaus Worm (Worm 1655: 383–384; Legati, 1677: 191). Legati, while compiling the Codex Cospi's description about minor 'letters' written at the margin as if they were brief comments or captions for the main figures (1677: 192), adds an interesting note about Valerio Zani being the donor of a precious Mexican *atlatl* (spear thrower) carved with figures and ideograms just obtained by the Aldrovandi Museum. This happened more than 60 years after Aldrovandi's death. Legati clearly wanted to show that it was Count Valerio Zani who gave the two refined items concerning Mexican hieroglyphics: that is the Mexican codex to the Cospi Museum and the spear thrower to the Aldrovandi Museum (Legati, 1677: 191–192). (The *atlatl* is now housed in the Pigorini Museum in Rome.)

Still unclear, nonetheless, are the reasons why Legati should include in the Cospi catalogue an item from the Aldrovandi Museum, Zani's name and the hieroglyphics. It is also unclear as to why the codex's cover was replaced with a seventeenth-century parchment from Bologna. It is possible that the new cover might have been due to Valerio Zani's electoral propaganda because the new parchment cover shows Zani as the donor. It should be noted that at the time, hieroglyphs were used in innovative research that aimed at discovering universal writing systems. These studies were led by the Bolognese friends of Marchese Ferdinando Cospi who joined the Academy of Gelati (hosted in Marchese Ferdinando's house), Count Valerio Zani being one of the promoters of this original research. At the same time, it appears that Zani was elected rector of the Gelati Academy between 1670 and 1671.

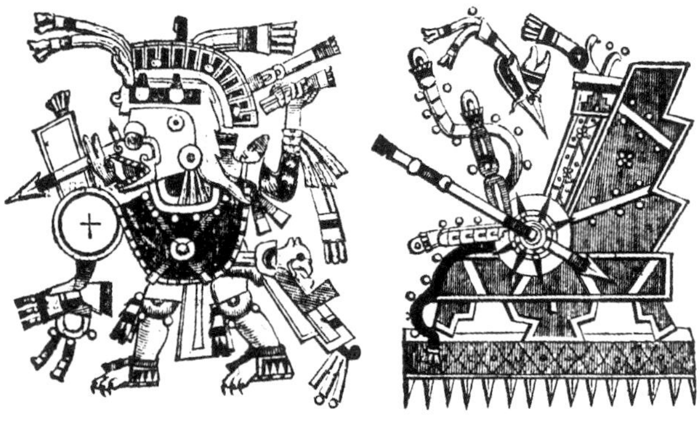

Figure 6. The lower part of Codex Cospi c. 10r (after Legati 1677: 192) (Biblioteca dell'Università di Bologna).

His rectorate is said to have been an energetic and innovative one (Bergamini 1988: 12–18). Perhaps this reputation may have emerged with reference to his outstanding present with Mexican hieroglyphs.

Legati, in spite of his interest in innovative investigations on ancient writing, demonstrated a peculiar way of interpreting objects that backed up his research (Legati 1677:191). An engraving from the lower part of c.10r of the Codex Cospi (Fig. 6) shows the god Tlahuizcalpantecuhtli reworked as if to show two 'punti di fuga' (as posited by Baroque aesthetic canons) in contrast to the one painted in the codex (Fig. 7).

In summary, Marchese Ferdinando Cospi, Dr Legati, and the Gelati academics saw the Cospi Museum's Mexican items as belonging to a laboratory of antiquities in which objects could be interpreted or 'read' according to the aesthetic rules of the time.

It is not yet clear where and how Valerio Zani obtained the Codex Cospi, nor can we reconstruct the codex's journey before it reached him. Recent research indicates that the Codex Cospi was possibly one of the books presented to Pope Clemente VII by Fray Domingo de Betanzos in Bologna during the ceremony on 3 March 1533. This is the same present that the pope shared with Friar Leandro after the ceremony. It is therefore possible to suggest that it could be the same book that Leandro Alberti subsequently gave to Achillini for his museum (Laurencich-Minelli 2012a).

Seven years after Clemente VII received the extraordinary present from Mexico in Bologna, and one year after Achillini's death, scholar Teseo Ambrogio degli Albonesi from Pavia visited the city. On this occasion the famous Renaissance orientalist reports that he saw in the hands of the canonist Paleotti an interesting book from the Indies (*ex India*) covered in 'tyger skin' that was painted on the *rectum* and on the *versum* with figures of men, animals and things 'clarified by a sort of captions which were both all around – the *tonalpohualli*(?) – and on the edge': figures that he describes as being 'really letters' (Albonesi 1539: 198r–v). This is true not only of the figures of men, animals and things but also for the ideograms that, as reported by Teseo Ambrogio, resemble captions to the figures (see Figs 6 and 7).

In short, Teseo Ambrogio's account of this manuscript indicates similarities with the *rectum* and *versum* of the Codex Cospi (cc.11–13, 21–31) – as well as of Laud and Vaticanus 3773 codices – which interestingly fit with Legati's description of the Codex Cospi. This suggests that the Codex Cospi might be the book that the erudite Teseo observed in the hands of the Bolognese canonist Paleotti that was among those given by Friar Domingo de Betanzos to Clemente VII on 3 March 1533, and which were kept in the town by Leandro Alberti. Canonist Paleotti, who at the time was a very young and promising student interested in collecting (Aldrovandi 1599: 636), later became Bologna's Cardinal Gabriele Paleotti. We cannot be sure whether or not he acquired the codex from Giovanni Achillini, Leandro Alberti, or from one of his friends. Further research is needed to clarify some of these issues. Certainly, after 132 years no one in Bologna remembers anything other than that the codex was a Mexican book. It was thought rather to be Chinese, as it appeared on its new cover on 26 December 1665.

Aztec objects in Bologna's collections during the eighteenth century

After Luigi Ferdinando Marsigli founded Bologna's Istituto delle Scienze in 1714, both the Aldrovandi and Cospi museums were moved into the Istituto's Stanza delle Antichità between 1742 and 1743 where they lost the obsolete connotations of *naturalia* and *artificialia*. These collections were no longer considered expressions of magnificent cultures flourishing in faraway lands, but rather as archaeological expressions of old civilisations. However, the Stanza delle Antichità continued to receive gifts, for example from Pope Benedetto XIV (the Bolognese Prospero Lambertini) who in 1745 donated various objects from Amazonia, and an Aztec musical instrument carved out of a human femur decorated with mosaic (Medica 1992: 42–43).

Jesuits, who came as exiles to Bologna from America in 1768, helped to revise the cultural affiliation of the objects kept in the Stanza delle Antichità, and confirmed the Aztec identity of some. Alongside the Istituto delle Scienze they presented scientific arguments for the complexity of Native American cultures that were generally disdained by Enlightenment scholars. The proximity to Mexican objects, and in particular to the Codex Cospi, stimulated Francisco Saverio Clavijero to write his monumental *Storia Antica del Messico* (Ancient History of Mexico) in Bologna during the years 1780–1781.

In 1803 both the Stanza delle Antichità of the Istituto delle Scienze and the Istituto itself became the responsibility of the University of Bologna, which also lent items to the newly founded Museo Archeologico. Unfortunately, prehistorian Luigi Pigorini's enthusiasm for antiquities was behind the 'theft' of the American items housed in Bologna for his new Museo Preistorico and Etnografico in Rome. In exchange for these items, the University of Bologna received a small prehistoric collection of the Villanova culture that had just been unearthed around the town's outskirts. As a result, the Codex Cospi and the humming bird of Figure 4 are the only Aztec items of the Istituto's Mexican collections left in Bologna today. The codex was left behind because it was housed in the Biblioteca Universitaria – which was forbidden to Pigorini – whereas the hummingbird was probably overlooked due to its small size as it is just 8.2 cm long (Laurencich-Minelli 1992, 2005).

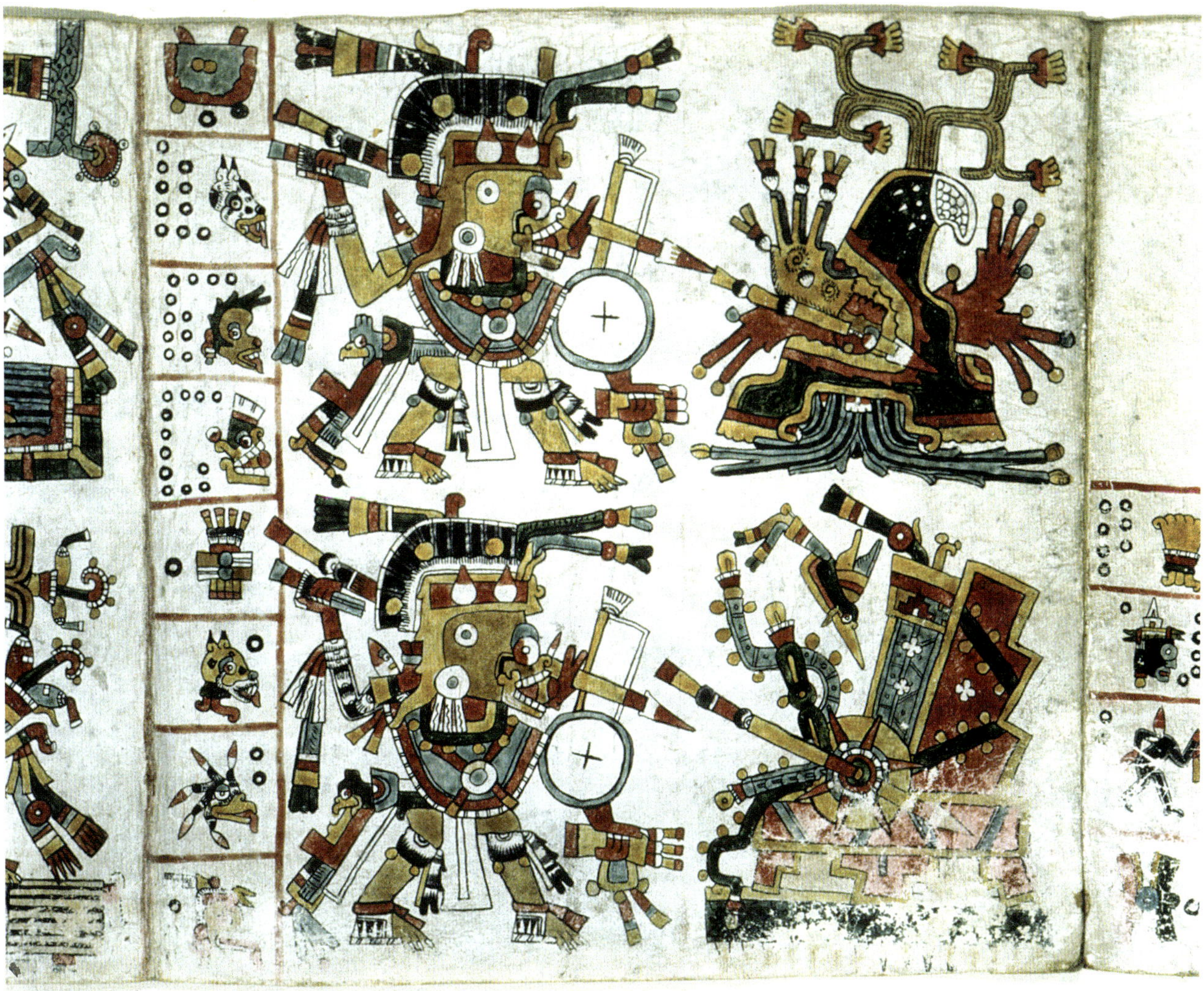

Figure 7. The Codex Cospi c. 10r (Biblioteca dell'Università di Bologna).

Conclusions

Between the sixteenth and the seventeenth centuries Bologna's museums housed several refined Mexican items that missionaries and travellers had brought back from the Americas as examples of unusual, yet complex, civilisations. These extraordinary objects give us a rare insight into the Mexican cultures thriving at the time of the Conquest (1519). They reached Bologna during the first half of sixteenth century not only because the city at the time was a hub of international activity, but also because of the close relationships enjoyed until the seventeenth century between Bologna's collectors, scholars, and the international world including the Medici family. At the same time, some Bolognese intellectuals and academics contributed to enlarge the collections and to prevent their loss.

Why Bologna's collections do not contain Inca objects (a civilisation that had been conquered around the same time as Clemente VII was receiving his gifts) is still puzzling. Reasons for this lack may be found in the political predicaments experienced at the time by Italy and Peru. By the second half of the sixteenth century Italy was about to be partially conquered by Spain, while Peru endured a long civil war between the Spanish Pizarristas and Almagristas, a fact that casts a shadow on Pizarro's victory at Cajamarca that was the result of his poisoning of Atahualpa's officers (Laurencich-Minelli 2007).

In conclusion, while the conquest of Mexico brought the first great pre-Columbian 'empire' of the Americas to the attention of Bologna's scholars and intellectuals, the conquest of Peru brought to the fore troubling questions about the legitimacy of the deeds accomplished in the name of conversion. What is more, the Medici had no interest in Peru because its harbours on the Pacific Ocean were inaccessible to them, whereas they continued to have access to the Atlantic harbours of Mexico and South America until the reign of Cosimo III de' Medici (1639–1723) (Laurencich-Minelli 2012b).

References

Alberti, L. [1541] (2006) *Historie di Bologna. Tomo II, 1511–1543*, A. Antonelli and M.R. Musti (eds). Bologna: Costa.

Alberti, L. (1568) *Descrittione di tutta Italia aggiuntavi la descrittione di tutte le isole.* Venezia: Ludovico degli Avanzi.

Albonesi (degli), T.A. (1539) *Introductio in Caldaicam Linguam, Syriacam atque Armenicam et Decem Alias Linguas (...).* Cremona: Simoneta Cremonensis.

Aldrovandi, U. (1599) *Ornitologiae. Hoc est de avibus historiae libri XII.* Bologna: Bellagamba.

Aldrovandi, U. (1648) *Musaeum Metallicum.* Bologna: Ferroni.

Anon. (1680) *Inventario semplice di tutte le materie esattamente descritte che si trovano nel Museo Cospiano. Non solo le notate nel libro già stampato e composto dal sig. Dottore Lorenzo Legati ma ancora le aggiuntevi dopo la Fabrica.* Bologna: Monti.

Bergamini, M.G. (1988) 'Dai Gelati alla Renia (1670–1698). Appunti per una storia delle Accademie Bolognesi', in M. Saccenti (ed.), *La Colonia Renia, Vol.II: Momenti e Problemi.* Modena: Mucchi, 5–52.

Clavijero F.S. (1780–1781) *Storia antica del Messico*, 4 vols. Cesena: Biasini.

Dávila Padilla, a. (1596) *Historia de la Fundación y Discurso de la Provincia de Santiago de México, de la Orden de Predicadores* (...). Madrid: Pedro Madrigal.

Giganti, A. [1586] (1984) 'Indice', in L. Laurencich-Minelli (ed.) 'L' indice del museo di Antonio Giganti. Interessi etnografici e ordinamento di un museo cinquecentesco', *Museologia Scientifica* **I** (3–4): 225–242.

Laurencich-Minelli, L. (1980) 'L'interesse americanista di Bologna attraverso i secoli', *Il Carrobbio, Rivista di Studi Bolognesi* **VI**: 236–237.

Laurencich-Minelli, L. (1982) 'Bologna und Amerika vom 16. bis zum 18. Jarhundert', in K.H. Kohl (ed.), *Mythen der Neuen Welt.* Berlin: Frölich und Kaufmann, 147–154.

Laurencich-Minelli L. (1983) 'Oggetti americani studiati da Ulisse Aldrovandi', *Archivio per la Antropologia e la Etnologia* **113**: 187–206.

Laurencich-Minelli L. (1984) 'L'indice del museo di Antonio Giganti. Interessi etnografici e ordinamento di un museo cinquecentesco', *Museologia Scientifica* **I**(3–4): 225–242.

Laurencich-Minelli, L. (1985) 'Museography and ethnographical collections in Bologna during the sixteenth and seventeenth centuries', in O. Impey and A. MacGregor (eds), *The Origin of Museums: The Cabinet of Curiosity in Sixteenth and Seventeenth Century Europe.* Oxford: Clarendon Press, 17–23.

Laurencich-Minelli, L. (ed.) (1992) *Codice Cospi: Calendario e Rituali Precolombiani.* Milan: Jaca Book.

Laurencich-Minelli, L. (2003) 'Some Mesoamerican items in Antonio Giganti's 'studio': a 16th century museum in Bologna', in D. Kop Jansen and E.K. de Bock (eds), *Colecciones Latinoamericanas/ Latin American Collections. Essays in Honour of Ted J. J. Leyenaar.* Leiden: Tetl, 27–42.

Laurencich-Minelli L. (2005) 'Bologna e gli Aztechi fra i secoli XVI e XVII', in A. Lupo, L. Lopez Lujan and L. Migliorati (eds), *Gli Aztechi tra Passato e Presente. Grandezza e Vitalità di una Civiltà Messicana.* Rome: Carocci, 117–130.

Laurencich-Minelli, L. (2007) *Exsul Immeritus Blas Valera Populo Suo e Historia et Rudimenta Linguae Piruanorum. Indios, Gesuiti e Spagnoli in Due Documenti Segreti sul Perù del XVII Secolo.* Bologna: CLUEB.

Laurencich-Minelli, L. (2012a) 'From the New World to Bologna, 1533: a gift to Pope Clement VII and Bolognese collections of the sixteenth and seventeenth centuries', *Journal of the History of Collections* (18 April 2011, http://jhc.oxfordjournals.org/content/early/2011/04/18/jhc.fhr014.short?rss=1) pp. 1–14).

Laurencich-Minelli, L. (2012b) 'Vuelo de plumas en colecciones italianas de los siglos XVI–XVII', in A. Russo, G. Wolf and D. Fane (eds), *El Vuelo de las Imágenes. Arte Plumario de Mexico y Europa.* Mexico City: Instituto Nacional de Bellas Artes, Mexico.

Laurencich-Minelli L. and Domenici, D. (eds) (2007) *Per Bocca d' Altri: Inca, Gesuiti e Spagnoli nel Perù del XVII Secolo. Tavola Rotonda sui Doc. Miccinelli,* Alma Digital Library, Università di Bologna. http://almadl.cib.unibo.it (ACTAS).

Laurencich-Minelli L. and Domenici, D. (forthcoming) 'Domingo de Betanzos' gifts to Pope Clemente VII in 1532–33: tracking the early history of some Mexican objects and codices in Italy'.

Legati, L. (1667) *Breve Descrizione del Museo di F. Cospi.* Bologna, Ferroni.

Legati L. (1677) *Museo Cospiano Annesso a Quello dell' Illustre Ulisse Aldrovandi.* Bologna: G. Monti.

Medica, M. (1992) 'Gli exotica americani donati da Benedetto XIV', in L. Laurencich-Minelli (ed.), *Bologna e il Mondo Nuovo.* Bologna: Grafis, 37–44.

Remesal Antonio (de) O.P. [1619] (1988) *Historia General de las Indias Occidentales y Particular de la Gobernación de Chiapa y Guatemala.* Mexico City: Editorial Porrúa.

Tugnoli-Pattaro, S. (1981) *Metodo e Sistema delle Scienze nel Pensiero di Ulisse Aldrovandi.* Bologna: CLUEB.

Worm, O. (1655) *Museum Vormianum seu Historia Rerum Rariorum.* Amsterdam.

Author

Laura Laurencich-Minelli, Università di Bologna, Bologna, Italy

Progress and evolution and changes to the practice of collecting in the mid-nineteenth century

Michael Thompson

Introduction

It is useful to consider briefly the nature of collecting just before the middle of the nineteenth century. The changes that took place then can then be assessed to see where Henry Christy's activities and his taste for turquoise fit in. Fortunately this can easily be done by studying the display in the fine Enlightenment exhibition in the Kings Library at the British Museum (in 2010). The 24 sections show collections, or parts of them, of the eighteenth and early nineteenth century. The assembled objects are 'static' in having the same single theme – exploration, natural history, science. They do not relate to each other in sequence to demonstrate development or evolution. It is the purpose of this article to show how this came about in mid-Victorian England.

Attention can be confined to three collectors, well known at the time, who were particularly influenced by some of these ideas: Henry Christy (1810–1865) (see Cook, in this vol., pp. 177–182), Lieutenant-General Augustus Pitt-Rivers (1827–1900) (Lane Fox before 1880, see Thompson 1977), and Sir John Lubbock (1834–1913), later Lord Avebury (Thompson 2009). Christy left his collection to the British Museum, Pitt-Rivers' earlier collection of the 1850–1870s is held at Oxford University (his later collection has been dispersed), while part of Lubbock's collection found its way to the British Museum. Christy and Lubbock were wealthy bankers, owners of private banks, while Lieutenant-General Pitt-Rivers had a military career but, although a younger son, he had the good fortune of inheriting an estate in Cranborne Chase, Dorset. The collections we are concerned with were ethnographic assemblages with a substantial prehistoric element, since what were then regarded as modern primitives were often viewed at that time as survivors from prehistoric times.

Progress

The idea of progress, a continuous movement forward for human beings in, at all events, their technology, is at least as old as the eighteenth century. It reached a climax in the next century but this is not a matter we can enter into here. Progress is implicit in events such as the French Revolution, for which the Scottish historian of that event writing in the 1830s put it neatly: 'Generations are the days of toilsome mankind: Death and Birth are the Vesper and Matin bells that summon mankind to sleep and rise refreshed for new advancement' (Carlyle 1833: 182). Each generation has a duty to advance over the previous one, or it happens naturally, so we are in a state of ceaseless movement of advancement.

In the 1840s it is no longer just the historian telling us of the movement but science itself reveals it, that is the fossil record revealed by the geologist. 'The inorganic has one final law gravitation. The organic ... rests in like manner on one law and that is development' (Chambers 1844: 360). Progress is now supported, indeed proved by science, and for Chambers the geological record demonstrated a form of evolution. This new role for natural history, particularly geology, was largely accepted by educated people. An unlikely witness to this was Alfred Lord Tennyson who requested a copy of Chambers' book. His well-known couplet in *In Memoriam* shows his interest in the conflict that existed for nineteenth-century man between Christian love and callous nature:

> Tho' Nature, red in tooth and claw
> With ravine shrieked against his creed (Tennyson 1850: 168–9).

In the 1850s real evolution emerged taking two very different forms: scientific biological evolution based on the natural selection of Darwin and Wallace and the rather more philosophic and all-embracing social evolution of Herbert Spencer (1862), both emerging about the same time at the end of the decade.

Before turning to our trio of collectors to see how each of them was affected by the different forms of evolution, a few words may be said about the much less known today – but very influential at the time – 'evolution' of Herbert Spencer. Although the numerous volumes of his *Synthetic Philosophy* (Spencer 1862–1893) are couched in scientific language, Spencer was not a scientist but first a railway engineer and later assistant editor of *The Economist*. His evolution was a philosophical concept that acted on all things in the organic and inorganic world including human institutions; all matter moved from the 'homogeneous to the heterogeneous' relentlessly (Spencer 1861, 1: 81, 275). It was not based on observation as with Wallace and Darwin but as it was comprehensive, for a collector it was easier to understand and to follow its consequences, so it was possible for Pitt-Rivers to try to arrange his collection of weapons using its principles (Pitt-Rivers 1874). It was said that the widespread optimism among scientists in the period emanated from Spencer (Sully 1884).

Taking the three collectors in order of age, Henry Christy, born 1810, is first and was undoubtedly influenced by the idea of Progress partly through the Great Exhibition of 1851 and from seeing the treatment of the remains as displayed by Thomsen in the National Museum of Antiquities in Copenhagen using the Three Age system: Stone Age, Bronze Age and Iron Age (see Cook, in this vol., pp. 177–182). His collection set out on a geographical basis was a large one. As he died young in 1865 he hardly had time to be influenced by Darwin's *Origin of Species* published in the autumn of 1859 unlike Pitt-Rivers, who lived to 1900 and found an explanation for his serial arrangement of his collection probably inspired by the Great Exhibition and Evolution whether in Darwin's or Spencer's form.

The reasons for seeing how Pitt-Rivers adopted his serial system are to be found in the Great Exhibition to be discussed below: at the time of the Bethnal Green catalogue (Pitt-Rivers 1874: xi) he already saw it in the evolutionary terms of Darwin (skulls) and Spencer (weapons), which seemed a ready explanation for what he had already been doing (Pitt-Rivers 1874; Thompson 1977: ch. 4). The sudden opening of his eyes, as he thought, to the truth of his serial arrangement of the objects tended to give a rather exaggerated adoption of evolutionary ideas in his collection. As the collection is now at Oxford the reader can judge for himself.

Sir John Lubbock, seven years younger than Pitt-Rivers, was indeed of the three the one most deeply imbued with evolutionary ideas since he acquired them directly from Darwin himself. He lived a mile and a half away from Downe, and taken from school by his father to help in the family bank he set himself on a course of intensive self-education. This was mainly in biology, helped by Darwin, who arranged for his father to give him a microscope that he had bought. Throughout his life, Lubbock's main interest was biology although evidently stirred by ideas of evolution as applied to man he studied prehistoric remains in Scandinavia, Switzerland and France that led to the publication of his first two books *Prehistoric Times* and *The Origin of Civilisation*, both running into six or seven editions up to the early years of the twentieth century. The first was primarily prehistoric as the title implies while the second was anthropological, a new form of study, but later publications were mainly biological. Clearly he influenced opinion and ideas of collecting, with evolution taken for granted in the background (Thompson 2009).

The Great Exhibition

So far we have merely mentioned the Great Exhibition of 1851 but its role in changing views deserves closer attention. The exhibition was held for six months in 1851 in Hyde Park and dismantled in the autumn to be removed to Sydenham. While in the park it attracted six million visitors paying a shilling a ticket as it was a commercial venture, the displayed objects being for sale with a special volume giving the prices (*Great Exhibition of the Works of Industry of All Nations 1851*). There were said to be 10 miles of exhibits but the visitor must have been deeply impressed by the building housing them both before and after entering. The Crystal Palace as it was called, built of glass and cast iron with a church-like ground plan with great lines of columns forming aisles on either side, was a veritable monument to the industry of the period.

That the object of the exhibition was to show progress at its height we are left in no doubt by the opening remarks of Prince Albert: 'Gentlemen – the exhibition of 1851 is to give a true test and living picture of the point of development at which the whole of mankind has arrived' (quoted in Ffrench 1950: 52). As important for us was its educational aspect. According to Henry Mayhew, the recorder of the London poor: 'the Great Exhibition, looked at in the true light is, we say once more a huge academy for teaching man the laws of the material universe by demonstrating the various triumphs of the material arts over nature' (Mayhew and Cruickshank 1851: 158). Mayhew wrote a kind of novel describing a family's journey south from the Lake District to see the Great Exhibition, from which the quotation is taken.

We may be confident that the three collectors of whom we speak were duly impressed by the exhibition. Christy said so, while Lubbock, then a youthful pupil of Darwin, had every opportunity to do so working in a City bank. There is some reason to suppose that Pitt-Rivers (Lane Fox) actually started collecting as a result of his experience in the Great Exhibition. In the Bethnal Green catalogue of his collection he stated that he started collecting in 1852, a few months after the Crystal Palace had been removed from Hyde Park (Pitt-Rivers 1874: xiii).

In 1851 Major Lane Fox was an instructor in the Army School of Musketry at Hythe where he wrote an official handbook on the subject (Pitt-Rivers ca.1850). The Great Exhibition contained many muskets and rifles, both British and foreign. Robert Hunt, in his handbook to the exhibition's catalogue in the section on this subject wrote:

> ... we refer back through history ... the inhabitants of this country were in a crude and barbarous state – they went nearly naked – their arms the club and stone hatchet, the spear and bow like those of the New Zealander, the Australian or African savage (Hunt 1851: 244).

Pitt-Rivers traditionally started his collection of objects arranged serially ('the evolution of culture' as he later called it) with muskets and rifles but changed (perhaps in 1852) to ethnographic and prehistoric material. Surely his very idiosyncratic collecting was prompted by thoughts or even the words along the lines of Hunt.

Seventeen years younger than Christy, Pitt-Rivers may have learnt a great deal from his larger collection although his motives for collecting and the arrangement differed markedly. Nevertheless, both tended to move away from simple collecting to digging up actual remains from the ground, no doubt in both cases influenced by what they had learnt from collecting. In the case of Christy, the association with Edouard Lartet, with Christy providing the finance, led to the excavations in the rock shelters in the valley of the Vezère, recorded in the *Reliquiae Aquitanicae* that initiated the study of the Upper Palaeolithic in France. With Pitt-Rivers it was not so much new material as his techniques that have made the four great volumes of *Excavations in Cranborne Case* so famous (Pitt-Rivers 1888–1898). His new museum at Farnham, Dorset was dominated by models of sites of these excavations since his earlier collection had already been given to Oxford.

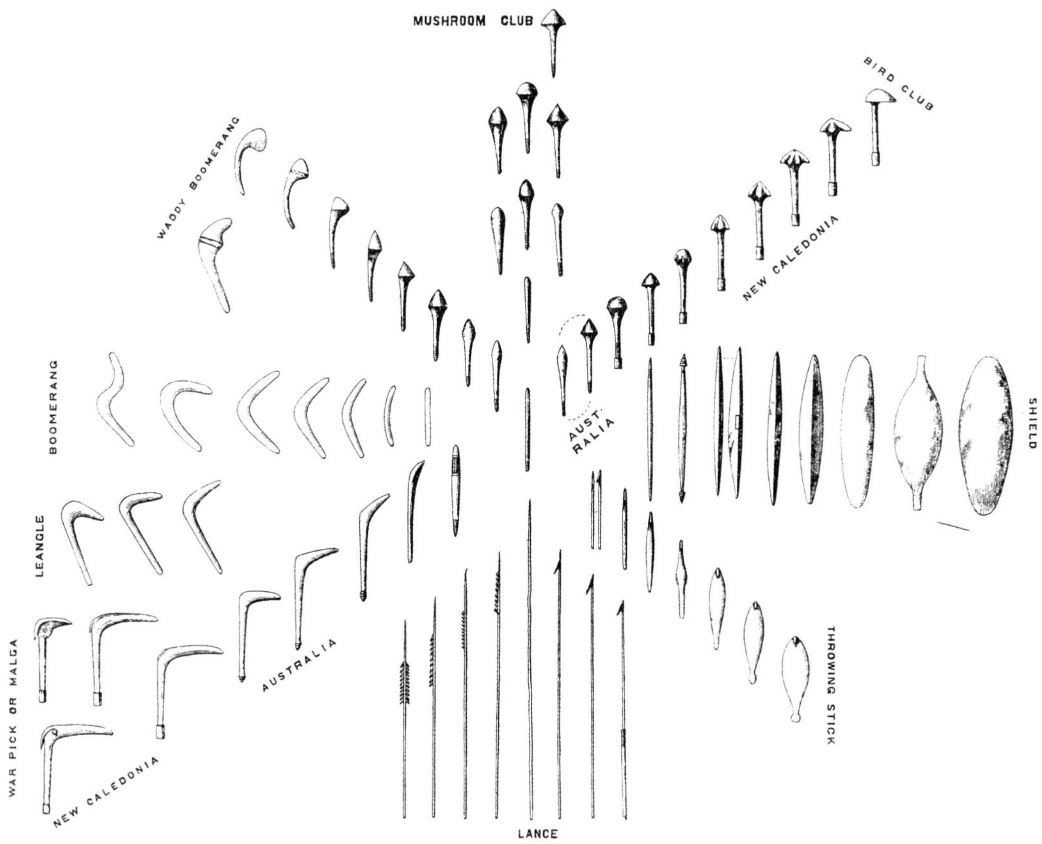

Figure 1. Diagram by Pitt-Rivers to demonstrate the derivation of a range of weapons in the South Pacific from an original stick (centre) by a process of evolution from his lecture on 'The Evolution of Culture'. (From *Proceedings of the Royal Institution of Great Britain* (1875),**7**: 496–520, used with permission.)

The last part of this paper will concentrate on Pitt-Rivers alone leaving the other two collectors behind, partly because his original collection survives more or less intact at Oxford and partly because he himself wrote in the Bethnal Green catalogue (Pitt-Rivers 1874) and in his lectures (Pitt-Rivers 1906) his reasons for and methods of collecting. It has to be remembered that his early collection had been given to Oxford in 1884 while the later collection at Farnham has been sold and dispersed and the building itself turned into dwellings (Thompson and Renfrew 1999).

In this later collection he had less interest in the earlier serial or evolutionary sequences, concentrating rather more on material from his excavations or purchasing ethnographic material on aesthetic grounds such as the Benin bronzes that he acquired at the end of his life (Pitt-Rivers 1900). In his thinking on the earlier material, the assumption was that all objects fitted into a series of development if one could establish the preceding and following objects in the series, and of course this series could be traced back to simple beginnings. The analogy with Darwin's tree of life is at once apparent. This can be seen more clearly in the simpler forms of ethnographic material: Figure 1 taken from his lecture in 1875 to the Royal Institution (reproduced in the 1906 reprint) shows clearly how he thought of weapons in the South Pacific area. From a stick in the middle developed clubs, shields, boomerangs, throwing sticks and so on. The weapons in the diagram are presumably based on actual examples. This kind of development would have been universal just as biological evolution embraced all life.

'The principles of variation and natural selection have established a bond of union between the physical and culture sciences which can never be broken. History is but another form of evolution' (Pitt-Rivers 1906: 24). It has been shown that his original idea that objects could be arranged to show a sequence of development was first prompted by the Great Exhibition of 1851, and that the notion of organic evolution came as a flash of light when it emerged at the end of the decade, apparently making plain that material culture and the organic world followed the same rules.

This extreme position may have been confined to Pitt-Rivers alone but there were others who thought that there might be something in it. Darwin was one of his four sponsors for the Fellowship of the Royal Society (Thompson 1977). The efforts by Huxley and others to persuade the government to accept the Bethnal Green collection for permanent display in the Natural History Museum implied some confidence in what it was intended to demonstrate. Indeed Oxford University surely only accepted the collection because it thought it had special significance, not just as any ordinary ethnographic collection. Since then of course there has been a loss of confidence in the theories of Pitt-Rivers and curators have had to face difficulties and demands for it to be reorganised on a geographical basis.

When the collection had outgrown his house, Pitt-Rivers placed it on loan in the Bethnal Green branch of the South Kensington Museum and set about writing a catalogue. However this was not completed – of the intended four volumes, only two were published: Part 1 Skulls and Part 2

Weapons (Pitt-Rivers 1874). There was an interesting introduction explaining the arrangement:

> Objects are arranged in sequence to show ... the successive ideas by which the minds of men in a primitive condition of culture have progressed in the development of their arts from the simple to the complex, from the homogeneous to the heterogeneous (p. xi).

Here he was quoting Herbert Spencer. He continued:

> Evolution and development are terms which, it is now beginning to be admitted are as applicable to the progress of humanity as to all other mundane affairs. Anthropology according to the more usual acceptation of the term deals with the whole history of human development.

His position is quite clear and with this sort of view the purpose of collecting evidently becomes the intention to demonstrate it. With his inheritance in 1880 of the vast Cranborne Chase estate overlapping four counties, he not only added Pitt-Rivers to his surname under the terms of inheritance but changed his collecting habits. Excavation, that had been an interest since Ireland in the 1860s, had now become his principal activity with ample resources to carry it out. The Farnham museum was dominated by the models of sites excavated: instead of theoretical considerations about the human past he could demonstrate what its remains actually revealed. It is upon this that his main claim to fame rests, and the skill with which he pursued it is vouchsafed for us in the four privately printed volumes of *Excavations in Cranborne Chase* (1887–1898). That his ethnographic interests had not weakened is clear from his acquisition of the Benin bronzes (Pitt-Rivers 1900), although his interest in these was more aesthetic than evolutionary. However, that he had not lost interest in the avid collecting of material to demonstrate the 'evolution of culture' is evinced by some of the surprising lists of purchases made on his travels as Inspector of Ancient Monuments (Thompson 1960: 119):

> Bought off George Sim, naturalist, 14, King Street, Aberdeen

	£	s	d
1 stone ball, Parish of Fyvie, Aberdeen	1	5	0
1 stone cup	1	10	0
Bronze brooch		15	0
1 gold brooch, harp (shape)		5	0
1 silver (brooch), heart shape		4	0
5 silver small (brooches)		13	9
5 flint arrowheads		19	0
1 iron (?bound) spade for turning cakes		2	0
1 Poor Man's candlestick		15	0

There are five other items on the list of equally miscellaneous nature with a total expenditure of £10 17s.The 'naturalist' selling the goods is the Victorian term for scientist, which Darwin or indeed Pitt-Rivers would have called themselves. Some of the objects might be regarded as folk material but we may suspect that they were purchased with a view to fitting into a series of some kind.

Although Pitt-Rivers differed considerably from Christy and Lubbock, who never adopted his serial development, both were certainly deeply influenced by ideas of progress, and evolution in the case of Lubbock. The search for sequence in the development of tools became normal for some years after until Radcliffe Brown brought the anthropologists closer to the people themselves. But we still see the idea of development deeply considered by those involved with ethnographic collections. In the period being considered, collectors were widely separated from the people who had made the objects collected and when the primitives themselves became the aim of study, inevitably theories based on the weapons were less relevant.

References

Carlyle, T. (1833) *Sartor Resartus*. London: n.p.

Chambers, R. (1844) *Vestiges of the Natural History of Creation*. London (published anonymously).

Ffrench, Y. (1950) *The Great Exhibition: 1851*. London: The Harvill Press.

Great Exhibition of the Works of Industry of all Nations 1851. Official descriptive and illustrated catalogue 1851. London: n.p.

Hunt, R. (ed.) (1851) *Handbook to the Official Catalogues*. London: n.p.

Mayhew, H. and Cruickshank, G. (1851) *1851: or The Adventures of Mr and Mrs Sandboys*. London: George Newbold.

Pitt-Rivers, A.H.L.F. (ca. 1850) *Instruction of Musketry*. London: n.p.

Pitt-Rivers, A.H.L.F. (1874) *Catalogue of the Anthropological Collection lent by Colonel Lane-Fox for Exhibition in the Bethnal Green Branch of the South Kensington Museum*, 2 vols. London: n.p.

Pitt-Rivers, A.H.L.F. (1887–1898). *Excavations in Cranborne Chase*, 4 vols (privately printed).

Pitt-Rivers, A.H.L.F. (1900) *Antique Works of Art from Benin* (privately printed).

Pitt-Rivers, A.H.L.F. (1906) *The Evolution of Culture and other Essays*, J.N.L. Myers (ed.). Oxford: Clarendon Press.

Spencer, H. (1862) *First Principles*, 2 vols. London: n.p.

Spencer, H. (1862–1893) *The Synthetic Philosophy*, 9 vols. London: n.p.

Sully, J. (1884) 'Scientific optimism', *The Nineteenth Century* **10**: 573–587.

Tennyson, A. (1850) *In Memoriam*, R. Ross (ed.) (1973). New York: Norton and Company.

Thompson, M.W. (1960) 'The first inspector of ancient monuments in the field', *Journal of the British Archaeological Association* **23**: 103–124.

Thompson, M.W. (1977) *General Pitt-Rivers: Evolution and Archaeology in the Nineteenth Century*. Bradford on Avon: Moonraker Press.

Thompson, M.W. (2009) *Darwin's Pupil: The Place of Sir John Lubbock, Lord Avebury (1834–1913) in Late Victorian and Edwardian England*. Ely: Melrose Press.

Thompson, M.W. and C. Renfrew (1999) 'The catalogue of the Pitt-Rivers Museum, Farnham, Dorset', *Antiquity* **73**: 377–393.

Author

Michael Thompson, Independent Researcher

In pursuit of the unity of the human race: Henry Christy and Mexico

Jill Cook

Introduction

In 1865 the British Museum acquired a large collection of ancient and modern artefacts from around the world bequeathed in the will of Henry Christy (1810–1865). By this time Christy was renowned for his archaeological research with the eminent French palaeontologist Edouard Lartet with whom he was excavating Palaeolithic sites in southwest France. This acclaimed work, which provided new evidence for the length of human antiquity and archaeological data in support of Darwin's theory of evolution, led to Christy's nomination to the fellowship of the Royal Society shortly before his death. However, although the posthumous display and publication of the French artefacts (Lartet and Christy 1865–1875) were important to the development of archaeology and public understanding of human evolution, it was the Mexican and Central American objects collected before, during and after a journey through Mexico in 1856 that captured the popular imagination when opened to the public at his home in Victoria Street, London before being moved to the Ethnographical Gallery of the British Museum (Weinberg 2009). Some authors have regarded this early collecting as little more than the by-product of business travel (Weinberg 2009: 89; Aguirre 2005: 180), but this review of Christy's life suggests that while commerce gave him the means to travel, his religious beliefs, humanitarian views and social network gradually determined his purpose.

Formative years

Henry Christy (Fig. 1) could afford to travel by virtue of his income from the family businesses and his own investments. He was born in Stockwell, South London in 1810, the second son of William Miller and Ann Christy who eventually had nine children. The family was prosperous thanks to William Miller's entrepreneurial skills as a manufacturer and retailer of hats (Beable 1927; McKnight 2000), as well as in banking, cotton manufacturing and property investments. The family were members of the Religious Society of Friends and Henry received a Quaker education pragmatically focused on literacy, numeracy and religious knowledge. An unpublished memoir, written by a family friend long after Henry's death (Baddeley n.d.), records that the religious education included study of Reverend William Paley's *Natural Theology* (1802). Paley (1743–1805) was an Anglican theologian who believed that biological organisms show such intricacy and complexity that they must be part of an intelligent design by a divine creator whom he famously likened to a watchmaker. From this he reiterated the view held by many through the long eighteenth century that knowledge of God could be achieved through the study of nature and, as for his contemporary Charles Darwin, this was probably the source of Henry Christy's lifelong interest in natural history, as well as the belief in the common origins of humanity in a single act of creation. Such beliefs also required humanitarian principles and toleration while begging the question of why people around the world existed in such varied states of economic and cultural development.

This question was one that Henry was to pursue later in life but for the young man a business career was inevitable.

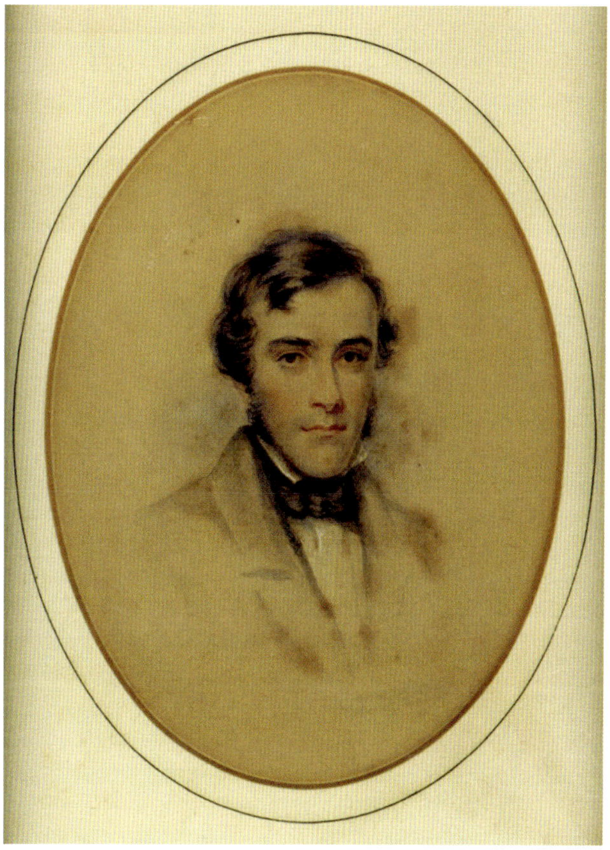

Figure 1. Watercolour portrait of Henry Christy as a young man (British Museum, Eu2006,Drg.1 © Trustees of the British Museum).

After working at the hat factory in Bermondsey and the shop in Gracechurch Street, City of London, he went to Cheshire to run the Stockport factory with his cousin Samuel Christy in 1831. Notebooks (Stockport Heritage Library, B/R/2/33) show him learning the trade on the shop floor and taking responsibility for the supply of materials and the quality of the hats. There are indications that Henry wanted more from life, but the death of his brother William in 1839 made him principal heir to the family fortune and 'circumstances forced me' (Christy 1852: 26 July) to remain in business. William's death affected him deeply (Christy 1852: 24 July).The brothers were close in age and shared an interest in botany. Through William, Henry met the eminent botanist William Hooker (1785–1865) with whom he corresponded until his death in 1865.

Natural history and humanitarian causes

The correspondence with Hooker reveals that Christy had supported an expedition to explore the geography and natural history of the Falkland Islands and wanted advice about sponsoring an expedition to the Rio Doce region of Brazil (Royal Botanic Gardens, Kew, DC16ff82, 6 August 1841). In both instances he had received encouragement from his friend, the eminent Quaker physician, abolitionist and social reformer Dr Thomas Hodgkin (1798–1866). In the case of the Brazilian expedition, Hodgkin was urging Christy's support so that the aims of the Aborigines Protection Society (henceforth APS) might be furthered. Evidence of this connection is important in understanding the origin and development of the ideas that led to Christy's collecting. By 1834 there was increasing concern for the terrible plight of aboriginal peoples in the British colonies. A House of Commons Select Committee on Aborigines in British Settlements led by abolitionist Thomas Fowell Buxton met from 1834 to 1837 to consider what might be done and Hodgkin, 'shocked at the loss of old traditions, with their pointers back to Noah's Flood', set up the APS (Desmond and Moore 2009: 45). The aim of this society was to save aborigines in British colonies from possible extinction and ruthless exploitation, as well as studying them for evidence about the early history of humanity.

Many who had supported the abolition of slavery were appalled that so many indigenous peoples were facing annihilation and joined the APS. The Christy family was part of a significant network of Quakers devoted to peaceful principles, abolition and social reform. Henry was sympathetic to the APS and the later development of his collecting was probably influenced by it. His views on the common origins and unity of humanity expressed in the account of his travel to Mexico (Tylor 1861: 18, 104) and the conclusion to his paper on the significance of his French discoveries with Lartet (Christy 1865: 372) certainly reiterate those of Hodgkin (Rosenfeld 1993: 191–196) and Dr James Cowles Pritchard (1786–1848), which were set out at the 1839 British Association for the Advancement of Science meeting in Birmingham where genocide and the gradual extinction of aboriginal peoples were presented as an irretrievable loss to science (Desmond and Moore 2009: 144–145). Pritchard spoke to the meeting about the unity and shared characteristics of all people. Later he published a book (Pritchard 1843) reiterating this point and adding that all races shared a common origin, probably in Africa, and their shared roots could sometimes be traced through their languages. At Birmingham, he and Hodgkin urged that travellers should collect data so that the common and varying features could be compared to show relationships and origins (Hodgkin 1841). This certainly influenced Christy and placed him among those who deplored opinion which, since the sixteenth century, had sought to bestialise aboriginal peoples and justify slavery or genocide by suggesting polygenetic origins around the world prior to the creation of Adam and the spread of God's chosen people after Noah's flood, based on theological or equally spurious assessments of racial characteristics (Livingstone 2008).

The APS gave rise to the Ethnological Society with its wish for a museum (Anon. 1944) but Christy did not immediately become a fellow. Throughout the 1840s he remained trapped by family and business obligations but found time for causes such as the Quaker Committee for the Relief of the Irish Potato Famine (Hatton 1993: 105–106, 178–179) and the repeal of the Corn Laws. With so much responsibility he did not travel abroad until 1850, when his motive for visiting the Middle East may have been as much personal as scientific.

Middle Eastern tour, 1850

A letter to Hooker dated 25 June 1850 (Royal Botanic Gardens, Kew, DC29ff162) indicates that Christy was about to leave on a journey that would take him to Lebanon, Syria, Turkey and Cyprus. In thanking Hooker for printed instructions on plant and seed collecting (Hooker n.d.), he promises to leave copies 'with consuls en route for the benefit of other travellers'. In botany as with antiquities, both the formal and informal growth of the British Empire provided opportunities for enhancing London's collections but it was essential to improve collectors' expertise (Endersby 2008: 55). Christy was modest about his own botanical abilities but clearly wished to do his best, although it is unlikely that botany was his only purpose in travel. Baddeley (n.d.) relates that despite an upbringing that discouraged interests in the fine arts and poetry, Christy was an admirer of Lord Byron and the proud owner of an autograph letter. Byron was a professional traveller and a man of taste and letters, who became immensely popular after the publication of *Childe Harold's Pilgrimage* published between 1812 and 1818 before the first complete edition in 1825. His works opened up new and exotic areas beyond the familiar lands of the Grand Tour where 'the scene was savage but the scene was new' (Bechler 1996). Bechler suggests that Byron tried to 'launch the souls of his readers, through sympathetic identification, into flight' (Becheler 1996: 59) and this may have worked for Christy who set out in his footsteps to escape from business and the parental home where he still lived unmarried.

No record of the journey exists but it is evident that Christy was a more pragmatic explorer than his romantic hero. While travelling he took some interest in antiquities and in Cyprus bought an important collection of votive sculpture, inscriptions and terracotta figures that he gave to the British Museum in 1852. He also collected artefacts that he saw in use in the villages and towns such as a threshing implement fitted with flint blades, pottery, carpets and textiles. The latter included loop pile towelling (Anon. 1951; Barber 1965). As drying the body with such Turkish towelling is much more effective and luxurious than with linen cloths, Christy sent the fabric to his brother Richard (1819–1898), who had been made a partner in the cotton business in 1843 (Anon. 1951). Machinery to manufacture towels was quickly developed and installed in the family's new factory at Fairfield, Stockport, originally intended for hat-making but not commissioned because of a downturn in trade.

Despite the vicissitudes of fashion and material resources, Christy wealth was increasing but the new product was even more momentous in that it took Christy to the 1851 Great Exhibition at the Crystal Palace in Hyde Park where the towels won royal approval and a first prize medal (Anon. 1951). More significantly, the exhibition showed Christy how objects from around the world revealed the varying state of technological development of different peoples. Such shows 'helped to define the nature and meaning of progress and made it part of the intellectual discourse of the age' (Keuran 1989: 26), placing an emphasis on the display of material culture. Christy recognised the value of such exhibitions for public interest and research. He saw how objects might be used as the empirical basis for studying past and present peoples. Here was the basis of a new subject that put museums in the forefront and was perhaps the catalyst for Christy to commit his activities to both the discourse of ethnography and the evidence on which it would be based.

Scandinavian influences

Early in the following year Christy resigned from the business now managed by his younger brothers, to whom his father had given partnerships to provide them each with a living. Henry could back out until he inherited. An obituary states that he wished 'to afford himself the unshackled leisure to follow his favourite pursuits ... to devote his time and ample means to the achievement of science' (Anon. 1865), but such clarity of purpose is not yet so evident. He was perhaps depressed following the death of his younger brother John (b. 1812) in April 1851, with news of which he wrote to Hooker (Royal Botanic Gardens, Kew, DC31ff103, 6 April) and a diary for 1852 opens gloomily on New Year's Day with the words 'a dull morning ... physically, mentally and atmospherically' (Christy 1852: 1 January). From 6 February to 8 June there are no entries, after which he is en route to Scandinavia where on his birthday at Scholt, Norway, he writes of his disappointment with his life 'although it had not been wholly in vain as regards the good of others' (Christy 1852: 26 July). He goes on to hope that his future will be 'better-directed and with more purpose!' and refers to 'the eventful step I took in retiring from the business while it cuts off many influential associations, limits and narrows one's field of action and shears one of much power, is to myself many ways so full of advantage' (Christy 1852: 26 July).

The steamer trip around the coast of Norway seems to have been a plant-collecting holiday in the footsteps of his late brother William (Christy 1852: 14 July), whose friends he was much moved to meet at Kaarfjord. After Norway, his journey continued to Sweden and Denmark where he had encounters that encouraged his interests in ethnographic collecting and suggested how he might do it.

In early October he visited the Stockholm museum of anatomist Professor Retzius, who was measuring skulls from different populations to ascertain their character, ancestry and relationships (Christy 1852: 7 October). He then went to Denmark and visited the Thorwaldsen and Ethnographic museums. At the former, he notes 'a collection of ravishing delight wonderful for its extent as the work of one man' (Christy 1852: 15 October) and if this showed him what could be done by an individual, his meeting with Professor Christian Thomsen (1788–1865) at the latter suggested how it should be done. Thomsen introduced Christy to the collection representative of Denmark's distant past organised chronologically and typologically under the heading Stone, Bronze and Iron Age following the Three Age system that he had helped to pioneer (Rowley-Conwy 2007: 37–47).

Christy's view of what he saw led him to think that all peoples advanced through these stages and their material culture would show which stage they had reached. Now he had an idea of how to document technological change through time and its stage of development in different places. This must have resonated with the imperative to preserve and record native peoples that he had acquired through Hodgkin, the APS and the Great Exhibition. It is apparent as the basis for the opinions about Mexica culture observed on the journey to Mexico where he saw the remains of a remarkable civilisation that had no use of iron, and despite some use of bronze had advanced by the use of stone tools (Tylor 1861: 102). He probably saw some of these tools and turquoise mosaic masks in Copenhagen and certainly made the acquaintance of the ethnographer Carl Steinhauer at this time. Now he began collecting artefacts seriously but through purchase rather than fieldwork and probably with the help of Steinhauer, whom he subsequently employed to catalogue his collection (Steinhauer 1862).

Developing a network

After a gentlemanly visit to Venice in 1853 to improve his drawing under the tutelage of William Cooke (1811–1880), the artist and gardener who became well known for his marine paintings and botanical illustrations, Christy was elected Fellow of the Ethnological Society and the Royal Geographical Society in 1854. He was proposed for both by Hodgkin. The Ethnological Society had emerged from the APS with the purpose of studying the natural history of humanity in order to address the question of whether the various peoples of the world had common origins and were simply in different states of development or had been created separately (Rosenfeld 1993: 195) and wanted to see a 'museum illustrative of the varieties of mankind, and the arts of uncivilized life' (Anon. 1944: 25). Christy was now beginning to make these objectives his own. In a letter to Hooker describing the cypresses he had seen in the gardens of Fontainebleau during a visit to Paris, which also took in a world trade fair, Christy says that he wishes Sir William 'could interest someone ... in matters non vegetable which relate to the less civilized countries especially of the Aborigines of the Southern Pacific', adding that in the

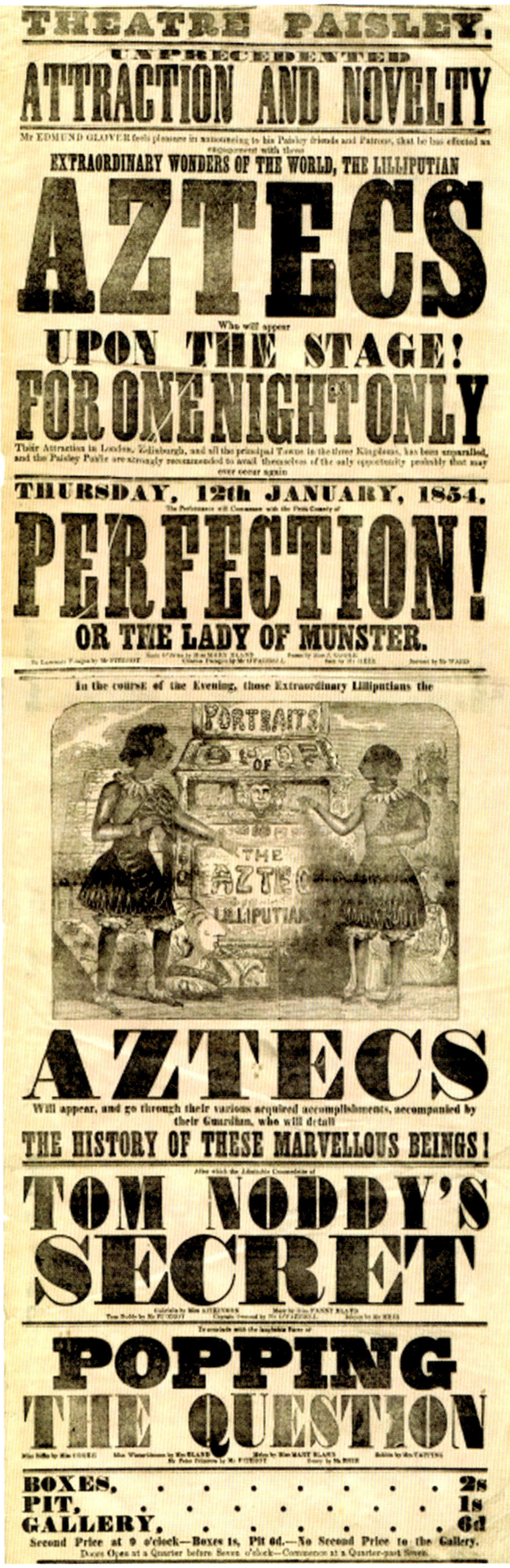

Figure 2. Poster for the *Aztecs* on tour around Britain in 1854 (British Museum, Am, EPH-AOA,Prt.9, © Trustees of the British Museum).

absence 'yet' of an Ethnographical Museum there was a risk that many objects would disappear as a result of 'the diffusion by our commerce of the productions of civilized Man' (Royal Botanic Gardens, Kew, DC35ff90, 15 October 1855).

At this date he was preparing for his visit to the Americas and may have seen the exhibition of two Mexican children billed as the *Aztec Lilliputians* (Fig. 2). The children became the focus of a heated debate about their origins and capabilities at the Ethnological Society (Aguirre 2005: 112–118; Altick 1978: 286). The characteristics of their skulls were variously described as deviant, atrophied or, more accurately, the consequence of dwarfism. The fervour of the discussions must have interested Christy. A letter to Anders Retzius shows that he was already acquiring Mexica antiquities with a view to serving scientific needs:

> I have for your museum 4 skulls of the Ancient Mexican race (Aztec) found with a number of Ancient idols and terracotta implements of Mexican worship which were buried at the time of the Spanish Conquest. I will send them to Dr Hodgkin's care for you and will ask you to be kind enough to allow the Secretary of the Ethnological Society to examine them before taking them to Sweden (Royal Swedish Academy of Sciences, Retzius archive incoming letters Vol.E1:2, 7 August, 1855).

Mexico

Christy left England in December 1855 for Cuba via New York on a journey that subsequently took him to Mexico before returning overland through the United States and further north through Canada to Hudson Bay. His diary for the voyage on the *North Star* to Cuba survives (Christy 1856) but notes only the weather, the state of the sea and related discomforts. What is known about the rest of the journey comes from letters to his parents and the book written by Edward Burnett Tylor (1832–1917) (1861), his companion on the journey who was later to become keeper of the University Museum at Oxford (1883) and first professor of anthropology at the university in 1896.

According to Tylor's account he met Christy by chance on an omnibus in Havana (Tylor 1861: 1). As Tylor was a Quaker whose family had a brass foundry in Bermondsey, it is unlikely that he was not already acquainted with the Christys. Later in the book, in commenting on the probable size of the Aztec population at the time of the Spanish conquest, he notes that 'when we left England, we both doubted the accounts of the historians' (Tylor 1861: 73), which suggests that they had both planned and prepared for the trip. Was it coincidence that brought them together? Tylor had been travelling in the USA before arriving in Cuba but agreed to go to Mexico, a country in the grip of civil strife, on an impulse. Christy does not mention their remarkable meeting to his mother until they are in Mexico but does not need to introduce him (Christy 1856). Whatever was behind this, Tylor states at the outset that the book is composed from notes and letters by both of them (Tylor 1861: 1). This is confirmed by the similarities between the opening chapter and letters to Christy's mother (Christy 1856). The nature of Christy's outward diary suggests it was intended for a travelogue, implying that a book was planned and suggesting that Tylor might have agreed to travel as his amanuensis but had needed for whatever reason to disguise the plan. The journey was certainly no spur of the moment decision, as Christy had a 'drawer-full of letters of introduction' that produced 'an immediate crop of acquaintances' (Tylor 1861: 40).

Anahuac

The choice of *Anahuac* as the main title for Christy and Tylor's journal of their Mexican journey immediately establishes their interest in the indigenous inhabitants by using a Mexica word meaning 'by the waterside'. This was the first name applied to the Valley of Mexico where the Mexica established their towns beside the lakes and homage to a people for whom the pair often express their admiration. Their intention to study them is confirmed early in the narrative when they arrive in the Highlands and are 'anxious to make sure that we have really got into the land of Aztec civilization, Mr Christy gets down from the diligence, and ... returns in triumph with a broken arrowhead of obsidian' (Tylor 1861: 35).

Tylor also notes that Christy was 'always on the lookout for a new seed or plant' (Tylor 1861: 52) and on the return journey they 'loaded the horses with the bulbs of some orchids as were still in a dry state, and would safely travel to Europe' (Tylor 1861: 314). Correspondence with Hooker shows that Christy germinated many of the seeds himself and gave specimens to Kew (Royal Botanic Gardens, Kew, DC38ff108, 5 June 1858).

In describing the Valley of Mexico, they write of how by inhabiting a group of swampy islands in the lake of Tezcuco, the Aztecs 'became a great and powerful nation' (Tylor 1861: 41) and subsequently note that

> they raised themselves, independently, to the extraordinary degree of culture which distinguished them ... The curious distribution of knowledge shows plainly that they found it for themselves, and ... not ... by transmission. We find a wonderful acquaintance with astronomy, even in such details as the causes of eclipses (Tylor 1861: 103).

This observation satisfied Christy's view that all peoples pass through the same stages of development at different rates and in different periods. The reference to the curious distribution of knowledge is amplified later when they reflect on the absence of iron tools and the fact that the civilisation advanced so far in its Stone and Bronze ages (Tylor 1861: 102, 138–139, 187). This is probably the first test for the Three Age system that Christy had learned in Copenhagen as a theory of development rather than a classification, and although evidently inadequate, he accepts it as a convenient rule of thumb. Similarly, the peopling of the earth as recorded in Genesis is accepted as the origins of the Mexica (Tylor 1861: 18, 104). In the mid-nineteenth century, this view indicated that Christy and Tylor did not subscribe to the view of polygenetic origins across the world before Adam, which was used by others to explain ethnic characteristics and justify views on racial superiority (Livingstone 2008). Arguing for monogenetic origins was not indicative of regressive Biblicism. Indeed, it was from such ideas argued out in the debates on slavery and race relations in the first half of the nineteenth century that the Darwinian concept of common descent was to emerge (Desmond and Moore 2009: 243). In this respect, although many of the views expressed in *Anahuac* may seem arcane to the modern reader, they reflect contemporary opinion that Christy had absorbed through his earlier travels and associations.

In addition to describing the various peoples of Mexico and Central America, as well as their archaeology and customs, Christy and Tylor report on the condition and material culture of the modern population. Not long after their arrival, they record Holy Week celebrations and evidently enjoy inspecting the representations of Judas as they were made, displayed, then fervently burned (Tylor 1861: 49–54). Elsewhere they are less tolerant of the Catholic Church and Spanish colonial rule when observing the condition of the people, noting vehemently the splendour of the churches and asking 'Is it possible that they can belong to these wretched, filthy cottages' (Tylor 1861: 36). Their Quaker ideals are evident in discussions about such matters as slavery (Tylor 1861: 13–14). They were also alert to injustice, such as chaining and parading all criminals as if murder and theft were equal crimes, and the prejudice revealed by 'the undisguised contempt with which the Indians have been treated for ages by the whites and mestizos' (Tylor 1861: 198). As in Copenhagen and Cuba, Christy made sure they visited a prison, his interest probably instilled by his mother's support for Elizabeth Fry, the champion of prison reform in Britain (Tylor 1861: 245–249). Free trade is also advocated as a means of relief for the poor, for whom the protection of the cotton industry made coarse cotton garments expensive (Tylor 1861: 1–6, 264, 300).

Conclusions

Weinberg (2009: 92–95) and Caygill (in this vol., pp. 183–196) have summarised the acquisition of Christy's Mexica and Central America collection that still attracts huge popular and academic attention. Discussion of these objects in *Anahuac* falls short of being a scholarly antiquarian study of the Mexica or Central American cultures, but shows the combination of religious, humanitarian, botanical and ethnographical ideas that formed the character and outlook of its authors, particularly Christy. It is perhaps best regarded as a report on the contemporary condition of the people and the need to preserve their past through collecting before modern life obliterated them.

This view of Christy's Mexican travel contrasts with Aguirre's (2005) argument that the period between 1820 and about 1860 represents a period when Britain attempted to extend its empire informally by developing free trade, diplomatic and cultural relationships in an effort to dominate Mexico following its independence from Spain in 1821. In a persuasive critique he describes how Victorian society acquired, presented, distorted and decontextualised the past cultures of Mexico and Central America between 1820 and about 1860, but relegates Henry Christy to a footnote (Aguirre 2005: 180, n. 85) on the grounds that he was a latecomer to the scene. More to the point, Christy does not fit Aguirre's stereotype of a wealthy businessman and collector complicit in an imperialist 'plot' (Aguirre 2005). Unlike Aguirre's main villain, William Bullock (1773–1849), who made a fortune from Mexican silver and exploited its people and antiquities (Aguirre 2005: 1–19), Christy made no money and brokered no business for himself or Britain during his Mexican expedition. He deplored the condition of the museums caused by war with America and civil strife rather than revelling in the opportunities this unregulated state might provide for the collector (Tylor 1861: 222), and shows a respect for the indigenous people not even found in Mexico at that time (Tylor 1861: 60–61). His fascination for Mexica culture and its implications for how complex societies develop probably encouraged his search for the turquoise that he rescued from oblivion in private European cabinets (Caygill, in this vol., pp. 183–196).

Christy's Mexican adventure undoubtedly marks the turning point in his career when his understanding of ethnography crystallised into the subject recognised today. Like many in his intellectual circle, his religious background and humanitarian views on slavery and race relations led him to try to prove the unity and common origins of all humanity. When Charles Darwin provided the mechanism of natural selection to show how this might have come about, Christy took his search for evidence to the remains of an even deeper past where he at last found his purpose and the confidence to publish in his own right.

Acknowledgements

I am grateful to my colleagues Marjorie Caygill, Colin McEwan, Jonathan King and Max Carocci at the British Museum, as well as Norman Hammond for their helpful comments on this paper. I am also indebted to the patience of the librarians and archivists from the Royal Botanic Garden, Kew, Gloucester Record Office, Stockport Heritage Library and the Quaker Library inter alia who have helped me find the disparate sources from which Christy's life has been constructed. Tony Gant from the Friends Meeting House at Wandsworth where Christy was a member provided insights into Quaker conduct and beliefs. Libs Bailey, a Christy descendant, helped with the family history and Chris Heal has been generous with information from his thesis about the hat business.

References

Aguirre, R.D. (2005) *Informal Empire: Mexico and Central America in Victorian Culture*. Minneapolis and London: University of Minnesota Press.

Altick, R.D. (1978) *The Shows of London*. Cambridge, MA and London: Belknap University of Harvard Press.

Anon. (1865) 'Obituary of Henry Christy Esq. FGS, FLS', *Proceedings of the Linnean Society* **1:** xxv–xc.

Anon. (1944) 'Foundation meeting of the Ethnological Society, 1843', *Man* **44**: 24–25.

Anon. (1951) *100 Years of the Royal Turkish Towel*. Manchester: Wm Christy & Sons Ltd.

Baddeley, W. St Clair (n.d.) *Memoir of Henry Christy, 1810–1865*. British Museum Centre for Anthropology archives (unpublished).

Barber, W. (1965) *The Chronicles of Canal Street from BC (before Christys) to 1968*. Manchester: J.Christie Miller.

Beable, W.H. (1927) *The History of Christys. The History of Hats*. Manchester: Christys.

Bechler, R. (1996) 'Lord Byron's grander tour', in C. Chard and H. Langdon (eds), *Transports: Travel, Pleasure and Imaginative Geography, 1600–1830*. Newhaven, CT: Yale University Press, 49–74.

Byron, Lord G.G. (1825) *The Childe Harold's Pilgrimage*. London: Dugdale.

Christy, H. (1852) 'Diary of a Voyage to Norway, Sweden and Denmark, 1852'. British Museum Centre for Anthropology archives (unpublished).

Christy, H (1856) 'Diary of the Voyage aboard the *North Star* from New York to Havana, 1856'. Gloucester Record Office, Baddeley Papers item 136 with letters to Ann Fell Christy (unpublished).

Christy, H. (1865) 'On the pre-historic cave dwellers of southern France', *Transactions of the Ethnological Society of London* **3**: 362–372.

Desmond, A. and Moore, J. (2009) *Darwin's Sacred Cause: Race, Slavery and the Quest for Human Origins*. London: Allen Lane.

Endersby, J. (2008) *Imperial Nature: Joseph Hooker and the Practices of Victorian Science*. Chicago and London: Chicago University Press.

Hatton, H.E. (1993) *The Largest Amount of Good: Quaker Relief in Ireland, 1654–1922*. London: McGill-Queen's University Press.

Hodgkin, T. (1841) 'On inquiries in the races of man', *British Association for the Advancement of Science Report, 11th Meeting* **52**: 53–54.

Hooker, W.J. (n.d.) *A Few Plain Instructions for Collecting and Transporting Plants in Foreign Countries*. Richmond: Royal Botanic Gardens, Kew.

Keuran, D.K. van (1989) 'Cabinets and culture: Victorian anthropology and the museum context', *Journal of the History of Behavioural Studies* **25**: 26–39.

Lartet, E. and Christy, H. (1865–1875) *Reliquiae Acquitanicae: Being Contributions to the Archaeology and Palaeontology of Périgord and the Adjoining Provinces of Southern France*. London: Williams & Norgate.

Livingstone, D.N. (2008) *Adam's Ancestors: Race, Religion and the Politics of Human Origins*. Baltimore, MD: Johns Hopkins University Press.

McKnight, P. (2000) *Stockport Hatting*. Stockport: Stockport Metropolitan Borough Council Community Services Division.

Pritchard, J.C. (1843) *The Natural History of Man; Comprising Inquiries into the Modifying Influence of Physical and Moral Agencies on the Different Tribes of the Human Family*. London: Balliére.

Rosenfeld, L. (1993) *Thomas Hodgkin: Morbid Anatomist and Social Activist*. London: Madison Books.

Rowley-Conwy, P. (2007) *From Genesis to Prehistory: The Archaeological Three Age System and its Contested Reception in Denmark, Britain, and Ireland*. Oxford: Oxford University Press.

Steinhauer, C.L. (1862) *Catalogue of a Collection of Ancient and Modern Stone Implements and of other Weapons, Tools and Utensils of the Aborigine's of Various Countries, in the Possession of Henry Christy*. London: n.p.

Tylor, E. B. (1861) *Anahuac: or Mexico and the Mexicans, Ancient and Modern*. London: Longman, Green, Longman & Roberts.

Weinberg, C.E. (2009) 'A collection within a collection: Henry Christy's ancient Mexican treasures at the British Museum', *Tribal Art* **52**: 84–97.

Author

Jill Cook, Department of Prehistory and Europe, The British Museum, London, UK

Henry Christy, A.W. Franks and the British Museum's turquoise mosaics

Marjorie Caygill

Introduction

Of a visit to the vast, jumbled, collection of Bram Hertz, the then owner of three Mixtec-Mexica turquoise mosaics, a columnist wrote in *Blackwood's Edinburgh Magazine* in May 1852:

> His room is like a pawnbroker's shop ... There are wardrobes richly inlaid, with a genealogy as carefully kept as the pedigree of a race-horse ... Then there are little ivory cabinets, and screens magnificently embroidered, all with their respective stories – there being no article that depends entirely on its intrinsic merits, but borrows a great part of its interest from the adventures it has gone through.

Like Hertz, the main aim in researching this paper has been to see if it was possible to find out more about the provenance (the 'adventures') of the British Museum's nine turquoise mosaics than is given in Elizabeth Carmichael's summary (1970: 35–36).[1]

Some of the lack of clarity in establishing the provenance of the Museum's mosaics can be accounted for by the involvement of the curator, collector and benefactor Augustus Wollaston Franks (1826–1897). Much of Franks's own vast collection, which he gave to the British Museum, and also many of the objects that he acquired in his official capacity as keeper (head) of the Department of British and Medieval Antiquities and Ethnography, are today lacking in provenance, among them the nine turquoise mosaics. In his will, Franks left his private notebooks, manuscripts and letters to his assistant and successor Hercules Read (1857–1929) who, he wrote, 'will know what best to do with the latter' (Caygill and Cherry 1997: 358). Although some, but not all, of the official papers have survived, virtually all of Franks's private papers have now disappeared.

First, the sometimes confusing connection between Henry Christy (1810–1865), A.W. Franks and the mosaics should be explained. Christy, a friend of Franks, died unexpectedly in 1865 and bequeathed his largely prehistoric and ethnographic collection, and a substantial sum of money (£5000) for its upkeep and expansion, to four trustees: the botanist Sir Joseph Hooker (1817–1911), the banker, politician and scientific worker Sir John Lubbock (later Lord Avebury) (1834–1913), the pharmacologist Daniel Hanbury (1825–1875) and the British Museum's A.W. Franks. Franks had been appointed in 1851 with the task of curating the Museum's somewhat meagre British collections. Fifteen years later he was determined fully to establish the worldwide scope of the British Museum's collections. Although the Museum had possessed ethnographical material since its foundation in 1753, and subsequently received important donations, no money was spent on purchases. Franks later wrote:

> When Mr Christy told me that he had appointed me a trustee he said that if I wished the main collection to go to the Museum I must fight for it. I did fight and the collection was secured for the National Museum (Caygill and Cherry 1997: 319–320).

So, persuaded by Franks, the other Christy Trustees rapidly agreed to the collection's transfer to the British Museum on certain conditions, including exhibition and the publication of a catalogue. If these conditions were judged not to have been met, the Christy Trustees retained the right to withdraw the bequest. Although the British Museum was managed by a group of Trustees answerable to parliament and largely funded through the Treasury, the 'Christy Collection' was to a large extent a separate entity, with its own funding, its own Board of Trustees and, until the mid-1880s, when space was released at Bloomsbury by the departure of natural history, its own building – Christy's home at 103 Victoria Street. Christy's original collection now began to expand rapidly by means of purchases made from the Christy Fund and also through donations from Franks and others. Thus, while all nine of the British Museum's turquoises are registered as part of the 'Christy Collection', the circumstances of their acquisition vary: three were bequeathed by Christy himself; three were bought with money from the Christy Bequest; and three were bought by Franks and donated.

Christy Bequest

Christy had a particular interest in Mexico following his travels there with Edward Tylor (1832–1917) in 1856 (Tylor 1861) (see Cook and Thompson, in this vol., pp. 177–182 and 173–176 respectively). It is not surprising, therefore, that when three turquoise mosaics from Mexico appeared in the saleroom in 1859 as part of the Bram Hertz collection, now owned and sold by Joseph Mayer of Liverpool, Christy was prepared to spend a total of £113 on their purchase. These, the three mosaics in the Christy Bequest, are: (1) a sacrificial knife with a flint

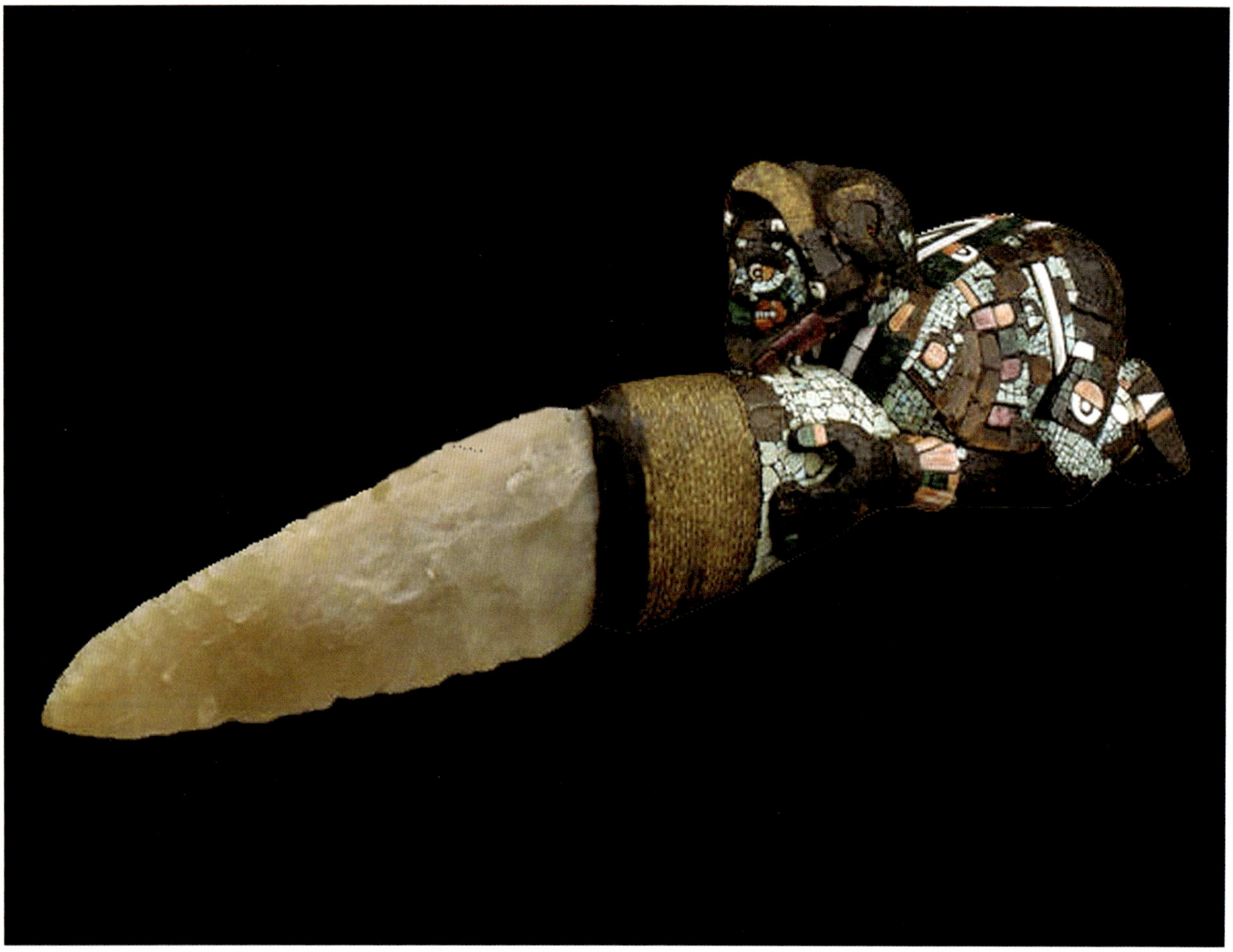

Figure 1. Sacrificial knife with a flint blade (British Museum, AOA, Am St.399, © Trustees of the British Museum).

blade (Fig. 1); (2) a mask, possibly of Xiuhtecuhtli (Fig. 2); and (3) a human skull, perhaps representing Tezcatlipoca (Fig. 3).

Bram Hertz is a shadowy figure. *Blackwood's Edinburgh Magazine* (1852: 603) includes a rather unflattering description – 'a little, round, oily-faced German ... remarkably fond of tobacco'. The writer comments:

> As we go shivering home, we will diverge for a moment into the most curious repository of nick-nacks the world contains – being the gatherings of thirty years, at a cost of thirty thousand pounds. We call in Argyll Street, and are civilly received by Mr Hertz, the proprietor of the collection ... Finally, he gives you a key, and sends you off, under the guardianship of his maid, to a house in Great Marlborough Street, which you will find filled, from cellar to garret, with works of a still more valuable description.

He was Jewish and the 1851 Census records that he was born in Hanover around 1794. He had connections with Frankfurt am Main, his wife's birthplace. He seems to have been operating in London in the early 1830s, possibly after the birth of his daughter, the educationist Fanny Hertz (1830–1908), in Hanover (*Oxford Dictionary of National Biography*), although Fanny's biographer gives 1837 as the date he settled in London. He is initially listed as a diamond merchant – he catalogued the collection of pearls and precious stones belonging to Henry Philip Hope (d. 1839) in 1839. He appears to have prospered, and, acting as both dealer and collector, was involved in a series of auction sales in the 1830s and 1840s. He published a catalogue of his collection (Hertz 1851) – possibly the first appearance of the mosaics in print since the seventeenth century – and in this (Hertz 1851: iii) reference is made to the collection having been acquired 'within the last twenty years'. In this catalogue the mosaics were numerically overshadowed by 3,500 other objects, and indeed they do not appear until the final page, but nevertheless it is evident that they were regarded by their owner as of particular importance. An undated summary catalogue of his 'museum', listing over 3,700 objects including the three turquoise mosaics, was produced in 1855 or later (Hertz n.d.). At some point there now enters Joseph Mayer (1803–1886) of Liverpool, rich silversmith and collector, a friend of Franks, who seems to have coveted the Hertz collection. There were three auction sales of Hertz material in the 1850s: 1854 (Sotheby's), 1857 (Phillips) and 1859 (Sotheby's).

The turquoise mask, knife and skull appear in the catalogue of Hertz's first Sotheby's sale (Sotheby and Wilkinson 1854), which took place in May 1854 and was described at some length in the *Morning Chronicle* of 29 May 1854. In the sale catalogue they were included with 650 classical objects, 24 Indian bronzes, seven Chinese antiquities and half a dozen Peruvian pots, most of which give the impression of not being of the first rank. The mosaics were, however, thought

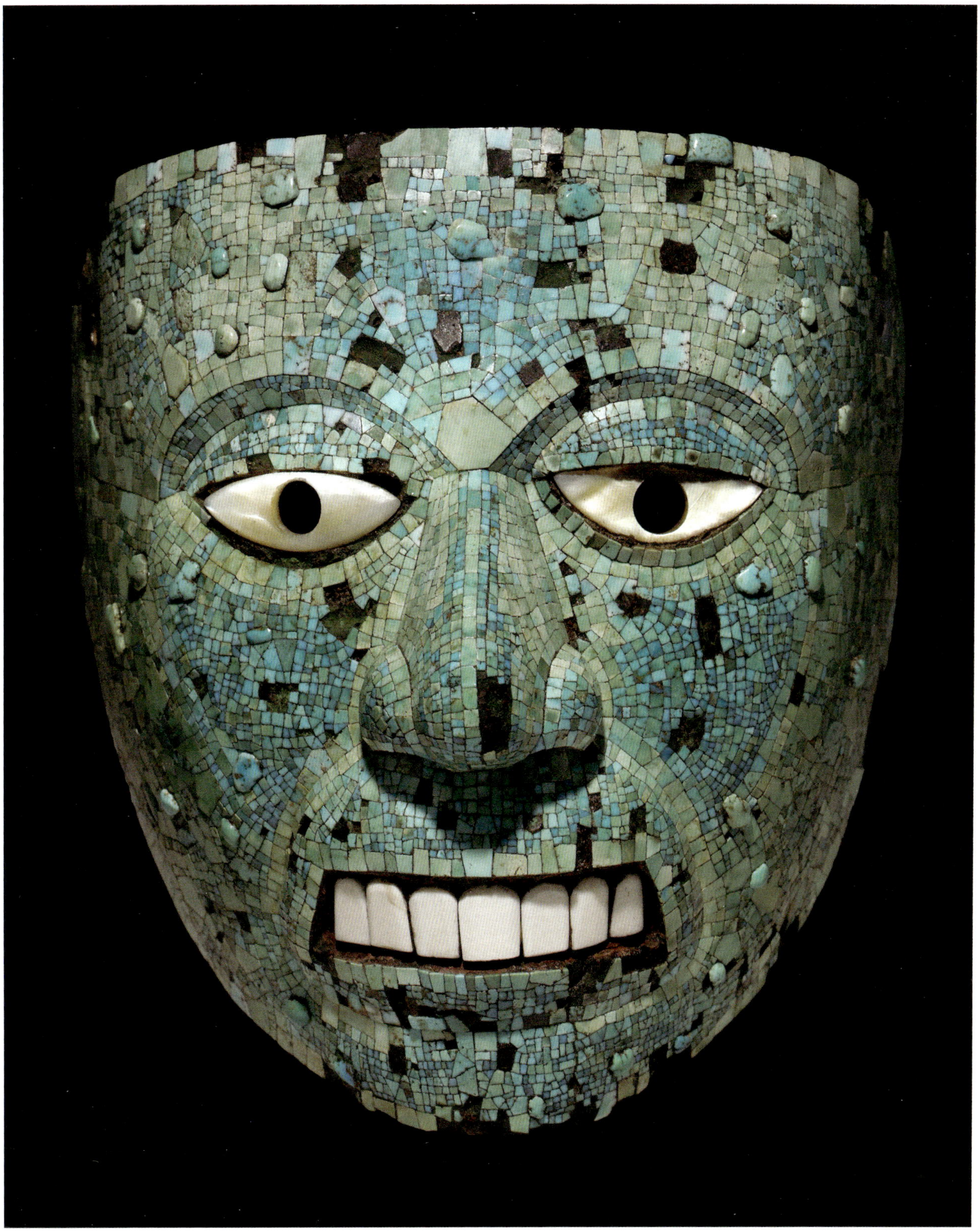

Figure 2. Mask, possibly of Xiuhtecuhtli (British Museum, AOA, Am St.400, © Trustees of the British Museum).

to be of sufficient importance to be emblazoned on the front cover of the catalogue, in ornate capital letters, described as 'Mexican Antiquities of the Highest Interest, Consisting of a Mask and Sacrificial Knife, and a Human Scull [sic] inlaid with Turquoise'. Hertz wrote:

> These three objects may be considered as unique, there is no record in any Catalogue of the great public museums, that such monuments of the ancient Mexican people are in existence (Sotheby and Wilkinson 1854: 19).

In an annotated sale catalogue in the British Library, the group of three mosaics is shown as having been acquired by what appears to be a Mr Roussey or Roussell for 14 guineas (the handwriting is difficult to read). However, quite who was

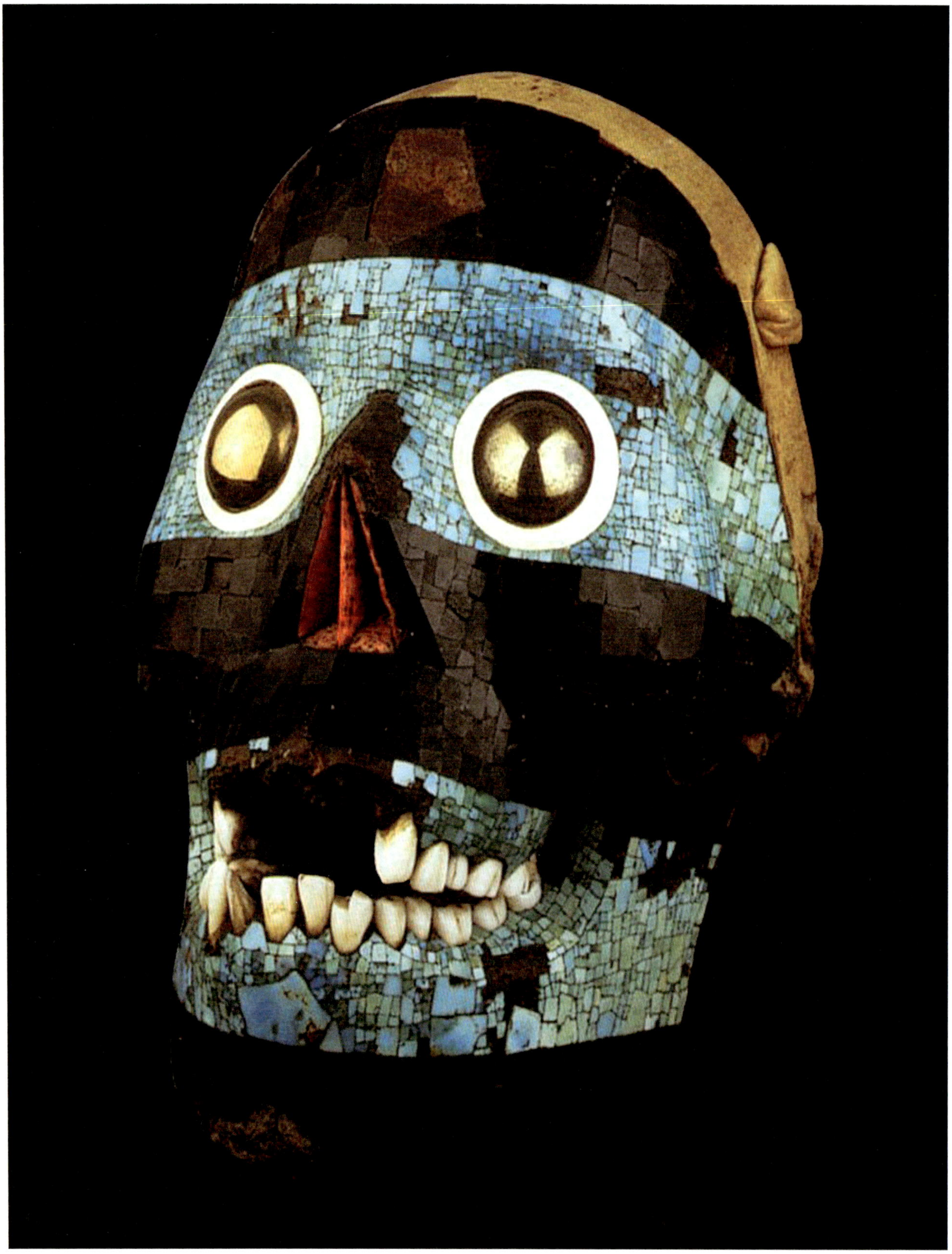

Figure 3. Human skull, perhaps representing Tezcatlipoca (British Museum, AOA, Am St.401, © Trustees of the British Museum).

buying for whom at this sale is now a little obscure. Although the names of over 30 buyers are given in this catalogue, it may be that some of the objects were bought in or were acquired by nominees buying on behalf of Mayer and a consortium of Liverpool businessmen said to be acting with the intention of donating the collection to their city (Gibson 1988: 11). The turquoises were in Hertz's possession in July 1856 when he wrote to Mayer offering his collection for sale, including 'extremely rare' 'specimens of Mexican, Chinese and Indian workmanship'. The deal was concluded by 15 January 1857 when removal arrangements were discussed (Liverpool Record Office).[2]

It has been suggested that Mayer may have over-reached himself financially and had to recoup his outlay. At all events, on 18 March 1857, *The Times* reported that after a 'quarter of a century' and because of ill health, Hertz was selling his collection prior to leaving England in search of a warmer climate. This second sale, of around 450 *objets d'art* and *vertu*, not including the mosaics, took place that month at Phillips with Mayer listed as the vendor (Phillips 1857). *The Times* had noted on 18 March 1857 that the archaeological part had already been sold by private contract.

Henry Christy now puts in an appearance, writing to Hertz on 2 February 1859, immediately prior to the sale, presumably asking about the provenance of objects that interested him, and receiving a reply on 5 February from Frankfurt setting out what Hertz could recall (Carmichael 1970: 37). Hertz's letter is rather vague but it is the best information we have about the provenance of the three mosaics. Although Hertz dated his letter 1858, a note in the Museum's archives states that two postmarks on its accompanying envelope are in fact for 1859 (BM, Asia).[3]

Hertz states that the mask and the sacrificial knife belonged to 'a celebrated collection at Florence', of which he had forgotten the name, the sale of which, he thought, 'took place some twenty odd years ago' – that is, in the 1830s. It has not so far been possible to establish which collection since, given the turmoil of the times, there is a long list of potential vendors among the Florentine nobility. Hertz said that he did not acquire the knife and mask direct from Italy but picked them up in London. He writes that the knife came from 'Mr Pratts in New Bond Street who brought this also from Venice' – which is a little confusing since in the previous paragraph of his letter he stated that both mask and knife 'belonged to a celebrated collection at Florence'. This is probably the infamous Samuel Luke Pratt of 47 New Bond Street, proprietor of an antique furniture business, today remembered particularly for his sales of fake antique armour (Watts 1992).

The mask, Hertz said, 'was acquired by a certain Descriever who at the time was travelling as courier with an English familly [sic] and settled afterwards as a curiosity dealer in London'. No 'Descriever' is listed in the 1841 Census. However, there is a Francis 'Deschryver', born abroad about 1801, then resident at 3 Great Newport Street, who is described as an 'antique furniture dealer'. He is almost certainly the source of the mask. Deschryver died some time between 6 June (the Census) and 16 September 1841 (*The Times* of 22 September recorded the birth of a posthumous daughter on that date). There is an earlier manuscript reference to a 'Deschryver' as the vendor of some 95 paintings 'just imported from the Continent' in the catalogue of a sale held by the auction house of Foster on 15–16 May 1834 (Foster 1834). If this is François/Francis Deschryver, it could indicate that he was by then established in London as a dealer. This might therefore place the sale of the Florentine collection prior to 1834. Further traces of Deschryver exist in two posthumous sales catalogues, one of which mentions in the title that he was 'for many years an importer' (Deschryver 1841a, b). Hertz mentions in his letter to Christy that he found a note in the case containing the mask stating that 'it belonged to a convent of nuns at Mozza [sic] and that it was of Egyptian origin'. He adds that the vendor of the mask (presumably Deschryver) asked £300 for it but 'I got it from him in exchange and it stood me about £80'.

The Minute books of the Royal Asiatic Society confirm that the mask and knife were definitely in Hertz's hands by 28 January 1843 when 'a sacrificial dagger, and a turquoise mask' were exhibited by him at a general meeting (RAS 1839–1845: fol. 124). Hertz writes of this event:

> Professor Wilson was in the chair, who doubted them to be of Mexican origin, as he argued that no Turquoise were found in Mexico, I replied that I was aware of that, but that Mexicans had communications with other countries, who might have supplied the material for these articles (Carmichael 1970: 37).

Since it was not displayed at the Royal Asiatic Society, it can be inferred that Hertz acquired the human skull after the beginning of 1843. He states in his 1859 letter to Christy:

> The skull was in a Collection at Bruges which was sold about 12 years ago, as also a wig which was described as a scalp. On the scull [sic] there is a material which is called obsidian, and which is found in great abundance in Mexico (Carmichael 1970: 37).

It is possible that Hertz could be referring to a private sale, but there is only one public auction listed in Art Sales Catalogues Online that took place in Bruges in the 1840s and which contained antiquities and miscellaneous objects.[4] This was the immense collection of the rich landowner and aristocrat Joseph van Heurne (1752–1844), built up over half a century, auctioned at Bruges in October 1844 after his death at the age of 91, which thus stands out as a possible source of Hertz's acquisition. A widower for 40 years, van Heurne had exhibited his collection in his residence at Bruges. The sale catalogue, which was distributed locally and in Berlin, Cologne, Paris and London, lists almost 2,000 items: 288 paintings, 198 drawings, 535 prints, 69 sculptures, 54 porcelains, 179 natural history objects, and 603 ivories, enamels, antiquities and miscellaneous objects. It was typical of the private collections of the period, the compiler of the catalogue commenting:

> on pourrait même la qualifier de Musée, tant par la diversité que par la rareté d'un grand nombre d'objets qui la composent ... ses Curiositiés rassemblées de façon à donner une idée des mœurs de la civilisation des différentes peuples (Vandervin 1844: 5).

The majority of the objects are European, but van Heurne did have more universal interests. There are, for example, ceramics from China, Japanese lacquer work, a Honduran hammock, a Senegalese princess's apron, an American peace pipe, 40 pairs of miscellaneous shoes and 12 kinds of East Indian tobacco, to name but a few. Listed in the catalogue, among 84 'Objets divers', is:

> 483. Un *Vanitas* d'une composition vraiment extraordinaire, consistant en une caisse sur la porte de laquelle on a représenté la portrait d'une jeune femme;

à l'intérieur se trouve un crane sur la face duquel on a appliqué des pierreries en mosaïque, les yeux sont en marcassite. A l'intérieur de la porte il y'a 8 vers latins relatifs au sujet. On croit que cette pièce date du milieu du l6eme siècle (Vandervin 1844: 106).

It seems very likely that this is the British Museum's human skull. The date and place are right, given that Hertz's letter is rather vague. If this is indeed the British Museum's skull one puzzle might be solved. In Hertz's 1859 letter to Christy and in both the 1854 and 1859 sales catalogues the skull was, for no obvious reason, said to be accompanied by a wig or scalp. It would make sense if this were part of a Vanitas ensemble, a portrait of a young woman outside but concealed inside a box a reminder of the transience of life – a skull with flowing locks. It might also explain the good condition of the leather strap attached to the skull (McEwan et al. 2006: 66–70). Vanitas painting, with its focal point of a human skull, was particularly popular in Northern Europe (Flanders and the Netherlands) in the sixteenth and seventeenth centuries. An assemblage of actual objects such as this is, however, unusual.

Edward Tylor in his book *Anahuac* (1861: 339) suggests that the Bruges origin of the skull indicates that it could have arrived in the Low Countries between 1521 and the expulsion of the Spaniards in 1579. It has, however, to be admitted that the scope of van Heurne's collection was wide and his sources of acquisition varied. Andrew van den Abeele (2009), an authority on Bruges and on van Heurne, notes that in the nineteenth century several ship owners operated from Bruges who might have brought back such a curiosity. He also mentions that around 1835–1840, for example, at least one Brugeois, Maximilien van Lede, travelled to Mexico and Latin America trying to found colonies. However, he also thought it possible that if the skull were already in Belgium in the sixteenth century it could have been in the collection of the governor, Margaret of Austria. Another possibility is that the Vanitas might perhaps have an ecclesiastical origin and van Abeele mentions that during the revolutionary period, when churches were closed or demolished and their treasures scattered and sold, van Heurne acquired all he could.

We do not know whether Hertz was present at the van Heurne sale. His reference to Bruges (Carmichael 1970: 37) is ambiguous so it is possible that he could have bought the skull later from a dealer. At all events, whatever their history, the three mosaics appeared in the Hertz/Mayer 1859 sale (Sotheby and Wilkinson 1859: 125–126), were bought by Henry Christy and were subsequently published by Tylor (1861) and in Steinhauer's catalogue of Christy's collection (1862: 33–34).

Christy Fund

Three mosaics were bought with funds from the Christy Bequest: (4) a shield (Fig. 4); (5) the mask depicting two entwined serpents (Fig. 5); and (6) the double-headed serpent, possibly a pectoral (Fig. 6). Little can be added to the provenance of the shield, which is described on its registration slip (BM, Department of Africa, Oceania and the Americas, henceforth AOA) as having been bought by the Christy Collection for £20 on 26 December 1866 from William Adams. Adams, a dealer, said that it had come from Turin (Read 1895: 395). It was registered by the Museum as 'A zodiac consisting of a perforated thin disk of wood ... In the centre are depicted the cardinal points' (BM, AOA). A sketch of what appears to be the British Museum shield occurs in a notebook used by Franks in 1870 (BM, Prehistory and Europe 1) during his travels on the Continent.

The skull mask with intertwined serpents emerged in the six-day sale in Paris in March 1870 of the collection of Anatole Demidov (1812–1870), first Prince of San Donato. This outstanding collection, of around 2,500 items, came from the prince's palatial villa at San Donato in Florence. The Demidov family were immensely rich through their interests in Russian arms manufacture and mining. Anatole's father, Nicholas (1772–1828) settled in Florence in 1822, following his appointment as Russian ambassador to the court of Tuscany, and in 1827 erected a grand Palladian palace on marshland to the north of the city. He is said to have 'passed most of his time in Florence and Rome among the relics of the arts, of which he was so enthusiastic a connoisseur' (*The Times*, 2 May 1870). Anatole, after the collapse of his disastrous marriage to Princess Mathilde Bonaparte in 1841, lived mostly in Paris where, it is said, he was latterly referred to as 'Prince Décomposition' owing to his 'bent and haggard form' (*The Times*, 2 May 1870). The mask was buried in the catalogue along with five other odds and ends – 'objets variés' – being described merely as:

475. Masque antique, couvert de turquoises et formé d'un serpent enroulé (Demidov 1870: 86).

The sale took place on Wednesday 30 March. Annotated copies of the sale catalogue available on Art Sales Catalogues Online record that the mask was bought by a M. Léman, presumably a dealer, for the modest price of 255 francs – about the same as the cost of four Sèvres plates or a Polish sabre.

How Franks learned of the mask is not recorded. In March 1870 he was otherwise occupied in the United Kingdom, but he did go on holiday to France at Easter, touring the west of the country. In a report dated 11 May 1870 (BM, Central Archive, henceforth CA) he informed the Trustees that he had spent a week in Paris (25 April–3 May) on his return. Here he had examined a large collection of Mexican antiquities that he could not afford. He did, however, succeed in buying from Léman, through an intermediary, 'a very remarkable Mexican mask made of a mosaic of turquoises' together with a New Guinea [pottery] stamp, for a total of £24 12s (BM, CA). Franks remarked to the British Museum Trustees in a report dated 8 June:

With respect to the purchase from M. Leman it did not appear to Mr Franks expedient that it should be known that the object was bought for the British Museum & the negotiation with M. Leman was very troublesome as he originally asked double the price paid (BM, CA).

It was suggested by Franks's assistant, Hercules Read, in 1894, that this mosaic mask (or the earlier Hertz piece) may be that described in a Medici inventory of 1640–1645 (Read 1895: 387).

A little more information can be added to the provenance of the two-headed serpent, which was registered by the

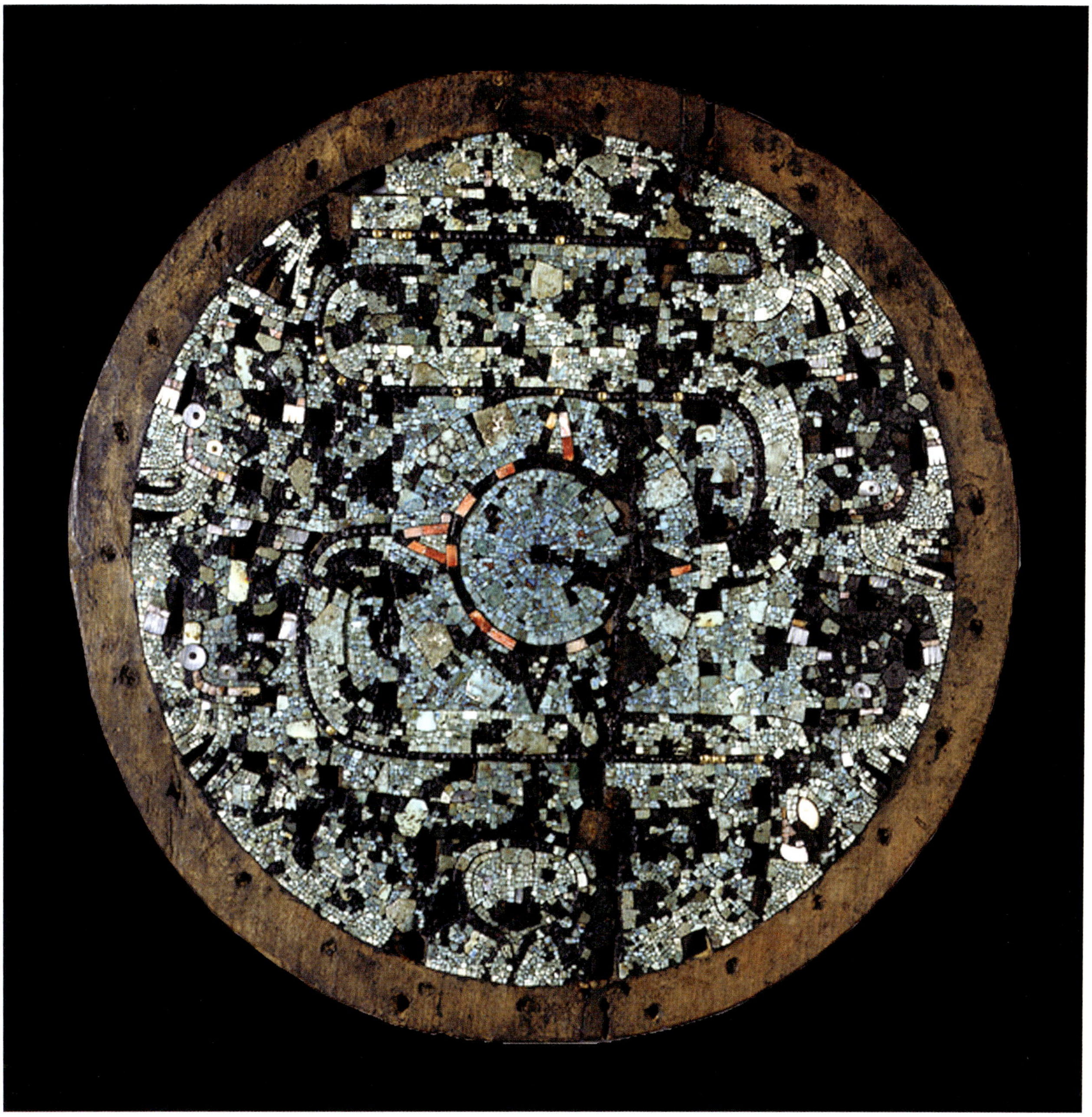

Figure 4. Shield (British Museum, AOA, Am St.397a, © Trustees of the British Museum).

Museum as 'Purchased. Christy Fund. September 1894. Duchessa Massimo through Lord Walsingham £100' (BM, AOA). It seems possible that Duchess Massimo (1840–1923) was not herself the vendor – certainly this is what she told Franks and Walsingham.

The serpent had already been offered to the British Museum on 3 September 1892 for £350 by a dealer, a Mr Ed. Joseph of 158 New Bond Street, who, having previously spoken to Read, wrote that he was acting on behalf of a 'Gentleman in Italy'. The purchase was declined on account of the high price (BM, Prehistory and Europe, henceforth P&E).

Quite how Walsingham and the duchess became involved is not apparent. Possibly the anonymous Italian owner may have tried to make further contact with the British Museum. Thomas de Grey, 6th Baron Walsingham (1843–1919), politician and entomologist, was a British Museum Trustee. He had a family connection with Duchess Massimo through her mother, Lady Mary Talbot (1815–1858) daughter of the 16th Earl of Shrewsbury, who had married Prince Filippo of Doria-Pamphilj-Landi. His wife, Leila, Lady Walsingham (d. 1906), also had contacts among the Italian nobility, having previously been Duchess of Sant' Arpino and later San Teodoro (Paget 1926: 265). There are letters from Walsingham in the British Museum archive (BM, P&E) describing the progress of sale negotiations, but quite who was negotiating for whom is not always clear.

On 23 July 1894 Walsingham wrote rather wearily to Franks:

> You will vote me a bore – If you do I shall only put the compliment back one step and mentally transfer it to the Duchesse Massimo who sends another photograph of a Mexican 'curio'. Please tell me about what to say to her. I know nothing about price but if we want it we could

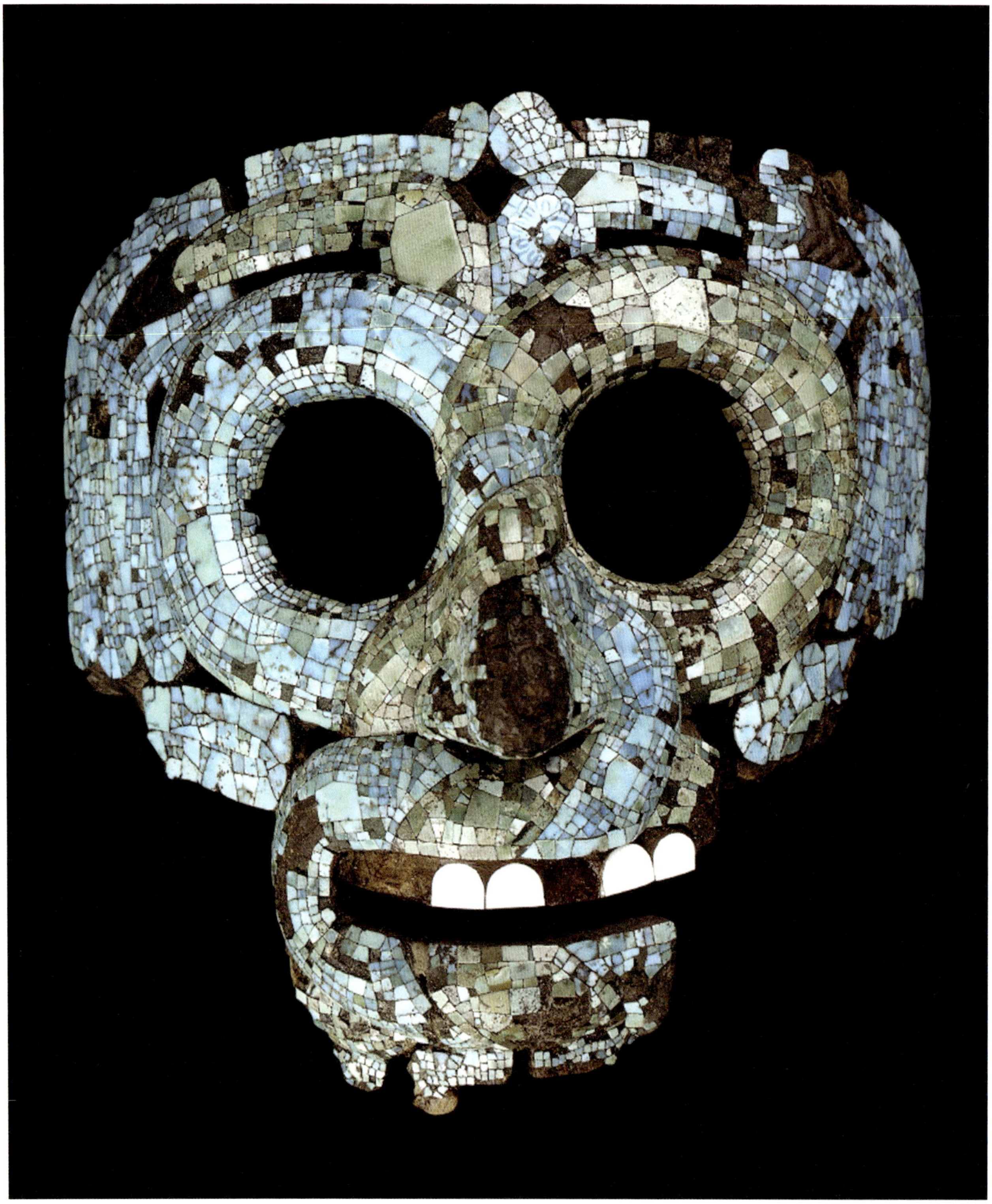

Figure 5. Mask depicting two entwined serpents (British Museum, AOA, Am 1987,Q.3, © Trustees of the British Museum).

> get it probably at its fair value ... the Duchesse will be staying with me next week (BM, P&E).

By 9 August, Walsingham had discovered that the price requested by Joseph in 1892 had been grossly inflated since, he writes, 'the owners had put the price at about £120'. In the same letter he states:

> Should you at any time think it worth while to make an offer for it for the Christy collection I should like to let the Duchesse Massimo know as she appears to be very anxious to obtain some help in this way for the family to whom it belongs (BM, P&E).

Walsingham, in Franks's absence, wrote to Read on 20 August:

> I will at once write to the Duchesse Massimo and ascertain whether the owner will take £100 ... I should think it would be better to have it sent on approval as Sir Wollaston Franks has only seen the photograph so

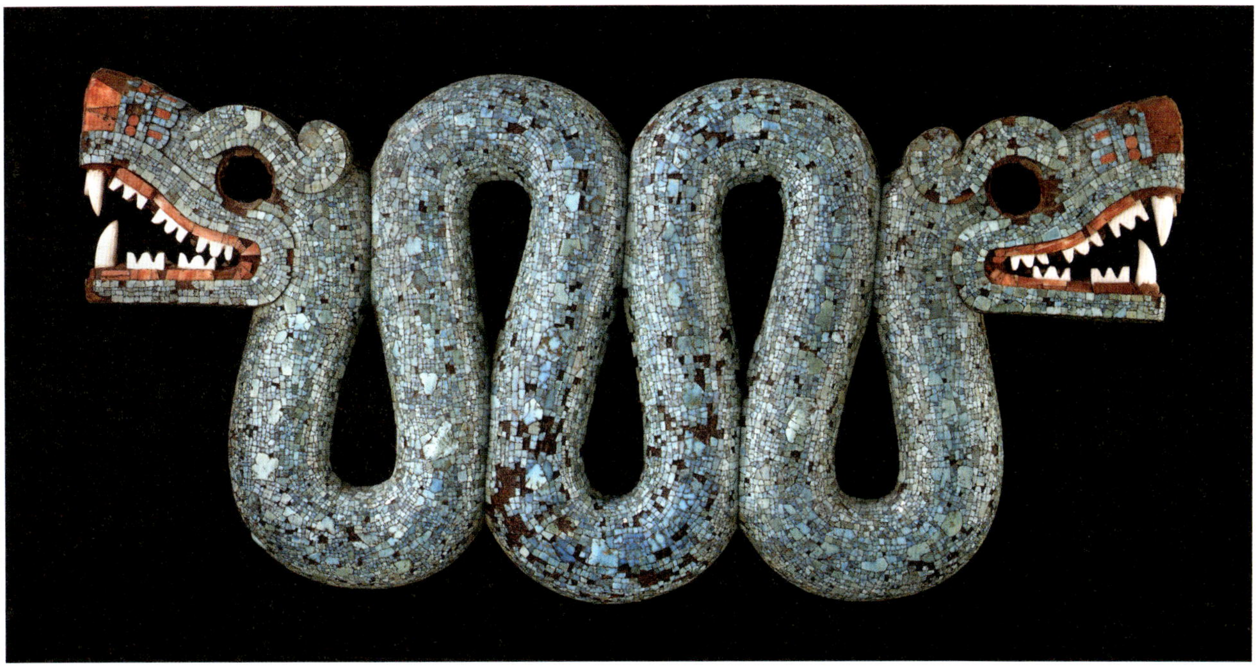

Figure 6. Double-headed serpent, possibly a pectoral (British Museum, AOA, Am 1894,-.634, © Trustees of the British Museum).

> far as I am aware. He may have seen the original some time ago (BM, P&E).

The offer was accepted, Walsingham writing to Franks on 14 September:

> I received a letter stating that the owner of the snake would accept £100 for it … although this was a slight disappointment. No details are given showing to whom the cheque should be drawn – If you will make the cheque payable to the Duchesse Massimo … I have no doubt it will reach its proper destination by being forwarded direct to her (BM, P&E).

The same day (14 September) Franks wrote to one of the Christy Trustees, Sir John Lubbock:

> The Duchess Massimo's wonderful Mexican mosaic has arrived and I hear from Lord Walsingham that she will accept the £100 for it, but grumbles (BM, P&E).

Franks, ever keen to strike a hard bargain, subsequently remarked in a letter to the duchess dated 28 September, announcing the arrival of the serpent: 'Il est très curieux, mais je ne dirais pas qu'il est beau' (BM, P&E). Because of the confidentiality of the negotiations Franks was still unsure to whom the cheque should be made out. It was agreed that the duchess's name could be inserted, but he requested a receipt. A letter from the duchess in Naples dated 5 October 1894 again makes reference to unknown owners:

> je vous envoie cijoint le reçu, je l'ai acquité de mon nom m'étant chargé (par les propriétaires du Serpent Méxicain) de toute la négotiation (BM, P&E).

The serpent was exhibited to the Society of Antiquaries on 22 November 1894.

So, for whom was the duchess acting? According to Society gossip, the duchess, née Doria-Pamphilj-Landi, and married to Emilio Massimo, 3rd Duca di Rignano (1835–1907), had had a difficult life. Around 1868 her friend Lady (Walburga) Paget (née von Hohenthal) (d. 1929), wife of the British ambassador Sir Augustus Berkeley Paget (1823–1896), wrote:

> Teresa Doria had been married at seventeen, almost at her mother's death-bed, to the Duc de Rignano, the only son of Duc Massimo, a young man not only weak, but almost wanting (Paget 1928: 135).

A.J.C. Hare also comments on the duchess's woeful marriage:

> As Duchess Rignano, Teresa Doria was wretched. We saw her afterwards at Genoa, in the old Doria Palace, with her mother, whose death was hastened by the sight of her daughter's woe and her own disappointed ambition. Before long the Duchess Teresa was separated from her husband (Hare 1896: 465).

In 1870 France had abandoned Rome and Italian independence was achieved. This was a turbulent period during which some noble families faced financial difficulties. In 1890 Lady Paget wrote, with reference to the duchess's brothers, Prince Giovanni (1843–1890) and Alfonso (1851–1914):

> A shocking thing has happened, poor Gianetto (Doria) died under the knife on Monday morning. All the Dorias are distracted. I am told Alfonso has been left very little, everything goes to his son, because he is such a spendthrift. For poor Duchess Massimo, who is completely ruined, it is a fearful blow, as he helped her much (Paget 1928: 489–490).

In the copy of Hercules Read's paper (1895) in the British Museum Department of Prehistory and Europe, the two

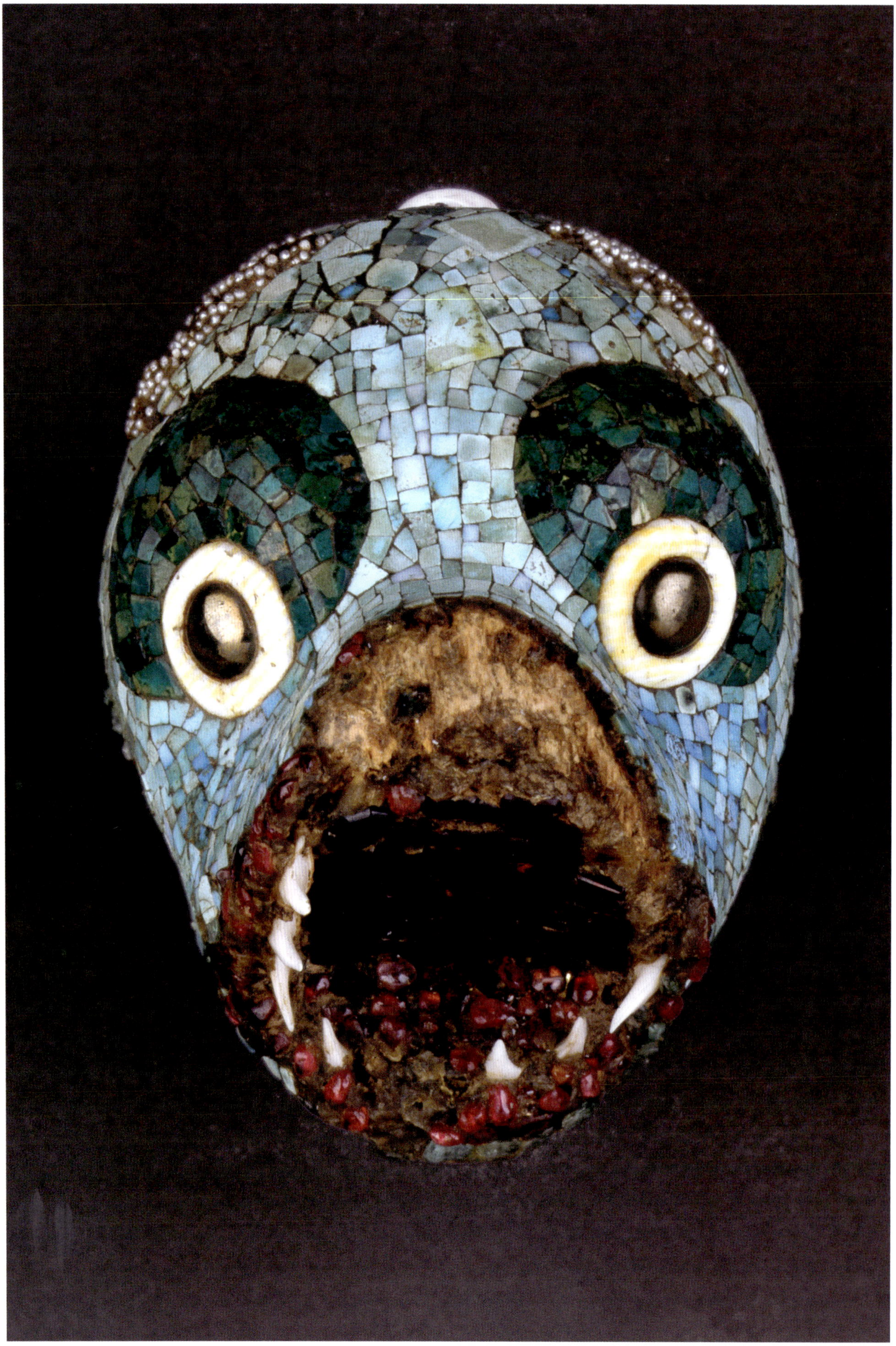

Figure 7. Animal-head pendant (British Museum, AOA, Am,St.400a, © Trustees of the British Museum).

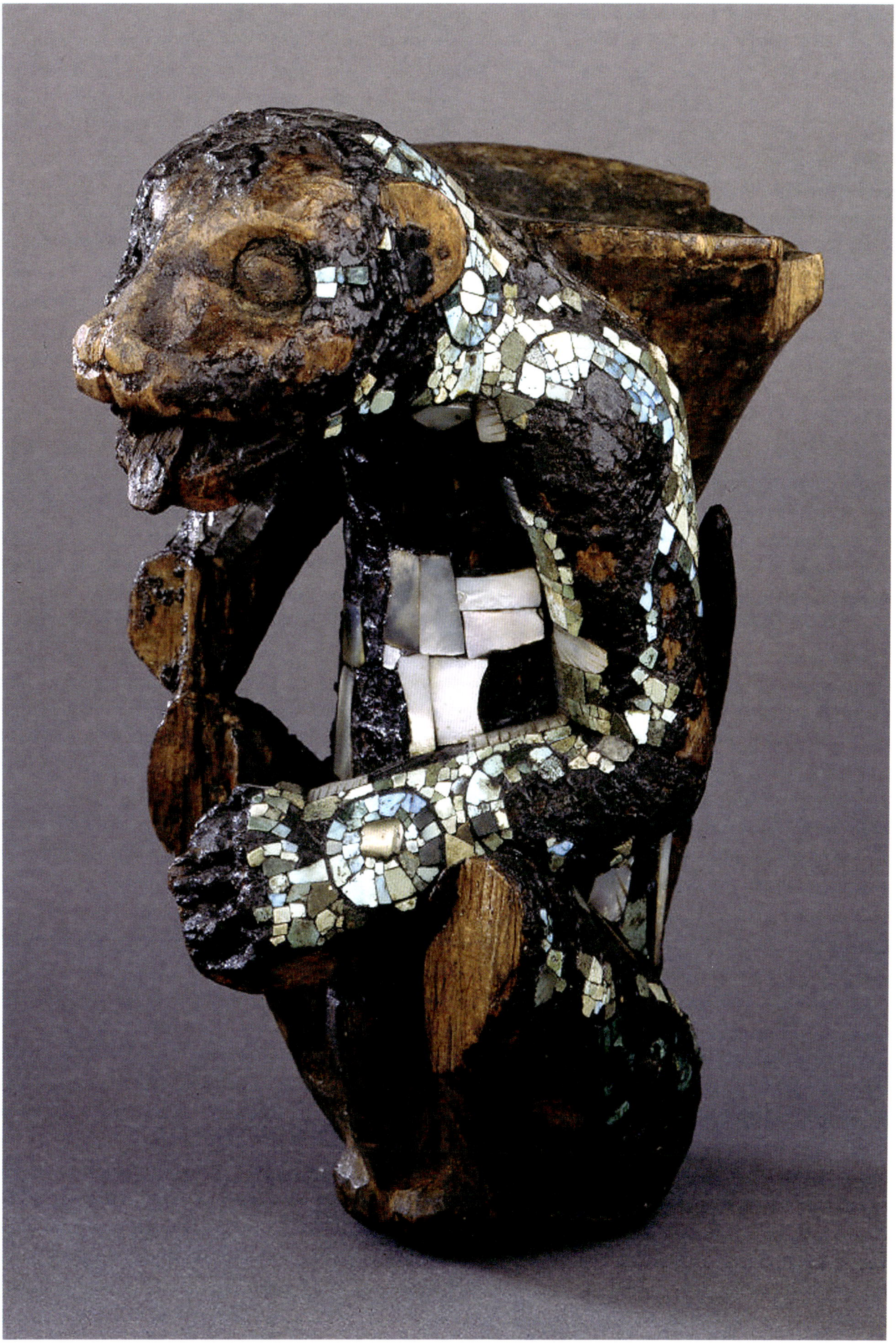

Figure 8. Seated animal and cup (British Museum, AOA, Am, +.165, © Trustees of the British Museum).

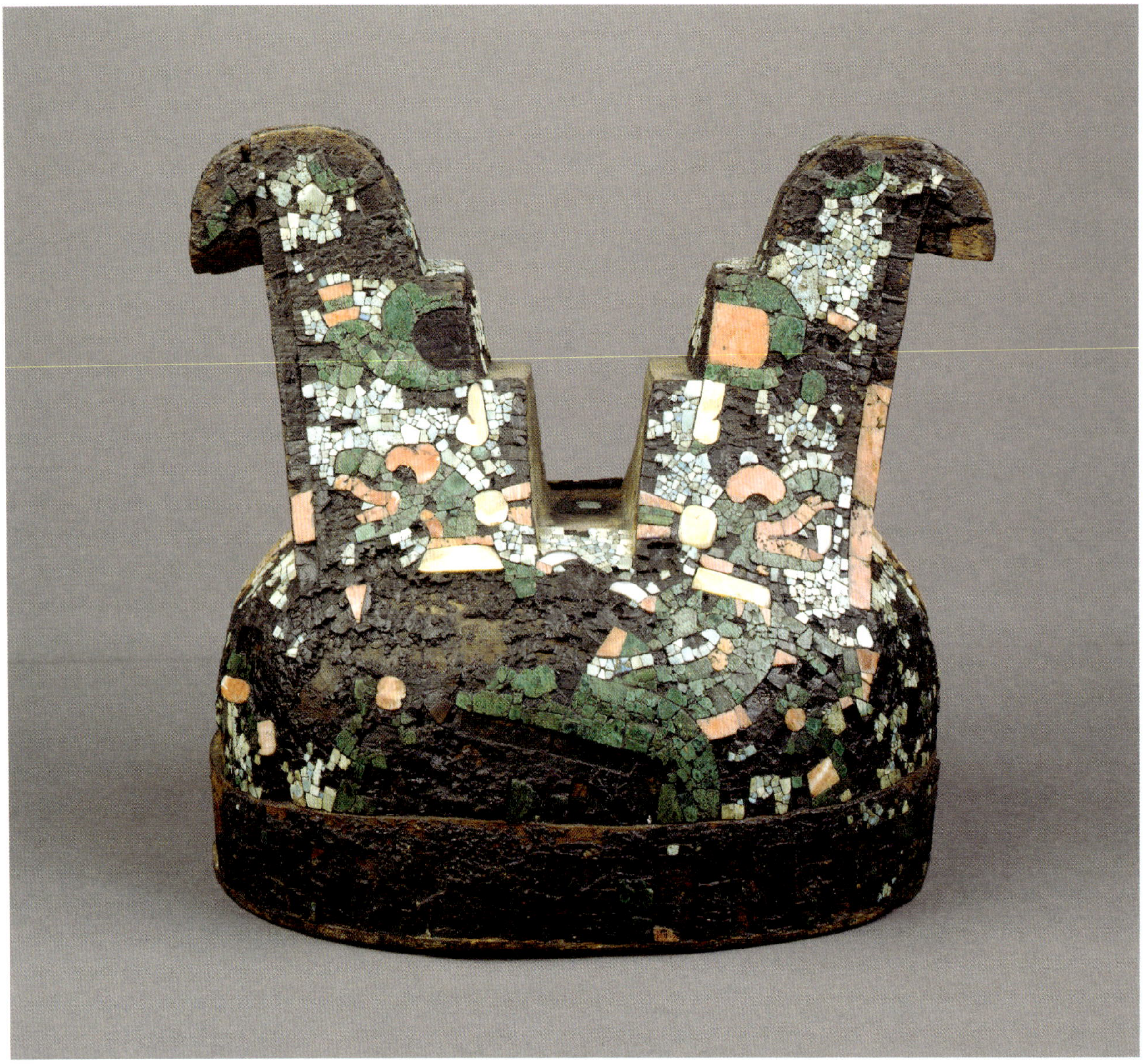

Figure 9. Helmet (British Museum, AOA, Am,+6382, © Trustees of the British Museum).

references to the serpent, which mention its source as 'an old collection in Rome' (1895: 389, 397) have been annotated in manuscript 'Prince Massimo'. The annotation may be by Read, but whether he had obtained confirmation of the provenance or was still guessing is not now known. According to the *Almanach de Gotha* (1893: 402–404) there were two lines of the Massimo family – one with two princely branches (Princes d'Arsoli) and the other (into which the duchess married) ducal (Ducs de Rignano). It has usually been assumed that the serpent could have come from the Massimo family but, given the dire state of the duchess's marriage, one might now speculate that the source could have been the Dorias. Or perhaps it was a family whose name is not recorded?

Augustus Wollaston Franks's donation

The final three mosaics are those bought by Franks himself and given by him to the Christy Collection: (7) animal-head pendant (Fig. 7); (8) seated animal and cup (Fig. 8); and (9) a helmet (Fig. 9).

All that is known about the animal-head pendant is that it was purchased by Franks in 1868 from William Adams, who had also supplied the shield, and presented by Franks to the British Museum. Its acquisition is recorded in the Christy Collection Register on 7 August 1868 (BM, AOA). Read (1895: 396) notes that Adams said that it came from northern Italy.

Franks's sketch of the mosaic shield in a notebook he used in 1870 (BM, P&E 1) has already been noted. In the same notebook and on the same page is a sketch of what might be an animal head. What is odd, however, is that it has two prominent ears – the British Museum's head has none and the 'dragon's head' mosaic from Schloss Ambras, now in Vienna, has only one (Heger 1892: 381). The latter appears in inventories from the sixteenth century onwards and was identified in 1891.

The seated animal (possibly a jaguar) supporting a cup was bought by Franks from Joseph Mayer for £20 and, as noted on the Christy Registration slip (BM, AOA) was presented to the British Museum in March 1877. In his younger days, Franks had acquired antiquities for Mayer and he had advised on the catalogue of Mayer's collection. Unfortunately Mayer gave no provenance for this piece (Read 1895: 398) and his papers are now scattered. Although Mayer acquired complete collections, he also travelled and made an uncounted multitude of lesser purchases, here and there in London and on the Continent (Gibson 1988: 12).

The helmet or headpiece at one time belonged to the archaeologist and antiquary Thomas Bateman (1821–1861) who, in 1844, built Lomberdale House, near Middleton, Derbyshire, to house his museum. Although Bateman's collection of 4,500 objects was particularly rich in British archaeology, it also included around 120 objets d'art and miscellaneous ethnographic curiosities. The latter covered a wide range, but the largest group, 22 items, came from Mexico. Bateman's published catalogue mentions the helmet, noting that it had been purchased in Paris by 'Mr Chaffers' in 1854 (Bateman 1855: 236). Bateman correctly identified his curio, describing it as an 'Ancient Mexican horned head-dress or helmet of wood, inlaid in mosaic with turquoise, malachite, coral (?) [sic], and mother-o'-pearl ... Probably unique, there being no specimen in any museums that I have hitherto visited' (1855: 236) and included it in 'Miscellaneous Ethnographical Objects'.

William Chaffers (1811–1892), is today quite well known as a leading nineteenth-century authority on gold and silver hallmarks and on makers' marks on pottery and porcelain. In contrast to his later scholarly reputation, he was described in the 1851 Census as a 'pawnbroker employing four men', so his activities may have included buying trips to the Continent. Bateman's collection survived his death in 1861 but a generation later, against his known wishes, it was broken up. At a sale on 14 April 1893 the annotated sales catalogue (BM, P&E) indicates that Franks acquired this 'interesting and very rare specimen' (Lot 127) for the middling price of £21 through 'Rollin', probably Charles Rollin of the Paris dealers Rollin & Feuerdant who had bid for him on other occasions (Sotheby et al. 1893: 22).

Conclusions

The story of the British Museum's nine turquoise mosaics has been taken a little further, at least in the nineteenth century, but often it seems that, as a solution presents itself, more loose ends appear, requiring further investigation. Some of this research would have been difficult a decade ago but today's electronic resources open up intriguing possibilities for future discoveries.

Notes

1. In reviewing the provenance of the British Museum's collection of nine mosaics, it is interesting to note how the known examples emerged in the nineteenth and twentieth centuries. To the initial three in the Christy Collection were added two in Copenhagen (Tylor 1861; Stevens 1870). The Christy collection enlarged to five (Stevens 1870). Bastian (1885) introduces two from the Ducal Museum, Brunswick. Pigorini (1885) lists five more in Rome, one in the Bateman collection, three in Berlin (including the two from Brunswick), and one in Gotha, producing a total of 19. Heger (1892) describes two from Schloss Ambras, Tyrol, transferred to Vienna which then had three. Read (1895) raises the Christy total to its final nine. Pogue (1912 and 1915) adds one from Honduras (then in Washington, DC). Saville (1922) doubles the total to around 45 by adding 18 in New York and four in Cambridge, MA (USA). Despite the loss of those in Berlin during the Second World War, which reduced the number, a further specimen from an old European collection, now in Dallas, was auctioned in London in 1978 (Sotheby's 1978). Shelton (1988, 1994) adds 10 from excavations such as those at Monte Alban, Oaxaca and the Templo Mayor, Mexico City and from caches discovered in caves in the Mixteca of Oaxaca and Puebla. The new finds were distributed in Mexico, New York (to which the Washington specimen was also transferred) and Brussels. This gives a total of around 55 plus smaller specimens.
2. Thanks are due to Louise Tythacott, University of Manchester, for identifying in the Liverpool Records Office five letters from Bram Hertz to Joseph Mayer, relating to the sale of Hertz's collection: 23 July 1856, 31 July 1856, 24 October 1856, 15 January 1857 and 2 March 1857.
3. Typed copy of Hertz/Christy letter reproduced in Carmichael (1970: 37). The original letter, according to an annotation on the typescript, accompanied the human skull when it was evacuated to an outstation during the Second World War. A typed note retained in the Wartime Archives of the BM Department of Asia states: 'Postmark on envelope reads: London DX Fe. 8. 59; also Frankf..t 5 Feb. 1859'.
4. A particularly helpful source in the compilation of this paper has been the database Art Sales Catalogues Online, the online edition of the *Répetoire des Catalogues de Ventes Publiques* by Frits Lugt. This includes full texts of a number of digitised catalogues from various sources. The National Art Library at the Victoria and Albert Museum has an extensive collection of sales catalogues as has the British Library, which holds the Sotheby's archival set. A small selection of sales catalogues relating to its collections is held by the library of the British Museum's Department of Prehistory and Europe.

References

Archives (unpublished)

British Museum, Central Archive (BM, CA)

British Museum, Department of Africa, Oceania and the Americas Archive (BM, AOA)

British Museum, Department of Asia Archive (BM, Asia)

British Museum, Department of Prehistory and Europe Archive (BM, P&E)

Liverpool Records Office, Mayer Collection 920 MAY, accession no. 2582

Royal Asiatic Society (RAS) (1839–1845). *Minutes of General Meetings*, December 1839–March 1845.

Published sources

Almanach de Gotha (1893) *Annuaire Généologique, Diplomatique et Statistique*. Gotha: Justus Perthes.

Bastian, A. (1885) 'Zwei altmexikanische Mosaiken', Verhandlungen der Berliner Gesellschaft für Anthropologie: 201.

Bateman, T. (1855) *A Descriptive Catalogue of the Antiquities and Miscellaneous Objects Preserved in the Museum of Thomas Bateman at Lomberdale House*. Bakewell: the author.

Blackwood's Edinburgh Magazine (1852) 'Our London Commissioner', *Blackwood's Edinburgh Magazine* **71**(439): 603.

Carmichael, E. (1970) *Turquoise Mosaics from Mexico*. London: British Museum.

Caygill, M. and Cherry, J. (eds) (1997) *A.W. Franks: Nineteenth-century Collecting and the British Museum*. London: British Museum Press.

Demidov, A.N. (1870) *Collections de San Donato, Objets d'art. Ordre des ventes ... Me Charles Pillet, commissaire-priseur ... M. Charles Mannheim, expert*. Paris: n.p.

Deschryver, F. (1841a) *The Ancient Collection of the Late Mr. F. Deschryver, Many Years an Importer ... A catalogue of Ancient Oak and Decorative Furniture, etc.* London: n.p.

Deschryver, F. (1841b) *A Catalogue of the Remaining Portion of the Stock ... of the Late Mr F. Deschryver ... viz. Ancient Furniture and Carvings, etc.*. London: n.p.

Foster, E. (1834) *Capital Pictures. A Catalogue of an Excellent Collection of Pictures, of the Italian, Dutch, and English Schools ... Just Imported from the Continent which will be Sold by Auction ... 15th of May 1834 & following day.* London: n.p.

Gibson, M. (1988) 'Joseph Mayer', in M. Gibson and S. Wright (eds), *Joseph Mayer of Liverpool 1803–1886.* Society of Antiquaries of London, National Museums and Galleries on Merseyside. Occasional Papers (new series), **X**, 11.

Hare, A. (1896) *The Story of My Life*, 6 vols, 1896–1900. London: George Allen.

Heger, F. (1892) 'Altmexikanische Reliquien aus dem Schloss Ambras in Tyrol', *Annalen des k.k. Naturhistorischen. Hofmuseum*, vii, **4**: 379–400.

Hertz, B. (n.d.) *A Catalogue of a Museum of Egyptian, Greek, Etruscan, Roman, Peruvian, and Mexican Antiquities Formed by B. Hertz ...* London: n.p. (published with a set of lithographic plates).

Hertz, B. (1851) *Catalogue of the Collection of Assyrian, Babylonian, Egyptian, Greek, Etruscan, Roman, Indian, Peruvian and Mexican Antiquities formed by B. Hertz.* London: n.p.

McEwan C., Middleton, A., Cartwright, C.R. and Stacey, R. (2006) *Turquoise Mosaics from Mexico.* London: British Museum Press.

Paget, Walburga, Lady (1928) *The Linings of Life*, 2 vols. London: Hurst & Blackett.

Phillips, Mr. (1857) *Catalogue of a Very Interesting Collection of Works of Art and Vertu, Formed by that Distinguished Connoisseur B. Hertz, Esq., ... which will be Sold by Auction ... 24th March 1857 and Two following Days ...* London.

Pigorini, L. (1885) *Gli antichi oggetti messicani incrostati di mosaico esistenti nel museo preistorico ed etnografico di Roma.* Rome: Reale Accademia dei Lincei.

Pogue, J. (1912) "The aboriginal use of turquoise in North America', *American Anthropologist*, new series, **14**(3): 437–466.

Pogue, J.E. (1915) 'The turquois: a study of its history, mineralogy, geology, ethnology, archaeology, mythology, folklore, and technology', in *Memoirs of the National Academy of Sciences*, Vol. XII, Part II. Washington, DC: National Academy of Sciences.

Read, H. (1895) 'On an ancient Mexican head-piece, coated with mosaic', *Archaeologia* **LIV**(II): 383–398.

Saville, M. (1922) *Turquoise Mosaic Art in Ancient Mexico.* New York: Museum of the American Indian Heye Foundation.

Shelton, A. (1988) 'Realm of the fire serpent', *British Museum Society Bulletin* **58**: 21–25.

Shelton, A. (1994) 'Cabinets of transgression: Renaissance collections and the incorporation of the New World', in J. Elsner and R. Cardinal (eds), *The Cultures of Collecting.* London: Reaktion Books, 177–203.

Sotheby, S. and Wilkinson, J. (1854) *Catalogue of a Very Interesting Portion of the Valuable and Important Collection of Antiquities, the Property of B. Hertz, Esq. ... which will be Sold by Auction ... 29th of May, 1854, and Two Following Days.* London: n.p.

Sotheby, S. and Wilkinson, J. (1859) *Catalogue of the Celebrated and well-known collection of Assyrian, Babylonian, Egyptian, Greek, Etruscan, Roman, Indian, Peruvian, Mexican and Chinese Antiquities, Formed by B. Hertz ... now the Property of Joseph Mayer, Esq;. ... which will be Sold by Auction ... 7th of February, 1859, and Fifteen Following Days ...* London: n.p.

Sotheby, Wilkinson and Hodge (1893) *The Bateman Heirlooms. Catalogue of the first portion of the very valuable collection of Works of Art and Antiquity formed by the late W. Bateman, Esq. and T. Bateman, Esq. of Lomberdale House, Youlgrave, Co. Derby ... which (by order of the Court of Chancery) will be Sold at Auction ... Friday, the 14th day of April 1893.* London: Dryden Press.

Sotheby's (1978) *Catalogue of Pre-Columbian, American, Indian, Oceanic and African Art. Day of Sale: Monday 11th December, 1978.* London: 44–45.

Steinhauer, C. (1862) *Catalogue of a Collection of Ancient and Modern Stone Implements, and of other Weapons, Tools, and Utensils of the Aborigines of various countries, in the possession of Henry Christy.* London: privately printed.

Stevens, E. (1870). *Flint Chips: A Guide to Prehistoric Archaeology as illustrated by the Collections in the Blackmore Museum, Salisbury.* London: Bell & Daldy.

Tylor, E. (1861). *Anhuac: or Mexico and the Mexicans, Ancient and Modern.* London: Longman, Green, Longman & Roberts.

van den Abeele, A. (2009) Personal communication, 6 September 2009.

Vandervin, H. (1844) *Catalogue des Collections de Tableaux, Dessins, Gravures, Antiquités, Curiosités, et Objets d'Histoire Naturelle, formant le Cabinet Van Heurne, Dont la Vente se fera à Bruges ... le 21 Octobre, 1844 et jours suivants ... sous la direction de A. Bollaert, Directeur des ventes, et ministère competent, assisté d'Henri Vandervin.* Gand [Ghent]: De Busscher Freres.

Watts, K. (1992) 'Samuel Pratt and armour faking', in M. Jones (ed.), *Why Fakes Matter: Essays on Problems of Authenticity.* London: British Museum Press, 100–107.

Author

Marjorie Caygill, Department of Africa, Oceania and the Americas, The British Museum, London, UK (retired)

On the eternal and the elusive: an appreciation of the work of Johnson and Bird

Henrietta Lidchi

Continuity and perspective

The question of turquoise and its role in the American Southwest is far from straightforward. In the hierarchy of valued materials turquoise may have the greatest prominence with ceremonial, aesthetic and economic values. This assertion, although it encapsulates many aspects of turquoise, does not tell the whole story.

In reviewing the indigenous history of Southwestern jewellery, there are a number of important aspects to acknowledge. There is the indisputable fact of the continuing habitation of the American Southwest from the pre-contact period to the present day, and the consistent production and use of turquoise beads and mosaic ornaments of a quality and quantity that fascinate (see Jernigan 1978). There are archaeological theories as to the rise to prominence of turquoise in the American Southwest, and the precise nature of the link to Mesoamerican consumption (see Weigand and Harbottle 1983). There are the Spanish accounts that describe the use of turquoise as gifts and payment. And then there is the more recent history, namely the documented bringing together of silver and turquoise in the late nineteenth century and the craft industry that later flourished under the influence and patronage of traders, dealers and collectors (see Adair 1944; Bedinger 1973). All these factors come together in a continuously evolving tradition which, since the early twentieth century, is as likely to include the output of individual artists and makers working out of a small individual or family-run workshop as that of a quasi-factory setting (see Ostler et al. 1996; Batkin 2008; Bsumek 2008). These shifts are the broad brushstrokes of history, and perhaps they play mostly with linear time, with a sense of the push and pull of things, and of things constantly changing.

What distinguishes Gail Bird and Yazzie Johnson's paper that follows (hereafter Bird)[1] is the sense of the constant, narrated through anecdotes, informed by deep knowledge and personal experience. Bird's paper (in this vol., pp. 203–212) comes from sustained conversations about (and much research on) this craft tradition that has *emic* and *etic* aspects: as seen from the inside and from the outside (something also addressed by Whiteley, in this vol., pp. 145–154). Bird's paper surveys the field from an insider's perspective, one very familiar with the ebb and flow of the art form, as well as the art market. It starts by addressing the work of contemporary Pueblo artists Angie Reano Owen (Santo Domingo, b. 1946) and Ray Lovato (Santo Domingo, b. 1946), whose mosaic and bead work graced the beginning and ending of the original lecture. This structure, starting and finishing with the work of important lapidary artists and turquoise tacitly reflects a circular view of time and tradition. It highlights that continuity, change and renewal are complementary and persistent aspects in the making of Native jewellery in the American Southwest. Within this overall narrative of continuity, Bird pinpoints moments of radical shift, such as the lasting innovation that was the emergence of Navajo silversmithing in the mid-nineteenth century.

Bird first analyses the technical and aesthetic distinctiveness of the Navajo silversmithing tradition through its most straightforward expression – the silver button. She describes those early techniques – bumping up (doming), stamping and filing – used to modify an American or Mexican silver coin into a versatile component of Navajo adornment, an ornament that might equally serve as currency. Bird moves from this button to survey some of the more complex pieces of Navajo jewellery that have become iconic Southwestern forms – the squash-blossom necklace, *jaclas* and *najas*. In so doing she celebrates the distinctive construction and design of Southwestern jewellery while acknowledging its syncretism. Bird then explores the collaborative work that arose in the early twentieth century under the patronage and influence of traders, particularly at Zuni. Considering figurative inlay work and the more sculptural three-dimensional carving tradition, she presents the work of Leekya Deyuse (Zuni, 1899–1966) and Leo Poblano (Zuni, 1905–1959). She equally showcases the work of the jeweller Charles Loloma (Hopi, 1921–1991) who, for many, transformed what was permissible within the definition of indigenous Southwest jewellery. Loloma pioneered novel materials, textures and sculptural forms, and his body of work is of enduring significance (see Struever 2007).

The lecture and paper are evidence of Johnson and Bird's readiness to be ambassadors for this dynamic tradition as artists and members of Native communities for whom turquoise and jewellery remain 'an important component of the living culture' (Bird, in this vol., p. 210). Bird's lecture, presented in London in 2009, was copiously illustrated and her choice of pieces exemplified the boldness and the balanced beauty of Southwestern jewellery. It emphasised the influence that makers and artists provided, and continue to provide, for each other.

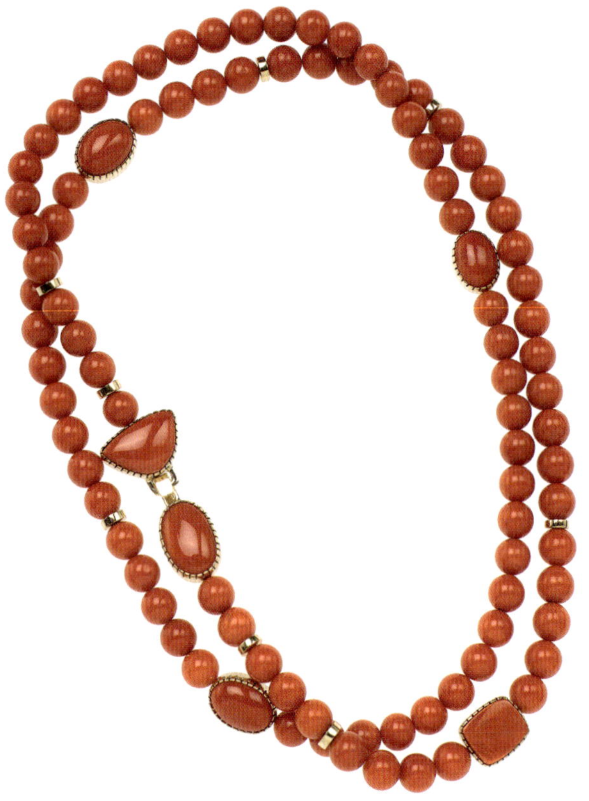

Figure 1. Necklace of coral beads and set coral cabochon, 2009. (NMS V.2009.245, © Trustees of the National Museums of Scotland.)

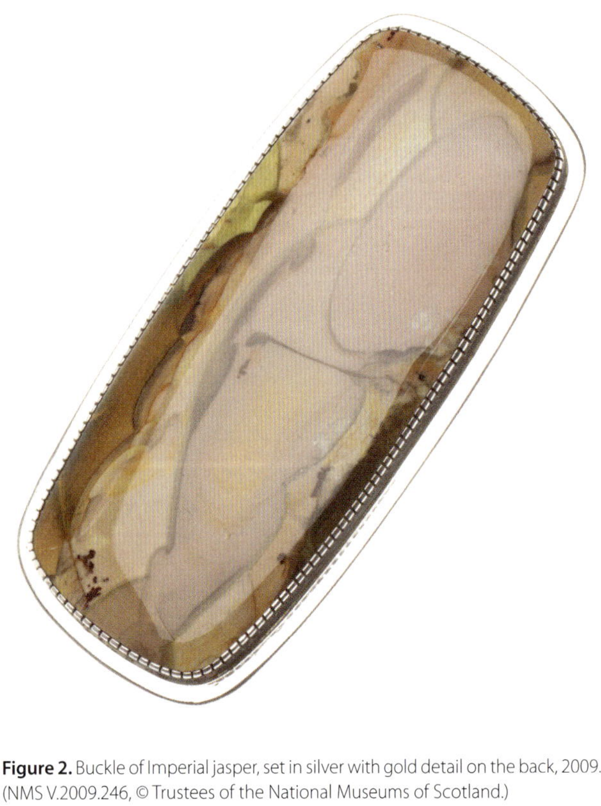

Figure 2. Buckle of Imperial jasper, set in silver with gold detail on the back, 2009. (NMS V.2009.246, © Trustees of the National Museums of Scotland.)

Materials and innovation

Today Johnson and Bird are acknowledged as key innovators and figures in the field, whose historical knowledge, technical skill and love of materials lend a richness and elegance to their work. In *Shared Images* (Pardue 2007), a catalogue dedicated to the work of Yazzie Johnson (Navajo, b. 1946) and Gail Bird (Santo Domingo/Laguna, b. 1949), Diana Pardue gives a sense of why Johnson and Bird might carry a unique authority and special insight when speaking about Southwestern jewellery. Pardue tracks Johnson and Bird's early friendship as teenagers, their training, collaboration and entry into jewellery-making. She charts their initial use of brass and unusual stones, the development of their distinctive style, as well as the manifold ways in which they have honed their skills. In considering the body of their work, *Shared Images* gives a sense of a sustained dedication to, and passion for, the craft of jewellery as artists, observers and curators. In recalling their shared career there is candour and clarity about times and places that contextualises the whole craft industry as something personally experienced. The layered illustrations, including working drawings and images of Johnson and Bird's workshop, provide an insight into the creative process. This emphasis on technique is picked up in Bird's discussion of history and design. *Shared Images* is mentioned here largely because it situates Johnson and Bird's own work within the wider field, something that Bird modestly does not do in the paper that follows.

In 2009, National Museums Scotland (Edinburgh) had the opportunity to buy two pieces by Johnson and Bird. These included a single stranded coral necklace of Mediterranean and Sea of Japan coral beads and cabochon set coral in gold (18ct) settings (Fig. 1) and a belt buckle of Imperial jasper set in silver (Fig. 2). These two new acquisitions are, in national terms, complementary to the Deschutes jasper and Yowah opal brooch set in gold (18ct) acquired by the British Museum (London) in 1998 (Fig. 3). Johnson and Bird are known for selecting and using precious or semi-precious stones of exceptional quality and unusual cuts that are collected as the opportunity arises, waiting to be placed into finished pieces. As Bird remarks, the Southwestern landscape has a particular pull; the record of human habitation is writ large upon it, architecturally and visually. A constant source of inspiration for their work (as are more quirky visual signifiers, such as road signs), it also explains the regular use of jasper as a stone, because different types of picture jaspers evoke the colours and contours of the American Southwest. The pastel colours of Imperial jasper (Fig. 2), for example, recall the large billowing clouds that build up in the summer time in Arizona and New Mexico.

In all three pieces, the undersides of the stone settings have overlay designs, or as Johnson and Bird like to call it, 'underlay'. Underlay is for Johnson and Bird a means to texture the back of a stone setting or to establish an additional and somewhat hidden narrative. On the brooch (Fig. 4) the lines and crosses reflect the geometric patterns of Navajo textiles, as well as the landscape and stars. On the buckle (Fig. 5), the figure is taken from a sketch made by Johnson in Petroglyph Park near Holbrook, Arizona and enhanced with a gold detail. On the necklace (Fig. 6), the underlay designs are of butterflies and dragonflies taken from pottery and petroglyph designs and can

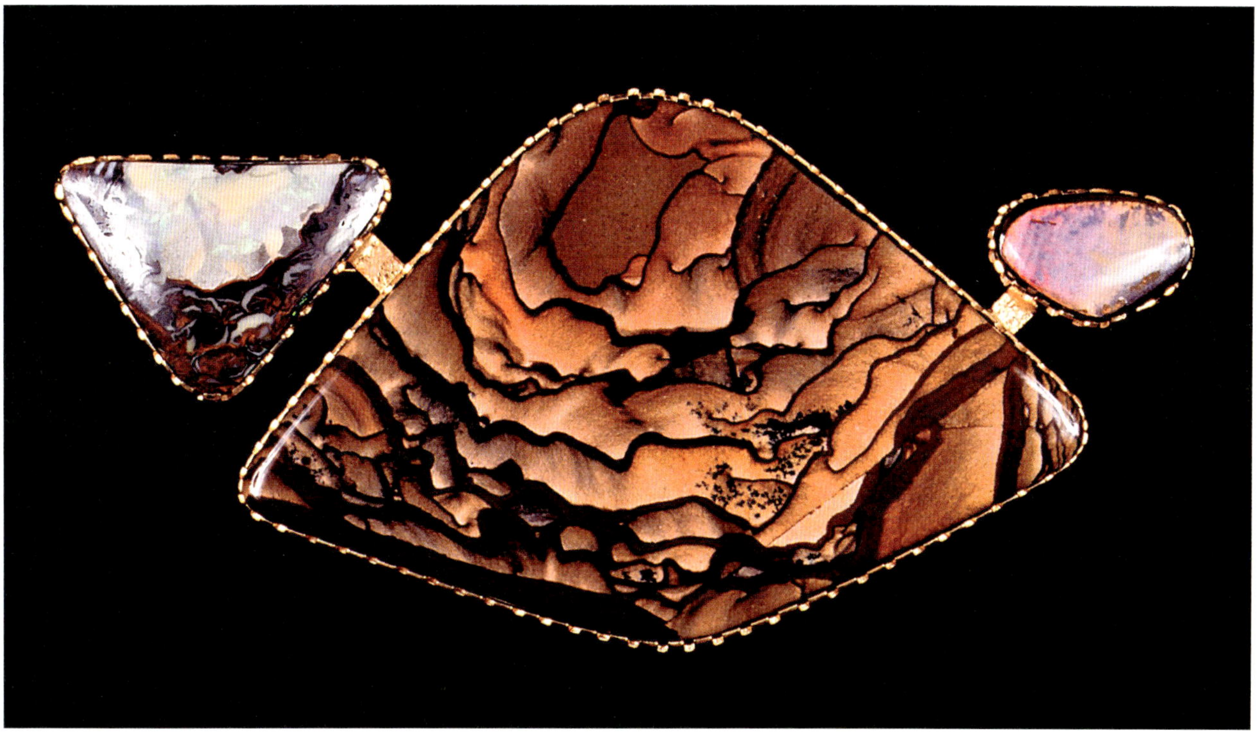

Figure 3. Brooch of Deschutes jasper and Yowah opal set in gold, 1998. (AOA 1998.06.1 © Trustees of the British Museum.)

be worn as alternate sides to the gold set coral cabochons. For the wearer the satisfaction of Johnson and Bird's jewellery is the slightly covert aspects of design: the underlay, the textures, the weight. In the British Museum's brooch only the wearer, or an acute observer, might pick out the texture that comes from the Navajo technique of tufa casting for the bridges between the painterly jasper and the exotic opals.

The arrival of Johnson and Bird in Edinburgh in December 2009 was an occasion to speak about the acquired pieces. The coral necklace was discussed at some length due to the distinct place of coral in jewellery. It substitutes for spiney oyster shell which, prior to the import and trade of coral, brought the colours of purple, orange and red into jewellery. Given its prominence coral can be understood as a correlative of turquoise.

> Coral in the Southwest is seen as a precious resource. It really has become part of the value system, and the culture of both Pueblo and Navajo people. It is integrated into the colour scheme and the belief system of the Navajo and it has been incorporated into the traditional apparel of Pueblo women. When you are preparing to take part in a ceremony, in many places, the traditional necklace that you wear is a necklace of coral beads … We tried to use it in the way that is a variation on a contemporary look, it's a fashion piece, but it also has a very traditional cultural link because of the colour and the materials (Bird 2009).

While the conversation addressed materials and design in general, the proportion of elements within the necklace was specifically discussed. This might seem an odd focus, but as Bird remarks, knowing how things are made yields new insights and connections. In the coral necklace there are eight gold discoidal beads, distributed asymmetrically, which add much to its visual rhythm. This is attributable to their weight and proportion as they evoke, in shape and size, the turquoise and shell discoidal beads used in *jaclas* or the beads of traditional shell and stone necklaces. Conceptually therefore, within this necklace, the gold beads can be seen as analogues of turquoise or shell. Looking again, the dynamic relationship between the coral and gold beads has another resonance: with squash-blossom necklaces (see Bird, in this vol., Fig. 7, p. 205). In the classic squash-blossom necklaces of

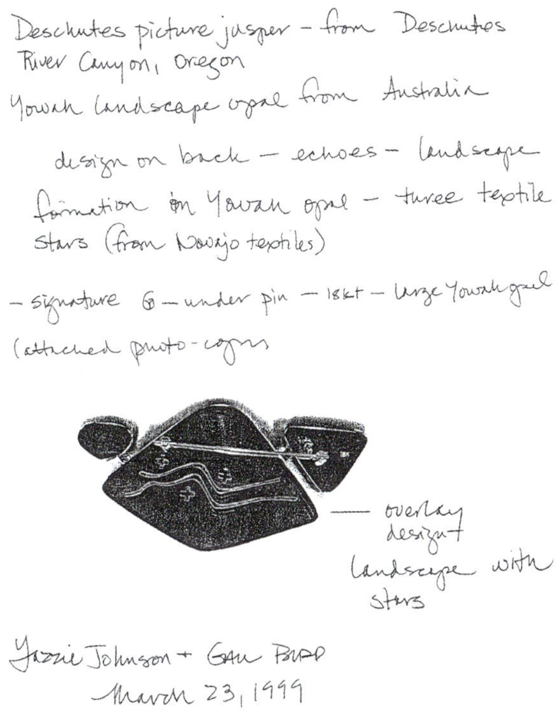

Figure 4. Underlay design of brooch, copied and annotated by Gail Bird in 1999. (AOA 1998.06.1 © Trustees of the British Museum.)

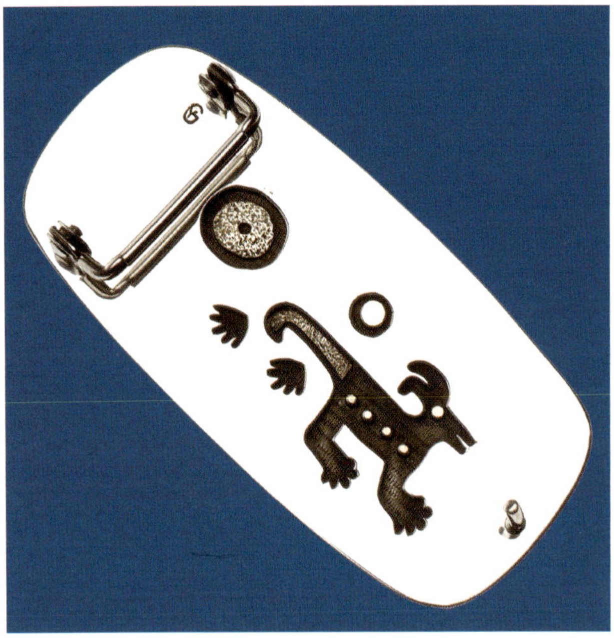

Figure 5. Underlay design of buckle (NMS V.2009.246, © Trustees of the National Museums of Scotland.)

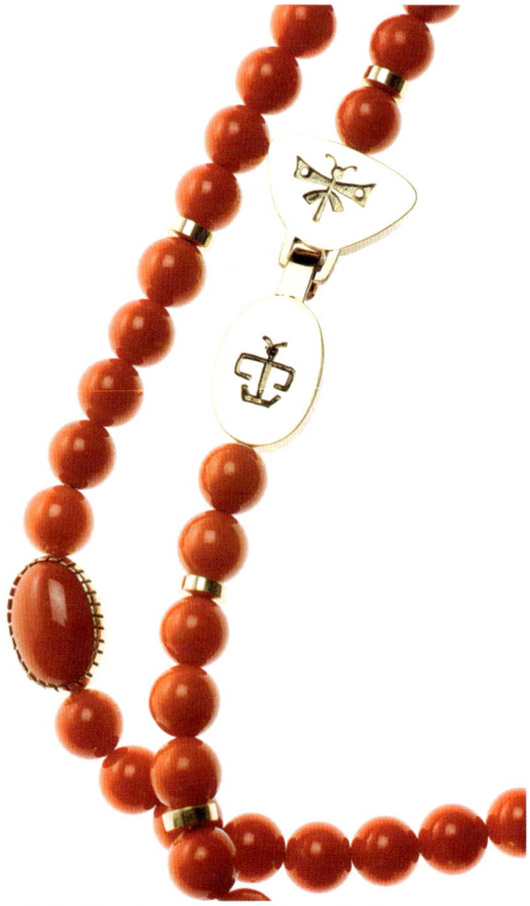

Figure 6. Detail of necklace showing underlay designs. (NMS V.2009.245, © Trustees of the National Museums of Scotland.)

the late nineteenth and early twentieth century, single strands of large silver beads are interspersed with large squash blossoms held by bars breaking the line of round beads. In later styles, prominent around the 1920s to the 1940s, cabochon-set turquoises were linked to strands of silver beads by textured bars, so that the roundness of one is interrupted by the edges of the other, very much like the Johnson and Bird necklace.

The evocations of contrast, form and proportion are tacit. They emerge on repeated viewing although they are wholly consistent with the aesthetics of balance, harmony and variation that is so representative of Southwestern jewellery design. The weight of the gold discoidal beads, like the textured bars in the British Museum brooch, is a consequence of tufa casting. Casting and the repeated working of metal are very much part of the aesthetic that comes from the American Southwest. This important but very subtle detail is achieved through Johnson's meticulous use of a traditional technique. This, in turn, is only perhaps apparent because the desired thickness and appropriate weight are not automatically supplied by mechanically milled sheet.

Mimicry and skeuomorphs

Good quality coral, like good quality American turquoise, is increasingly hard to find. While Johnson and Bird may collect stones with a regard to quality and colour, they also collect and document the materials that have come onto the market masquerading or substituting for natural turquoise. Turquoise is a soft stone and the hard and lustrous 'gem quality' American turquoise, with its wide range of blue and green hues, is desirable and rare. Once one moves away from the standard supply stores, buying turquoise that is good quality seems to be a process veiled with some mystique. Accounts talk about the selective purchasing of older jewellery where the value of the stones exceeds that of the piece, or elusive sources known through individual networks. There are stories of old mined turquoise cached in deposit boxes, waiting to be released onto the market.

The discrepancy between supply and demand, and the price of good American turquoise, has created a series of alternatives, suitable for local consumption and for selling to the average consumer unable to afford collector prices. In the American Southwest, at jewellery supply stores, most of these alternatives can be purchased. Turquoise can be found that is enhanced or chemically treated to improve the colour, or stabilised – treated with epoxy under pressure – to give it permanent hardness. Turquoise is also compressed or reconstituted from small chips. There are good reasons why using stabilised, treated and compressed turquoise might appeal: it is cheaper, more reliable and more versatile. Untreated low grade turquoise, on the other hand, has poor colour, it can be chalky, unpredictable and wasteful of time and labour if used for fine work. Consequently makers of turquoise beads or turquoise carvings often choose to use enhanced materials and this is frequently an acceptable material for regulated art shows. Imitating turquoise is not a novel phenomenon (it is present in the archaeological record) but the mimicry of natural stone and Native-made jewellery remains an issue of concern for those who rely on it for their income.

In addition to those materials mimicking natural stone there are skeuomorphs. The chemical composition of turquoise means that it forms a series with other, equally attractive, natural stone, most notably chalcosiderite, where the tipping point comes through increasing substitution of iron for copper. The stone known as New Landers turquoise,

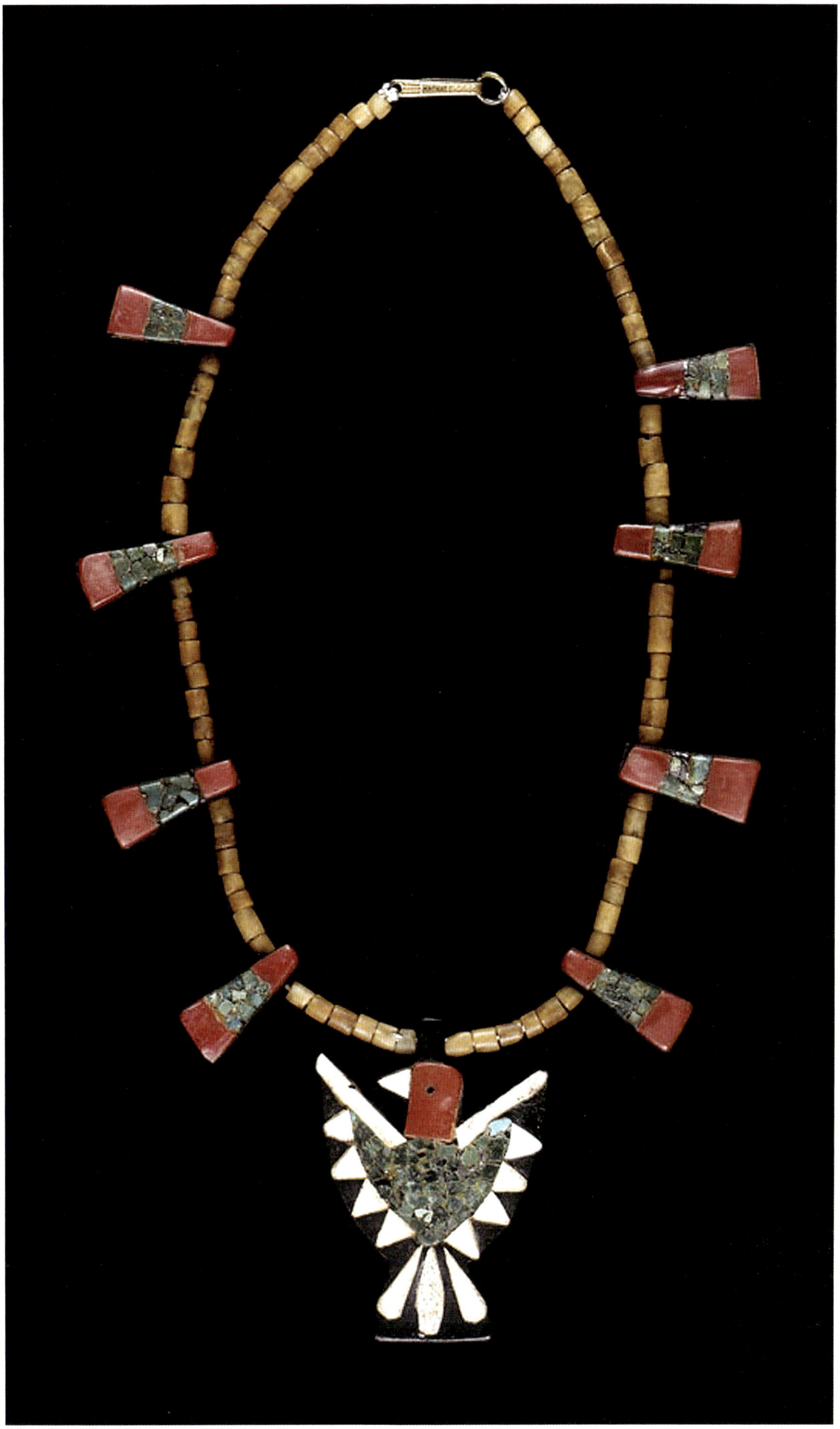

Figure 7. 'Depression' necklace with thunderbird central pendant, Santo Domingo, 1930s–1950s. (AOA. 2001.16.4 © Trustees of the British Museum.)

for example, occurs in a chemical series: when it is light blue-green it contains more aluminium and is mostly turquoise but when it is yellow or a dark moss green it contains mostly iron and is chalcosiderite. Variscite is also attractive and used in jewellery as a substitute for turquoise, though chemically it lacks the necessary copper. What is interesting about mimicry as opposed to skeuomorphs is that on the basis of colour and 'naturalness' they cater to different communities of consumption. Much locally worn jewellery contains stabilised stone, whereas New Landers, being a natural stone, is often sold to collectors.

The relative conformity of styles, the consistent use of stones for colour, the mimicry of natural materials to allow for greater abundance, are all signs of the continuing salience of jewellery to the pattern of social and ceremonial life of Native communities in the American Southwest. Indeed the continuing relevance has created a certain tolerance and playfulness in relation to materials, not always reflected in collecting. Seemingly absent from museum collections are the plastic necklaces made of red and blue plastic Clorox buckets in the 1970s. These buckets were cut up to make 'coral' and 'turquoise' bead and tab necklaces. Johnson, in a conversation in the 1990s, reminisced fondly about seeing these worn in abundance at Shiprock Fair, on the Navajo Reservation. Walking around Shiprock he saw elaborately adorned peopled walking towards him and marvelled momentarily at the wealth on display until the wind changed, blowing the light necklaces hither and thither. Bird also alludes to this when she writes about the dyed corn necklaces made in Cochiti (Bird, in this vol., Fig. 25, p. 211). This is more particularly the case with the 'Depression' necklaces made in Santo Domingo (Fig. 7) which, while collectible, are still not sufficiently well understood in terms of their composition or makers. These necklaces show a desire to modify form and style, and to select certain kinds of material. Recognition that these not only express individuality in terms of the maker, rather than simply conformity or a response to commercial opportunity, is opening up research bringing these less well-understood forms into view (see Bird, in this vol., Fig. 27, p. 211 for a variety of earrings in this style).

Bird, in writing a brief history of turquoise through ornament, explores the complexity of Southwestern adornment as a language in which makers, wearers and viewers are important constituencies, repeatedly drawn into a visual debate (for this tripartite categorisation see Goring 2006). In the American Southwest, artist and wearer are not separate identities: making and using jewellery for traditional reasons is part of life. Sometimes jewellery is so descriptive and figurative that the wearer can literally be wearing a miniaturised and idealised image of him/herself (see Bird, in this vol., Fig. 15, p. 208). As Bird remarks (in this vol., p. 212), in the Southwest '[y]ou have to have the jewellery. That's why the art continues.' As a transacted object of art and culture jewellery in the Southwest is a vehicle for self-reflection and projection: it has many purposes and multiple destinies.

Note

1. I shall use Bird as a convention when referring to the paper by Bird and Johnson in this volume, and the lecture given in 2009, for reasons of readability. I shall refer to the artistic partnership as Johnson and Bird.

References

Adair, J. (1944) *The Navajo and Pueblo Silversmiths.* Norman: University of Oklahoma Press.

Batkin, J. (2008) *The Native American Curio Trade in New Mexico.* Santa Fe, NM: Wheelwright Museum of the American Indian.

Bedinger, M. (1973) *Indian Silver: Navajo and Pueblo Jewelers.* Albuquerque: University of New Mexico Press.

Bsumek, E. M. (2008) *Indian-Made: Navajo Culture in the Marketplace, 1868–1940.* Kansas: University of Kansas Press.

Bird, G. (2009) From Y. Johnson and G. Bird, taped interview, National Museums Scotland, Edinburgh (9 December 2009).

Goring, E. (2006) '"Jewelry and communication: breaking the code", juried exhibition in print', *Metalsmith* **26**(4): 6–9.

Jernigan W.E. (1978) *Jewelry of the Prehistoric Southwest.* School of American Research Southwest Indian Arts Series. Santa Fe, NM: School of American Research.

Ostler, J., Rodee, M. and Nahohai, M. (1996) *Zuni: A Village of Silversmiths.* Zuni: A:shiwi Publishing.

Pardue, D.F. (2007) *Shared Images: The Innovative Jewelry of Yazzie Johnson and Gail Bird.* Phoenix, AZ: Heard Museum Press.

Struever, M.H. (2007) *Loloma: Beauty is his Name.* Santa Fe, NM: Wheelwright Museum Press.

Weigand, P.C. and Harbottle, G. (1993) 'The role of turquoises in the ancient Mesoamerican trade structure', in J.E. Ericson and T.G. Baugh (eds), *The American Southwest and Mesoamerica: Systems of Prehistoric Exchange.* New York: Plenum Press, 159–177.

Author

Henrietta Lidchi, Department of World Cultures, National Museums of Scotland, Edinburgh, UK

Shared images: continuity and innovation in turquoise jewellery

Gail Bird and Yazzie Johnson

Introduction

Yazzie Johnson and I are self-taught jewellers from New Mexico. Our work is considered innovative and contemporary because of our use of gold and unusual stones. We consider it an extension and continuation of the jewellery arts of the American Southwest. We have both studied older jewellery, learning design and construction in the process. In addition, we have both read extensively and been fortunate to have had access to museum and private collections, older jewellers, scholars and historians. The techniques Yazzie uses are hand techniques based on traditional Navajo and Hopi metalsmithing. My role is to plan, design, critique and explain our work. Over the course of our 35-year career, we have given demonstrations and been asked to speak about our work. We have developed presentations detailing our learning and experience. We have photographed many of the places and art ourselves. We have met and interviewed others. Our approach is a personal, yet historical, view of the evolution of jewellery production in the American Southwest. Here in talking about jewellery, we are going to stress the use of turquoise, its importance in the lives and culture of the people of the Southwest and the reasons for its viability as an art form.

Commentary

Figure 1 is a great piece of contemporary mosaic work by a woman named Angie Reano Owen from Santo Domingo Pueblo. It is new, as is the necklace illustrated in Figure 2 of incredibly beautiful Bisbee turquoise nuggets strung on rolled shell beads. These pieces have a long evolutionary history. Work like this has been produced in the American Southwest for centuries. The techniques have endured, but the tools and technology have changed the quality of materials used. That an art form could survive in almost an unaltered form is testament to the resilience of the people living and working within their environment.

This jewellery comes from the American Southwest, a land of limited resources and very little water. I like to think that those resources are used in the most imaginative and careful way; that people have always handled nature and natural

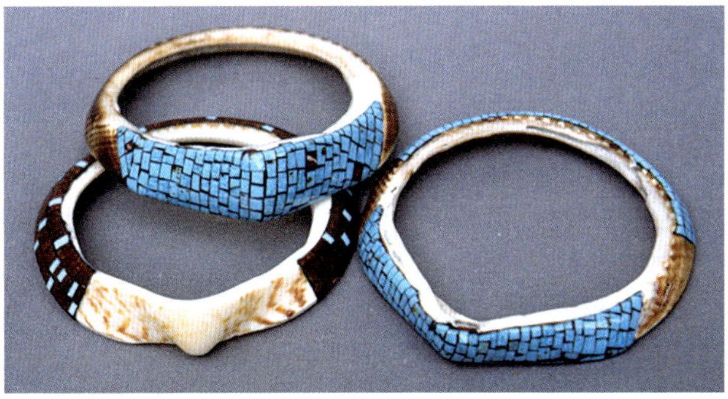

Figure 1. Shell and inlay bracelets by Angie Reano Owen. Shell, turquoise, pipestone and coral, 15 × 45 cm diameter. Private collection. (Photo: Gail Bird and Yazzie Johnson, used with permission.)

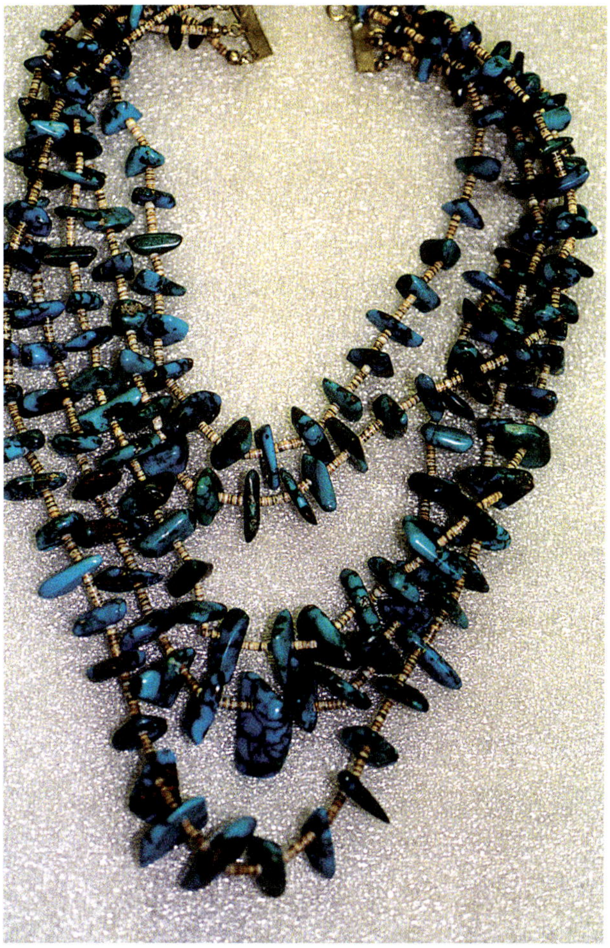

Figure 2. Anonymous. Five-strand turquoise necklace: graduated lengths, 56–64 cm. Bisbee turquoise nuggets, rolled *Olivella* shell beads. Heard Museum, Phoenix, Arizona (gift of Maureen Nichols). (Photo: Gail Bird and Yazzie Johnson, used with permission.)

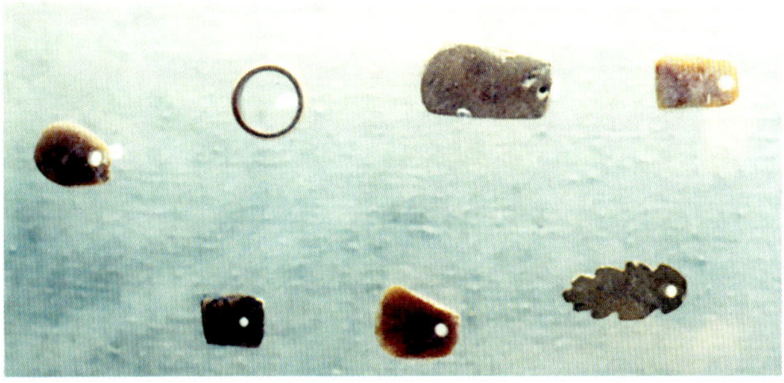

Figure 3. Various pendants, stone, shell, Hohokam culture (ca. AD 1300–1450). Private collection. (Photo: Gail Bird and Yazzie Johnson, used with permission.)

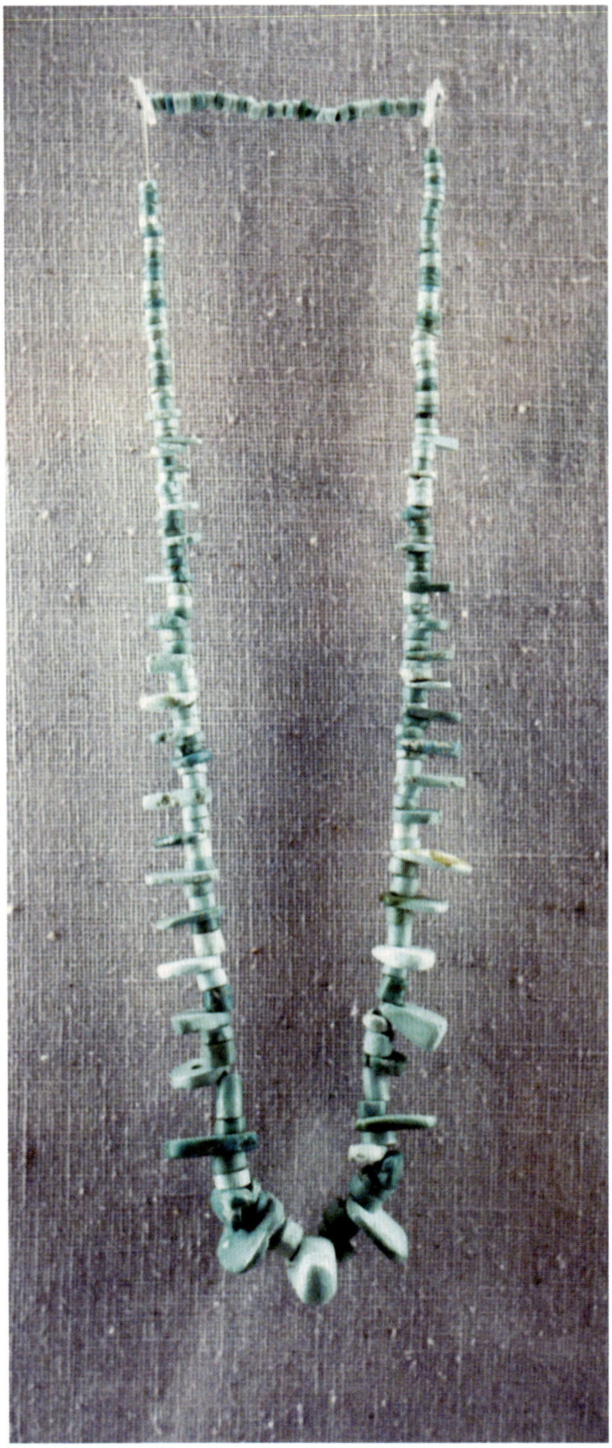

Figure 4. Rolled turquoise tab and bead necklace, Hohokam culture (ca. AD 1300–1450). Private collection. Photo: Gail Bird and Yazzie Johnson, used with permission.

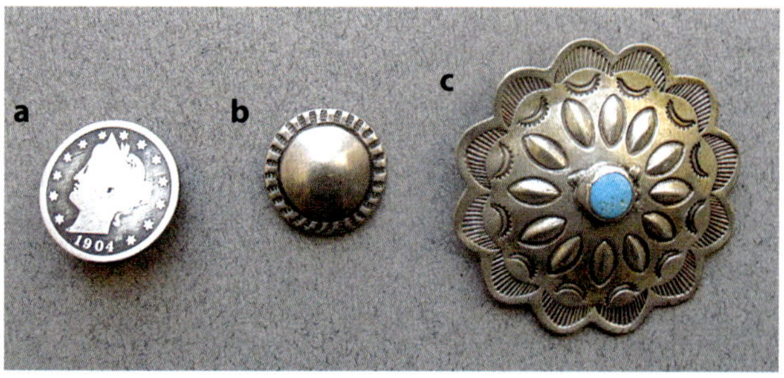

Figure 5. (a): Anonymous, probably Navajo, 1904. Liberty Head dime, mounted on copper shaft to make a button. Diameter 72.22 cm. Catalog number 47/1191c. (b): Silver button, no date. Diameter 1.9 cm. Catalog number 47/974. (c): Stamped silver *concha* with turquoise set, ca. 1900. Diameter 4.44 cm. Catalog number 47/232a. Wheelwright Museum of the American Indian, Santa Fe, New Mexico.

assets in the Southwest. If you look at the Native landscapes you can see that use was made of the cliffs by carving into the rock walls. People created homes by digging out living spaces, making use of the natural qualities of the land and the natural physical beauty of their surroundings. Cliff dwellings are juxtaposed with natural tufa formations.

Other structures begin to develop. Ideas develop and art in the form of self-depiction evolves on the rock walls. When you begin to think of self, the next step is self-ornamentation. The early pieces are very simple stone beads and pendants made of shell, bone, stone and clay (Fig. 3). Hohokam mosaic work, earrings and pendants were made by affixing turquoise and *Spondylus* shell to bone, stone and shell surfaces. There are examples of rolled bead necklaces: materials are broken or cut into small flat pieces, individually drilled, strung on a strong cord (now wire), then hand-rolled over a series of large flat stones to create a uniform diameter bead. Figure 4 is an example of a tab or nugget and rolled bead necklace. The tabs are made by taking pieces of turquoise, smoothing them on a flat stone then drilling. The necklace has a soft appearance and colour, because the easiest turquoise to work is the softer lighter material. In early work you see shades of blue and green. It is not until metal tools are introduced that you begin to see dark, hard turquoise being used.

There has been discussion about trade and where materials come from. The ideas of trade, adaptation and change are instrumental in the development of the jewellery of the American Southwest. As shells come inward, inland, they gain value by the series of trades that take place before reaching their final owner. When you see the finished piece, you have to wonder about the time and tools. How were the pieces sliced, formed, drilled? There were no metal tools. Simple stone and bone instruments were used to cut, smooth and polish, creating simple yet elegant objects.

In 1540 things change. The Spanish come to the Southwest and when they come not only do they bring a new religion, they bring domestic animals, metal and different arts. The jewellery starts to change, but not immediately. It is not until the 1800s that the art of silversmithing begins to develop and in its earlier stages it is not silversmithing but blacksmithing that is learned and taught. It begins with copper and brass, very simply. Bracelets are made by the hammering, pounding and filing techniques that are the blacksmith's craft.

When people look at a finished piece, they do not always understand its complexity. When I talk about jewellery, I like to explain the process; I feel the simplest way is to show how something is made. To most modern people buttons are a functional item. In the hands of a Navajo they become a highly decorative ornament.

The button illustrated in Figure 5a is a Liberty Head dime. It is already a self-contained ornament. It has the profile of a woman with small stars surrounding her head. Someone has learned to solder, to make a loop of copper, solder it to the reverse and it becomes a wearable ornament. The button in the centre (Fig. 5b) involves advanced tool-making and two steps. It is made by taking a ball-peen hammer and using it to create a concave depression in a wooden stump. Then you set your circle of silver over it and using the same hammer you pound to get a half-domed cap. It is very simple – using basic tools you create more tools. Taking a simple stamp and using it over and over again creates the crenellated edge on the outside. This is a simple button. Navajo women wear blouses of velvet and velveteen and they will take dozens, sometimes hundreds of buttons and sew them along the collar, the sleeve line, the seams and the cuffs. Varying in size, very few if any serve a function, but an idea is extended, creating a very complex whole.

Two more buttons made by very simple techniques; they look very similar but they are not (Fig. 6). The one on the left with the crenellated edges involves two stamps. The button on the right looks very similar but instead somebody has taken a very thick piece of silver, made the dome, and then using a straight file has carved away the edge. When you look at it in relief you see the series of triangles, a series of jagged lines going around the edge. Very simple tools though very complex, and very time consuming.

The beginning of simple technology leads to an elaboration of design and style. The idea of a simple bump-up leads to the creation of male and female dies to produce *repoussé*-like ornamentation. Stampwork becomes more complex as smiths and toolmakers gain access to finer tools. They can saw away metal and create curves as this scalloped edge button indicates (see Fig. 5c). The fine-lined stampwork is possible because the fine files are able to cut into tool steel and create greater intricacy. Soldering has become a little more refined. Someone has learned to make a bezel, a thin frame or holder for a stone, and has soldered the frame to the button's surface. Turquoise now gains an importance in embellishing a plain ornament.

Adaptation and change are very important parts of what makes Southwest jewellery. These are two tufa-cast *najas*, made and worn by Navajos. They were probably worn as pendants (Fig. 7, top and centre). The *naja* is a Navajo design adapted from the Spanish. When the Spanish come, they come highly ornamented; they even ornament their horses. The horses wear silver head stalls, and at the front there was this crescent-shaped pendant. It is an adaptation of a Moorish symbol, to ward off the evil eye, but to a Navajo it doesn't mean the same thing. It has a pleasing shape that looks like a crescent moon.

Then things begin to come together (Fig. 7, bottom). You take the simple half-domed button, drill a hole, file the edges, solder the two halves and you have a silver bead. You take the *naja*, refine it and add fine turquoise. You take another item you've seen the Spanish wear; silver pomegranate blossoms sewn on the sides of their pants. What is a pomegranate?

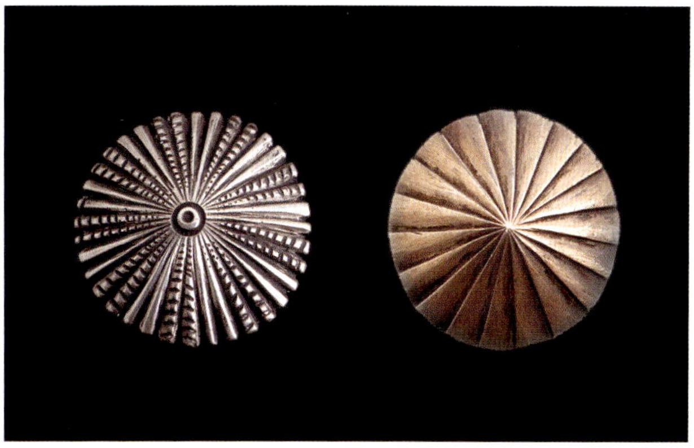

Figure 6. Navajo stamped and filed silver buttons. Anonymous. Heard Museum, Phoenix, Arizona.

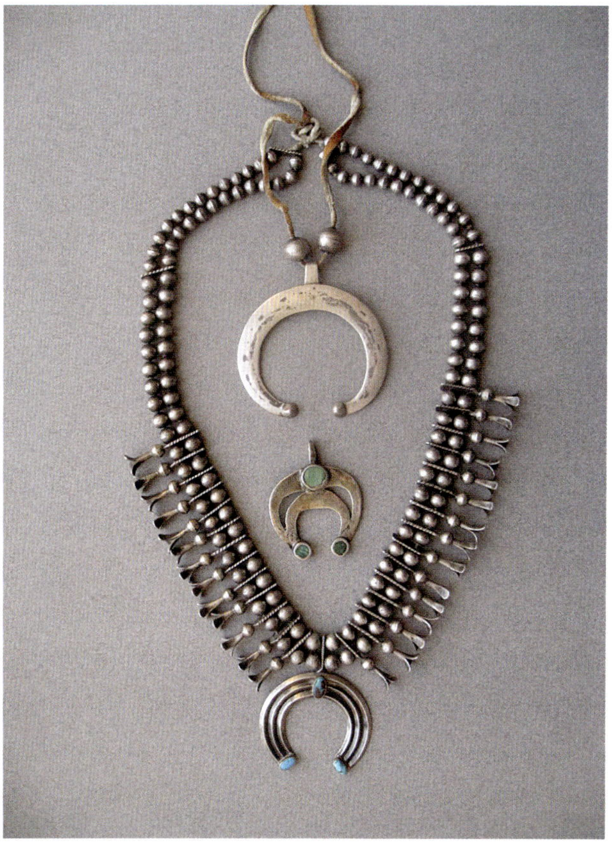

Figure 7. *Top*: Anonymous, probably Navajo. *Naja* on leather thong, no date. 8.25 cm wide × 8.25 cm long. Catalog number 47/678. *Centre*: Silver *naja* with three turquoise sets, ca. 1885. 6.35 cm high × 5.08 cm wide. Catalog number 47/677. *Bottom*: Silver squash-blossom necklace and *naja* with three turquoise sets, ca. 1900. *Naja* is 7.62 long × 6.35 wide. Total length of necklace is 40.64 cm. Catalog number 47/290. Wheelwright Museum of the American Indian, Santa Fe, New Mexico.

Its blossom looks like a squash blossom, but the petals are too short. If the petals are elongated it looks like something we know, so it changes. How do you put this all together? Functional bars to extend the blossoms away from the beads, but why leave them plain? If I take my file and run the edge over the bar, I can create lines. The lines add nothing to the stability or strength of the bar but they give a dimension and strength to the finished piece. This is where technology, adaptation and change all come together; the creation of a uniquely Navajo piece marked by simplicity, strength and balance. When you become familiar with the descriptive style, you think you can recognise and identify work with

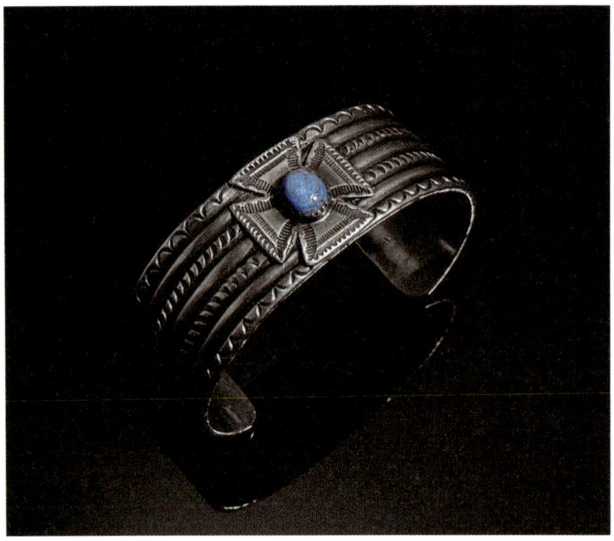

Figure 8. Anonymous, Navajo. Stamped, appliqué silver bracelet with turquoise. Heard Museum, Phoenix, Arizona.

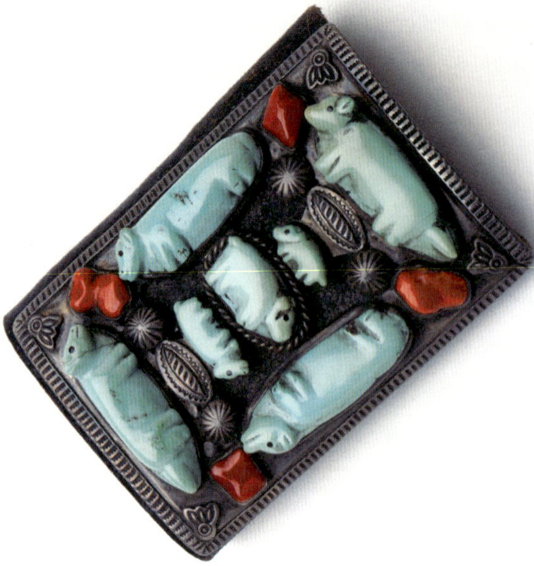

Figure 10. Silver bow guard by Zuni artists Leekya Deyuse (turquoise bears), and Dan Simplicio (coral). Heard Museum, Phoenix, Arizona.

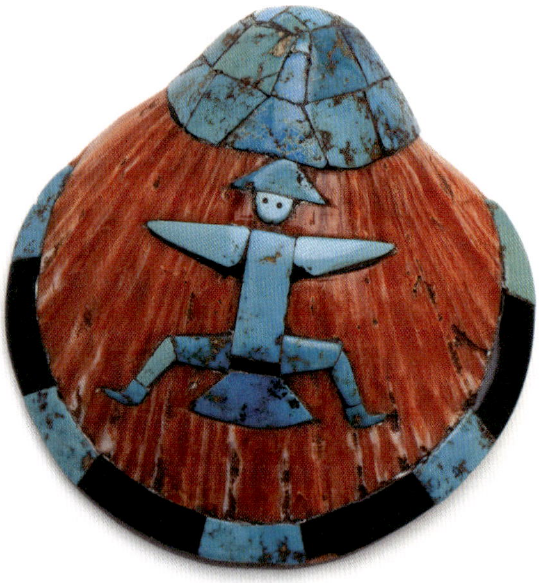

Figure 9. Zuni knife wing figure. Blue Gem turquoise, *Spondylus* shell. Heard Museum, Phoenix, Arizona.

authority. I love this necklace. I have used it as an example for years although several years ago I noticed it had been strung incorrectly. To me, it represents the use of heavy silver, the simplicity of design, the sparing use and importance of fine turquoise balance and symmetry.

Figure 8 is a classic example of Navajo silver work: heavy metal, enhanced by the application of a second layer of silver, deep stampwork, and with the emphasis on a single central stone. Later there is use of much larger stones, the embellishment of twisted wire, and multiple symmetrical matched set stones. The balance and symmetry and clean solid metalwork in bracelets combine with the scale of the turquoise in perfect balance.

This piece (Fig. 9) from the late 1940s is distinguished by the dramatic interplay of natural Blue Gem turquoise and *Spondylus* shell from the Gulf of Mexico. The *Spondylus* shell is inlaid with a turquoise knife-wing figure. Artistically the Zunis have the freedom to use symbols adapted from both their secular and religious oral traditions and they embellish their work with them.

Collaborative work was orchestrated in the late 1940s at Zuni by a trader named C.G. Wallace. It shows how much importance and influence a trader could have providing he had the right materials at the right time and was there to supply and support artists. The next images feature work from this period.

The turquoise bears in this silver bow guard (Fig. 10) are more great examples of Zuni stonework. They are also the work of Leekya Deyuse. Small coral branches have been set at each corner. The overall scale of the bow guard matches the scale of the carvings making it almost sculptural.

The large turquoise and coral butterfly inlaid in jet (Fig. 11) is surrounded by small turquoise stones set in silver. It is one of the finest examples from this or any period.

Another area in which the Zunis excel is in their depiction of human figures, both secular and ceremonial (Fig. 12). These pink shell carvings of a man and woman were carved by Leo Poblano. They are about 15 cm high. The woman is wearing a turquoise necklace. Around the base of her skirt are inlaid small rosettes of turquoise that are meant to represent silver and turquoise pins worn down the right side and above the hem of a woman's *manta*. The man has a necklace of rolled beads and a *concha* belt set with turquoise.

There is a lot of debate about who makes what: Navajos, Zunis, men, women. Navajos are the first to learn; they develop a style then teach others. Men make jewellery. If women help, there are no records. Families work together. In the late 1940s names start to emerge and some are recognised.

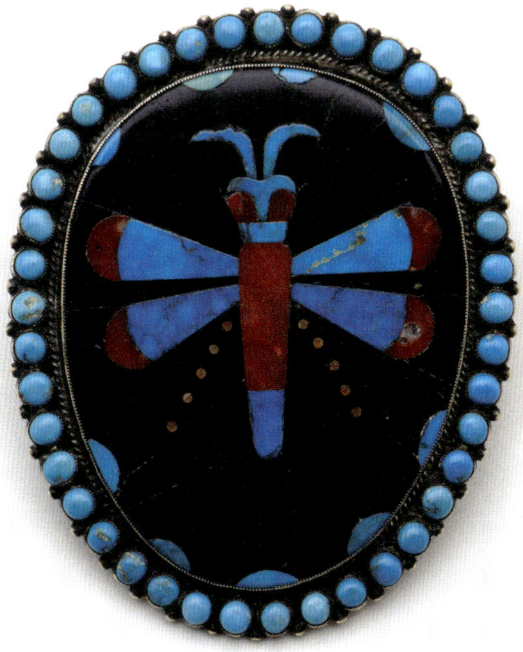

Figure 11. Butterfly inlaid in jet, by John Gordon Leak (Zuni). Heard Museum, Phoenix, Arizona.

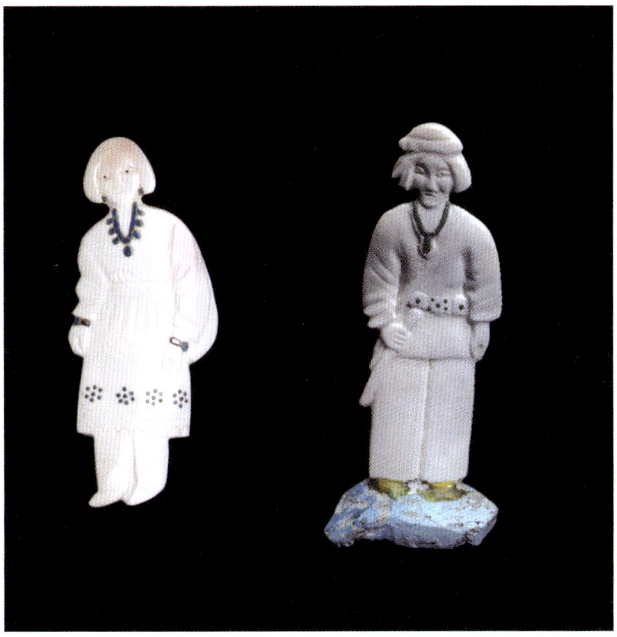

Figure 12. Pink shell and turquoise figures by Leo Poblano (Zuni). Heard Museum, Phoenix, Arizona.

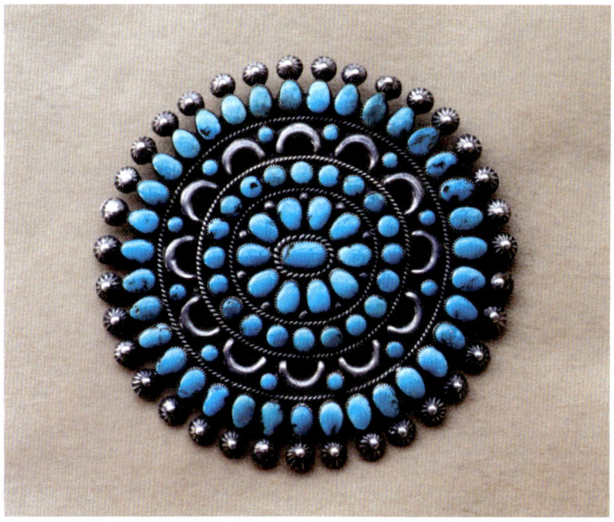

Figure 13. Turquoise and silver cluster pin. Zuni. Private collection. (Photo: Gail Bird and Yazzie Johnson, used with permission.)

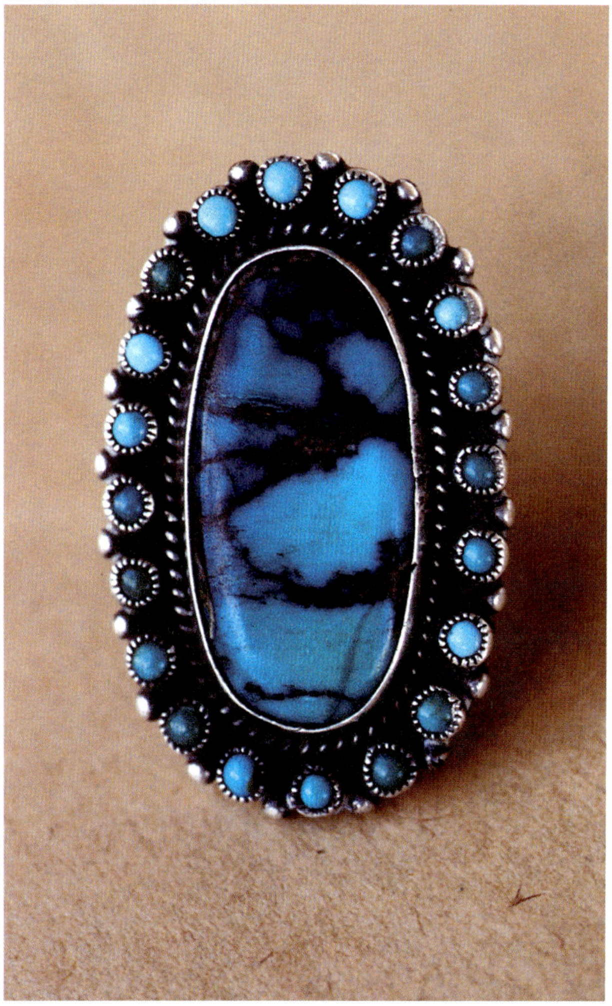

Figure 14. Bisbee turquoise ring. Anonymous artist, Navajo. Private collection. (Photo: Gail Bird and Yazzie Johnson, used with permission.)

Individual styles start to appear. When those styles mature, design increases in complexity and the distinction starts to blur. Zuni work is supposed to be small stones set in complex silver designs, Navajo is larger and heavier. The distinctions are there, but categorising meshing styles becomes harder.

You might think this piece is Navajo (Fig. 13). We know that it is Zuni. It was made by a person who developed this style in the 1950s and it was identified by the dealer from whom it was purchased. It is a very complex piece. When you make comparisons between jewellery you look at the finished product, you look at how the silver work is put together, how finely the turquoise is cut, the quality of the stones and then the piece tells you what it is.

Figure 14 is a great Bisbee turquoise ring set with tiny turquoise surrounding the major stone. Turquoise can be named in many ways. It can have a place name, a geographical location; sometimes it is named for the person who discovered

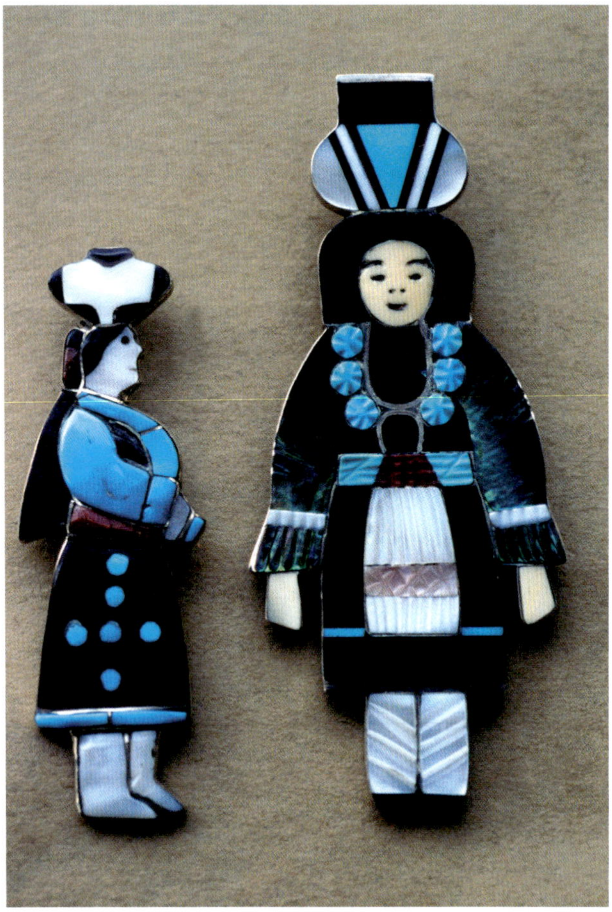

Figure 15. Two female inlay figures. Anonymous artist, Zuni. *Left*: Silver, jet, mother-of-pearl, coral, turquoise, lapis lazuli, 70 × 25 cm. *Right*: Silver, jet, pink and white mother-of-pearl, turquoise, coral, abalone, 90 × 45 cm. Private collection. (Photo: Gail Bird and Yazzie Johnson, used with permission.)

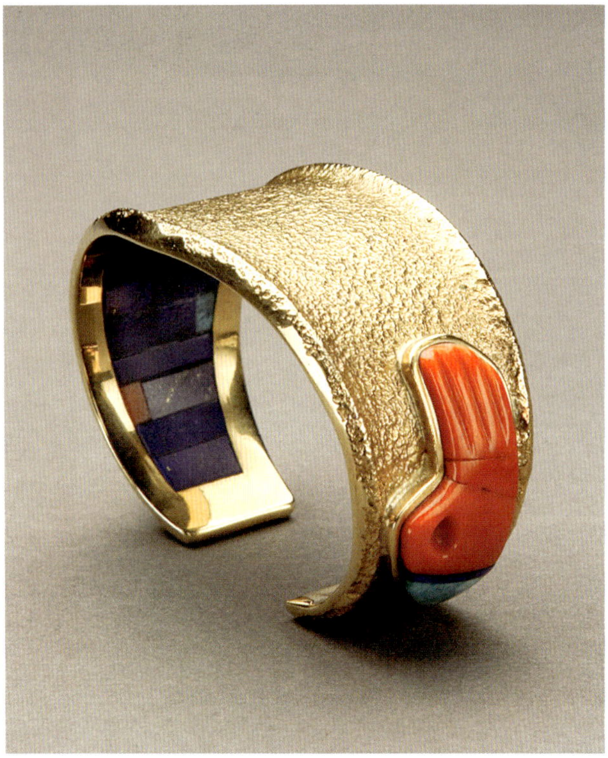

Figure 16. Charles Loloma, Hopi. Tufa-cast bracelet with badger paw motif and interior inlay, ca. 1980. Gold, lapis lazuli, turquoise and coral. 6.35 cm wide × 3.175 cm high. Promised gift. Wheelwright Museum of the American Indian, Santa Fe, New Mexico. (Photo: Addison Doty.)

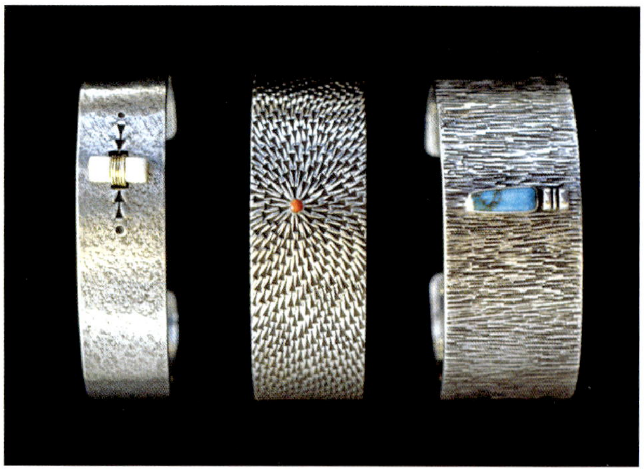

Figure 17. Three bracelets by Norbert Peshlakai. *Left*: Silver, ivory, 14K gold wire. 15 × 152 cm. *Centre*: Silver and coral. 25 × 152 cm. *Right*: Silver and turquoise. 30 × 152 cm. Private collection. (Photo: Gail Bird and Yazzie Johnson, used with permission.)

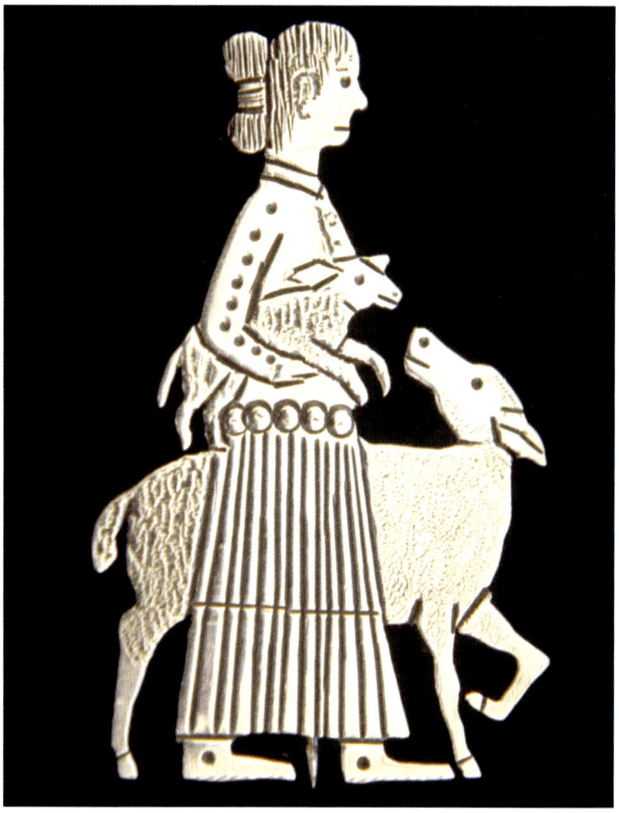

Figure 18. Woman and sheep. Pin. Clarence Lee, Navajo. Silver. 52 × 32 cm. Private collection. (Photo: Gail Bird and Yazzie Johnson, used with permission.)

or owned the mine or sometimes it is given a romantic or descriptive name, such as Blue Gem or Royal Blue. Bisbee turquoise is from Bisbee, Arizona. It is distinguished by a very deep dark blue with a chocolate brown matrix that swirls through it.

The Zuni art of depicting human figures extends into today (Fig. 15). Many jewellers specialise in creating stylised images of traditionally attired figures. The larger pin can range in size from 2.5 cm to 12.7 cm in height. They have commonalities. You can see the carved contours of the dress; each article and layer of clothing is delineated by the change of material. The *manta* or top layer is made of jet, the underdress is pink mussel shell; abalone, turquoise and white shell detail the apron. Her jewellery is prominently presented in turquoise.

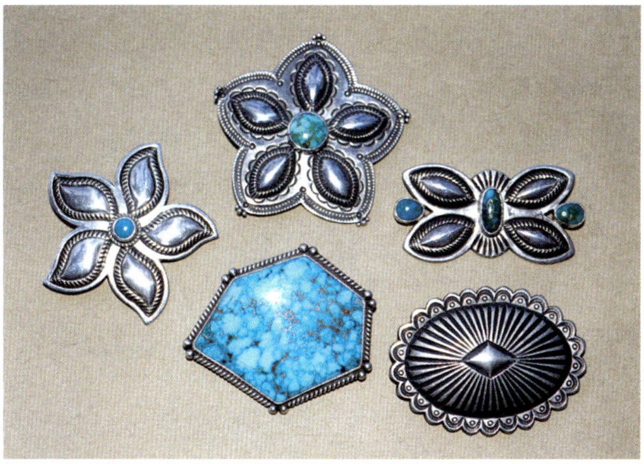

Figure 19. Stamped silver and turquoise pins. Perry Shorty. Navajo. Private collection. (Photo: Gail Bird and Yazzie Johnson, used with permission.)

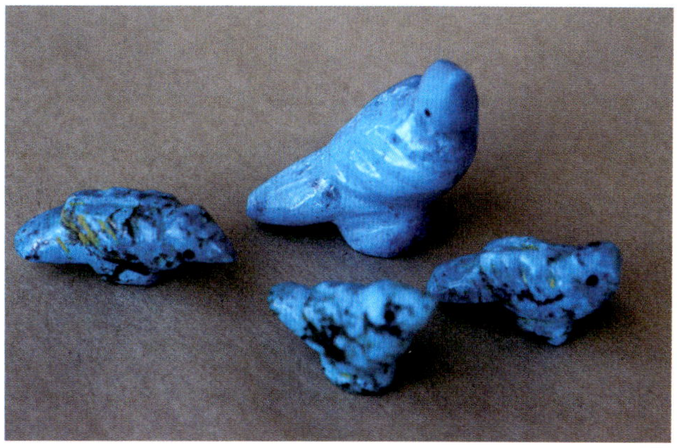

Figure 20. Four turquoise fetish birds by Sara Leekya. Various types of American turquoise. From top: 45 × 62 cm; 15 × 42 cm; 32 × 50 cm; 25 × 50 cm. Private collection. (Photo: Gail Bird and Yazzie Johnson, used with permission.)

Though there is a standardisation of styles by the 1960s, in the late years of that decade everything begins to change largely because of Charles Loloma. He is traditionally raised and educated at Hopi, then leaves to pursue an education as an artist. He studies ceramics and painting. He is aware of his surroundings, his religion and background have impacted upon his views and it is recalled in his work. When he discovers jewellery, he explores the medium using what he can afford in building ideas. This bracelet (Fig. 16) of tufa-cast gold has an interior lined with turquoise and other stones reflecting a different sense of values and awareness. The additional colour and value is known to the wearer and hidden from the viewer.

The idea of texture can be seen in the work of Navajo jeweller Norbert Peshlakai (Fig. 17). Now in his mid-50s, he uses a complex style of hammering and stamping, basic Navajo techniques – not to create balance and symmetry, but to create texture. These bracelets are made by taking a single stamp and repeatedly pounding it into silver, creating a textured surface rather than a lined or designed piece. If you look at the stones, the turquoise, the coral, the ivory, you see that he has given that single stone the importance that early smiths gave to stones except that he has done it in a uniquely different way.

Sheep: they are very important to the Navajo. Figure 18 is the work of Clarence Lee. I like this woman, stamped out in silver, carrying a little goat. You can see that she is very proudly dressed in her Navajo outfit, wearing a *concho* belt. The jewellery is important because it identifies her as a well-dressed Navajo woman, just as the Zuni pieces identified Zuni women.

Figure 19 illustrates stamped silver and turquoise pins by Perry Shorty, a Navajo man probably in his early 40s. When he started work, he worked in a store in Gallup that sold supplies to Navajo silversmiths. As Perry watched them work he kept thinking to himself, 'I can do this, but I just need to get started. How am I going to do this and make it different?' He started at the very beginning; making his own stamps, going back to traditional styles, casting his own silver, pouring coin silver, pounding it out so that he made his own silver sheet and wire. He began to work in a style that is very evocative of the old and very simple, very plain but in keeping with the highest style of Navajo work.

These carved turquoise birds (Fig. 20) were made by Sara Leekya, the daughter of Leekya Deyuse. We met about five years ago. She works by herself in the tradition of her father carving individual fetishes. Small fetishes used to be carved for the use of the individual. Most men would carry one in a pouch or they would be placed at shrines. The idea of an art object for a table or shelf was an alien concept but the Zuni now make them for sale. These are rather small; they are less than 7.6 cm high.

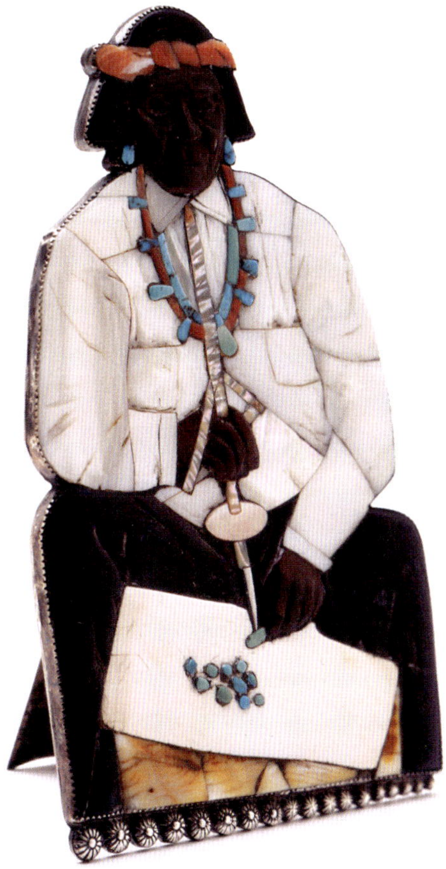

Figure 21. Leo Poblano. Shell, turquoise. Heard Museum, Phoenix, Arizona.

Figure 21 is a large inlaid Zuni piece from the 1950s. It was probably carved by Leo Poblano and shows a man drilling beads using his hand-drill placed vertically over a stone on a stationary surface. He is rather grandly dressed to be working, but the image created was probably at the instigation of C.G. Wallace to inform, impress and ultimately woo the collector.

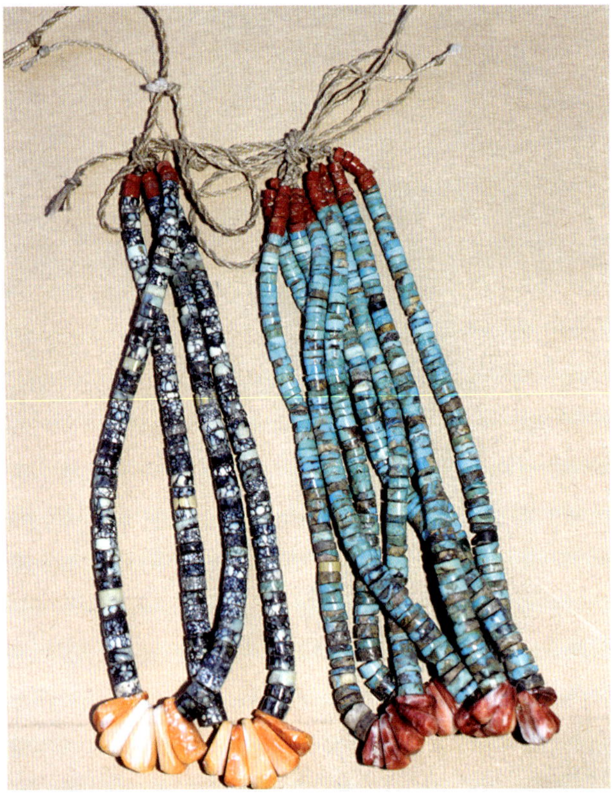

Figure 22. Two *jaclas*, both by Ray Lovato (Santo Domingo Pueblo). *Left*: Coral, turquoise and shell. 126 cm length × 4 cm diameter. *Right*: Double pair: coral, Blue Gem turquoise, *Spondylus* shell. 126 length × 4 cm diameter. Private collection. (Photo: Gail Bird and Yazzie Johnson, used with permission.)

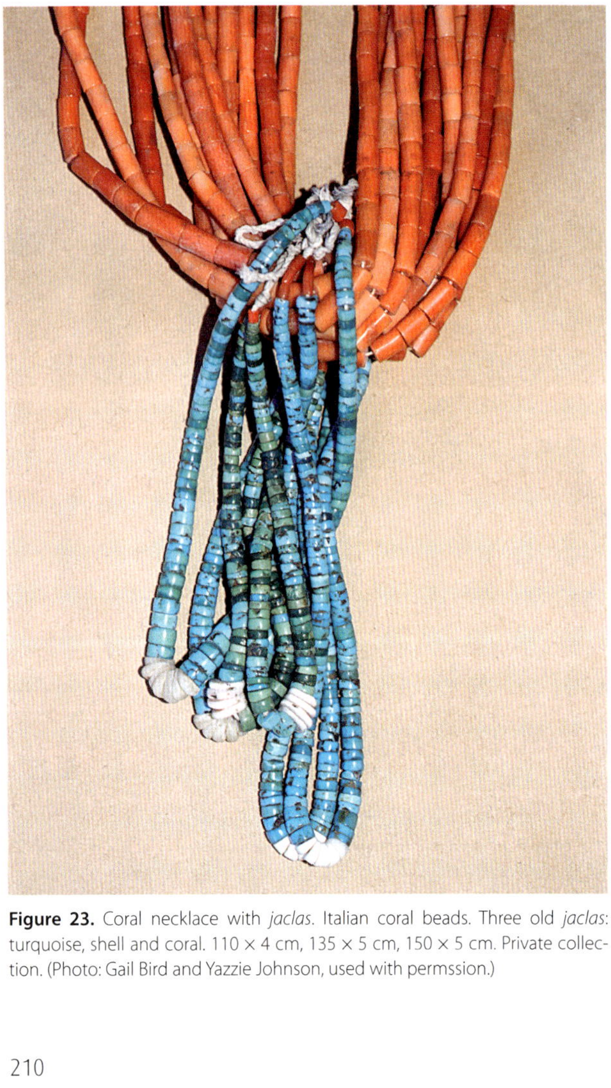

Figure 23. Coral necklace with *jaclas*. Italian coral beads. Three old *jaclas*: turquoise, shell and coral. 110 × 4 cm, 135 × 5 cm, 150 × 5 cm. Private collection. (Photo: Gail Bird and Yazzie Johnson, used with permssion.)

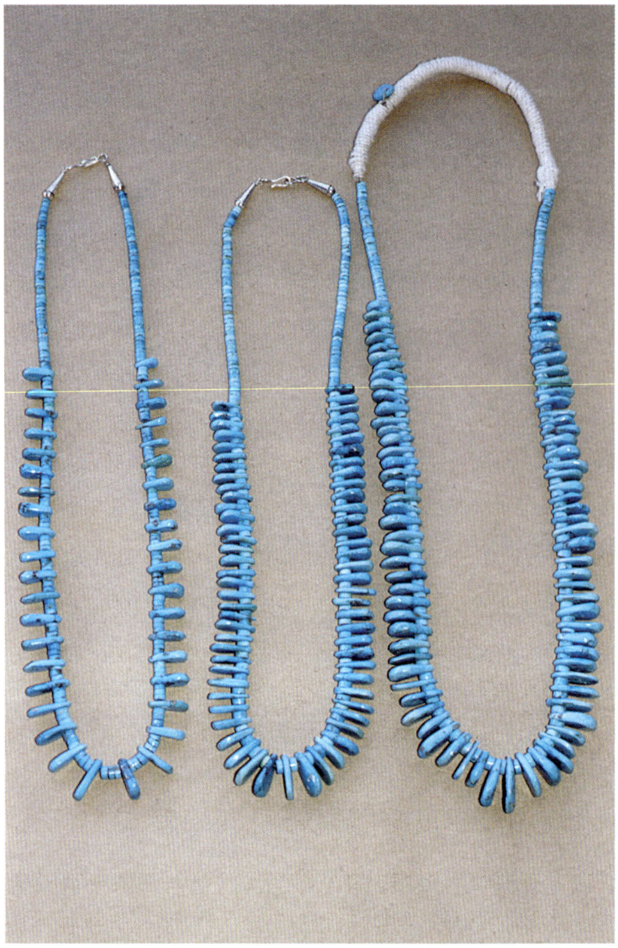

Figure 24. Three bead and tab turquoise necklaces by Ray Lovato. 110 × 4 cm, 135 × 5 cm, 150 × 5 cm. Private collection. (Photo: Gail Bird and Yazzie Johnson, used with permission.)

No discussion of turquoise can ignore *jaclas*. They are worn and used by most Pueblo and Navajo people of the American Southwest. They are also highly prized by non-Natives. Traditionally they were primarily made at Santo Domingo Pueblo and traded with other tribes. Figure 22 illustrates fine examples of natural turquoise *jaclas*, one with carved shell corn at the centre bottom (left); the double pair (right) has shaped coral. *Jaclas* can be worn as earrings, short necklaces or attached to the bottom of a necklace (Fig. 23). The impact of colour is of primary importance, just as the Zuni turquoise and *Spondylus* shell pieces seen earlier. The colour red does not appear naturally or in an easily workable material in the Southwest. In the 1920s coral is imported from Italy in bead form and is introduced by C.G. Wallace in Zuni. It becomes an important trade item readily accepted into the material culture of Pueblo and Navajo people. A coral necklace with *jaclas* like this would be worn by women at one of the many Pueblo dances along the Rio Grande.

There has been discussion about materials; what their use might have meant, how the juxtaposition of different colours brings or creates light, and brings alive the mosaics. All are good points, but do not take into account one of the major characteristics of turquoise: it is a soft material, it absorbs oils, and will change colour when worn, so sometimes that play of colour is pure happenstance and luck.

Wearing and using rolled bead and tab necklaces and *jaclas* is an important component of the living culture of

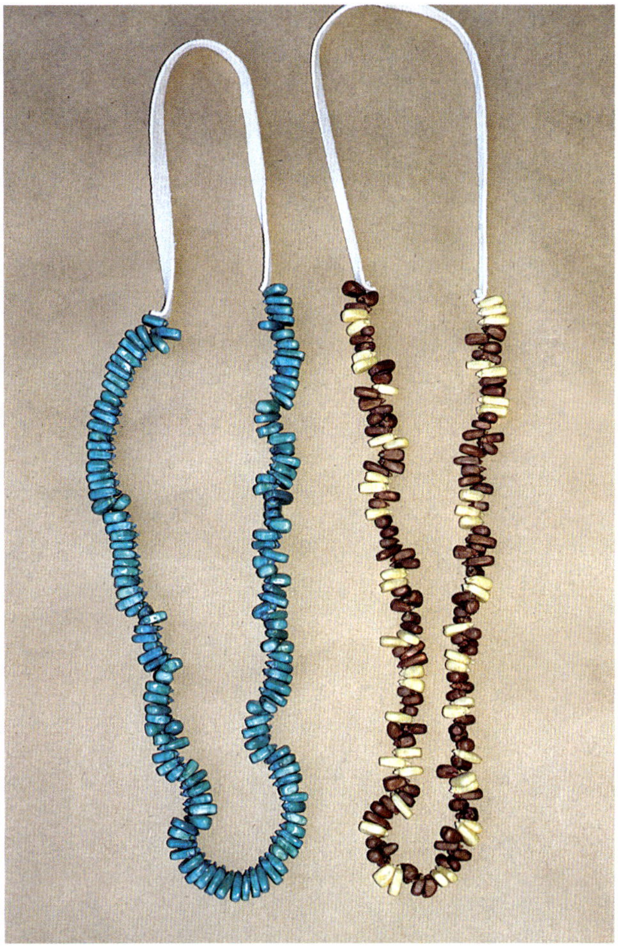

Figure 25. Two dyed corn bead necklaces, Santo Domingo Pueblo. Length: ca. 30 cm. Private collection. (Photo: Gail Bird and Yazzie Johnson, used with permission.)

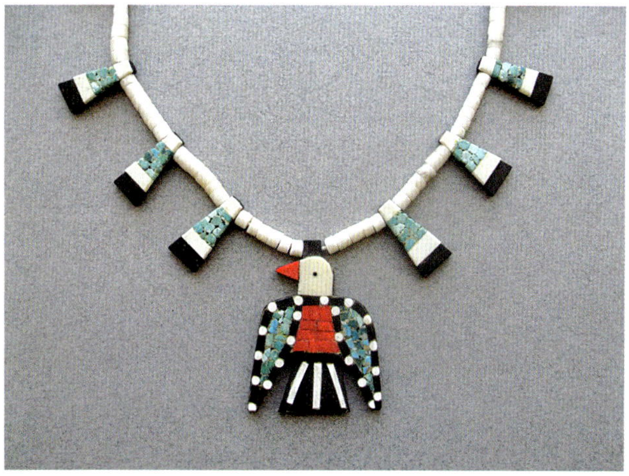

Figure 26. Anonymous, Santo Domingo Pueblo. Thunderbird necklace, ca. 1945. Hard rubber, plastic or celluloid, turquoise, gypsum beads. Pendant is 6 cm high × 4.76 cm wide. Necklace is a total of 40 cm long. Catalog number 47/759. Wheelwright Museum of the American Indian, Santa Fe, New Mexico

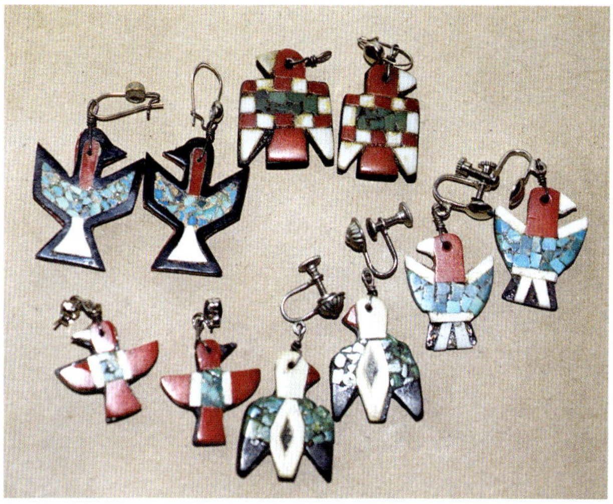

Figure 27. Anonymous, Santo Domingo Pueblo. Mosaic, plastic turquoise and shell, various sizes. Private collection. (Photo: Gail Bird and Yazzie Johnson, used with permission.)

many American Southwest tribes. Work of this type (Fig. 24) is still being produced for sale, trade and personal use. Very fine contemporary work incorporating a diverse array of stone, shell, silver and gold is being made.

For traditional jewellers, what they do is a necessary part of their purpose in this world. To participate, you have to possess and wear the material goods that define life. In ceremonies and often in secular life, the costuming, the clothing and the jewellery identify the wearer as belonging and being part of the larger community. There are many who feel strongly that their work helps to continue their cultural lineage.

One such man is Ray Lovato from Santo Domingo Pueblo. His entire family participates in the fabrication and sale of these necklaces. He is known for his use of natural materials. In a time when the use of stabilised and dyed turquoise is more prevalent, Ray continues to produce old-style beads that are highly prized. The family also fabricates a line using lower quality materials that meet the demands of the commercial market. These are two necklaces that look like shell and turquoise nuggets (Fig. 25). They are white Indian corn that has been dipped in fabric dye baths, sold to tourists along the highways and offered in trading posts and gift shops. They are sold as what they are. They provide an outlet and an income for those who do not have the skills or who can't afford to work with stone. They also provide a purchasable souvenir. This reliance on supplemental income has its roots in the early development of the tourist trade. Over the years we had heard stories from a Santo Domingo friend about her family making and selling tourist jewellery. The work was seen as a way to feed the family and learn the craft. Who would have thought that cut-up plastic hair ornaments and combs could be reconfigured in such a way?

There is tremendous growth and variation in the production and sale of Native arts. In all areas a new class of worker-artist evolves. The Depression of the 1930s affects this income base well into the 1960s. The shell and turquoise workers of Santo Domingo Pueblo respond to this decline by creating objects based on old ideas and techniques but using less valuable, though equally colourful, materials. Gypsum, plastic and low-grade turquoise replace the original material and new designs emerge. Unlike the Zuni, Santo Domingo artists have severe restrictions on what symbols they can use. Luckily birds are allowed and natural forms evolve into fantasy (Fig. 26). Then later, what looks like turquoise, shell, coral and jet is actually cut-up plastic glued onto black battery casing. This type of work has been known for years, and pieces are included in museum collections as oddities. Recently Yazzie and I were privileged to see and photograph a private collection whose range, playfulness and craft surpassed anything we had ever seen (Fig. 27).

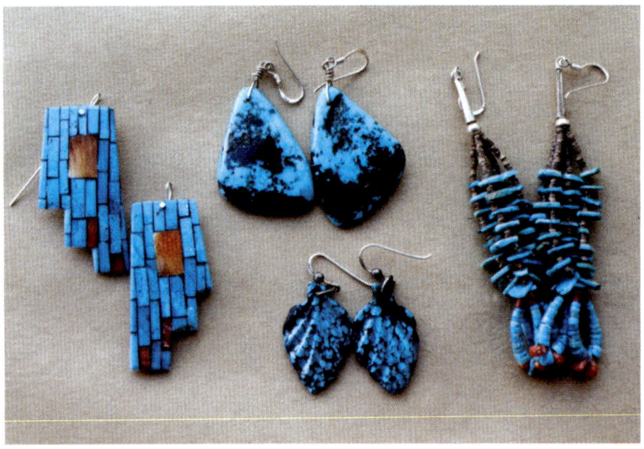

Figure 28. Four turquoise earrings. *Left*: Turquoise and *Spondylus* shell by Carlotte and Percy Reano. 4.4 × 2.2 cm. *Centre top*: Natural Tyrone turquoise by Ray Lovato, 3.8 × 2,5 cm. *Centre bottom*: Stabilised Chinese turquoise by unknown Zuni artist. 2.8 × 1.9 cm. *Right*: Miniature necklace: turquoise, olivella shell, coral by Ray Lovato. 1.5 × 6.8 cm. Private collection. (Photo: Gail Bird and Yazzie Johnson, used with permission.)

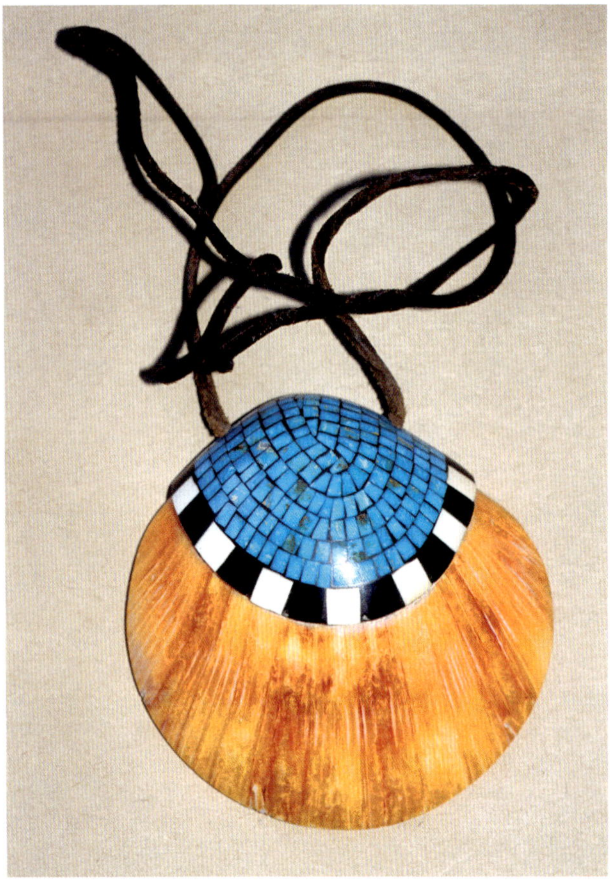

Figure 29. Mosaic *Spondylus* shell pendant by Angie Reano Owen. Turquoise, shell, jet. Private collection. (Photo: Gail Bird and Yazzie Johnson, used with permission.)

When the economy improved and Indian jewellery regained popularity in the early 1970s, quality materials and craft resurfaced. The historical connection between the Santo Domingo tourist jewellery and the prehistoric Hohokam forms were made and the revival of fine Santo Domingo shell work began. These last illustrations show contemporary work by Ray Lovato and Angie Reano Owen to compare and illustrate those connections. The first pair of earrings (Fig. 28, left) recalls Hohokam mosaic style, but instead of the regularity of block on block placement, the turquoise pieces are longer and the finished form has been shaped to almost resemble a human profile. There is a pair of very fine dark blue tab earrings with dark black matrix by Ray Lovato (Fig. 28, centre top). The carved leaves are Zuni, the turquoise is stabilised (Fig. 28, centre bottom). Many contemporary carvers often use stabilised material to reduce loss when carving intricate items. The miniature necklace/earrings (Fig. 28, right) were made by Ray Lovato and his family. He probably made them for same reason he once made a replica of the enormous Leekya Deyuse Blue Gem necklace he saw at the Millicent Rogers Museum. He wanted to try to excel at something within the confines of his defined world.

The dance shell (Fig. 29) was made by Angie Reano Owen. While the tradition of making rolled shell and turquoise beads and dance shells for personal use had continued side-by-side with the gypsum and plastic made for sale, the makers were unaware of many of the prehistoric roots of their art. When Angie saw a photograph of a plain *Gycymeris* (cockle shell) pendant, she envisioned it partly in mosaic and worn as a bracelet (see Fig. 1). She made many variations partially decorating using an assortment of materials before fully covering the surface. She found a shell with a large natural curve and sliced and formed a cuff, which she then decorated with turquoise and coral. She changed and revived the art by altering the shapes and using new natural materials. She encouraged her extended family members to also work to revive the family and village tradition in their own individual styles. Eventually she experimented with her mother's old squash-blossom patterns, using natural materials instead of plastic. She continues to work, inspiring her children and grandchildren to make items to be worn and to be accessible to all. She may be one of the best-known 'names' in the world of Indian jewellery, but the act of providing beautiful jewellery for her people is of utmost concern. She lives the traditional ideal; jewellers are here to serve a purpose.

In my talk at the British Museum my last slide was a photograph of a Santo Domingo woman dressed as a Navajo woman. There are dances at Santo Domingo in which the participants dress and dance as Navajos. Out of respect for their neighbours, everything must be correct. You have to have the jewellery. That's why the art continues.

Authors

Gail Bird and Yazzie Johnson, Artist Farmers, New Mexico, USA

We wear our wealth

Shane Hendren

Reflections

I am Shane R. Hendren, a Navajo (Dine), from Tohatchi, New Mexico, on the Navajo Nation. I currently reside in Albuquerque, New Mexico. I am Tlogi, my mother's clan, born of Irish, Comanche, Choctaw, and Cherokee, my father's heritage. My maternal grandfather is Nakai Dine of the Toh haglinee clan. My paternal grandfather is Choctaw, Cherokee and Irish. A paternal great-grandfather is Comanche and Irish descended from Quanah Parker. I am a direct descendant of Manuelito, one of the Navajo head men who signed the treaty of 1868 that returned the Navajos from the concentration camp at Bosque Redondo to their traditional homeland in the four corners region of the southwest USA.

I am the current President of the Indian Arts and Crafts Association (henceforth IACA), which was formed in 1974 by artists, retailers and wholesalers with the purpose of preserving and protecting authentic American Indian arts and crafts. Governor Bill Richardson has also appointed me to the State of New Mexico's Native American Arts & Crafts Authenticity Task Force. These associations offer opportunities to ensure that authentic indigenous art continues to be a viable livelihood for the makers of this art. Their purposes are directly connected to the preservation and maintenance of my culture for future generations around the globe.

In the same manner as my ancestors, I have lived a semi-nomadic life, moving to where the resources necessary to sustain life are more easily accessed. As a result, I have lived on and off the Navajo Reservation spending a great deal of my life on ranches throughout the American Southwest. As livestock producers, we Navajos are tied directly to the land. One can only raise as many calves for market as the grass

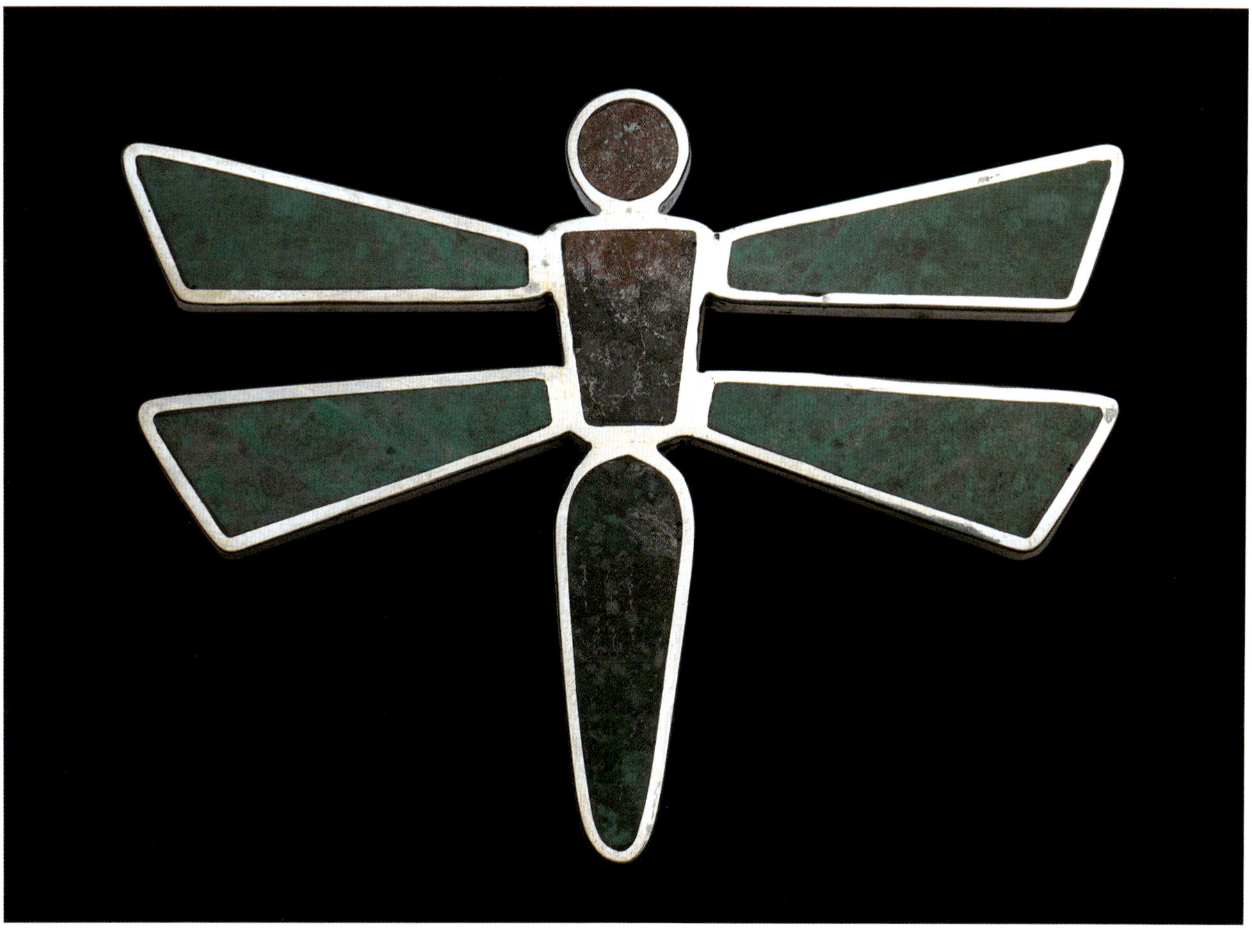

Figure 1. Brooch by Shane Hendren. Silver, turquoise (British Museum, AM, 2010,2002.1, © Trustees of the British Museum).

will allow. A keen knowledge of the environment is directly linked to traditional beliefs that support and encourage living in balance and harmony with the universe.

Navajos are raised to walk in beauty; to be in balance and harmony within the universe is one's ultimate life goal. These morals are buttressed by stories that reinforce and illustrate how and why this is important. Part of the Navajo Creation story that addresses when we emerged into this present world discusses how the ants broke off pieces of the turquoise sky of the fourth world before leaving it to come into this current world. They took these turquoise pieces and distributed them around the earth so that we could use them in the future. Turquoise provides protection and ensures health and happiness. Therefore, when a Navajo infant is born, the infant is given turquoise to wear.

Throughout one's life, a Navajo is given turquoise, often in the form of jewellery, to mark significant changes like puberty, or accomplishments such as graduation from school. Wearing one's turquoise and jewellery demonstrates to others that the wearer is wealthy and comes from a successful, productive and strong family. The individual serves as a representative of their immediate and extended family, whether that family is biological or clan related. Both turquoise and jewellery are a symbolic display of the collective accomplishments of a person's family.

In modern times, turquoise has proven to be a commodity more stable than money. Navajo artists have used it in increasingly creative ways to provide for their families' needs. Turquoise has been so widely used by Native artists that it has become synonymous with Indian art, particularly Navajo jewellery. For many Native individuals, the making of art, jewellery in particular, is their primary vocation. Navajo art is rooted in traditional beliefs that make it a vehicle for the maintenance and retention of culture. Turquoise in jewellery is a portable representation of this culture. By this means, many people, from around the world, are drawn to acquire it. Turquoise that is creatively and beautifully made into jewellery is highly prized by collectors.

This popularity has become a double-edged sword. As the popularity of Native art, and especially jewellery, has grown so have the instances of fakery. In a recent *Wall Street Journal* article about such unscrupulous practices it was reported that as much as 75 percent of the art being sold today is fake. Considering that the same article reports that the Indian art business is a billion dollar industry, that is a substantial amount of money that is being diverted from Native artisans whose ideas and creativity are being stolen from them. When you also consider that Native Americans as an ethnic group in the United States have the highest rates of unemployment, and live in abject poverty, these numbers become increasingly unsettling.

This is not a new phenomenon; in fact it is the very reason why IACA was formed in 1974. Today IACA is actively working to educate the public about authentic Indian art and change the laws that apply to the sale and production of Native-made products. These changes benefit all those involved in the industry, from the artisan to the retailer and on to the collector, ensuring that what is being sold as genuine Indian art is in fact genuine. This also includes ensuring that the materials being used are properly identified, and this also applies to turquoise. Like the art itself, turquoise has been faked and sold fraudulently to unsuspecting buyers. When misrepresentation is discovered, it not only damages the legitimacy of the art, but also those like me who have dedicated themselves to producing art with this material.

Artists are the ambassadors of their people. Through their work they provide a vehicle for the rest of the world to gain a better understanding of who we are. Collecting of Native American art gives one a direct link to our people and our culture – a culture that is vibrant and alive. Most of the art and jewellery produced by Native artists is done by hand. For this reason it is an actual extension of that person, it is their legacy.

The making of art is ceremonial by its very nature. Artists produce their work when they are in a good place mentally, emotionally and spiritually. This state of being ensures that the traditional values we have been raised with are perpetuated. Our art serves as an indicator to the world and our children that our beliefs have value to us and those who collect our work. This value is intellectual, spiritual and commercial. Most of all, it is our voice in a visual language that is appreciated globally.

Author

Shane Hendren, Master Metal Smith/Indian Arts and Crafts Association, USA

Eighteenth-century economy, twentieth-century merchandising: the market for turquoise in the American Southwest, 1900–1940

Cheri Falkenstien-Doyle

Introduction

On 2 November 1938 John Adair, researching his book *The Navajo and Pueblo Silversmiths*, interviewed Herman Schweizer, an executive with the famous hotel chain the Fred Harvey Company and manager of its Indian Department in Albuquerque, headquarters of the firm's curio business.

> We used to sell a little [Navajo jewellery] ... But a lot of [it] was too heavy for the tourists' taste, so at that time I started in having silver made up to order. It was in 1899 that the turquoise from the north – Nevada and Colorado – began to come in here. I was the one who got a mine owner by the name of [William Petry] in Nevada ... to cut the stones for Indian use, cutting flat square and oblong stones. The first trader to make silver for me was a trader at Thoreau [New Mexico]. From there we branched out to Sheep Springs and the surrounding regions (John Adair Collection, Field Notes, Book 6).

When Herman Schweizer began farming out silver and turquoise to be made into souvenir jewellery, Navajos had been silversmiths for fewer than 50 years. They were introduced to the rudiments of working silver as early as the 1850s by the same Hispanic New Mexican blacksmiths who taught them to make and repair horse gear, but the craft did not flourish until after 1870. At that time many men who had recently been released from internment at Fort Sumner in New Mexico Territory – where between 1864 and 1868, in an effort to curtail raiding and other violence among Native Americans, Hispanics and Anglo-American settlers, the United States government imprisoned more than 7,000 Navajos – began working as wage labourers at military installations such as Fort Wingate, the Navajo Agency at Fort Defiance, or railroad towns such as Thoreau. Their wages were paid in silver coin, which could be melted and cast into ingots and then laboriously hand-wrought. The first Navajo silversmiths used blacksmithing tools, improvised implements and fine files to make bracelets, rings, concha belts, and bridles for Native use, as well as watch fobs and tableware on commission for American soldiers, traders and other outsiders (Matthews 1883).

But while silver had always been a medium of exchange and commerce with the various occupants of the American Southwest, the value of turquoise rested in the ceremonial and religious significance it held for indigenous peoples throughout the region. In Gary Witherspoon's explanation of the relative importance of silver and turquoise among Navajo people, he says 'Navajos view silver as a very attractive metal that lends itself well to jewelry making and sculpture.' Turquoise on the other hand:

> is a symbol of the grand cosmic scheme of 'hózhó' (beauty and harmony); and ... it functions as a lubricant to enhance this scheme when it gets disrupted. Turquoise is a symbolic means of communication that expresses social harmony and universal beauty. Its natural attractiveness makes it an ideal symbol of the dominant concept in Navajo world view ... It is special because it is a means of harmonious communion with the other beings in the universe (Witherspoon 1981: 48–49).

So while Schweizer's scheme to commercialise Native American jewellery may have represented a logical step in the development of a market for silver, it marked a profound change in the way that turquoise was obtained, handled and used.

Navajo, silver and turquoise

When anthropologist John Adair conducted his initial investigations into the origins of Navajo and Pueblo silversmithing, he interviewed men who were among the first generation of smiths. Many of them remembered when turquoise stones were first set into silver jewellery – about 1880 among Navajos and by 1885 at Zuni – and they remarked on its scarcity and value. On 6 June 1938 a smith called Keneshde recalled that people from Santo Domingo Pueblo, who traditionally worked the turquoise deposits at Cerrillos, New Mexico, brought rough stone to Zuni and traded it for buckskins and black wool cloth from which women's dresses were made. Zunis cut and polished the turquoise and traded it to Navajos, who set it into silver bracelets that they traded back to Zuni. According to another smith, Juan

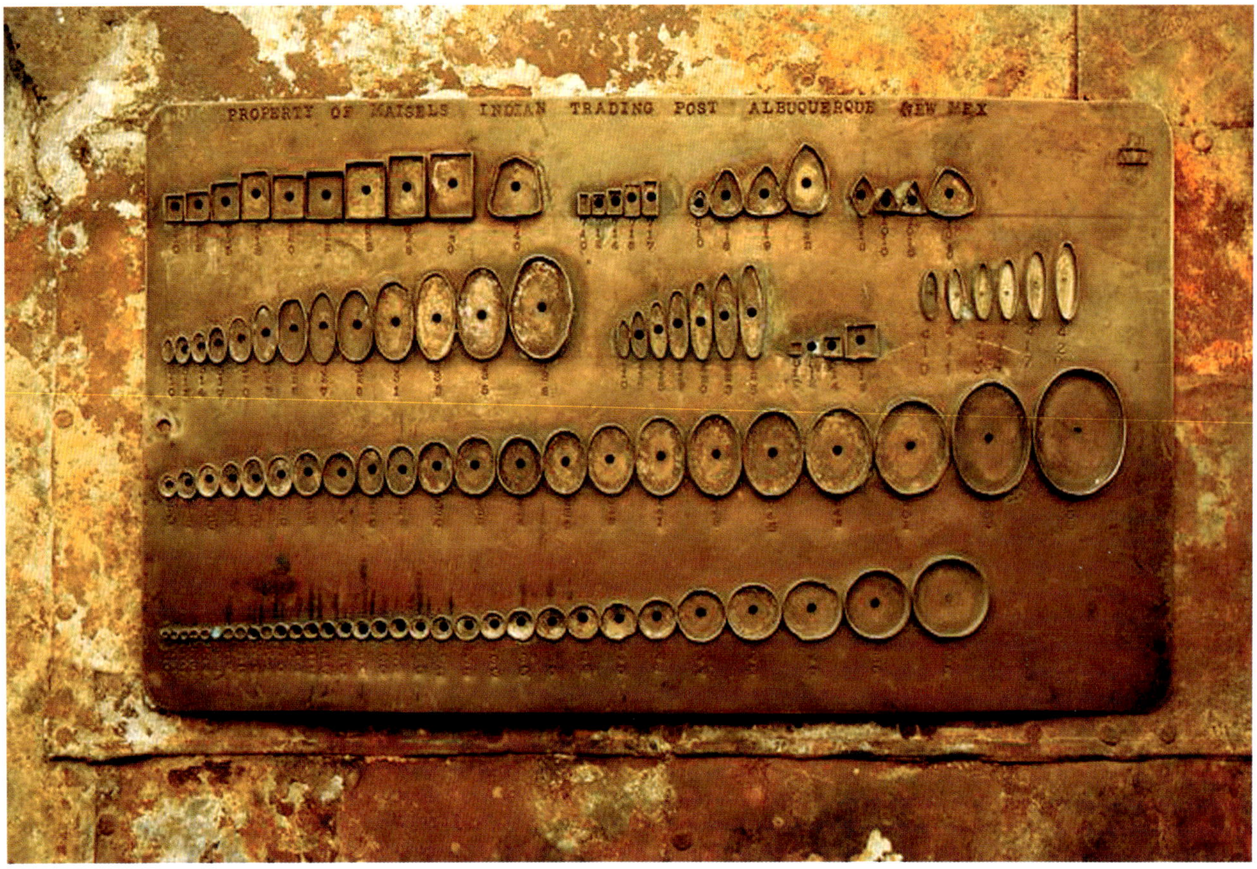

Figure 1. A master gauge for standardising commercially cut turquoise, used by Maisel's Indian Trading Post of Albuquerque, New Mexico during the 1930s. Width 35.56 cm. Collection of Armen Chakerian. (Photo: Addison Doty.)

Deleosa, 'the Zunis would give two good pieces of turquoise for a good silver bracelet'. By 1890, when Zunis became interested in obtaining their own stone from the Cerrillos deposits, they first had to secure permission and pay a fee to an American mine owner (John Adair Collection, Field Notes, Book 1).

Large-scale mining in Nevada commenced in 1859 with the discovery of the Comstock Lode, the first important silver strike in US territory. Until then the United States imported all its silver. With the outbreak of the Civil War, Nevada's production of gold and silver became crucial to the survival of the Union, and the territory was made a state in 1864, at least in part with the hope that its mineral wealth would help to bring down the country's tremendous war debt. As mining activity escalated, a variety of minerals, including turquoise, was discovered as well.

When the commercial mining of turquoise started in Nevada around 1889, among the first to become involved was Herman Schweizer's collaborator William Petry, a 20-year-old lapidary who had immigrated to the United States from Veitsrodt, Germany in 1886 (United States Passport Applications, 1795–1925). Before settling in Los Angeles, California, he lived in New York and in Colorado, where his association with a lapidary named William Kley may have provided his entry into the turquoise trade. In about 1890 a prospector named George Simmons went looking for copper and discovered turquoise fragments near his campsite. When he eventually located the vein of turquoise and began to mine it, he sent some of the rough stone to Kley. Simmons met William Petry through Kley, and hired him to set up a lapidary shop at the mine. According to journalist James R. Morrissey, 'After that, no more rough turquoise was sold. The bulk of the finished gems were sold to Woods & Lamont, wholesale gem dealers in New York City' (Morrissey 1968: 7).

By 1915 Petry was living with his extended family in a house on Rogers Avenue in Los Angeles. With him were gem cutters Ernest and Ludwig Petry, and several members of his married sister's family, Albert, Fritz, Harold and Walter Zimmer. Fritz was also a lapidary and in 1917 Walter, a businessman, founded the Walter C. Zimmer Company, a wholesale jewellery firm still operating today in downtown Los Angeles. Most of Petry's family worked with him at his firm in the Broadway Central Building, about 2½ miles from home in what became the city's jewellery district (Los Angeles Directory Company 1915).

Petry became a mine owner in 1902 or 1903 when he purchased the Royal Blue mine in Esmeralda County, Nevada for $3000. The Royal Blue became one of Nevada's major producers of high-quality turquoise. In 1907, Petry sold a portion of the mine to the Himalaya Mining Company (who sold it back to him several years later), but in 1908 continued to work the area, and in fact patented his claim (Morrissey 1968: 25). The flat blue stones that Herman Schweizer sought for the Harvey Company's souvenir jewellery could easily have come from Petry's mine. The United States Geological Survey for 1907 called it 'good blue turquoise in seams an eighth of an inch or more in thickness in a dark gray chalcedony matrix' (Sterrett 1908: 827).

William Petry became one of a growing number of gem dealers from large cities – primarily Los Angeles and Denver – who peddled Nevada turquoise to trading posts and mercantile companies throughout Navajo country. On 25 June 1938

Father Berard Haile of St. Michaels Mission, Arizona, told John Adair about meeting a turquoise peddler – the representative of a Los Angeles wholesale jewellery company – as early as 1907. Mike Kirk, the trader at Manuelito, New Mexico recalled buying commercially cut Nevada stones from a Denver firm in about 1912 (John Adair Collection, Field Notes, Book 1). Ruth Kirk, the wife of wholesaler John Kirk wrote that in about 1912 'F.A. Marcher of Los Angeles opened up turquoise mines in Nevada, had the stones cut, and on an annual trip through the reservation sold them to traders who in turn traded them to the Indians' (Kirk 1945: 31).

In addition to traders, urban gem dealers found some of their best customers among curio shops that specialised in the mass production of Indian (or 'Indian style') jewellery. Jonathan Batkin's book *The Native American Curio Trade in New Mexico* (2008) traces the commercialisation of Native American jewellery as it developed in shops such as Julius Gans's Southwest Arts and Crafts in Santa Fe, which opened in 1916, and Maisel's Indian Trading Post of Albuquerque, which opened in 1923. These shops, and others throughout the American Southwest, employed young Native American men to make lightweight bracelets, rings, pins, and novelty items for sale to tourists and mail order customers. Julius Gans's employees made jewellery essentially by hand, but used shortcuts such as pre-designed patterns and sheet silver (rather than ingots). Maisel's operation was completely mechanised, except for a small amount of hand finishing (Fig. 1). Both companies placed bench workers in full public view: watching an Indian silversmith at work was a popular part of the retail experience.

Companies outside the American Southwest produced Indian-style novelty jewellery using no Native American labour whatsoever. Southern California manufacturers including the Pacific Jewelry Company of Santa Monica and the Turquoise Indian Trading Company of Santa Barbara flooded the market with imitations, and were large consumers of low-grade turquoise that they routinely dyed and processed. They also introduced methods for standardising turquoise that became widespread in the curio jewellery trade. On 25 October 1940, E. Tobiasson, an owner of the Pacific Jewelry Company of Santa Monica, told John Adair that:

> He had managed to standardize the shape of the stones, and he was the first one to do this. He said that in the old days it was impossible to get turquoise in quantity to fit the cups [manufactured bezels]. But he furnished the [turquoise] cutters ... with a blank just the size of the cup, and the finished stone had to be of such a size that it could be pushed through this blank. The turquoise dealers give him first choice of their stones and those that he rejects go over to Maisel's and the Indian country and as they are already cut to fit his cups he has in this way standardized the size and shape of the stones used in manufactured silver (John Adair Collection, Field Notes 1940).

By 1930, reservation traders, Native American jewellers, arts patrons and museum professionals were united in the belief that mechanised jewellery shops posed a serious threat to the survival of traditional Navajo and Pueblo silversmithing. Most thought that mass-producers of Indian-style jewellery competed unfairly with independent Native American smiths and with the traders who sold their work; some maintained that all curio shop methods, particularly the use of sheet silver, were inauthentic and that Native smiths should use only the most laborious hand methods. When in 1936 Congress established the Indian Arts and Crafts Board (henceforth IACB) to revitalise and promote traditional Native American arts, one of its most urgent matters of business was to regulate the production of Native American jewellery. In 1937 Commissioner of Indian Affairs John Collier issued standards by which jewellery would be considered authentic. These included the prohibition of power-driven machinery, and the requirement that all work be 'individually processed and ... entirely hand-made', and that 'all stone used, including turquoise, is to be hand-cut and polished. This permits the use of hand or foot-driven wheels' (Batkin 2008: 199) (Fig. 2).

In 1940 Adair began a survey of traders and silversmiths on the Navajo reservation that culminated in a report on the state of the craft to the IACB. Adair was particularly interested in the problem of turquoise. On 28 June he interviewed Alfreda Ward of New Mexico's Santa Fe Indian School, whose silversmithing program was known for turning out well-trained craftsmen. She told him that it was 'impossible for them to get good turquoise, rough cut, these days. She said that they would pay up to $50 a [pound] for good stones. She said they might even have to buy polished stones and rough them up.' The following day the trader at Pinehaven Trading Post in Arizona criticised the IACB for 'trying to maintain standards which are too high to be practical', and recited a litany of reasons why he thought stones by non-Native lapidaries using modern commercial methods should be permitted: mine owners made less money selling rough turquoise than they did selling cut and polished stones; traders could weigh pre-cut stones before supplying them to craftsmen, and thereby ensure that the correct amount of turquoise came back in the finished

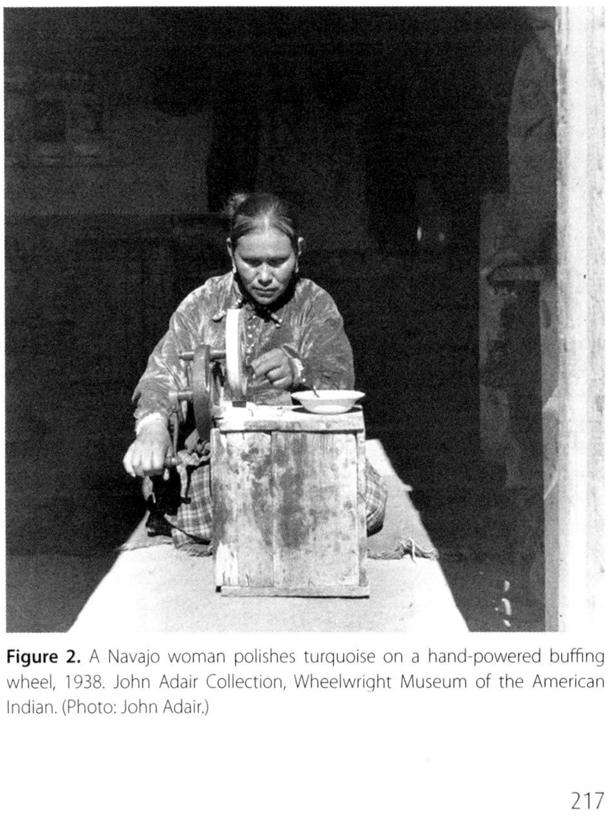

Figure 2. A Navajo woman polishes turquoise on a hand-powered buffing wheel, 1938. John Adair Collection, Wheelwright Museum of the American Indian. (Photo: John Adair.)

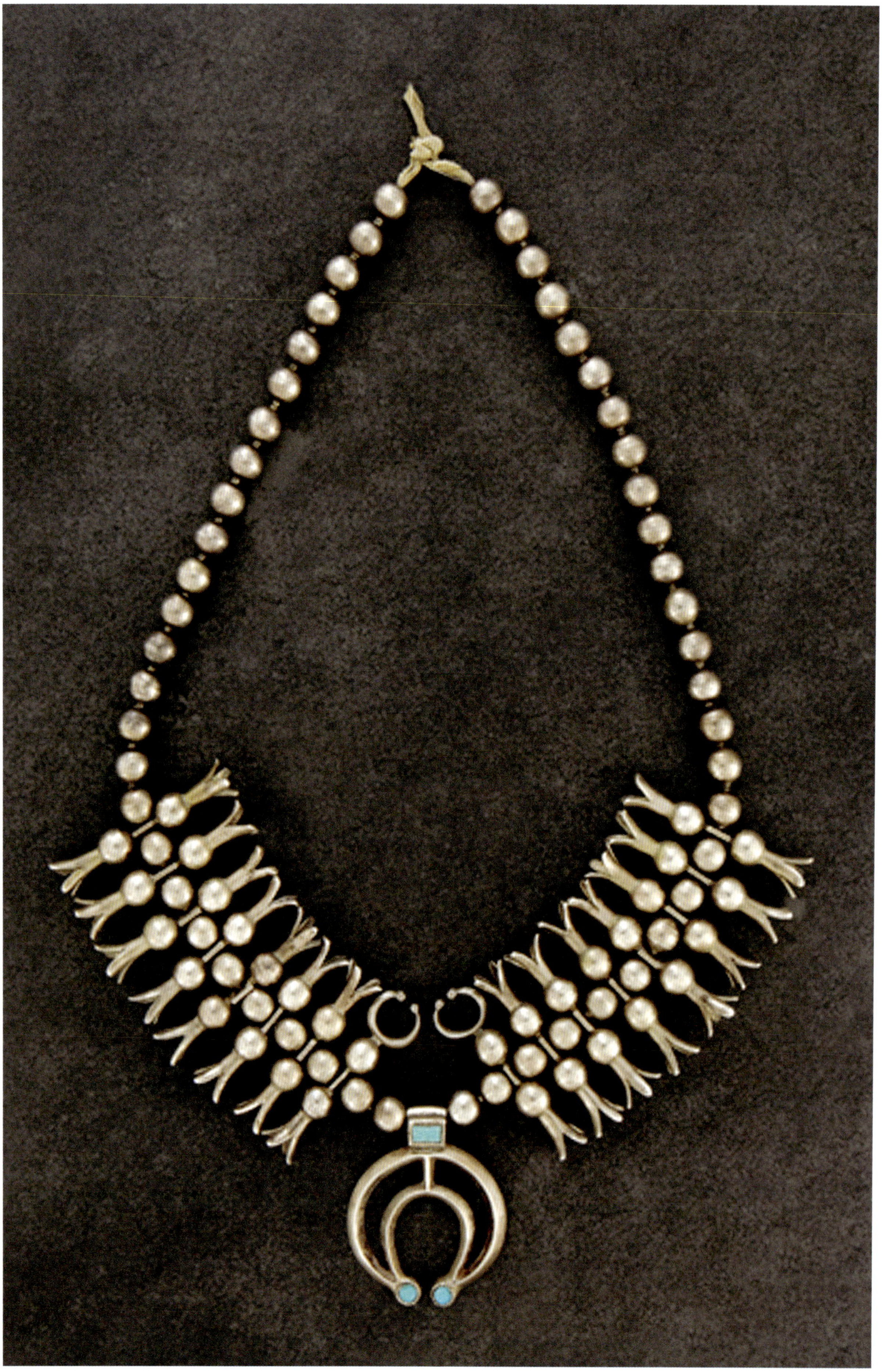

Figure 3. Squash-blossom necklace by Slender Maker of Silver, ca. 1885. The pendant contains flat, hand-polished turquoise. Wheelwright Museum of the American Indian. (Photo: Addison Doty.)

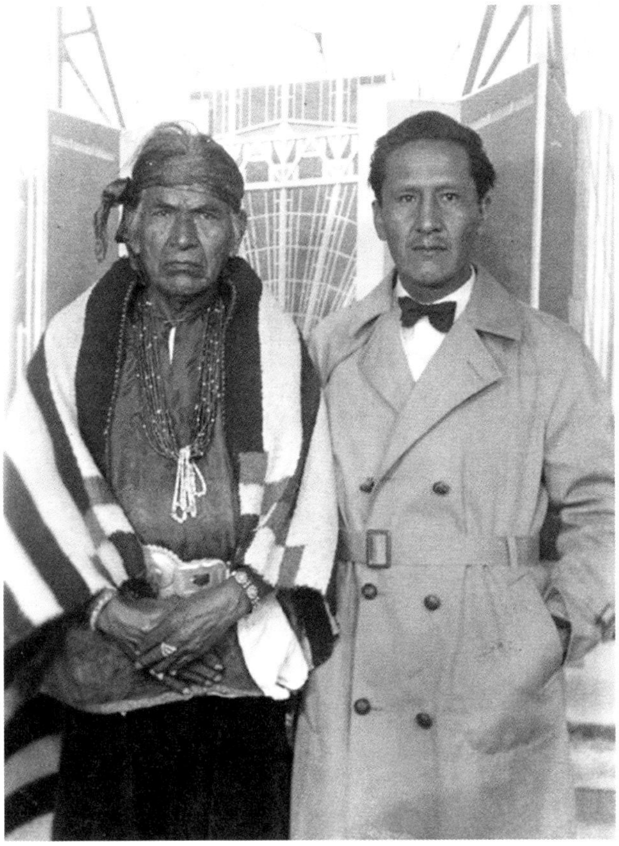

Figure 4. Fred Peshlakai (right) with Navajo medicine man Hastiin Klah at the Century of Progress Exposition, Chicago, Illinois in 1933 or 1934. Harriett Williams Collection, Wheelwright Museum of the American Indian.

jewellery; and the IACB inspectors could not tell the difference between turquoise hand-cut by highly skilled craftsmen at Zuni and Santo Domingo Pueblo, and commercially cut stones. At the 1939 Gallup Ceremonial, jewellery submitted by the Pinehaven Trading Post was rejected for containing commercial turquoise, when in fact it had been cut by lapidaries at Zuni (John Adair Collection, Field Notes 1940).

Trader after trader expressed the opinion that the use of rough turquoise was not only impractical but also that the material was nearly impossible to obtain. Out of the 80 silversmiths that Adair surveyed, only 23 said 'they polished any turquoise at all' (1941: 35). Adair discovered that while turquoise came from a number of sources, the best came from two southern California suppliers who held 'a virtual monopoly on gem turquoise'. These were Walter Godber of Los Angeles, who operated the Yerington mines in central Nevada and owned the Western Gem Company of Los Angeles, and the most prolific, Doc Wilson of Wilmar, California (Adair 1941: 75).

Wilson got into the turquoise trade in 1903. When Adair interviewed him on 28 October 1940 he owned full or part interest in at least seven Nevada mines, including Lone Mountain in Esmeralda County and Blue Gem in Lander County, and he may have owned the Number 8 mine in Eureka County, which he purchased in 1935 and later sold (Morrissey 1968: 12). Wilson travelled to New Mexico every 60 days to sell gem quality, pre-cut turquoise to traders and curio dealers who farmed it out to Navajo and Zuni smiths, or had it set into silver by bench workers in their shops. Before 1937 he refused to sell to mass-producers of Indian-style jewellery, but the IACB ruling against commercially cut stones so angered him that he began selling scrap and low-grade turquoise to them (Harry Boyd to Adair, 30 August 1940). In his report to the IACB Adair wrote:

> Mr. Wilson ... told me that the manufacturers here in Los Angeles will not pay more than $5.00 a pound for turquoise, and they prefer to pay $2.00. (Good gem turquoise sells for $35.00 to $50.00 a pound.) He sells his rejected stones and scraps to these companies, all of whom soak the turquoise in paraffin to give it a deeper color, and some, like the Pacific Jewelry Company, dye it (1941, 75).

From their home in a suburb east of downtown Los Angeles, Wilson's wife Hazel managed the family's Navajo jewellery business. They sold authentic old jewellery collected on the reservation, and employed a Navajo smith named Frank Peshlakai, and occasionally his brother Fred Peshlakai, to set their best turquoise into silver (John Adair Collection, Field Notes, 1940).

John Adair met Fred Peshlakai in 1938. At that time Peshlakai made and sold most of his work from a small shop on Olvera Street, a tourist destination in the city's El Pueblo de Los Angeles historic district. Born in 1896, Fred was one of at least three children born to the second wife of Slender Maker of Silver, a first-generation master smith, who late in life became known as Old Peshlakai (Dodge n.d.). From the family's home near Whiskey Creek, in the backcountry northwest of Crystal, New Mexico, Old Peshlakai ran what Adair referred to as 'a regular silver shop', paying as many as ten men to help him produce beautifully crafted silver jewellery (Adair 1944: 23) (Fig. 3). Fred learned the trade from his father when he was about 18 years old.

By about 1929, Fred had become a Christian evangelist and was working with the Presbyterian mission at Ganado, Arizona (Dodge n.d.; Ganado Mission n.d.), but in 1930 or 1931 he moved to Gallup, New Mexico, where he operated a curio store and learned commercial metal casting techniques from a local German jeweller. On 1 October 1931, Fred was hired as the first instructor of silversmithing at the Charles H. Burke School in Fort Wingate, New Mexico – one of at least three Indian schools to introduce silversmithing programs in that year. In 1933 he demonstrated his craft at the Chicago Century of Progress Exposition (John Adair Collection, Field Notes, Book 4, 20 August 1938) (Fig. 4). Fred Peshlakai left the Burke School in 1935, following criticism that his experience as a commercial jeweller in Gallup rendered him unsuitable as an instructor of Native American students, and the false accusation that he was encouraging the use of modern mass-production methods in his classroom (Batkin 2008: 194–195). Before settling in Los Angeles he worked briefly for Reese Vaughn of Vaughn's Indian Store in Phoenix, Arizona (Pardue 2005: 65).

Fred opened his Los Angeles shop in the summer of 1938, and through the work he produced there over the next 30 years earned a reputation as an expert in the selection and use of turquoise (Fig. 5). He rarely worked for others, and insisted upon managing his own career for most of his life – a situation that was uncommon among Native American jewellers until well after the Second World War (Batkin 2010). Unlike most smiths, whose access to materials was closely controlled by

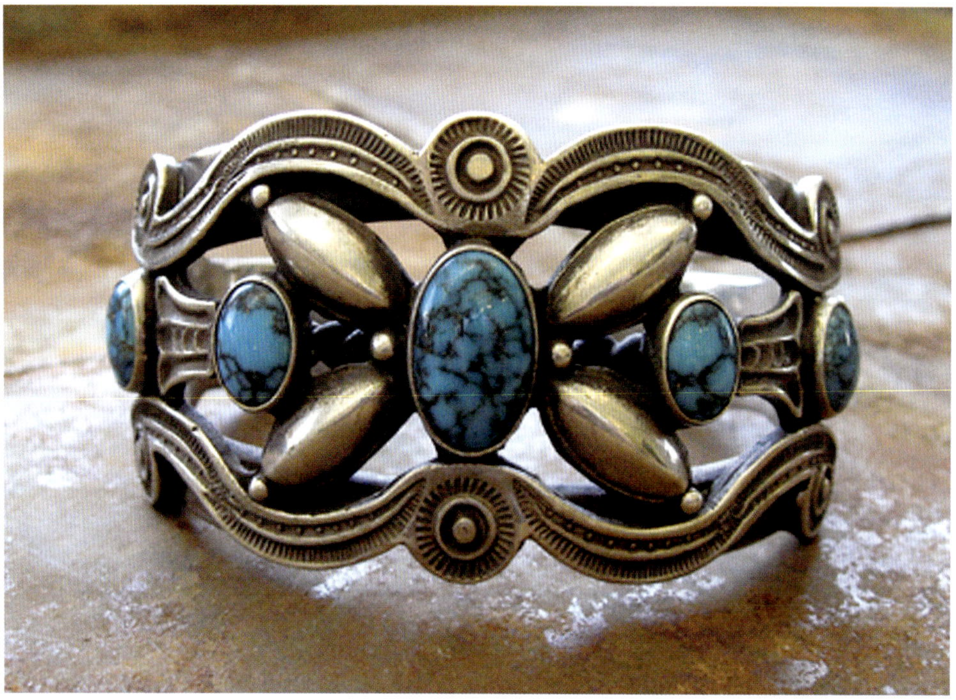

Figure 5. Sterling silver and turquoise bracelet by Fred Peshlakai. Width 3.5 cm. Collection of Lauris Phillips. (Photo: Cheri Falkenstien-Doyle.)

the traders or shop owners for whom they worked, Peshlakai obtained his materials directly. He bought his silver from the Wildberg Brothers Smelting and Refining Company in Los Angeles (John Adair Collection, Field Notes, 24 October 1940). He worked for Doc Wilson in exchange for some of the finest stones on the market, and visited Wilson's mines to select rough turquoise that he cut and polished himself, using methods and equipment that would have been available to any modern lapidary (Phillips 2010).

Peshlakai's career signalled the decline of a system in which reservation traders, shop owners, and others struggled to maintain control not only over the output of Native jewellers, but also over the resources necessary for the perpetuation of their craft – a situation Adair described as 'an eighteenth century system of economy that is adapted only to frontier conditions ... trying unsuccessfully to compete with twentieth century merchandising techniques' (1941: 15). Fred Peshlakai was one of a growing number of Native American silversmiths who through their own initiative joined the ranks of modern studio jewellers.

References

Adair, J. (1938) 'Field Notes, John Adair Collection', Wheelwright Museum of the American Indian (unpublished).

Adair, J. (1940) 'John Adair Collection', Wheelwright Museum of the American Indian (unpublished).

Adair, J. (1941) 'Report on Silver Production of the Southwest', John Adair Collection, Wheelwright Museum of the American Indian (unpublished).

Adair, J. (1944) *The Navajo and Pueblo Silversmiths*. Norman: University of Oklahoma Press.

Batkin, J. (2008) *The Native American Curio Trade in New Mexico*. Santa Fe, NM: Wheelwright Museum of the American Indian.

Batkin, J. (2010) Personal communication, 16 July 2010.

Dodge, C. (n.d.) [ca. 1929] Unpublished typescript, Chester Faris Collection, Wheelwright Museum of the American Indian.

Ganado Mission to Navajo Indians (n.d.) [ca. 1930] Board of National Missions of the Presbyterian Church in the USA, New York.

Kirk, R. F. (1945) 'Southwestern Indian jewelry', *El Palacio* **52**(2): 22–32.

Los Angeles City Directory Company (1915) *Los Angeles City Directory 1915 Including San Pedro and Wilmington* (http://rescarta.lapl.org:8080/ResCarta-Web/jsp/RcWebBrowse.jsp, accessed 2010).

Matthews, W. (1883) 'Navajo silversmiths', *2nd Annual Report of the Bureau of American Ethnology for the Years 1880–1881*. Washington, DC: Government Printing Office, 171–178.

Morrissey, F.R. (1968) 'Turquoise deposits in Nevada: Nevada Bureau of Mines and Geology Report 17'(http://www.nbmg.unr.edu/dox/r17/r17text.pdf, accessed 5 August 2009).

Pardue, D. (2005) 'Native American silversmiths in the Southwest', *American Indian Art Magazine* **30**(3): 62–69.

Phillips, L. (2010) Personal communication, 21 June 2010 and 16 July 2010.

Sterrett, D.B. (1908) 'Precious stones', in *Department of the Interior, United States Geological Survey, Mineral Resources of the United States, Calendar Year 1907*. Washington, DC: Government Printing Office, 795–842.

United States Passport Applications (1795–1925) Ancestry.com [database on-line]. Provo, UT, USA: Ancestry.com Operations, Inc., 2007 (accessed 12 October 2009).

Witherspoon, G. (1981) 'Silver and turquoise in the Navajo universe', in C.J. Davies (ed.), *White Metal Universe: Navajo Silver from the Fred Harvey Collection, Heard Museum, Phoenix*, 47–50.

Author

Cheri Falkenstien-Doyle, Wheelwright Museum of the American Indian, Santa Fe, New Mexico, USA

The turquoise mines of the Cerrillos Hills Mining District, gemstone materials, and their usage: ancient and modern

Douglas Magnus

Introduction

The turquoise mines of the Cerrillos Hills Mining District, located in Santa Fe County, New Mexico, are known to be among the most ancient sources for turquoise in all of North America. Archaeological evidence found at these sites dates back to at least AD 900. The gemstone materials extracted here figured prominently in the existence and development of regional indigenous cultures, and their connections to distant peoples throughout the Americas. There is little doubt that the massive amount of material mined, prior to the arrival of the Spanish in the sixteenth century, was used by Mesoamerican peoples, including Zapotec, Toltec, Mixtec and Aztec (Mexica).

Located 100 miles southwest of the Cerrillos mines, the ancient Puebloan (Anasazi) people of Chaco Canyon (ca. AD 900–1200), were major participants in the trading and use of turquoise. Evidence points to Cerrillos as a major source. Modern research has not proved the extent of these speculations, but the physical evidence is irrefutable. These ancient mines are a testament to the industry of their makers, and the volumes of turquoise surely went far south, very likely through Anasazi trade networks, among others, although earlier systems remain mysterious (Fig. 1).

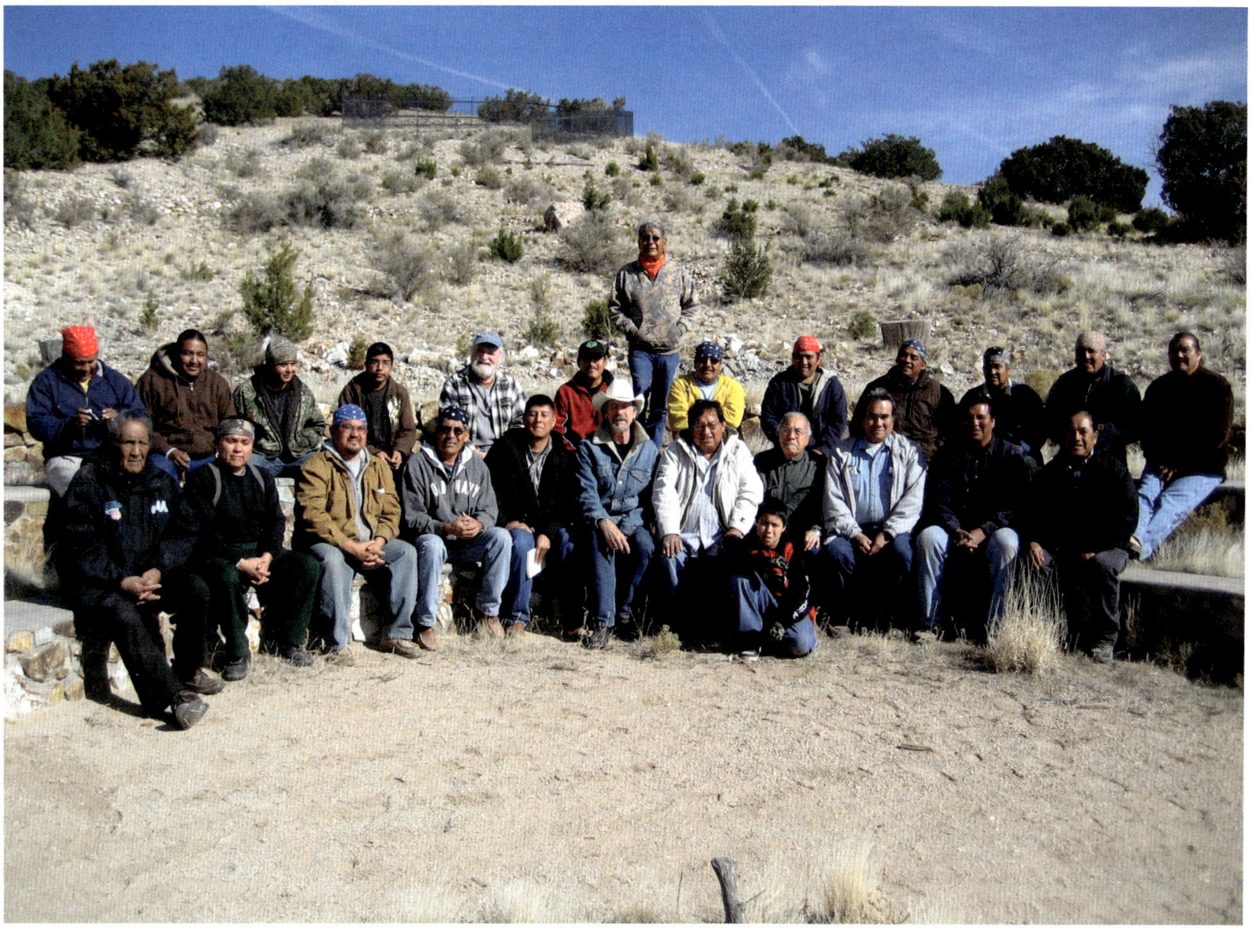

Figure 1. Santo Domingo Pueblo officials and the author, 2009. (Photo: Douglas Magnus.)

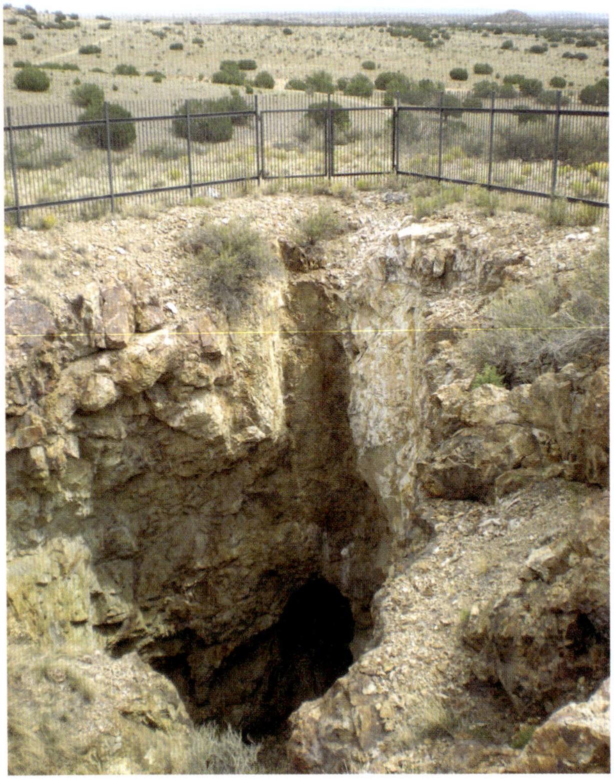

Figure 2. Cerrillos mines in 2009. (Photo: Douglas Magnus.)

Because the mines, as we see them today, have been altered by modern developments, it is remarkable that many ancient traces remain. One can still marvel at the visible workings of the ancients, and what they accomplished with simple stone tools. The ancient mine shafts are still visible (Fig. 2). Actually, one can marvel at the stone tools themselves, and wonder how they were shaped, transported and used. Even the broken fragments were used, and many of those tools litter the grounds surrounding the mines, even now.

The geology of the Cerrillos Hills Mining District produced relatively small deposits of turquoise and other precious materials, and thus, no major modern mining has erased the ancient workings entirely. It is noteworthy that these small deposits produced a wide variety of colour and quality – in veins, nuggets and nodule forms – and in generally smaller proportions than are found at some of the other turquoise mining locations in the southwestern United States. Nevertheless, the ancients recognised special quality considerations of Cerrillos turquoise. Something about the region perhaps; the colour range of the stone, hardness and translucency of the best of the stone, and of course, sacred attachments to the land and its brilliant blue sky. The region around the area of these turquoise mines is famous for some of the most luminous skies in the world. It is difficult to say exactly what were the attractions of the ancient peoples to these sources, but they were of great wonderment to modern miners who rediscovered these mines in the 1870s. They marvelled at the extent of the workings created by the ancients. One of the mysteries is that the ancient shafts at Turquoise Hill, and others in the district, had been entirely filled in, possibly during the period following the Pueblo revolt, or possibly before, in an effort to conceal their treasure from the Spanish. The mines were purposely sealed with earthen material in an apparently ceremonial manner. The shafts were packed with dirt and stone, not necessarily from the mines themselves, and a hard layer at about every 8 ft was packed, probably by dancing, and the layer would be ringed with stone tools.

These mines would remain closed until the American period following the 1872 General or National Mining Act (42nd Congress Sess. II Ch. 152. 1872), which encouraged prospecting and mineral development, and settlement in the expansion of the American West. The first of the mines to reopen at Turquoise Hill was likely the Muniz Lode, which would later become the American Turquoise Company's famous 'Tiffany' mines. This company would also reopen the Castilian Lode, located at the west end of the formation, half a mile from the Tiffany. Because the Muniz/Tiffany mine was perhaps the first turquoise strike in modern America (though it could perhaps have been in the Mount Chalchihuitl mine, 3 miles south in the Cerrillos Hills), and because a blue stone of great quality (attributed to the Muniz/Tiffany mine) was to receive gemstone acclaim at the Paris exhibition of 1878, the modern era of interest in American turquoise began. Tiffany Company's inclusion of turquoise in the award-winning designs of George Paulding Farnham, incorporating turquoise into American Indian inspired pieces of silverwork, were a major breakthrough at subsequent Paris exhibitions in 1889 and 1900.

The Cerrillos turquoise of Turquoise Hill was proclaimed equal in beauty and quality to that of Persian turquoise (the world standard and primary source), by Tiffany's eminent gemologist and vice-president, George Frederick Kunz, who went on to become a shareholder in the original American Turquoise Company, formed in 1890–1891. With the purchase of the claims at Turquoise Hill, the quest for turquoise in the Los Cerrillos began in earnest. Soon after, Irish miner James P. McNulty was hired to supervise the mining, and he remained and mined the Los Cerrillos for 40 years, a story well described by his descendant Patricia McGraw in *Tiffany Blue* (McGraw 2006). The work under his supervision constituted the bulk of the significant mining at Turquoise Hill in the modern era. The American Turquoise Company was beset with problems as it attempted to control US turquoise mining/marketing, cartel-fashion, as new turquoise discoveries continued to be made in the southwestern United States.

Turquoise as a gemstone in modern times has ridden the tides of fashion and popularity, except among the Native Americans, where it is a perennial favourite. The 1970s was a noteworthy period of demand for turquoise, particularly American Indian jewellery, but turquoise enjoyed a burst of glory in the early years of the twenty-first century, which has at last established the stone as a universal favourite. Markets and availability however, continue to fluctuate.

Cerrillos turquoise in the twenty-first century is rare. Active mining is limited to a hobby level of prospecting. The mines are mostly exhausted and not commercially viable. It is unlikely that serious future turquoise mining will exist in the Cerrillos Hills Mining District. The current Santa Fe County ordinances prohibit or discourage mining on a commercial scale. The turquoise mines at Mount Chalchihuitl are (2009) to become part of the Santa Fe County owned,

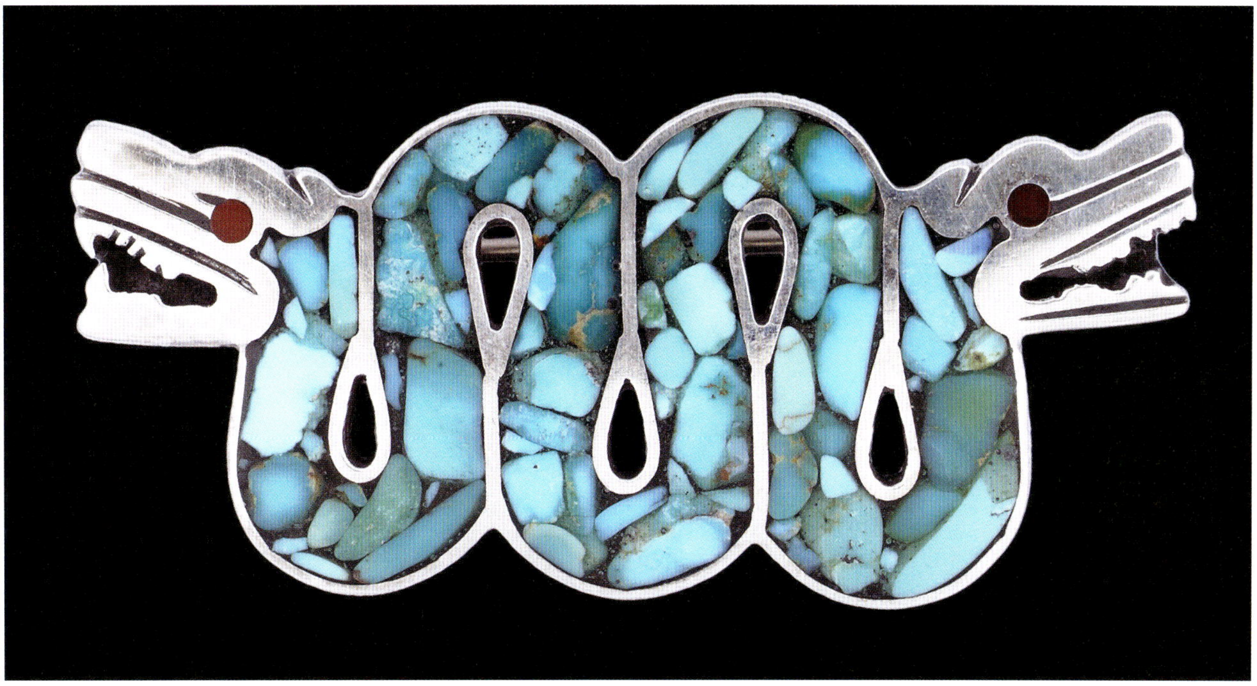

Figure 3. Brooch made in Sante Fe, ca. 2008, after the Mexica double-headed serpent pectoral, using Old Castilian turquoise. Made and presented by the author (Department of Africa, Oceania and the Americas, British Museum. 2009,2004.9. ENA109623, © Trustees of the British Museum).

Cerrillos Hills Park (New Mexico 2012). Also, the mines of Turquoise Hill have been fenced and remain preserved under their current ownership – Millennium Turquoise Mines Inc. and me. It is likely these mines will remain as they are for the foreseeable future.

Personal observations and experience with Cerrillos turquoise

Since the early 1970s I have worked extensively at making jewellery, and turquoise has been my primary gemstone material (Fig. 3). Though I have experience with other gemstones, it is precisely because of my experience with other materials that I have chosen to stay close to turquoise. Other gems certainly hold their own appeal, but turquoise is for me the most interesting. It is a vast and endlessly fascinating realm. My personal journey in lapidary work with gem materials for inclusion into my jewellery creations actually began with Cerrillos turquoise, when a friend showed me the mines in 1973. There, I was able to gather a multitude of small stones left behind from earlier mining activities, from which I was able to create many jewellery items, in both traditional cabochon cuts and inlays. Most of the knowledge I have gained these many years is through a passionate drive of intuitional self-teaching, and just plain work. I am not an academic. There has never been a disconnection between myself and the stone. I understand it from the matrix of the earth, to its extraction, and the various processes rendering it into gems and jewellery hands-on. My story may differ from others by virtue of my connection to the region in which I live and work, its history as regards to turquoise, and also because of the extraordinary good fortune I have had to come in contact with the Cerrillos mines, and now because I have the privilege of ownership of some of them (I own the mines at Turquoise Hill, in the northern part of the Cerrillos Hills Mining District) (Fig. 4); other turquoise mines exist a few miles to the south). All of this combines for a unique relationship with the stone.

Of course, through the years I have met others with a passion for turquoise including miners, cutters, geologists, jewellers, and traders. Exchanges with them have helped me to gain a perspective on the subject. The stone as a part of nature is subject to extremes in variation. The basic properties of turquoise's geology remain the same no matter where on earth it is found, but there is much to consider as to the unique

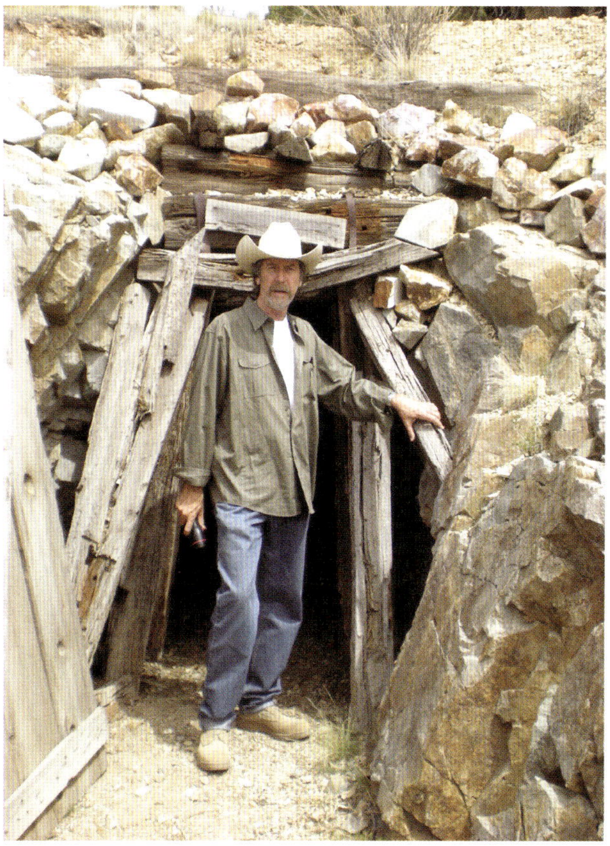

Figure 4 The author in front of one of the mines. (Photo: Douglas Magnus.)

circumstances of its formation. In other words it involves the 'when and how' that may affect the material as concerns the gemstone producer. Hardness and depth of colour, consistency, matrix inclusions and porosity are key anxieties. Many questions remain within the field of geological study as to the formation of turquoise, and other gemstones, and likely they always will. To the lapidary, jeweller, connoisseur or customer, there is no question – it's all about beauty, durability and source. This rule applies mostly to natural turquoise and not to treated, stabilised or synthetic materials. It is my experience that many people will seek out the turquoise of a specific source, for a variety of reasons mostly of a personal nature, but often because of historical connections or spiritual perceptions and interests.

About the old and the new

In considering the use of the turquoise, I have harvested from the grounds (dumps and scatterings) at Turquoise Hill, and have been faced with many problems. I imagine these to be the same issues that all ancient and modern people are faced with when considering a variety of possible design ambitions. First of all, it is necessary to possess a significant amount of raw materials from which to choose the appropriate candidates for the desired work. In order to accomplish this, it may be necessary to possess 100 to 1000 or more times the necessary number of stones for the intended work, so as to provide proper selection in matching colours and characteristics. Then the lapidary must work each stone to reveal its potential for inclusion in the desired creation. Any problems within the stone must be exposed prior to further work. These problems include fractures, porosity, undesirable matrix, soft spots, colour changes, and possibly other issues.

The process of working with these materials is time consuming and difficult. It involves cutting, grinding, sanding and polishing. The experienced modern lapidary will feel the stone's quality before seeing its colour and characteristics, as he touches it to the grindstone. The sound the stone makes during the cutting process will reveal much about the material's hardness. If a whitish streak develops upon the grindstone, one knows the material is soft. Cerrillos turquoise can range in hardness from a chalky 3 or 4, to a superior translucent 6, or possibly a rare 7, on the Mohs scale of mineral hardness. During the process of converting raw material to finished gemstone, fully 50–75 percent of the material is lost. The creation of items containing many individual stones, such as in mosaic inlays, may require extreme amounts of stone to be shaped and fitted together, creating a seamless flow of colour and characteristics, into the desired effect. Much of the Cerrillos turquoise was found in thin veinlets (1–2 mm thickness) with one side of the matrix still attached. The stone can be cut as a veneer of colour attached to the rhyolite/monzonite (host rock) matrix. The ancients frequently did this in pendant forms for earrings and necklaces. This is a difficult enough process using modern equipment, but was especially challenging in ancient times. A modern individual with contemporary tools can accomplish the work of many men or women in a fraction of the time, but the ancients accomplished the job admirably. Considering their methods they completed the work as well as any modern person could. Most ancient turquoise was sanded smooth by hours of rubbing against a flat lapidary stone often of mine host rock, rhyolite or monzonite.

For my own work with Cerrillos turquoise, I have collected perhaps millions of pieces of stone; from 1 mm sized chips to fingernail size and larger, and also a wide selection of matrix varieties of turquoise. In order to work with these, it has been necessary to spend countless hours cutting, trimming and shaping the raw materials into configurations to reveal the stones' best potential. At this point, the thinner, flat pieces could be considered tesserae. The stones are sometimes backed with epoxy steel for support and strength, and to level out irregular surfaces, then placed on dowel sticks using lapidary dop wax. The stones are thus ready to be handled for final cutting and polishing. These I transfer to a series of trays containing foam florists' blocks, which support the stones on their sticks facing up, allowing me to see each stone for selection for any given purpose. At this stage, it is all about potential. Some pieces are set aside as possible bead-making material. Matrix material may become fetish, or other carvings, and small sculptures. When the time is right, I will search through the selection of prepared materials in the hope of finding the right stones to match my designs. Even then it is often difficult to find the best materials for the job – much of the material contains fractures, which I try to avoid for my fine jewellery pieces. Almost no two stones are ever able to be matched in colour, characteristics or size! My observations of the ancient American, and Mesoamerican turquoise jewellery that I have seen convinces me that this has always been true.

Ancient people were good at finding uses for all the turquoise materials, including what nowadays might be considered waste or scrap. They would incorporate it into all manner of objects. I once heard tell of the practice by Acoma Pueblo Katsina makers of grinding the chalk turquoise material into a fine dust. The maker would then blow the dust onto a sticky surface upon the Katsina figure as a sort of paint. Early Spanish records tell of Zuni Pueblo houses with turquoise inlaid into the lintels above doorways.

In the making of jewellery and other objects, it is my experience that little has changed from ancient to modern, except the tools. Today's tools utilise steel and diamond abrasives. We have motors and rapidly turning wheels, speeding the process from cutting to final polish. Nevertheless, many of the concepts, the handling of each stone, and assembly of parts remain the same.

Reference

McGraw, P. (2006) *Tiffany Blue: The True Story of Turquoise, Tiffany and James P. McNulty in Territorial New Mexico 1892–1933.* Santa Fe, NM: Lone Butte Press.

Author

Douglas Magnus, Artist/Designer, Magnus Studio/Cerrillos Mines, Santa Fe, New Mexico, USA

Index

Page numbers in *italics* are for illustrations, e.g. humming birds 84, 85, 143, *168*, 169, 170. Page numbers followed by (n) are for notes, e.g. Arts Sales Catalogues Online 187, 188, 195(n 4). The following abbreviations are used in the index:

BM (British Museum, London)
EMB (Ethnologisches Museum, Berlin)
KC (Kislak Collection, Washington, DC)
MC (Museo de las Culturas, Oaxaca)
MNA (Museo Nacional de Antropología, Mexico City)
MRAH (Musée Royaux d'Art et d'Histoire, Brussels)
MV (Museum für Völkerkunde, Vienna)
NMAI (National Museum of the American Indian, Washington, DC)
NMS (National Museums Scotland, Edinburgh)

Aborigines Protection Society (APS) 178
Achillini, Giovanni 165, 166
Acosta, Jorge 42, 47
Adair, John, *The Navajo and Pueblo Silversmiths* 215, 217, 219, 220
Adams, William 188, 194
adhesives 21, 25, 48–49, 52, *63*, 79
aesthetics 143
  approaches to collecting 155–162, 170
  see also colour; iconography; poetry; scientific approaches
Aguirre, Porfirio 17
Aguirre, R.D. 181–182
Ah Canul tombs 22, 23
Albert, Prince Regent 174
Alberti, Friar Leandro 165, 166, 170
Albonesi, Teseo Ambrogio 170
Aldrovandi, Ulisse 166
Aldrovandi Museum 166–167, 168, 169
amazonite 17, 18, 101
Ambras collection 104, 107, 110
American Turquoise Company 222
Anahuac (Tylor) 181, 188
analysis *see* materials analysis; provenance
Anasazi people (Ancestral Puebloans) 31, 150, 152, 221
animal, seated, and cup (BM) *56*, *193*, 194–195
animal-head mirror frame (MV) 103, 110–114, *111*
animal-heads
  formerly in EMB 103, 114–115(n 1)
  pendant(?) (BM) 11–12, *13*, *112*, 114, *192*, 194
APS (Aborigines Protection Society) 178
aristocracy, Aztec (*tecuhtli*) 96, 101
art market 197, 214
  see also sales catalogues; trade
Arts Sales Catalogues Online 187, 188, 195(n 4)
atequilizcuicatl (song of water pouring) 140–141
atlatl (spear thrower) 122–124, *123*, 129, 130, 169
Azcatitlan Codex 138, *142*, 143
Aztec exhibition (1854) 180, *180*
Aztecs (Mexica) 91–101
  Calendar Stone 129, 133
  cosmology 84–86
  deities, ruler and nobility 93–96
  ethnographical studies of (19th-century) 180–182
  mosaics and objects in Bolognese museums 165–171
  mosaics in BM 3–4, 7–13, 55–63, 183–195
  mosaics in MV 3–4, 103–115, 194
  poetry and thought 135–143
  society, economy and religion 91–92
  symbolism 83–86, 128–132, 135–143
  tributary states and trade 96–101
  uses of turquoise 92–93

Baakal, masks from kingdom of 24–26
'backing' materials *see* supports
Baddeley, W. St Clair 178
Bassani, Ezio 155
Bateman, Thomas 195
Batkin, Jonathan 217
beauty (concept of) 143
Bechler, R. 178
belief systems
  Navajo 214, 215
  Pueblo 148–150, 153
  see also poetry; ritual; symbolism
Bennett, E.M. 66, 67
Bernal, Ignacio 120
Bethnal Green catalogue (Pitt-Rivers) 174, 175–176
Bird, Gail and Johnson, Yazzie 197–202, 203
birds
  bird's head ornament from Oxkintok 23, *24*
  cotinga 124–125, 143, 147
  fetish birds 209, *209*
  humming birds 84, 85, 143, *168*, 169, 170
  in O'odham myth 153
  roseate spoonbill 85
  thunderbird necklace *211*
  tsooro (mountain bluebird) 147, *147*
birth, mining (extraction) as 135–136
Bisbee turquoise jewellery *203*, 207–208, *207*
Blackwood's Edinburgh Magazine 183, 184
Blake, William P. 66, 67, 152
blue (colour) 47
  Maya Blue 124, 128–129
Blue Gem turquoise 206
'bluebird seeds' (*tsorposi*) 147, 153
Bolognese museums, 16th- and 17th-century collections 165–171
Borbonicus Codex 131, 133, *139*
Borgia Codex 125, 128, 132
botanical collecting 178, 179, 181
Boturini Codex 96, 137
bracelets
  Navajo silverwork 206, *206*
  shell and inlay (Owen) *203*
  silver and turquoise (Peshlakai) *208*, 209
  tufa-cast gold (Loloma) *208*, 209
breast pieces *see* pectorals
British Museum
  brooch, Deschutes jasper and Yowah opal 198, *199*
  Mixtec-Mexica mosaics 3–4, 7–13, 55–63, 183–195
    animal-head (pendant?) 11–12, *13*, *112*, 114, *192*, 194
    double-headed serpent (pectoral?) *9*, 11, 188–194, *191*
    eagle warrior knife handle (sacrificial knife) 56, *56*, 183–184, *184*, 185–188
    helmet *56*, *194*, 195
    seated animal and cup *56*, *193*, 194–195
    shield 11, 57, *57*, 108–109, *108*, 110, 188, *189*

skull (representing Tezcatlipoca?) 56–57, *56*, 184, 185–188, *186*
Tlaloc skull mask with entwined serpents *56*, 57, 188, *190*
'warty' mask, possibly of Xiuhtecuhtli *56*, 57, 61, *63*, 184–188, *185*
see also Christy, Henry
brooches
Deschutes jasper and Yowah opal set in gold (BM) 198, *199*
double-headed serpent (Magnus) 223, *223*
eagle brooches (19th-century, BM) 2–3, *2*, *3*
silver, turquoise (Hendren) *213*
buckle, Imperial jasper set in silver (Johnson and Bird, NMS) 198, *198*
building-offerings 152
Bullock, William 182
Burgkmair, Hans (drawing attrib. to), black youth holding a shield 107–108, *107*, 110
Burial 6, Moon Pyramid, Teotihuacan 19–21
Burland, C.A. 114
butterflies
jewellery inlaid in jet (Leak) 206, *207*
symbolism of 117–118, 126, *126*, *131*, 133
buttons 197, *204*, 205, *205*
Byron, Lord (George) 178

cabochon turquoises 61
Cabrera, Rubén 19
Calakmul, masks from 23
Calendar Stone, Aztec 129, 133
Cantares Mexicanos 140–141
cape, turquoise 96, *142*, 143
Carlyle, Thomas 173
Carmichael, Elizabeth 183, 187
carving, Zuni 197, 206–208, 215–216
Caso, Alfonso 120, 121
catalogues and inventories
Bateman collection 195
Bethnal Green catalogue (Pitt-Rivers) 174, 175–176
Christy collection 179, 188
Cospi Museum 168
Giganti and Aldrovandi collections 166–167
Hertz collection 184
sales catalogues 184–185, 187–188, 195
Settala Museum 158–160
see also exhibition catalogues
caterpillars, symbolism of 117, 118, 129–130, *131*
cave (*also* mine *or* womb: *oztotl*) 135–136
Caygill, M. and Cherry, J. 183
Cedrela odorata (Meliaceae family) 7–11
optical microscope images of *8*, *9*
censers, Aztec-style 132
centipedes, symbolism of 127–128
centzonmimixcohua (400 stars of the north) 84
ceremony *see* ritual and ceremony
Cerrillos mines 35–38, 66–67, 151–152, 153, 221–224
Chaco Canyon 29(table 1), 31, 35–38, 146, 152, 221
Chaffers, William 195
Chalchihuites workshops 83
chalchihuitl (greenstone/jade/jadeite) 93, 117, 137, 138–139, 151–152
see also greenstone; jade
Chalco glyph *138*, 139, *139*
chalcosiderite 200–202
Chambers, R. 173
chemical properties 29–30, 32–35, 68–71
Chichen Itza 124, 125, 128
disk from Temple of the Warriors 46, 47
Chilam Balam of Chumayel 142
Christy, Henry 2, 3–4, 173, 174, 177–182
and Hertz collection 183, 187
Christy Bequest 183–188
Christy Fund 188–194
see also British Museum, Mixtec-Mexica mosaics
chrysoprase 23
Cihuacoatl 93
Cihuateteo (warrior women) 132
cinnabar 25–26
Cinteotl 93
Clavijero, Saverio, *Storia Antica del Messico* 170
Clemente VII, Pope 165, 167, 170
cliff dwellings 204
clinochlore 18
Coayxtlahuacan 100
Cobean, Robert 43
Codex Azcatitlan (*Códice Azcatitlan*) 138, *142*, 143
Codex Borbonicus 131, 133, *139*
Codex Borgia 125, 128, 132
Codex Boturini 96, 137
Codex Cospi 169–170, *169*, *170*, *171*
Codex Dresden 126
Codex Fejérváry-Mayer 130
Codex Magliabechiano 130–131, 133, 141
Codex Mendoza 61, 91, 96, 129, 135, 142
illustrations from *94–95*, *98–99*, *139*
Codex Vaticanus 130, *156*
Codex Vindobonensis *136*, 169
Códice Florentino *see* Florentine Codex; Sahagún
Códice Telleriano-Remensis 85–86
Cola de Palma, Oaxaca 124, *124*, 128
collections and collectors, European 3–4
16th- and 17th-century Italy 155–162, 165–171
mid-19th century 173–176, 177–182, 183–195
see also British Museum; Christy, Henry; Museum für Völkerkunde; *names of collectors and museums*
colour
of *E. coralloides* 11–12
and etymology of glyphs 139, 142
of fire 132–133, 140
of greenstones 15, 19, 24
in jewellery design 210
in mask from Malinaltepec 17
in masks from kingdom of *K'aan* 23, 24
Maya Blue pigment 124, 128–129
in Pueblo thought systems 146–147, 148, 150
red paint (cinnabar) 25–26, 47, 49
of shell and coral 12, 199
of turquoise 1, *30*, 57, 67, 135, 222
comets and meteors 122, 124, 129, 130, *140*, 141, *141*
see also cosmology; sky; star symbolism; sun
common descent, concept of 181
Comstock Lode 216
conservation and restoration
disk from Tula 42, 45–52
masks from Maya area 23–26
mosaics in MV 106, 111, 114
Offering 99, Templo Mayor 75–79
reuse of tesserae 19, 25, 59, 61, 124
Cooke, William 179
copper 24, 101
copper isotopes 31, 33, 38
coral 199, 210
coral necklace (Johnson and Bird, NMS) 198, *198*, 199
coral tree (*Erythrina americana*) 11–12, *11*
corn (maize)
dyed corn necklaces 202, *211*
symbolism of 84, 85
Coronado, Francisco Váquez de 151
Cortés, Hernán, 'Moctezuma's gifts' to (1519) 4, 80, 92, 93, 107, 114
cosmology
Aztec 84–86
Pueblo 146–147
see also comets; sky; star symbolism; sun
Cospi, Marchese Ferdinando 168
Cospi Codex 169–170, *169*, *170*, *171*
Cospi Museum 168–170
costume
and jewellery, present-day Navajo 211–212
ritual Tupinambá 159–160, *161*
see also diadem; headdresses; helmet; masks; regalia
cotinga 124–125, 143, 147
craftspeople *see* jewellery makers; lapidaries; manufacture; materials; silversmithing; *names of individual makers and craftspeople*
creation story, Navajo 214
crown *see xiuhuitzolli*
cuicatl (song) 135, 140–141
Culhua Mexica (Aztecs) *see* Aztecs
cultural (substitute) turquoise 3, 18, 200–202, 212
culture
and choice of materials 7
poetry and etymology 135–143
and traditional jewellery 197, 203–212, 213–214
see also belief systems; poetry; ritual; symbolism
cut (polished) stones
archaeological analysis of 19–21, 24, 48, 81–83
and rough-cut 217–219, 220

dance shells pendants (Owen) *151*, 212, *212*
Darwin, Charles 174, 175
Darwinian evolution 173, 174, 177, 181
dating 18, 101
see also provenance
De Bry, Theodore, *Historiae Antipodum* 160, *161*

De Michele, Vincenzo 159
deities 83, 84–86, 93
see also names of individual deities
Deity with Two Horns (MV) 103, *113*, 114
Demidov, Anatole 188
'Depression' necklaces *201*, 202
Deschryver, Francis 187
Deschutes jasper and Yowah opal brooch set in gold (BM) 198, *199*
design *see* iconography; jewellery makers; symbolism
Deyuse, Leekya 197, 209
silver bow guard 206, *206*
Di Peso, C.C. 67
diadem (*xiuhuitzolli*) 83, *92*, 95–96, 128, 129, 130, 131, *132*
directional associations, Hopi 146–147
disks
Chichen Itza 46
Offering 99, Templo Mayor, Tenochtitlan 75–86
Tula 41–52
dog pectoral (*xolocozcatl*) 131
dolomite 21
Domingo de Betanzos, Friar 165, 170
double-headed serpent (pectoral?, BM) *9*, 11, 188–194, *191*
Dresden Codex 126
Durán, Fray D. de 85
Durand, K.R. and Durand, S.R. 35–36
dyed corn necklaces 202, *211*
Dzibanché, masks from 23–24

eagle brooches, 19th-century (BM) 2–3, *2*, *3*
eagle warrior knife handle (sacrificial knife, BM) 56, *56*, 183–184, *184*, 185–188
Early Postclassic Toltec 41–52, 117, 124–128
earrings (Lovato) 212, *212*
Easby, E.K. and Scott, J.F. 113
economy, Aztec 91–92
see also trade; trade networks
Ehecatl-Quetzalcoatl (Aztec wind deity) 126
Eisen, Gustav 66
Erythrina americana (*Erythrina coralloides*) 11–12, *11*
VP-SEM micrograph of *10*
ethnographical collecting, 19th-century 174–176, 177–182, 183–195
Ethnological Society 178, 179
etymologies 4, 137–139, 146–147
of jade 117
of turquoise 1, 3, 129
see also languages; poetry; symbolism
European collections *see* collections and collectors
evolution, concepts and theories of 173–176, 177
Three Age System 174, 179, 181
excavations, archaeological
Burial 6, Moon Pyramid 19–21
disk from Tula 41–45
funerary masks, Maya area 22–26
Malinaltepec mask 15–17
Offering 99 75–77, *76*
Pitt-Rivers 174, 176
see also mines and mining
Excavations in Cranbourne Chase (Pitt-Rivers) 174, 175, 176
exhibition catalogues 198
exotica, collection of 155–162

face paint 83–84
fake jewellery 3, 214
Farnham Museum, Dorset 174–175, 176
feather paintings 157–158, *157*, *158*
Feest, C. 111, 113
Fejérváry-Mayer Codex 130
female myth and ritual 24, 148–150
Cihuateteo (warrior women) 132
Ferdinand, Archduke of Tyrol 104, 107
fertility, myth and ritual 47, 48, 148–149, 150
festival of *quecholli* 85–86
see also ritual and ceremony
fetish birds (Leekya) 209, *209*
figurines, Teotihuacan 118, *118*
see also human figures
Fiore, Cesare, *View of the Settala Museum* 158, *158*
fire and smoke 132–133, 136, 139–141
fire god *see* Xiuhtecuhtli
fire serpent *see* Xicani; Xiuhcoatl
Florentine Codex (*Códice Florentino*) 130, 135, 136, 137, 138, 140, 141, 143
see also Sahagún
Fontaine, Françoise 128
Foord, E.E. and Taggart, J.E. 68
Foshag, W.F. 117
Franks, Augustus Wollaston 3, 183, 188, 190–191, 194–195
FTIR analysis 17, 18
funerary deposits 17, 19–26

Gamio, Manuel 130
Garibay, K.Á.M. (ed.), *Romances de los Señores de la Nueva España* 137, 143
Gelati Academy 169–170
gem dealers, 20th-century urban 216–217
gender
and Navajo jewellery making 206–207
myth and ritual 148–150
see also women
geological deposits 18, 23, 30–31, 32–33, 223–224
see also mines and mining
geological properties 33, 67–71, 223–224
mineral samples, Burial 6, Moon Pyramid 19, 21
see also chemical properties
Giganti, Antonio 165–167
glyphs 137–139
Xi glyph *120*, 121–122, 125–126
Yax glyph 126
see also hieroglyphs
Godber, Walter 219
gold beads, and coral 199–200
Graulich, Michael 86
Graz, Archduke Karl of 114
Great Exhibition, The (1851) 174, 179
great-house system 146, 152–153
Greater Southwest *see* Southwest
greenstones 15–26
in BM mosaics 59
from Burial 6, Moon Pyramid 19–21
on funerary masks, Maya area 22–26
as supports 18, 19
on Teotihuacan-style mask, Malinaltepec (MNA) 15–19, *16*, 101
Guadalupe community 29(table 1), 31, 35–38
Guerrero, Teotihuacan-style mask from (MNA) 15–19, *16*, 101

Hanbury, Daniel 183
hand-made jewellery, standards for *216*, 217
Harbottle, G. and Weigand, P. 97
hardness, of turquoise 222, 224
Hare, A.J.C. 191
headdresses 83, 84
see also diadem; helmet
Heger, Franz 104, 105, 110, 111, 113, 114
helmet, Mixtec-Mexica (BM) *56*, *194*, 195
helmet headdresses, Teotihuacan 118–124
helmet masks, War Serpent symbolism in 118–124
hematite 17, 23
Hendren, Shane, brooch *213*
herbs, and meaning of *xihuitl* 137, *137*, 138
Hertz, Bram, collection of mosaics 183, 184–188
Heyden, Doris 126
Hidden, W.E. 66
hieroglyphs, study of in Bologna 167–168, 169–170
see also glyphs
Historiae Antipodum (De Bry) 160, *161*
Hochstetter, Ferdinand von 104
Hodgkin, Dr Thomas 178
Hohokam culture
jewellery 204, 212
village (Redtail) 69
Hooker, William
Christy's correspondence with 178, 179–180, 181
trustee of Christy bequest 183
Hopi 146–147, 148, 150
jewellery makers 197
Huehueteotl (Old God) 130, 136
huehuetlahtolli (moral speeches) 135
Huexotla 97
Huitzilopochtli (god of war) 83, 85, 91, 92, 93, 130
human figures
Burial 6, Moon Pyramid 19
disk, Offering 99, Templo Mayor 79–80, *81*, 83–84, 110
panel (MC) *109*, 110
Teotihuacan figurines 118, *118*
Zuni depictions of 206–207, 208, *208*, 209, *209*
see also skulls; warrior figures
humanitarianism 177, 178
humming birds 84, 85, 143, *168*, 169, 170
Hunt, Robert 174
hunting, ritual 85–86
hydrogen isotopes 31, 33, 38

IACA (Indian Arts and Crafts Association) 213, 214
IACB (Indian Arts and Crafts Board) 217, 219
Ichikawa, Schoichi 46
iconography

royal masks from kingdom of *K'aan* 23, 24
shields 57, *105*, 106–107, *106*, 110
warrior figures 79–80, *81*, 82–83, 110
see also symbolism
identity, cultural 202, 211
idols, Aztec, in Giganti Museum 166–167, *167*
ilhuicatl (sky) 141
imagery *see* iconography; symbolism
imaging technologies 7–12, 55–56, 59, 77–78
Imperial jasper buckle, set in silver (Johnson and Bird, NMS) 198, *198*
in situ isotopic measurements 30, 33–35, 36, 38
in situ lifting methods 44, 49
Indian Arts and Crafts Association (IACA) 213, 214
Indian Arts and Crafts Board (IACB) 217, 219
'Indigenists' and 'Mexicanists' debate 31
inlay jewellery *203*, 208, *208*
insignia *see* regalia; rulers; warrior figures
intelligent design, theories of 177
Inventario semplice (Anon. [Cospi Museum inventory]) 168
inventories *see* catalogues and inventories; exhibition catalogues
Isías, Alejandra Quintanar 19
isotopic measurement
copper and hydrogen 30, 31, 33–35, 36, 38
lead and strontium 66, 69–71
Istituto delle Scienze, Bologna 170
Italy, 16th- and 17th-century collections 155–162, 165–172
Ixtapantongo, rock painting 124, 128–129
Izeki, M. 96

jaclas 210–211, *210*
jade masks from Maya area 23–26
see also chalchihuitl; greenstones
jaguar-like characteristics of War Serpents 119–120
Jarrilla Mountains 68
jasper 198
Jay I. Kislak Collection 132
Jesuit scholars 170
jewellery
in 19th-century Europe 2–3
20th century merchandising 215–220
design, present day Navajo 197–202, 203–212
jewellery makers 197, 203–212, 215, 223–224
see also lapidaries; manufacture; materials; silversmithing; *names of individual makers and craftspeople*
Johnson, Yazzie 202, 203
and Gail Bird 197–202
Judge, W.J. 31, 35

K'aan, masks from kingdom of 23–24
Keres 146, 148
Keuran, D.K. van 179
Kewa 150, 151–152
Kim, J. 32
Kirchner, Norbert 106, 111
Kirk, R.F. 217
Kislak Collection *see* Jay I. Kislak Collection
Kley, William 216
knife wing figure, Zuni 206
knives 166, *166*, 168
sacrificial knife with flint blade (eagle warrior knife handle, BM) 56, *56*, 183–184, *184*, 185–188
Köhler, Ulrich 129, 130
Kolowisi (Zuni plumed water serpent) 149
Kunstkammer of Archduke Karl of Graz 114
Kunz, George Frederick 222

Lane Fox, Augustus *see* Pitt-Rivers, Augustus
languages
Pueblo 145–146
Tojolabal 130
see also etymologies; Nahuatl; poetry
lapidaries
Aztec 93–96
Pueblo 197, 203–212
Toltec 47
see also jewellery makers; manufacture; materials; silversmithing; *names of individual makers and craftspeople*
Lartet, Edouard 177
lead isotopes 66, 69–71
Leak, Gordon, butterfly inlaid in jet 206, *207*
Lee, Clarence, pin with woman and sheep *208*, 209
Leekya, Sara, turquoise fetish birds 209, *209*
Legati, Dr Lorenzo 168, 170
Lehmann, W. 114
Lenkersdorf, C. 130
Léry, J. de, *Navigatio in Brasiliam* 160
Lévi-Strauss, C. 145
Liberty Head dime *204*, 205
Libro di piume (Minaggio) 157–158, *157*
life-cycle/lifestyle
and traditional jewellery 210–211
Navajo 213–214
Pueblo rituals 148
Lightbown, Ronald 155
lime paste 23, 25
limestone 21
Loloma, Charles 197
tufa-cast gold bracelet *208*, 209
Los Angeles, jewellery trade 216–217, 219–220
Lovato, Ray 197
earrings 212, *212*
rolled bead and tab necklaces and *jaclas* 210–211, *210*
Lubbock, John, Lord Avebury 173, 174, 183
luminescence imaging 55–56, 59

macaw 153
machiotlahtolli (proverbs) 135
Magliabechiano Codex 130–131, 133, 141
Magnus, Douglas 223–224
mahogany 23
maize (corn) symbolism 84, 85
makers *see* jewellery makers; lapidaries; manufacture; materials; silversmithing; *names of individual makers and craftspeople*
malachite 24, 59
male myth and ritual 148–150
Malinaltepec, Teotihuacan-style mask from (MNA) 15–19, *16*, 101
manufacture 19–21, 24–25
BM mosaics 59, 61
disk from Tula 48
disk, Offering 99, Templo Mayor 80–83
traditional jewellery-making 203, 204, 205, 209, 211, 224
see also jewellery makers; lapidaries; materials; materials analysis; silversmithing; technology; tools; *names of individual makers and craftspeople*
market *see* art market; economy; trade
Marsigli, Luigi Ferdinando 170
masculinity, turquoise as symbol of 148–150
masks 114, 167
funerary masks, Maya area 22–26
helmet masks, with War Serpent symbolism 118–124
Teotihuacan-style, Malinaltepec (MNA) 15–19, *16*, 101
Tlaloc skull mask with entwined serpents (BM) *56*, 57, 188, *190*
'warty' mask, possibly of Xiuhtecuhtli (BM) *56*, 57, 59, 61, *63*, 184–188, *185*
mass sacrifices 85–86, 117–118
Massimo, Duchess Teresa 115(n 2), 189–194
Mastache, Guadalupe 43
materials 15
traditional jewellery-making 203, 204, 205, 208, 211
see also copper; coral; greenstones; malachite; resin; sandstone; shell; silversmithing; turquoise; wood
materials analysis 7–13, 29–31, 32–35
and sources of artefacts 67
BM mosaics 55–56, 59, 62
Burial 6, Moon Pyramid 19–21
disk from Tula 42, 48–49, 51
funerary masks from Maya area 23–24, 25
Offering 99, Templo Mayor 77–79
Scorpion Village and Redtail site 70–72
Teotihuacan-style mask, Malinaltepec 17–19
matlalxihuitl (blue turquoise) 92, 96
Matrícula de Tributos 129
Maya
funerary masks 22–26
symbolism 119, 126, 127–128
Maya Blue pigment 124, 128–129
Mayer, Joseph 183, 184, 186–187, 194–195
Mayhew, Henry 174
McNulty, James P. 222
meanings *see* etymologies; languages; poetry; symbolism
Medici collections 168, 171, 188–189
Melgar, E. and Bautista, P. 81
men, male myth and ritual 148–150
Mendoza Codex *see* Codex Mendoza
merchandising *see* art market; economy; trade
metals, Early Postclassic 124

metalwork, silversmithing 197, 203, 204–206, 209
meteoric beings 122–124, 129, 130
see also comets; sky; star symbolism; sun
Mexica *see* Aztecs (Mexica)
'Mexicanists' and 'Indigenists' debate 31
Mexiquito, War Serpent monument 122
Milanese collections 157–162
Settala Museum 155, 158–162, *158*
military accoutrements *see* helmet; knives; pectorals; regalia; shields; warrior figures; weapons
mimicry 200–202
mimixcohua 85, 86, 137
Minaggio, Dionisio, *Libro di piume* 157–158, *157*
mine (*also* cave *or* womb: *oztotl*) 135–136
mineralogy of turquoise *see* chemical properties; geological properties; materials analysis
mines and mining 18, 66–67, 151–152, 216, 219
see also geological deposits; provenance regions
mining (extraction), as 'birth' 135–136
mirror frames 114
animal-head frame (MV) 103, 110–114, *111*
mirrors, Toltec *tezcacuitlapilli* 127–128, 132
Mitelli, Giuseppe 168
Mixcoatl (god of fire and war) 83, 84, 85, 86
Mixtec
mosaic shield *108*, 110
Yahui (meteoric being) 122
see also British Museum, Mixtec-Mexica mosaics
Moctezuma I (Motecuhzoma Ilhuicamina)
'gifts' to Cortés (1519) 4, 80, 92, 93, 107, 114
royal emblem of 96
monuments, Cola de Palma 124, *124*, 128
Moon Pyramid, Teotihuacan 19–21
Mopan Maya 130
moral speeches (*huehuetlahtolli*) 135
Morandotti, Alessandro 159
Morning Green and Parrot (O'odham myth) 153
Morris, Earl 46
Morrissey, James R. 216
Mortimer, Cromwell 1
mortuary bundles 118–120, *118*, 126, 130–131
mortuary song (*atequilizcuicatl*) 140–141
mosaics 15
see also tesserae
Motagua River region 23
Motecuhzoma Ilhuicamina *see* Moctezuma I
mother-of-pearl (*Pinctada mazatlanica*) 17, 23
Munich Kunstkammer inventory 113, 114
Musée Royaux d'Art et d'Histoire, Brussels, shield *108*, 110
Museo Nacional de Antropología (MNA) Mexico City
disk from Tula 52
Teotihuacan-style mask 15–19, *16*, 101
Museum für Völkerkunde, Vienna
Mexican mosaics in 3–4, 103–115, 194
animal-head mirror frame 110–114
Deity with Two Horns *113*, 114
shield 103, *103*, 104–110
museums *see* collections
myths (*tlamachiliztlahtolzazanilli*) 135
see also belief systems; poetry; ritual; symbolism

nahuallahtolli (shamanistic texts) 135
Nahuatl, poetry and culture 130, 135–143
najas 205, *205*
Nalda, Enrique 24
National Museums Scotland (NMS) 198
Nauhyohuehue 139
Navajo 148–149, 213–214
jewellery and silversmithing 197–202, 203–212, 215–220
The Navajo and Pueblo Silversmiths (Adair) 215, 217, 219, 220
Navigatio in Brasiliam (Léry) 160
necklaces, Navajo 198, 199–200, 202, 203, 204, 210–211, *218*
New Landers turquoise 200–201
Nicholson, H.B. 130
Nizza, Friar Marcos de 151
nobility, Aztec (*tecuhtli*) 96, 101
Nowotny, K.A. 104, 105, 106, 107, 111, 113, 114
nugget (*or* tab) and rolled bead necklaces 203, 204, *204*
Nuttall, Zelia 104

Oaxaca, Museo de las Culturas, mosaic panel *109*, 110
obsidian 21, 23
Offering 99, Templo Mayor, Tenochtitlan 75–86, *80*, 110
offerings 46–47, 85, 152
Olivier, G. and López Luján, L. 130
Olmos, Fray Andrés de 143
Oppel, A. 103, 108
optical microscopy 7–11, 55, 59, 81
organic evolution, concept of 174
Owen, Angie Reano 197, 203, 212
shell and inlay bracelets *203*
Spondylus princeps pendants (dance shells) *151*, 212, *212*
Oxkintok, Yucatán, funerary mosaics 23–24
oyohualli 83, 86
oztotl (mine, cave *or* womb) 135–136

Paalölöqangw (Hopi feathered serpent) 149
Paget, Walburga, Lady 191
paint
face paint 83–84
red (cinnabar) 25–26, 47, 49
turquoise dust used as 224
paintings *see* feather paintings; pictorial expression; rock painting
Pakal, funerary mask 24–26, *25*
Palacio Quemado 41, 43, 46–47
Palenque, Chiapas, masks from 24–26
Paleotti, Gabriele 170
Paley, William, *Natural Theology* 177
Paquimé 128
paraffin wax lifting method 44, 49, 50–51
Pardue, Diana, *Shared Images* 198
Parrot, and Morning Green (O'odham myth) 153
Paynal 93
pectorals
dog pectoral (*xolocozcatl*) 131
Mixtec-Mexica, double-headed serpent (BM) *9*, 11, 188–194, *191*
Toltec 125–126, 131
pendants
animal-head (BM) 11–12, *13*, *112*, 114, *192*, 194
Hohokam culture 204, *204*
Peshlakai, Fred 219–220
Peshlakai, Norbert, bracelets *208*, 209
Petry, William 216
pictorial expression 135
Pignoria, Luigi 156, 170
Pinctada mazatlanica (mother-of-pearl) 17, 23
pins, Navajo jewellery *208*, 209, *209*
Pinus, used as support 19
VP-SEM micrograph of *9*, *10*, 11
Pitt-Rivers, Augustus Henry Lane Fox 173, 174–176
planerite 68
plastic jewellery *201*, 202, 211–212, *211*
Las Plazuelas, Guanajuato 122, *122*
Poblano, Leo 197, 206
man drilling beads 209, *209*
pochteca (Aztec merchants) 96, 97
poetry, culture and etymology 135–143
Pogue, Joseph 66
Pohl, John 122
Pollard, H. 100
Postclassic period (AD 900–1521) 66
Pratt, Samuel Luke 187
pre-Columbian trade networks *see* trade networks
Primisser, Johann 113
Pritchard Dr James Cowles 178
procurement strategies, trade networks 29, 31, 35–38
progress, concept of 173–174, 179
provenance
BM turquoise mosaics 59–62, 183–195, 195(n 1)
Giganti and Aldrovandi collections 166–167
MNA greenstone mosaics 18
mosaics in MV 103, 104, 107–108, 110, 114
tesserae from Offering 99, Templo Mayor 77–79
provenance regions 29, 31–33, 35–38, 47, 66–67, 69–72
proverbs (*machiotlahtolli*) 135
Pueblo Bonito 83, 152
Pueblo, modern day 145–154
cosmology 146–147
lapidary artists 197, 203–212
myth and ritual 148–150, 152
and turquoise in history 150–153
Pueblo Revolt (1680) 151–152
Purpus, Carl A. 110
pyrite 68
adhesive 49
plaques 48, 49, 52

Quakerism, and social reform 178, 181
quecholli 85–86
Quetzalcoatl (feathered serpent) 47, 93, 126
Quetzaltepec 99–100
Quiauhteopan 97, *98–99*, 99–101
quincunx 83, 85, 136, *136*

Raman spectroscopy 55, 59
Ramusio, Giovanni Battista 156
rare earth element patterns 31–33
raw materials, properties of 7–13
see also materials analysis
Read, Hercules 183, 188–189, 191–195
reconstruction of disk from Tula 50–52
see also conservation and restoration
red paint (cinnabar) 25–26, 47, 49
Red Queen, mask of from Palenque 24, *24*
Redtail site (Hohokam village) 69–72
regalia
Aztec 79–80, 128–132
royal 96, 128, *142*, 143
Hopi 150, *150*
Toltec 125–127, 132
religion *see* belief systems; ritual; society; symbolism
resin
in base of mosaic shield (MV) 104, 115(n 4)
in disk from Tula 48–49
in VP-SEM micrographs *8*, *10*, 11
restoration *see* conservation and restoration
Retzius, Professor Anders 179, 180
riddles (*zazanilli*) 135, 141
ring, Bisbee turquoise 207–208, *207*
ritual and ceremony
Aztec 85–86, 141
funerary deposits 17, 19–21
offerings 46–47, 85, 152
present day 202, 211–212, 214
Pueblo 148, 149–150, 152
Toltec 46–47
Tupinambá costume and dances 159–160, *161*
rock painting, Ixtapantongo 124, 128–129
rolled shell bead and tab necklaces 203, 204, *204*
Rollin, Charles 195
Romances de los Señores de la Nueva España (Garibay) 137, 143
Rome, museums and collections in 155, 170
roseate spoonbill 85
rough-cut turquoise 217–219, 220
Royal Asiatic Society 187
Royal Blue mine, Esmeralda County 216
royalty *see* regalia; rulers, Aztec
Ruiz de Alaecón, Hernando 139, 141
rulers, Aztec 93–96, 128, 140, 142–143
Ruvalcaba Sil, José Luis 17
Ruz, Alberto 24

sacrifices 85–86, 117–118
sacrificial knife with flint blade (eagle warrior knife handle, BM) 56, *56*, 183–184, *184*, 185–188
Sahagún, Fray Bernardino de 48, 95, 96–97, 100
Códice Florentino 130, 135, 136, 137, 138, 140, 141, 143
sales catalogues 184–185, 187–188, 195
see also art market; trade
San Francisco Call 66
Sánchez, H. and Robles Camacho, R. 21
sandstone support 43, 45, *47*, 48, 52
Santo Domingo Pueblo
jewellery-makers 203–212, 215
officials *221*
see also Kewa
Saville, M. 100
Scandinavia, Christy in 179
scanning electron microscopy (SEM) 7–11, 21, 70–71, 81
Schele, Linda 124–125
Schlosser, J. von 158
schools, silversmithing 217, 219
Schweizer, Herman 215
scientific approaches 168, 170, 173–176
properties of raw materials 7–13
Scorpion Village 69 72
seasons 148
see also temporal cycles
seated animal and cup (BM) *56*, *193*, 194–195
secondary ion mass spectrometer (SIMS) analysis 30, 33–35
Seler, Eduard 124, 126
SEM (scanning electron microscopy) 7–11, 21, 70–71, 81
serial evolution 174–176
serpent symbolism
Classic period 118–124, 127–128, 129–130, 132
Pueblo myth and ritual 149–150
Toltec myth and ritual 47
serpentine 19
serpents
double-headed (pectoral?, BM) *9*, 11, 188–194, *191*
Tlaloc skull mask with entwined serpents (BM) *56*, 57, 188, *190*
Settala, Manfredo 158, 159–160
Settala Museum, Milan 155, 158–162, *158*
sexuality, myth and ritual 148–150
shamanistic texts (*nahuallahtolli*) 135
Shared Images (Pardue) 198
sheep, as Navajo motif 209
shell
bird's head ornament from Oxkintok 23, *24*
mother-of-pearl (*Pinctada mazatlanica*) 17, 23
in Navajo jewellery *151*, 203, *203*, 204, *204*, 212, *212*
see also Spondylus princeps
shields
Aztec 99
in BM 11, 57, *57*, 108–109, *108*, 110, 188, *189*
in NMAI 109–110
in MV 103, *103*, 104–110
disk from Tula identified as 47
in MRAH *108*, 110
shooting stars 122, 124, 129, 130
see also comets; sky; star symbolism; sun
shops, jewellery 217, 219–220
Shorty, Perry, stamped silver and turquoise pins 209, *209*
Sierra Nahuat 130
Silver Bell Mountains 69
silver bow guard (Deyuse and Simplicio) 206, *206*
silversmithing, Navajo 197, 203, 204–206, 209, 215–220
Simmons, George 216
SIMS analysis 30, 33–35
Sinkankas, J. 67
skeuomorphs 200–202
skull mask with entwined serpents (BM) *56*, 57, 188, *190*
skulls, human
formerly in EMB 103, 114–115(n 1)
representing Tezcatlipoca? (BM) 56–57, *56*, 184, 185–188, *186*
study of 179, 180
sky (*ilhuicatl*) 130, 141
see also comets; cosmology; star symbolism; sun
Slender Maker of Silver (Old Peshlakai) 219
squash-blossom necklace *218*
Sloane, Sir Hans 1
Smith, Mary Elizabeth 122
smoke and fire 132–133, 136, 139–141
Snow, C.H. 66
social evolution 173–174
society
Aztec 91–92
present-day Southwest communities 202
Pueblo 148
social reform, 19th-century 178, 181
and value and status of turquoise 1–3, 31, 57
see also belief systems; myths; ritual; symbolism
song (*cuicatl*) 135, 140–141
Sotheby and Wilkinson, sale catalogues 184–186, 188
Southwest
jewellery design 197–202, 203–212
pre-Columbian trade networks 29, 31, 35–38
Spanish explorers 92, 96, 97–99
and silversmithing 204
and turquoise trade 150–151
see also Cortés
spear thrower (*atlatl*) 122–124, *123*, 129, 130, 169
Spencer, Herbert 173–174, 176
'sphinxes', Aztec, in Cospi Museum 168–169
Spondylus princeps (thorny oyster) 18, 150, 154(n 2), 204
coral as substitute for 199
optical images of 12, *12*
pendants (Owen) *151*, 212, *212*
Zuni knife wing figure 206
squash and squash blossom 145, 148, 150
squash-blossom necklaces 199–200, 205, *218*
stampwork 205, *206*, *209*
standards, for hand-made jewellery *216*, 217
star symbolism 84, 85, 86, 120, 122–124, 129, 130

in poetry 141
Steinhauer, Carl 179, 188
Stone, Andrea 122
stonework, Zuni 197, 206–208, 215–216
Storia Antica de Messico (Clavijero) 170
Strombus gigas 12, 22
strontium isotopes 66, 69–71
Sugiyama, Saburo 19
summer and winter 148
Sun
in Aztec myth 85, 86, 142–143
in O'odham myth 153
in Zuni belief 148, 149–150
supports 15, 23, 137
greenstone 18, 19
provisional, during restoration 50–52
sandstone 43, 45, *47*, 48, 52
wood 19, 23, 25, 45, 48, 51–52, 104
identification of properties 7–12
symbolism
Classic period 117–124
Early Postclassic Toltec 46–47, 124–128
Maya 24
Mexica (Aztec) 128–132, 135–143
deities and nobility 93–96, 101
mosaic from Offering 99 83–86
Pueblo 146–150
Santo Domingo artists 211–212, 214, 215
and social value and status of turquoise 1–3, 31, 57
see also iconography

tab (*or* nugget) and rolled bead necklaces 203, 204, *204*
Taube, Karl 24
technology *see* manufacture; materials analysis; tools
tecuhtli (Aztec nobility) 96, 101
Tecutepec 96
Telleriano-Remensis Códice 85–86
Temple of Inscriptions, tomb from 24–26
Temple of Quetzalcoatl, Teotihuacan 118–124
Temple of the Warriors, disk from 46, 47
Templo Mayor, Tenochtitlan, Offering 99 75–86, *80*, 110
temporal cycles 85, 129, 141–142, 147, 148
Tennyson, Alfred, *In Memoriam* 173
Tenochtitlan 91
Offering99, Templo Mayor 75–86, *80*, 110
tenontzalli (verbal relation) 135
Teotihuacan 118–124, 126, 127–128
Burial 6, Moon Pyramid 19–21
teoxihuitl (fine turquoise) 92, 95–97, 99–100, 137–138
tesserae 136–137
disk (Tula) 48, 51–52
reuse of 19, 25, 59, 61, 124
shield (BM) 57
shield (MV) 104–106, *105*
see also manufacture; materials analysis
Tewas 146, 148
Texcoco 91
tezcacuitlapilli (back mirror) 127–128, 132
Tezcatlipoca (god of war) 47
skull representing (BM) 56–57, *56*, 184, 185–188, *186*
Tezozomoc, Alvarado 95–96
Thomsen, Professor Christian 179
Three Age System 174, 179, 181
thunderbird necklace (Santo Domingo Pueblo) *211*
'Tiffany' mines 222
time *see* temporal cycles
The Times 187, 188
Tlacopan 91
Tlahuizcalpantecuhtli (Aztec deity) 83, 84, 170
Tlaloc (rain god) 75, 120, 126
Tlaloc skull mask with entwined serpents (BM) *56*, 57, 188, *190*
Tlaltecatzin of Cuauhchinanco 143
Tlaltecuhtli (mother earth) 136, *136*
tlalxiuhco (or *tlalxicco*, 'in the place of turquoise') *139*, 140, 141
tlamachiliztlahtolzazanilli (myths) 135
Tochpan, tribute from *94–95*, 96(map), 97–99, 97(map)
Tojolabal (language) 130
Toltecs (Toltecayotl) 42, 46–47, 96, 117, 124–128
tombs *see* funerary deposits
Tombstone Prospector 66
tools
analysis of traces left by 19–21, 24, 48, 81–83
ancient mining tools 222
traditional jewellery-making 205, 224
Toltec, *tezcacuitlapilli* (mirror) 127–128, 132
tourist trade 211, 215, 217
trace element profiling 31–33
trade
16th-century 150–151
20th-century 206, 215–220
present-day 197, 202, 204, 211, 214
see also sales catalogues
trade networks, pre-Columbian 15, 29, 31, 35–38
Aztec 96–101, 117
provenance regions 29, 31–33, 35–38, 47, 66–67, 69–72
Toltec 47
see also tribute
traditional jewellery and jewellery-making 197–202, 203–212, 214
travel, and collecting, 19th-century 178–179, 180–181
travel literature, 16th-century 155–156
tribute
and provenance 59, 61
Aztec tributary states *94–95*, 96–101, *98–99*
see also trade networks
Triple Alliance, Aztec 91
tsooro (mountain bluebird) 147, *147*
tsorposi ('bluebird seeds') 147, 153
tufa casting 199, 200, *208*, 209
Tula, Early Postclassic Toltec 117, 124–128
disk found at Palacio Quemado 41–52
Tupinambá ritual costume 159–160
Turin collections, exotica in 157
'turkey' (fake turquoise) 3
turquoise 1–3
chemical properties of 29–30, 32–35, 68–71
definitions of 1
misidentification of 32, 115(n 5)
and non-turquoise materials on BM mosaics 56–62
quality of 216, 217, 222
social and status value of 31, 57
sources of 65–66
see also mines and mining; provenance; trade networks
substitute (cultural turquoise) 3, 18, 200–202, 212
types of 92
see also greenstones; tesserae
Turquoise Hill mines 222
Tyler, H.A. 152
Tylor, Edward Burnett 180–181, 182
Anahuac 181, 188

uel popoca teuxiuitl (very smoky fine turquoise) 92
UIL imaging 55–56, 59
ultraviolet-induced luminescence (UIL) imaging 55–56, 59
underlay designs 198–199, *199*, *200*
Underworld 85, 86
Urcid Serrano, Javier 122, 124

van den Abeele, Andrew 188
van Heurne, Joseph 187, 188
Vandervin, H. 187
Vanitas 187–188
variscite 202
Vaticanus Codex 130, *156*
Velázquez Castro, A. 17, 81
Venetian travel literature, 16th-century 155–156
Venus (planet) 84, 85, 86
verbal relation (*tenontzalli*) 135
Victoria, Queen 2
Vindobonensis Codex *136*, 169
virtues of turquoise 143
Vokes, A.W. and Gregory, D.A. 69
VP-SEM micrographs, *Cedrela odorata* 7–8

Wallace, C.G. 206, 210
Walsingham, Thomas de Grey, 6th Baron 189–191
war deities 47, 84, 150
War Serpent 118–124, 128
Ward, Alfreda 217
warrior figures 122–124, *123*, 129, 130, 169
on Aztec-style censer (KC) 132, *132*
on disk from Offering 99 79–80, *81*, 83–84, 85
on mosaic panel (MC) *109*, 110
on shield (MV) *105*, 106–107, *106*, 110
Toltec 47, 125, *125*, 126
warrior knife handle (sacrificial knife, BM) 56, *56*, 183–184, *184*, 185–188
warrior mortuary bundles 118–120, *118*, 126, 130–131
'warty' mask, possibly of Xiuhtecuhtli (BM) *56*, 57, 59, 61, *63*, 184–188, *185*
weapons 85, 174, 175, *175*
see also knives; shields; warrior figures
weathering 33, 68
Weigand, P.C. 67, 152
Weigand, P.C. and Weigand, A. 97
Weinberg, C.E. 181
Wilson, Doc 219, 220

winter and summer 148
Witherspoon, Gary 215
womb (*also* mine *or* cave: *oztotl*) 135–136
women, myth and ritual 24, 148–150
Cihuateteo (warrior women) 132
wooden supports 19, 23, 25, 45, 48, 51–52, 104
identification of properties 7–12
workers *see* jewellery makers; lapidaries; silversmithing; *names of individual makers and craftspeople*
workshops *see* manufacture
Worm, Olaus 169
worms *see* caterpillars
writing systems *see* glyphs; hieroglyphs

X-ray diffraction (XRD) *17*, 18, 48, 77
X-ray fluorescence (XRF) 17, 55, 59, 77–78
Xaltepec 100
Xi glyph *120*, 121–122, 125–126
Xicalanco 97
Xicani (Zapotec fire serpent) 122, 124
xihuitl (common turquoise) 92, 129, 135, 137, 138
also *xiuhtli*, the comet *140*, 141
Xiuhcoatl (fire serpent, *also* turquoise serpent) 47, 117, 120–121, 128, *128*, 129–130, *131*
xiuhcoatl-shaped comet *141*
xiuhpilli (sun) 142–143
Xiuhtecuhtli (god of fire) 47, 93, 117, 125, 126, 130, 132–133, 136
mask of(?) 184, *185*
xiuhtezcatl (decorated mirror) 100
xiuhtli (or *xihuitl*, the comet) *140*, 141
xiuhtomolli (half-round, half-flat turquoise) 92
xiuhtomoltetl (green and white turquoise) 92, 97
Xiuhtototl (cotinga) 125, 126
see also cotinga
xiuhuitzolli (or *xihuitzolli*: diadem) 83, *92*, 95–96, 128, 129, 130, 131, *132*
xixitl (low-grade turquoise) 92
xolocozcatl (dog pectoral) 131
Xolotl 93
XRD *17*, 18, 48, 77
XRF 17, 55, 59, 77–78

Yacatecuhtli 93
yacaxihuitl (nosepiece) 126, 128, 131, 132
Yahui (Mixtec meteoric being) 122
Yax glyph 126
year *see* temporal cycles
Yoaltepec 97, *98–99*, 99–101

Zani, Count Valerio 169–170
Zapotec, Xicani (fire serpent) 122, 124
zazanilli (riddles) 135, 141
Zimmer, Walter C. 216
Zuni 146, 149, 150–151
carving and stonework 197, 206–208, 215–216